Nonlinear Water Waves with Applications to Wave-Current Interactions and Tsunamis

CBMS-NSF REGIONAL CONFERENCE SERIES IN APPLIED MATHEMATICS

A series of lectures on topics of current research interest in applied mathematics under the direction of the Conference Board of the Mathematical Sciences, supported by the National Science Foundation and published by SIAM.

GARRETT BIRKHOFF, *The Numerical Solution of Elliptic Equations*
D. V. LINDLEY, *Bayesian Statistics, A Review*
R. S. VARGA, *Functional Analysis and Approximation Theory in Numerical Analysis*
R. R. BAHADUR, *Some Limit Theorems in Statistics*
PATRICK BILLINGSLEY, *Weak Convergence of Measures: Applications in Probability*
J. L. LIONS, *Some Aspects of the Optimal Control of Distributed Parameter Systems*
ROGER PENROSE, *Techniques of Differential Topology in Relativity*
HERMAN CHERNOFF, *Sequential Analysis and Optimal Design*
J. DURBIN, *Distribution Theory for Tests Based on the Sample Distribution Function*
SOL I. RUBINOW, *Mathematical Problems in the Biological Sciences*
P. D. LAX, *Hyperbolic Systems of Conservation Laws and the Mathematical Theory of Shock Waves*
I. J. SCHOENBERG, *Cardinal Spline Interpolation*
IVAN SINGER, *The Theory of Best Approximation and Functional Analysis*
WERNER C. RHEINBOLDT, *Methods of Solving Systems of Nonlinear Equations*
HANS F. WEINBERGER, *Variational Methods for Eigenvalue Approximation*
R. TYRRELL ROCKAFELLAR, *Conjugate Duality and Optimization*
SIR JAMES LIGHTHILL, *Mathematical Biofluiddynamics*
GERARD SALTON, *Theory of Indexing*
CATHLEEN S. MORAWETZ, *Notes on Time Decay and Scattering for Some Hyperbolic Problems*
F. HOPPENSTEADT, *Mathematical Theories of Populations: Demographics, Genetics and Epidemics*
RICHARD ASKEY, *Orthogonal Polynomials and Special Functions*
L. E. PAYNE, *Improperly Posed Problems in Partial Differential Equations*
S. ROSEN, *Lectures on the Measurement and Evaluation of the Performance of Computing Systems*
HERBERT B. KELLER, *Numerical Solution of Two Point Boundary Value Problems*
J. P. LASALLE, *The Stability of Dynamical Systems*
D. GOTTLIEB AND S. A. ORSZAG, *Numerical Analysis of Spectral Methods: Theory and Applications*
PETER J. HUBER, *Robust Statistical Procedures*
HERBERT SOLOMON, *Geometric Probability*
FRED S. ROBERTS, *Graph Theory and Its Applications to Problems of Society*
JURIS HARTMANIS, *Feasible Computations and Provable Complexity Properties*
ZOHAR MANNA, *Lectures on the Logic of Computer Programming*
ELLIS L. JOHNSON, *Integer Programming: Facets, Subadditivity, and Duality for Group and Semi-Group Problems*
SHMUEL WINOGRAD, *Arithmetic Complexity of Computations*
J. F. C. KINGMAN, *Mathematics of Genetic Diversity*
MORTON E. GURTIN, *Topics in Finite Elasticity*
THOMAS G. KURTZ, *Approximation of Population Processes*
JERROLD E. MARSDEN, *Lectures on Geometric Methods in Mathematical Physics*
BRADLEY EFRON, *The Jackknife, the Bootstrap, and Other Resampling Plans*
M. WOODROOFE, *Nonlinear Renewal Theory in Sequential Analysis*
D. H. SATTINGER, *Branching in the Presence of Symmetry*
R. TEMAM, *Navier–Stokes Equations and Nonlinear Functional Analysis*

Miklós Csörgo, *Quantile Processes with Statistical Applications*
J. D. Buckmaster and G. S. S. Ludford, *Lectures on Mathematical Combustion*
R. E. Tarjan, *Data Structures and Network Algorithms*
Paul Waltman, *Competition Models in Population Biology*
S. R. S. Varadhan, *Large Deviations and Applications*
Kiyosi Itô, *Foundations of Stochastic Differential Equations in Infinite Dimensional Spaces*
Alan C. Newell, *Solitons in Mathematics and Physics*
Pranab Kumar Sen, *Theory and Applications of Sequential Nonparametrics*
László Lovász, *An Algorithmic Theory of Numbers, Graphs and Convexity*
E. W. Cheney, *Multivariate Approximation Theory: Selected Topics*
Joel Spencer, *Ten Lectures on the Probabilistic Method*
Paul C. Fife, *Dynamics of Internal Layers and Diffusive Interfaces*
Charles K. Chui, *Multivariate Splines*
Herbert S. Wilf, *Combinatorial Algorithms: An Update*
Henry C. Tuckwell, *Stochastic Processes in the Neurosciences*
Frank H. Clarke, *Methods of Dynamic and Nonsmooth Optimization*
Robert B. Gardner, *The Method of Equivalence and Its Applications*
Grace Wahba, *Spline Models for Observational Data*
Richard S. Varga, *Scientific Computation on Mathematical Problems and Conjectures*
Ingrid Daubechies, *Ten Lectures on Wavelets*
Stephen F. McCormick, *Multilevel Projection Methods for Partial Differential Equations*
Harald Niederreiter, *Random Number Generation and Quasi-Monte Carlo Methods*
Joel Spencer, *Ten Lectures on the Probabilistic Method, Second Edition*
Charles A. Micchelli, *Mathematical Aspects of Geometric Modeling*
Roger Temam, *Navier–Stokes Equations and Nonlinear Functional Analysis, Second Edition*
Glenn Shafer, *Probabilistic Expert Systems*
Peter J. Huber, *Robust Statistical Procedures, Second Edition*
J. Michael Steele, *Probability Theory and Combinatorial Optimization*
Werner C. Rheinboldt, *Methods for Solving Systems of Nonlinear Equations, Second Edition*
J. M. Cushing, *An Introduction to Structured Population Dynamics*
Tai-Ping Liu, *Hyperbolic and Viscous Conservation Laws*
Michael Renardy, *Mathematical Analysis of Viscoelastic Flows*
Gérard Cornuéjols, *Combinatorial Optimization: Packing and Covering*
Irena Lasiecka, *Mathematical Control Theory of Coupled PDEs*
J. K. Shaw, *Mathematical Principles of Optical Fiber Communications*
Zhangxin Chen, *Reservoir Simulation: Mathematical Techniques in Oil Recovery*
Athanassios S. Fokas, *A Unified Approach to Boundary Value Problems*
Margaret Cheney and Brett Borden, *Fundamentals of Radar Imaging*
Fioralba Cakoni, David Colton, and Peter Monk, *The Linear Sampling Method in Inverse Electromagnetic Scattering*
Adrian Constantin, *Nonlinear Water Waves with Applications to Wave-Current Interactions and Tsunamis*

ADRIAN CONSTANTIN
University of Vienna
Vienna, Austria

Nonlinear Water Waves with Applications to Wave-Current Interactions and Tsunamis

siam.
SOCIETY FOR INDUSTRIAL AND APPLIED MATHEMATICS
PHILADELPHIA

10 9 8 7 6 5 4 3 2 1

Library of Congress Cataloging-in-Publication Data
Constantin, Adrian.
Nonlinear water waves with applications to wave-current interactions and tsunamis / Adrian Constantin.
p. cm. -- (CBMS-NSF Regional conference series in applied mathematics ; 81)
Includes bibliographical references and index.
ISBN 978-1-611971-86-6
1. Nonlinear waves. 2. Nonlinear wave equations. 3. Tsunamis--Mathematical models. I. Title.
QA927.C596 2011
551.46'3--dc23

2011023012

To Professor Lars V. Hörmander
and Professor Henry P. McKean
with grateful admiration.

Contents

Preface **xi**

1 Introduction **1**
1.1 General considerations 1
1.1.1 The role of mathematics 1
1.1.2 Modelling 2
1.2 Aims of the lectures 3

2 Preliminaries **5**
2.1 Simplifying assumptions 5
2.2 The governing equations 11
2.2.1 The equation of mass conservation 11
2.2.2 The equation of motion: Euler's equation 12
2.2.3 The boundary conditions 13
2.2.4 Well-posedness 15
2.3 Vorticity 18
2.3.1 Irrotational flows 26
2.3.2 Currents 27
2.4 Appendix for Chapter 2 29
2.4.1 Surface tension 29
2.4.2 Dissipation of energy 31
2.4.3 Flow-invariant sets 33
2.4.4 The compression rate of a fluid 37
2.4.5 Sobolev spaces 38
2.5 Notes for Chapter 2 40

3 Wave-current interactions **41**
3.1 Steady periodic waves of small amplitude 46
3.1.1 The linearization 47
3.1.2 Existence of nonlinear waves of small amplitude 53
3.1.3 The dispersion relation 63
3.2 Rotational steady waves of large amplitude 72
3.2.1 The global continuum 72
3.2.2 Nodal pattern and bounds 79
3.2.3 Numerical simulation of waves of large amplitude 87

3.3 Symmetry of rotational steady water waves . . . 90
3.4 Regularity of the streamlines . . . 94
3.5 Appendix for Chapter 3 . . . 97
3.5.1 Fredholm operators . . . 97
3.5.2 Local bifurcation . . . 99
3.5.3 Global bifurcation and degree theory . . . 100
3.5.4 Elliptic boundary value problems . . . 114
3.5.5 Maximum principles . . . 125
3.6 Notes for Chapter 3 . . . 126

4 Fluid kinematics for wave trains **131**
4.1 Particle paths beneath a Stokes wave . . . 132
4.2 Pressure beneath a Stokes wave . . . 152
4.3 Appendix for Chapter 4: Gerstner's wave . . . 155
4.4 Notes for Chapter 4 . . . 160

5 Solitary water waves **163**
5.1 Particle paths beneath an irrotational solitary water wave . . . 166
5.2 Pressure beneath an irrotational solitary water wave . . . 170
5.3 Solitons . . . 174
5.4 Appendix for Chapter 5 . . . 180
5.4.1 Integrable systems . . . 180
5.4.2 Alternative model equations within the same regime . . . 235
5.5 Notes for Chapter 5 . . . 238

6 Breaking waves **241**
6.1 Long waves of small amplitude . . . 243
6.2 Wave breaking for long waves of moderate amplitude . . . 248
6.3 Appendix for Chapter 6 . . . 257
6.3.1 Semigroup theory approach toward well-posedness . . . 257
6.3.2 The evolution of extrema . . . 265
6.4 Notes for Chapter 6 . . . 266

7 Modelling tsunamis **275**
7.1 General considerations . . . 276
7.2 The December 2004 tsunami . . . 285
7.3 The 1960 Chile tsunami . . . 287
7.4 Appendix for Chapter 7 . . . 289
7.4.1 Nondimensionalization and multiple scales approach . . . 289
7.4.2 The highest wave ever measured . . . 295
7.4.3 The March 2011 Japan tsunami . . . 296
7.5 Notes for Chapter 7 . . . 297

Bibliography **299**

Index **319**

Preface

"There is no subject so old that something new cannot be said about it."
F. Dostoevsky

This book is the written and somewhat expanded version of the author's lectures delivered at the NSF-CBMS international research conference organized at the Mathematics Department of The University of Texas-Pan American, Edinburg, Texas, USA, during the week May 17–21, 2010. The aim of the conference was to provide an overview of some of the main results and recent developments in nonlinear water waves. Accordingly, these lecture notes represent, in part, a survey of fundamental aspects of the field and, in part, a discussion of several important topics of current research interest. The material presented contains a selection of aspects of water-wave motion that is believed to be of intrinsic mathematical interest and of physical relevance, and where a mathematical study can be pursued to an advanced stage, enabling us to derive conclusions that explain, to some extent, observed phenomena. The mathematical considerations are preceded by a discussion of the underlying physical factors, and we discuss the physical relevance of the mathematical results that are presented. No attempt is made to cover, in any complete sense, all the important topics of water-wave theory; indeed, we do not even exhaust the topics discussed here! The material described in these lectures is not intended, in any sense, to provide a systematic treatise of the theory of water waves. The exposition aims more at conveying a feel for the scope and usage of the main ideas and techniques than at proving all the relevant theorems in detail. However, many important results are proved in full, while for others the proofs are merely sketched, in which case we strive to provide either sufficient background to enable the details to be completed without difficulty, or appropriate references. In an effort to make the book accessible to a wide audience we do not aim for the most general formulation, thus avoiding an excess of pure technicalities, nor do we have any qualms in attempting, occasionally, to present a lot of details, even at the risk of boring those able to proceed faster—such readers may indulge in judicious skipping. We hope that mathematicians, physicists, and engineers will find that the interplay between physical concepts/insights and mathematical ideas/methods provides a convincing description/explanation of some specific water-wave phenomena encountered in nature. We also believe that the reader without advanced expertise in mathematics/physics will be able to find some passages of interest: we have tried to present mathematical results in conjunction with physical interpretations, in order to provide some acquaintance with the subject.

I am very grateful to Professor Lokenath Debnath, the chair for this particular NSF-CBMS conference, for inviting me to be the principal speaker and for his efforts to assemble a successful NSF-CBMS application. Many thanks to the faculty and staff members of the Department of Mathematics at The University of Texas-Pan American for their gracious and warm hospitality, offering a perfect environment for scientific interaction and making the stay so enjoyable for all the participants. I also wish to thank all of the participants for providing stimulating discussions and contributing talks, keeping the lectures and workshop sessions so lively. I am grateful to the National Science Foundation (USA) for its generous support and to SIAM for publishing this book.

Throughout my activities related to the study of water waves I was fortunate to benefit from many conversations and collaborations with a number of researchers, profiting a lot from their generous sharing of ideas, insights and enthusiasm, as well as from their support and encouragement. I am especially indebted to Professor Joachim Escher, Professor Robin S. Johnson, and Professor Walter A. Strauss. I also wish to thank them for their advanced lectures on the occasion of this CBMS conference. Professor Joachim Escher, Professor David Lannes, Professor Walter A. Strauss, Dr. Mats Ehrnström, Mrs. Anna Geyer, Dr. David Henry, Dr. Boris Kolev, Mr. Raphael Stuhlmeier, Dr. Eugen Varvaruca, and Dr. Erik Wahlèn read portions of the draft manuscript. I have greatly profited from their comments. This book would have been far less than it now is without the invaluable service rendered by the thoughtful suggestions of Professor Robin S. Johnson on large parts of the manuscript. I am also indebted to several copyright owners for permission to reprint photographs and figures. The author would also like to thank Ann Manning Allen, Sara J. Murphy, Gina M. Rinelli, and Ronald C. Rosier for their help with the production of the book.

I cannot express adequately how much pleasure I have gained from consulting various publications (research/survey papers as well as lecture notes and books) related to these notes. Throughout the text I have tried to refer to these texts as an acknowledgement of their inspirational role but also as an indication of possible further reading. Nevertheless it is likely that I neglected to mention many other sources from which I have profited; if so, I sincerely apologize.

Adrian Constantin

Vienna, February 2011

Chapter 1

Introduction

"Remember, when discussing about water,
to induce first experience, then reason."
Leonardo da Vinci

Roughly 70 % of the Earth's surface is covered by water and waves can develop whenever the surface of the water is free to move, providing us with beautiful—or terrifying—sights. The study of water waves has had practical applications throughout history; its recent significance was demonstrated dramatically by the tsunamis of 2004 and 2011. The motion of water waves is very complex, due to the fundamental property that water, unlike solid structures which allow a limited degree of deformation without breaking, is a fluid capable of unlimited deformation and of distorting under even small disturbance forces. The great complexity of water-wave phenomena makes progress in their understanding depend on a fruitful cooperation between theory and experiment.

1.1 General considerations

1.1.1 The role of mathematics

Mathematics is essential as we seek to understand this fascinating field of physical reality, and adds immensely to the enjoyment obtained from the observation of the wealth of water-wave phenomena nature confronts us with: "Hydrodynamics is one of those fundamental areas where progress at any moment can be regarded as a standard to measure the real success of mathematical science" [17]. The study of water waves draws freely from various mathematical disciplines, but one cannot expect to use mathematical tools mechanically. Even at simple levels, subtle use of mathematical concepts is needed and many important achievements in the study of water waves are based on profound theories, some of which were pioneered in this context and "... stimulated developments in the domains of pure mathematics, such as complex analysis, topology ..." [17]. Also, cf. [250]: "water waves present sufficiently challenging problems that much of the development of the mathematical theory of wave motion over the past 200 years has been pioneered on their theory." The theory of water waves has profound and fascinating connections to many areas of classical and modern mathematics, with the added advantage that very often the behavior of the waves may be analyzed by direct observation. Without mathematics, the study of water waves, though possibly informative, leads to a very limited understanding: the absence of a mathematical formulation means that there is little indication of the justification/degree of credibility of a theory and it diminishes the possibility of extending observations to other, but different, scenarios on one's own account. A superficial treatment cannot offer the sense of satisfaction and involvement produced by genuine understanding, while, at the other extreme, focusing on advanced topics can produce a false sophistication which emphasizes technique rather than ideas—abstruse theoreticians eventually lose themselves in the arid labyrinths of pure technique.

1.1.2 Modelling

Given a physical situation, under certain simplifying assumptions, established laws from physics can be applied to obtain a model of the underlying processes. Physical insight determines the crucial factors governing a given phenomenon and helps to identify the factors that can safely be neglected. The conclusions reached will reflect reality (that is, specific situations which may be observed in nature or experimentally) only insofar as the accuracy of the model permits: the value of a model depends on the number of physically useful deductions which can be made from it. The quest is to capture the salient features of the phenomenon under investigation, based on the simplest possible mathematical model. The "truth" of the model is meaningless, just like perfection in a model is unattainable, as all experiments contain inaccuracies of measurement and, in addition, effects other than those accounted for cannot be totally excluded. One fascinating feature of the study of water waves is that their motion can exhibit qualitative differences at different scales: deep water versus shallow water, large wavelength versus small wavelength, and so on. Real-world observation of such qualitative differences establishes the credibility of mathematical models for water waves. The observation that the idealization involved in the mathematical technique of modelling, whereby some facts which are only known with some degree of accuracy are considered to be exact and from them mathematical statements (theorems) are deduced, often leads to a very accurate description of real-world phenomena is sometimes referred to as "the inconceivable effectiveness of mathematics in natural sciences" [316]. However, one should beware of purely abstract studies: "the fine poison of a mathematical education for the physicist" [179] leads, sometimes, to a theory that separates from reality and is no longer relevant to it. Gathering field data and performing experiments are key processes in the quest to understand and predict water waves, offering invaluable insight and providing motivation for further studies. Theory and experiment interlace: theory interprets, to some extent, experiments and builds upon them, while experiments confirm or discard theories. However, portraying scientific discovery as resulting from the confrontation of theoretical predictions with experiments, with perfect matching establishing unassailable scientific truth, is misleading (we will see a concrete example when discussing particle trajectories beneath Stokes waves). Last but not least, many theoretical developments were stimulated and supported by numerical computations (for example, solitons—nonlinear waves that can recover their shape after colliding—were discovered as a result of numerical simulations) and often such simulations are compelling even if they do not yet relate to rigorous theory. While numerical simulations are important to unveil possibilities that guide the subsequent analysis, rigorous theory is necessary to validate such investigations and their interpretation; without theory the computations are capable of producing reams of indigestible information that is practically useless. Theory is a conceptual tool, rather than merely a device to express and interpret numerical calculations.

1.2 Aims of the lectures

Nonlinear systems are disproportionate in cause and effect; they are not subject to the principle of superposition, so their behavior cannot be understood as a sum of the behaviors of their parts. Most predictions for water-wave motions use linear approximations, but this is adequate only for waves of small amplitude—for waves of moderate and large amplitude nonlinear effects cannot be neglected. We will discuss some recent important developments in the study of water waves of moderate and large amplitude.

- We investigate **wave-current interactions**, more precisely the existence theory of steady two-dimensional water waves propagating at the surface of water with a flat bed in a flow with a general vorticity distribution, providing also qualitative insight (e.g., the symmetry and regularity of the wave profiles and of the streamlines).
- We elucidate **the flow pattern beneath irrotational steady water waves** propagating at the surface of water with a flat bed (particle trajectories, properties of the fluid velocity field, behavior of the pressure), discussing both **Stokes waves** (periodic traveling waves) as well as **solitary waves**.
- We describe the insight into **wave breaking** provided by some integrable model equations arising in the shallow-water regime.
- We address some aspects of **tsunami modelling** with emphasis on the possible relevance of **soliton theory**.

Each of the above themes offers a wide range of open questions, for some of which conjectures supported by numerical or experimental evidence were previously formulated and are described here. The thrust of the presentation is analytical, but throughout the discussions we take advantage of input from experimental evidence and numerical simulations. Our main motivation for the present study starts from the observation that ocean waves are too untidy to fit into a general descriptive framework, and a direct numerical solution of the equations is often impossible, and even if computationally within reach, no significant quantitative or qualitative information about the behavior of solutions can be obtained in this way. Since understanding wave motions cannot derive solely from reading off nature, as even observations rest on a conceptual framework, to make progress there are practically no alternatives but to use the basic approach of mathematical physics: the specific water-wave motion under study is viewed as a perturbation of a known solution found in advance by other methods. By perturbation we understand that the equations governing the motion are close in an appropriate sense to equations that admit exact solutions, either available explicitly or, if implicit, providing in-depth qualitative information. Starting with these exact solutions one can work (analytically or numerically) with expansions of increasing accuracy and complexity toward extracting the relevant information about the specific problem under study. This approach toward describing the physical reality motivates us to investigate very regular waves that serve as starting points for more complicated wave patterns. In choosing the specific equations for which we seek exact solutions, one has to trade off between the desire for generality and the possibilities opened up for in-depth studies. For example, the first level of approximation is that of linear theory, in which case classical in-depth studies are available. Taking into account nonlinear effects, even at the level of weakly nonlinear theory, increases the phenomenological and the mathematical complexity. We try to motivate physically our simplifying assumptions (e.g., neglecting viscosity and surface tension) and, whenever possible, we will try to point out experimental studies that lend support to our theoretical considerations. While no mathematical expression of natural phenomena can be considered to represent the last word, certain have such overwhelming evidence in their support that it makes no sense to regard belief in them as provisional. Of course, it must be reckoned a mistake if we take them for granted in too wide a variety of circumstances, way beyond their original intent.

While a considerable amount of the material in the lectures is primarily appropriate for experts, introductory material and discussions intended to motivate further studies and to develop insight are included for advanced undergraduate students, for graduate students, and for newcomers to the field (in an attempt to provide a bridge between different mathematical levels). To make the presentation more suitable for independent study we aimed to provide a detailed exposition of the ideas involved. Knowledge of advanced calculus and linear algebra is assumed and a smattering of complex function theory, ordinary differential equations, partial differential equations, and functional analysis would be useful. The mathematical concepts are introduced as needed, and prior experience with fluid mechanics, though beneficial, is not expected. The methods come from diverse areas of mathematics, and, since the book is designed to serve both as an introductory text and as a reference for researchers, whenever more advanced material is required for our considerations we try to offer some background information by describing the principal ideas in that particular field. We hope that the material illustrates the interplay between mathematics (pure as well as applied), physics, and engineering which pervades the whole subject and gives it a special fascination.

Chapter 2

Preliminaries

"Scientific theories can be thought of as constituting maps of the physical world. No map ever represents everything that might be said."

J. Polkinghorne

In the study of water-wave flows, just like in most scientific endeavors, one uses theories that are fairly simple to specify but have complex implications: they are able to explain accurately a large class of observations and to make definite predictions about future observations. In this chapter, we present the framework that underpins our later developments by specifying the underlying physical assumptions and deriving from them the governing equations of the motion.

2.1 Simplifying assumptions

We now describe the fundamental simplifying assumptions, indicating the grounds for neglecting certain factors; we shall also define some notions and present the governing equations.

The macroscopic viewpoint: We consider the behavior of matter in the large (on a scale large compared to the distance between molecules) in order to evade the difficulties inherent in the modelling of a fluid as consisting of a granular structure of discrete molecules. This idealization amounts to the assumption that the water is distributed continuously (the continuum hypothesis). This hypothesis is in complete accord with the everyday observation of water where the idea of regarding its structure as anything other than continuous would be unnatural. This inevitably implies that it is possible to attach a definite meaning to the notion of the value of some property of the water at a point.

The assumption of constant density: The density of water is effectively a function of depth, temperature, and salinity [206] so that the following numerical data are relevant: temperature variations in the ocean, varying from the freezing point of seawater, $-2°C$, to values around $27°C$, are responsible for density changes of up to 0.5%, and salinity changes account for variations in density of up to 0.2%, while a descent of 1 km in depth increases the density by no more than 0.5%. This great resistance to changes in density allows us to regard water for most purposes as homogeneous to a very high degree of accuracy, with density $1000 kg/m^3$. Of course, at great depths of the ocean the change in density of the water is significant, but the variation of water density in the ocean is limited to about 4%, the maximum being attained at depths of around $10 km$.

The assumption of inviscid flow: We will neglect viscosity. Although water is a viscous fluid, "an inviscid approximation is often realistic since the velocity profile in the water, whether due to laminar viscosity or turbulent mixing, is usually established over length scales which are long compared with a wavelength" [89]. For a discussion of the rationale for neglecting the viscous action leading to wave decay see the Appendix to this chapter.

The motion of the air: The influence exerted by the air on the water's surface, by virtue of the Earth's atmosphere, is in the form of pressure acting on the surface, taken to be a constant. (The viscous action of the air at the surface is ignored.) This assumption decouples the motion of the air from that of the water so that in analyzing the motion of the water, we neglect the motion of the air above. While any disturbance of the free surface implies some motion of the air, the argument is made that the change in the pressure in the air due to this motion is negligible, implying that the air pressure may be approximated by its undisturbed value P_{atm} (constant atmospheric pressure). We comment that the mechanism mainly responsible for the most rapid stage of growth of sea waves, under the action of the wind, still remains to be found. For this reason, we suppose that at some moment in the past a disturbance was created on the surface of the still water in some fashion (by the wind, for example, but other causes are also possible, e.g., earthquakes in the case of tsunamis), and we investigate the subsequent motion of the water.

Gravity water waves: The balance between a restoring force and the inertia of the system governs the evolution of the disturbance of the surface of water—this is the so-called free surface (as its form is not prescribed a priori) and its determination is our primary objective. We will study **gravity waves** arising through the restoring action of gravity on water displaced from the equilibrium level (still water). We shall not investigate internal waves (occurring at the interface between two layers of water of different density), planetary waves (tides related to the effects of the Moon or of the Sun on oceanic movements, effects of the Earth's rotation), capillary waves (for which the flatness-restoring force is not gravity but surface tension), or capillary-gravity waves (for which the effects of surface tension are comparable with those of gravity). The effects of surface tension can be neglected for waves that are not very small (see the discussion in the Appendix to this chapter).

The assumption of a rigid flat bed: The topography of the bed has a great influence on the surface waves. In view of the present limited understanding of the fundamental processes involved, we will restrict our attention to a rigid flat bed. While a moving bed is of interest in the generation of tsunami waves by an undersea earthquake and, for sandy beds, the water flow induces a motion of the sand (as well as the percolation of the water through the bed), these phenomena are not investigated.

The assumption of two-dimensional flow: We will mainly consider regular wave trains of plane waves—**steady periodic two-dimensional waves**—for which the motion is identical in any direction parallel to the crest line and which propagate at constant speed in a fixed direction. For such a wave train, the **crest** is the most elevated part of the wave about the undisturbed water level, the lowest part of the wave that is below this surface being called the **trough**, while the vertical distance from the wave crest to wave trough is called the

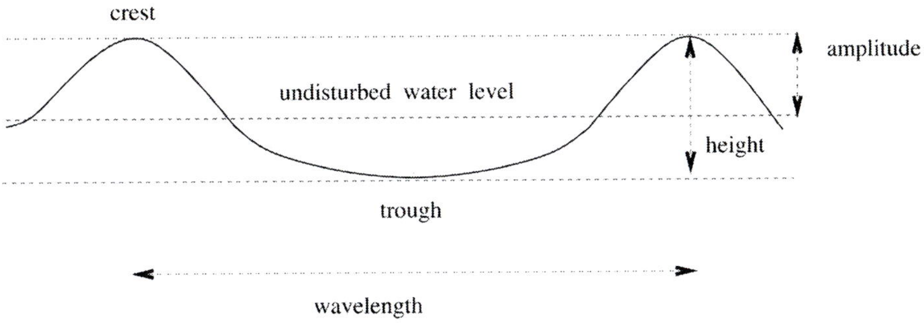

Figure 2.1. *The main characteristics of swell.*

Photograph 1. *Swell.* [From "Waves" by Steve Hawk; © 2005 by Steve Hawk. Used with permission from Chronicle Books LLC, San Francisco]

Linear wave theory predicts sinuosoidal wave profiles for regular wave trains with profiles that are monotone between crest and troughs but in the photograph taken from the Great Ocean Road (Australia), the crest is higher and narrower and the trough broader and less deep. Only nonlinear theory can capture this lack of crest/trough symmetry.

Photograph 2. *A regular wave train.* [From "Waves" by Steve Hawk; © 2005 by Steve Hawk. Used with permission from Chronicle Books LLC, San Francisco]

This wave train in water of small depth with a flat bed near the coast of California is even further from a sinusoidal wave pattern, being almost flat near the trough and presenting a pronounced elevation near the crest.

wave height. The **amplitude** of the wave is the maximum deviation of the wave from undisturbed water surface, being thus either the distance from the wave crest or the distance from the wave trough to the undisturbed water level; for sinusoidal profiles these two distances are the same, but periodic sea waves tend to have sharper elevations and flatter depressions so that the maximal elevation above the sea level usually exceeds the maximal depression below it. Wave amplitudes, measuring a few mm at the lower end, can attain a few tens of m at the upper end. The **wave period** is the time required for two successive crests (or two successive troughs) to pass a fixed point in space, while the horizontal distance between two successive crests (or troughs) is the **wavelength**, with range from cm to hundreds of km. The ratio of wavelength to wave period is the **wave speed**, with the fastest waves traveling at hundreds of km/h whereas the slowest travel below $1\,km/h$; cf. [206].

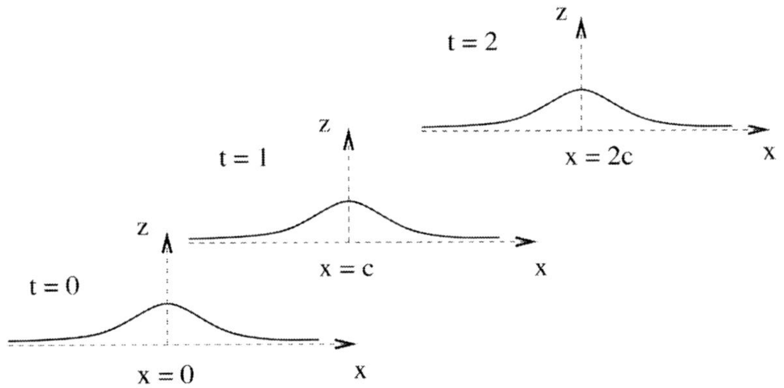

Figure 2.2. *Propagation of a solitary wave with speed c.*

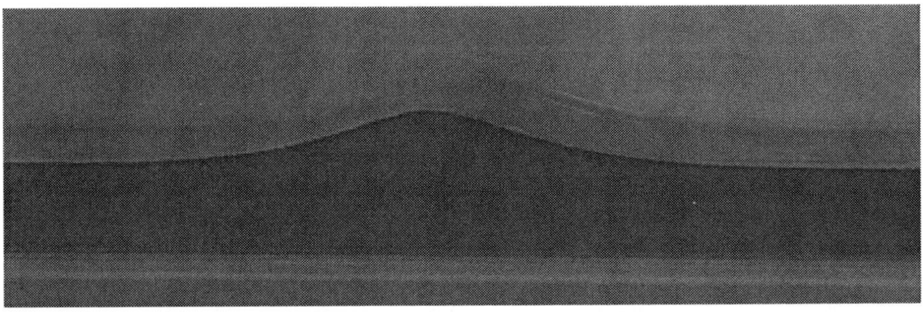

Photograph 3. *A solitary wave in the laboratory.*
[Reproduced by permission of Dr. S. Weber, Bayreuth University, Germany]

This wave of elevation (the entire surface is above the asymptotic water level far out), with a profile that is symmetric and localized around the wave crest, was generated in a $12m$ long rectangular wave tank, of width $10cm$, with mean water level $7cm$.

The occurrence of such waves can be observed in many different scenarios. Indeed, ocean waves are classified as either sea or **swell**. Irregular patterns made up of various waves with different speeds, wavelengths, and heights are called **sea**. When these waves move past the area of influence of the generating winds, they sort themselves into groups with similar speeds and wavelengths. This process produces swell that is characteristically a regular pattern of undulation of the ocean surface and serves as a prototype for steady periodic two-dimensional water waves. Swell often moves thousands of miles away from a storm to a shore somewhere (for example, swell originating from Antarctic storms has been recorded close to the Alaskan coast after traveling more than $8000km$).

Solitary waves: In addition to regular wave trains, we will investigate solitary waves—localized two-dimensional waves of a permanent form propagating at constant speed in a fixed direction. Solitary waves are less ubiquitous than wave trains but nevertheless represent observable and beautiful wave patterns.

Photograph 4. *A solitary wave in Molokai (Hawaii).*
[Copyright February, 1993 Robert I. Odom]

Photograph taken in February 1993 on the south shore of the Molokai island in the Hawaiian archipelago. Molokai has a $60\,km$ long reef located about $1.6\,km$ offshore, which runs almost the full length of the island. The inshore lagoon has almost constant depth of about $1\,m$ at high tide, the bottom being very uniform and mostly smooth with sand farther offshore and silt closer to shore. In the winter, distant Pacific storms produce waves which are almost normally incident on the reef where they break. Near-perfect solitary waves separate from the initial disturbance caused by the breaking waves and are observed to propagate shorewards almost from the reef edge.

Photograph 5. *A solitary wave in Maui (Hawaii).*
[Copyright February, 2003 Robert I. Odom]

Photograph taken in February 2003 at Maalea Bay on Maui island in the Hawaiian archipelago. Maalea Bay in central Maui has a very shallow bottom slope, which allows the incoming ocean waves to steepen without breaking into near-perfect solitary waves.

2.2 The governing equations

We now present a derivation of the basic equations for water waves and discuss the boundary conditions which select the water-wave problem from all other possible solutions of these equations. Throughout the presentation, we suppose that initially (at time $t = 0$) a disturbance is created on the surface of the still water in some fashion (by the wind, for example), and we wish to determine the subsequent motion of the water wave. Several assumptions (discussed in the previous section), that are relevant for water waves, will be made; we regard them as a mathematical convenience since the adequacy of any assumption can only be judged when results which it predicts accurately reflect what happens in reality. Most of the approximations that will be described can be traced to the use of plausible arguments, which have proved most fruitful in the physical understanding of these phenomena and are made partly for simplicity (even with them, one cannot solve the mathematical problems). There are certain mathematical conditions to be satisfied throughout the derivation of the governing equations, such as differentiability hypotheses; that these hold in all cases where their necessity arises will be assumed on grounds of physical continuity. It might be thought that the study of water waves in this idealized form could throw but little light on the behavior of actual water waves. We shall see that this is so far from being the case that the theory can, in important instances, explain not only qualitatively, but at times also quantitatively, observed phenomena.

2.2.1 The equation of mass conservation

Imagine a volume V, bounded by a C^1-surface S, within water of constant density ρ. If $\mathbf{n}$ is the outward unit normal on S and $\mathbf{u} \in C^1$ is the fluid velocity at a point in the water, the outward velocity component across S is $\mathbf{u} \cdot \mathbf{n}$. Decomposing $\mathbf{u}$ into normal and tangential components, the tangential ones do not contribute to the outflow, so that the rate at which mass flows out of V is

$$\int_S \rho \mathbf{u} \cdot \mathbf{n} \, dS.$$

The rate of change of mass in V is

$$\frac{d}{dt} \int_V \rho \, d\mathbf{x} = \int_V \frac{d\rho}{dt} \, d\mathbf{x} = 0$$

in view of the homogeneity assumption (constant density). Since matter (mass) is neither created nor destroyed anywhere in the water, the rate of change of mass in V is brought about only by the rate of mass flowing into V across S. But we know that this rate of change is zero, so

$$\mathbf{0} = -\int_S \rho \mathbf{u} \cdot \mathbf{n} \, dS$$

or

$$\int_V \nabla \cdot \mathbf{u} \, d\mathbf{x} = 0,$$

by the divergence theorem. This has to be true for any volume V, yielding the **equation of mass conservation** (for an incompressible fluid)

$$\nabla \cdot \mathbf{u} = 0. \tag{2.1}$$

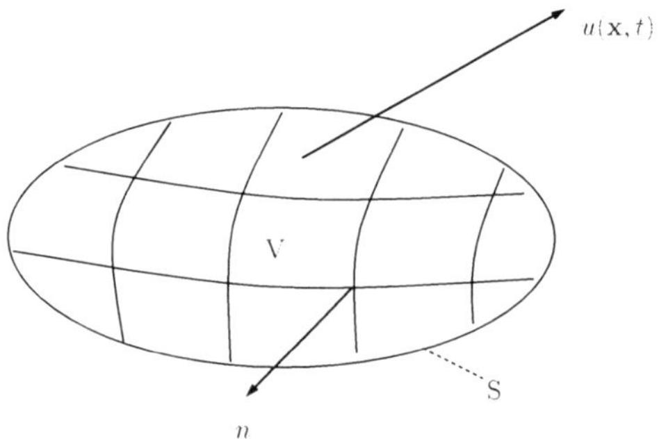

Figure 2.3. *The mass crosses the boundary S of the water volume V at a rate equal to the surface integral of* $\rho\,\mathbf{u}\cdot\mathbf{n}$ *over S.*

2.2.2 The equation of motion: Euler's equation

There are two types of forces that are relevant in fluid mechanics: these distinguish between body forces (external forces), having their source exterior to the fluid and being practically the same for all particles, and internal forces (local forces) exerted on a fluid element by other elements nearby; the latter are the forces which obey Newton's third law (equality of action and reaction). For gravity water waves, the body force is due to gravity so that, in Cartesian coordinates $\mathbf{x} = (x, y, z)$ with z measured upward, $\mathbf{F} = (0, 0, -g)$, where $g > 0$ is a constant. The internal forces here are pressure forces, whereby a lump of water is acted on across its surface by the rest of the continuum in a purely normal direction to the surface (there are no tangential components): the internal forces acting on an infinitesimally small area δS within the water region take the form of the thrust due to the pressure (a scalar quantity P depending on the position of the point at which it is measured and on time)—a force of magnitude $P(\delta S)$ in the direction of the inward normal. This characterization of the internal forces is valid in an **ideal fluid**, a concept supposing the matter to be continuously distributed and the fluid to be inviscid. The concept of pressure in an incompressible fluid presents certain difficulties: in a compressible fluid, given an adiabatic relation $P = P(\rho)$, for example, the pressure determines the density, but here the variable of state, P, is denied any influence upon the way the fluid ocupies the space. The **hydrodynamic pressure** P has to be regarded as a reaction to the constraint of incompressibility (2.1).

Consider now a volume V having as boundary the C^1-surface S within the water. We assume that $\mathbf{u}$ and P are of class C^1 throughout the fluid. The total force (body + local) acting on the water in V is

$$\int_V \rho\,\mathbf{F}\,d\mathbf{x} - \int_S P\,\mathbf{n}\,dS,$$

$\mathbf{n}$ being the outward unit normal on S. Using the divergence theorem, we transform the above expression into

$$\int_V (\rho\,\mathbf{F} - \nabla P)\,d\mathbf{x}.$$

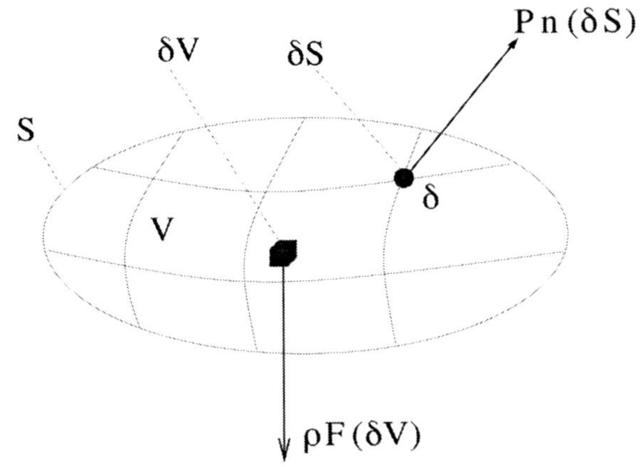

Figure 2.4. *Pressure force across the surface element δS and the external force (gravity) acting on the water volume element δV.*

We now apply Newton's law of motion on the body of water surrounded by S:

$$\int_V \rho \frac{D\mathbf{u}}{Dt} d\mathbf{x} = \int_V (\rho \mathbf{F} - \nabla P) d\mathbf{x}. \tag{2.2}$$

The left side is the sum of the acceleration times mass (recall that density is mass/volume) for each particle in the volume V, while $\frac{D}{Dt}$ is the total differentiation following the particular particle. Indeed, the velocity vector of a particle is $\mathbf{u} = \mathbf{u}(\mathbf{x},t)$, so that the particle follows the path $\mathbf{x}(t)$ on which $\frac{d\mathbf{x}}{dt} = \mathbf{u}(\mathbf{x}(t),t)$, and thus the acceleration of the particle is

$$\frac{d^2\mathbf{x}}{dt^2} = \frac{\partial \mathbf{u}}{\partial t} + (\mathbf{u}\cdot\nabla)\mathbf{u} = \frac{D\mathbf{u}}{Dt},$$

as used in (2.2). Since the V in (2.2) is arbitrary, we obtain the **Euler equation** of motion

$$\frac{D\mathbf{u}}{Dt} = -\frac{1}{\rho}\nabla P + (0,0,-g). \tag{2.3}$$

The pressure P appears as the fourth unknown besides the components of the velocity $\mathbf{u}$. Note the nonlinear character of Euler's equation caused by the presence of the term $\mathbf{u}\cdot\nabla\mathbf{u}$. This prohibits the use of the principle of superposition and makes the analysis of the equations a considerably difficult mathematical problem.

2.2.3 The boundary conditions

There are two boundaries of the fluid region: the rigid flat bed and the water's free surface (assuming that the fluid extends to infinity in all horizontal directions). The **kinematic boundary conditions** (called this because they do not involve the action of forces) express the fact that a particle on the boundary remains there at all times.

Let $S(\mathbf{x},t)=0$ be the equation of the C^1 free surface, where $S:\mathbb{R}^4\to\mathbb{R}$ is a C^k-function (with integer $k\geq 1$) with nonvanishing spatial gradient,[1] $(S_x,S_y,S_z)\neq\mathbf{0}\in\mathbb{R}^3$. Choosing an arbitrary particle $\mathbf{x}_0$ lying at time $t=t_0$ on the free surface, the subsequent location $\mathbf{x}(t)$ is given by the unique solution of the differential equation $\mathbf{x}'(t)=\mathbf{u}(\mathbf{x}(t),t)$ with initial data $\mathbf{x}(t_0)=\mathbf{x}_0$, and which satisfies $S(\mathbf{x}(t),t)=0$ for $t\geq t_0$. Differentiation with respect to t shows that a necessary condition is $\frac{DS}{Dt}=0$ along the surface. To see that this condition is also sufficient is less obvious. We now have to deal with a so-called flow-invariance problem [261]: if the initial data lies on the free surface, does this property persist for the resulting solution of the differential equation $\mathbf{x}'=\mathbf{u}(\mathbf{x},t)$? Formally, the answer should be in the affirmative as the condition that $\frac{DS}{Dt}=0$ all along the surface expresses precisely the fact that the flow $(\mathbf{x}(t),t)$ is everywhere tangential to the time-dependent free surface, and the absence of a normal component should prevent particles from leaving the free surface. To be more precise, this physical interpretation relies on the fact that, when water is in contact with a rigid surface (bottom) or with another fluid (air) with which it does not mix, the water and the surface with which it is in contact must have the same velocity normal to the surface if contact is to be preserved; the surface is defined by the property that water does not cross it. If $\mathbf{n}$ is the normal unit vector at the point $\mathbf{P}$ of the surface of contact, $\mathbf{u}=(u_1,u_2,u_3)$ is the velocity of water, and $\mathbf{q}$ is the velocity of the point $\mathbf{P}$ of the surface, we must have $\mathbf{q}\cdot\mathbf{n}=\mathbf{u}\cdot\mathbf{n}$ or $(\mathbf{u}-\mathbf{q})\cdot\mathbf{n}=0$ (see Figure 2.5). If $S(x,y,z,t)=0$ is the equation of the surface, then

$$\mathbf{u}\cdot\mathbf{n}=\frac{u_1S_x+u_2S_y+u_3S_z}{\sqrt{S_x^2+S_y^2+S_z^2}},\qquad \mathbf{q}\cdot\mathbf{n}=-\frac{S_t}{\sqrt{S_x^2+S_y^2+S_z^2}}.$$

The equation $\frac{DS}{Dt}=0$ therefore expresses the condition that the velocity of the water relative to the surface is perpendicular to the normal; that is, the velocity of water relative to the surface is everywhere tangential to the surface. These considerations have a firm mathematical basis (see the Appendix to this chapter).

On the free surface $z=h(x,y,t)$ the kinematic boundary condition is

$$u_3=h_t+u_1h_x+u_2h_y \quad \text{on} \quad z=h(x,y,t), \tag{2.4}$$

while on the fixed, horizontal, flat bed $z=-d$ the kinematic boundary condition is

$$u_3=0 \quad \text{on} \quad z=-d. \tag{2.5}$$

For gravity waves we ignore the effects of surface tension; the influence exerted by the atmosphere on the water surface is therefore[2] in the form of pressure acting on the surface, taken to be the (constant) atmospheric pressure P_{atm}. This gives the **dynamic boundary condition**

$$P=P_{atm} \quad \text{on} \quad z=h(x,y,t), \tag{2.6}$$

[1] Imposing this gradient condition amounts to excluding waves with rough free surfaces since if, at some instant t, the level set $\{\mathbf{x}\in\mathbb{R}^3:\ S(\mathbf{x},t)=0\}$ fails to be a hypersurface (a C^k-submanifold of dimension 2 in $\mathbb{R}^3$), then the gradient vector must be zero at all singular points [135].

[2] If surface tension is taken into account, then the difference of pressure on the two sides of the free surface is balanced by the effects of surface tension.

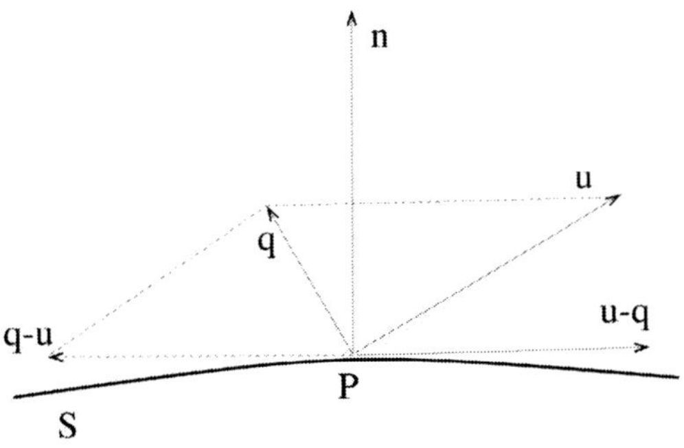

Figure 2.5. *The physical interpretation of the kinematic boundary condition.*

which reflects the external actions on the free surface (at the bottom we do not encounter this type of condition as no interactions occur there) and so decouples the motion of the air from that of the water; we may then concentrate on the motion of the water without considering the motion of the air. While any disturbance of the free surface clearly implies some motion of the air, the argument is made that the change in the air pressure due to this motion is negligible because its density is small,[3] so the air pressure may be approximated by its undisturbed value P_{atm}.

The general description of the water-wave problem for gravity waves is encompassed by Euler's equation (2.3) and the equation of mass conservation (2.1) together with the boundary conditions (2.4)–(2.6): these are the **governing equations for water waves**. A distinctive feature in water-wave theory is that one of the boundaries—the free surface—is not known and must be determined as part of the solution. This is a **free boundary problem**: the shape of the surface waves and the motion of the fluid are interdependent.

2.2.4 Well-posedness

The first issue that needs to be addressed in connection with the governing equations for water waves is their **well-posedness**. If these equations are a genuine mathematical model of physical phenomena, they should have the properties that, if appropriate data are specified initially (say, at time $t = 0$), then, at least in some small time interval $[0, T)$ with $T > 0$:

(i) a solution exists;
(ii) the solution is unique;
(iii) the solution depends continuously on the data, in some reasonable topology.

The necessity of the existence of a solution, condition (i), is motivated by physical reality. It is not the only requirement that is necessary. Experience suggests the hypothesis of determinism: if the same initial scenario repeats itself exactly, we should get the same outcome, that is, (ii) must hold. The last requirement, (iii), is also needed for a satisfactory model. Since one can never be certain that particular initial conditions can be repeated exactly, then, if the data are almost the same, we expect the outcome to be almost identical,

[3]Air near sea level has a density that averages $1.275\,kg/m^3$, while that of the water is about $1000\,kg/m^3$.

at least for some time. The reason why the formulation of well-posedness is local in character (that is, for some small time interval) is that in the context of water waves one has to make provisions for wave breaking. While no physical and little qualitative insight is to be gained from the study of well-posedness—and for this reason, rather than pursuing an in-depth study of this problem, we will only point out the essential features without going into technical details—this aspect of the problem is of fundamental importance. Lack of well-posedness invalidates the model.

The mathematics behind the well-posedness of the governing equations is deep and interesting, and still not completely resolved, despite remarkable recent advances. While a satisfactory general well-posedness result remains elusive, it is known that the governing equations, for flows without vorticity (see below), are well-posed.

Let us first formulate the well-posedness problem in more precise terms before briefly discussing the most relevant results. Given the rigid bed $z = -d$, the data (at time $t = 0$) consists of the initial surface $z = h_0(x, y)$ and of the initial divergence-free irrotational velocity field $\mathbf{u}_0(\mathbf{x})$ with $(h_0, \mathbf{u}_0) \in H^{s+1}(\mathbb{R}^2) \times H^{s+1}(\mathbb{R}^3)$ with the Sobolev[4] index $s \geq 3$. Then there is a unique solution

$$(h, \mathbf{u}) \in C^1([0, T); H^s(\mathbb{R}^2) \times H^s(\mathbb{R}^3))$$

with initial data $(h_0, \mathbf{u}_0)$, and such that $\mathbf{u}(\cdot, t) : \mathbb{R}^3 \to \mathbb{R}^3$ is divergence free and irrotational at any fixed instant $t \in [0, T)$. Notice that $s \geq 3$ ensures that both h and $\mathbf{u}$ are C^1 by Sobolev embeddings [111]. In order to include the largest variety of possible water flows, one would like to ask for the minimal regularity for these functions; continuously differentiable functions come naturally to mind. The mathematical technicalities that are encountered suggest the choice of Sobolev functions as the well-posedness discussion is more elegantly described in the setting of these Hilbert spaces.

Notice that we have not specified anything about the pressure. The reason for this is that the mathematical role of the pressure inside the fluid domain is to keep the evolution of the velocity within the space of divergence-free vectors: the role of the pressure within the fluid is to guarantee that the left side of the Euler equation (2.3) is a gradient, a condition that can be expressed exclusively (but in a less elegant form) in terms of the components of the velocity field. The dynamic boundary condition (2.6) can also be expressed as a condition on $(h, \mathbf{u})$: at any fixed instant $t \in [0, T)$, knowing the free surface $z = h(x, y, t)$ and the velocity field $\mathbf{u}(\mathbf{x}, t)$, the condition that the pressure is constant on $z = h(x, y, t)$ can be expressed exclusively in terms of $\mathbf{u}$ and h. This is accomplished by using Euler's equation (2.3) to write the condition that any derivative of $P(\cdot, t)$ in a tangential direction to the free surface vanishes, and the possible time dependence of the constant value $P_0(t)$ of P restricted to the free surface is eliminated by requiring $P_0'(t) = 0$, a relation that again can be expressed as a condition on $\mathbf{u}$ and h. A hint that the pressure P is not to be regarded as an independent variable in the governing equations, being determined by the configuration of the fluid velocity and by the shape of the free surface, is given by observing that in the governing equations P is defined up to an additive irrelevant constant: for any constant $P_0 \in \mathbb{R}$, the function $P_0 + P$ becomes the pressure provided that we allow the adjustment of (2.6). Despite all this, the pressure P is not merely an ornamental device for the purpose of elegant formulations: it is physically relevant and of invaluable use in certain mathematical

[4]See the Appendix to Chapter 2 for a brief discussion of Sobolev spaces.

considerations, as we shall see in Chapter 3 when we discuss the symmetry of periodic traveling waves.

A water flow without swirls is called **irrotational**, the mathematical expression of this condition taking the simple form of imposing that the velocity field is curl free. Irrotational water waves are common physical occurrences and there are considerable mathematical advantages in the analysis of these flows, as we shall see. For our present discussion the most relevant information is that an initially irrotational flow remains so at later times (see the discussion of vorticity below). Lannes [194] proved the well-posedness of the governing equations for irrotational water waves without imposing the condition that the motion is a small perturbation of still water, as was the case in earlier work. As shown by Wu [318], the result goes well beyond our framework, allowing even for initial wave profiles that are overhanging (and therefore not representable in the form $z = h(x,y,t)$ for some function h), provided the water's surface is non-self-intersecting. Thus the governing equations for irrotational water waves have a firm mathematical foundation.

For general flows, that is, allowing for nonzero vorticity, the well-posedness issue for the governing equations for gravity water waves is not yet settled. Nevertheless, if initially all along the water's free surface the exterior normal derivative of the pressure $\frac{\partial P}{\partial n}$ is negative, that is, if there is some $c_0 > 0$ such that

$$\frac{\partial P}{\partial n}(\mathbf{x},0) \leq -c_0 < 0 \quad \text{on} \quad z = h(x,y,0),$$

Lindblad [207] and Coutand and Shkoller [80] proved that the governing equations are well-posed (and throughout the existence time a condition of this type on the pressure will hold). This condition is not required in the irrotational case because in this setting one can prove that it holds automatically. However, for water flows with vorticity this condition has to be imposed initially, as failure to do so might result in the governing equations being ill-posed, as proved by Ebin [104].

We conclude our discussion of the well-posedness of the governing equations for water waves by highlighting an important gap in the theory. This concerns the issue of breaking waves: characterizing how a solution fails to exist after some finite time. There are solutions that exist for all times. Indeed, for three-dimensional flows Germain, Masmoudi, and Shatah [124] proved recently that if the irrotational data have small size in a certain weighted Sobolev norm, then the solution exists for all times $t \geq 0$. The key ingredients of their proof are high-order energy estimates and dispersive estimates which are obtained by the analysis of the spatial and temporal resonances (the three-dimensionality being advantageous because the basic decay rate is faster and this contributes to improved dispersive estimates). As for two-dimensional flows, in Chapter 3 we will prove the existence of steady traveling waves, allowing even the presence of nonzero vorticity in the flow. However, the wave-breaking phenomenon is commonly observed for waves near the shore or in the open sea. It seems intuitively obvious that solutions do not exist for very long after they overturn (that is, after the slope of the free surface becomes infinite so that viewing it as the graph of a function is no longer possible) but a rigorous mathematical proof of this fact is not available. Nor are there special examples to guide us, and the physical contention is not even supported by some partial geometrical results (for example, of the type that describe how the curvature of the surface might diverge at some point). This remains a challenging problem of great importance.

2.3 Vorticity

A fundamental property of a fluid flow is the **vorticity**,

$$\boldsymbol{\omega} = \operatorname{curl} \mathbf{u} = \nabla \wedge \mathbf{u},$$

measuring the local spin or rotation of a fluid element (thereby producing a rotational or swirling motion).

To see that the vorticity $\boldsymbol{\omega}$ is a measure of the local rotation of the particles, consider for simplicity a time-independent flow with a smooth velocity field $\mathbf{u}(\mathbf{x})$; we show that, to leading order (i.e., the linear approximation), the flow in a neighborhood of some fixed point, $\mathbf{x_0}$, of the fluid domain can be viewed as a combination of infinitesimal translation, rotation, and deformation, with the vorticity accounting for the rotational part. Indeed, the smooth velocity field has a Taylor series expansion near $\mathbf{x_0}$:

$$\mathbf{u}(\mathbf{x}) = \mathbf{u}(\mathbf{x_0}) + [(D\mathbf{u})(\mathbf{x}_0)](\mathbf{x} - \mathbf{x_0}) + O(|\mathbf{x} - \mathbf{x_0}|^2)$$

for $|\mathbf{x} - \mathbf{x_0}|$ small, where $(D\mathbf{u})(\mathbf{x}) = \left(\frac{\partial u_i}{\partial x_j}\right)_{1\le i,j\le 3}(\mathbf{x})$ is the differential of $\mathbf{u}$ at $\mathbf{x}$. We write $(D\mathbf{u})(\mathbf{x_0})$ as the sum of its symmetric and antisymmetric parts,

$$(D\mathbf{u})(\mathbf{x_0}) = D_S(\mathbf{x_0}) + D_A(\mathbf{x_0})$$

with

$$D_S(\mathbf{x_0}) = \frac{(D\mathbf{u})(\mathbf{x_0}) + [(D\mathbf{u})(\mathbf{x_0})]^T}{2}, \qquad D_A(\mathbf{x_0}) = \frac{(D\mathbf{u})(\mathbf{x_0}) - [(D\mathbf{u})(\mathbf{x_0})]^T}{2},$$

where J^T denotes the transposed matrix of J. A straightforward calculation yields

$$2\,D_A(\mathbf{x_0})\mathbf{h} = \boldsymbol{\omega}(\mathbf{x_0}) \wedge \mathbf{h}, \qquad \mathbf{h} \in \mathbb{R}^3,$$

so that to leading order in $|\mathbf{h}|$, where $\mathbf{h} = \mathbf{x} - \mathbf{x_0}$, the velocity field is the (unique) sum of three terms:

$$\mathbf{u}(\mathbf{x_0} + \mathbf{h}) \approx \mathbf{u}(\mathbf{x_0}) + D_S(\mathbf{x_0})\mathbf{h} + \frac{1}{2}\,\boldsymbol{\omega}(\mathbf{x_0}) \wedge \mathbf{h}. \tag{2.7}$$

The three terms have a natural physical interpretation in terms of translation, deformation, and rotation, as we now demonstrate.

If $\mathbf{x}(0)$ is a particle location near $\mathbf{x_0}$ at time $t = 0$, then its trajectory is $\mathbf{x}(t) = \mathbf{x_0} + \mathbf{h}(t)$, so that in the linear approximation we are interested in understanding the behavior near $\mathbf{0}$ of the solutions to the differential equation

$$\frac{d\mathbf{h}}{dt} = \mathbf{u}(\mathbf{x_0}) + D_S(\mathbf{x_0})\mathbf{h} + \frac{1}{2}\,\boldsymbol{\omega}(\mathbf{x_0}) \wedge \mathbf{h} \tag{2.8}$$

with initial data $\mathbf{h}(0) = \mathbf{x}(0) - \mathbf{x_0}$. We think of each of the three terms in the sum appearing on the right-hand side of (2.8) as a vector field generating a flow near $\mathbf{0}$.

(I) Retaining only the first term in (2.8), we solve the differential equation

$$\frac{d\mathbf{h}(t)}{dt} = \mathbf{u}(\mathbf{x_0})$$

with the solution $\mathbf{h}(t) = \mathbf{h}(0) + \mathbf{u}(\mathbf{x_0})t$ describing an **infinitesimal translation**.

(II) The flow generated by the second term in (2.8) represents a **stretching** of the flow along the eigenspaces of the symmetric matrix $D_S(\mathbf{x_0})$, called the deformation or rate-of-strain matrix; cf. [216]. Indeed, this diagonalizable matrix has three real eigenvalues $\lambda_1 \leq \lambda_2 \leq \lambda_3$, not necessarily distinct, such that the corresponding eigenvectors $\mathbf{e}_i$ $(i = 1, 2, 3)$ span $\mathbb{R}^3$, the stretching along these directions being simply the information encoded in the defining relations

$$D_S(\mathbf{x_0})\,\mathbf{e}_i = \lambda_i\,\mathbf{e}_i \quad \text{for} \quad i = 1, 2, 3.$$

To see this, observe that, if we ignore the first and third terms in (2.8), we obtain

$$\frac{d\mathbf{h}}{dt} = D_S(\mathbf{x_0})\,\mathbf{h}.$$

Writing $\mathbf{h}(t) = h_1(t)\mathbf{e}_1 + h_2(t)\mathbf{e}_2 + h_3(t)\mathbf{e}_3$, the above equation is equivalent to three elementary linear equations

$$\frac{dh_i}{dt} = \lambda_i\,h_i, \qquad i = 1, 2, 3.$$

The rate of change along the $\mathbf{e}_i$-axis is λ_i, and so we obtain

$$h_i(t) = h_i(0)\,e^{\lambda_i t} \quad \text{for} \quad t \geq 0.$$

Thus the local effect of the vector field $D_S(\mathbf{x_0})$ is merely an expansion or a contraction along each of the axes $\mathbf{e}_i$ (according to whether λ_i is positive or negative), hence the name "stretching" or "deformation." Furthermore, if $\mathbf{u}$ satisfies the equation of mass conservation (2.1), then $0 = \text{trace}\,(D\mathbf{u}(\mathbf{x_0})) = \text{trace}\,(D_S(\mathbf{x_0}))$ so that

$$\lambda_1 + \lambda_2 + \lambda_3 = 0.$$

Consequently, the deformation associated to $D_S(\mathbf{x_0})$ is volume preserving, because a fluid flowing with velocity $\mathbf{u}$ is being compressed at the rate $\nabla \cdot \mathbf{u}$ (see the disccusion in the Appendix to this chapter) and the velocity field of an incompressible fluid is divergence free.

(III) The third flow, obtained by solving

$$\frac{d\mathbf{h}}{dt} = \frac{1}{2}\,\boldsymbol{\omega}(\mathbf{x_0}) \wedge \mathbf{h}, \tag{2.9}$$

defines a rigid **rotation**[5] with angular velocity $\frac{1}{2}\,\omega(\mathbf{x_0})$. Indeed, assuming $\boldsymbol{\omega}(\mathbf{x_0}) \neq \mathbf{0}$, the solution $\mathbf{h}(t)$ to (2.9) describes a motion at constant angular speed $\theta = \frac{|\omega(\mathbf{x_0})|}{2}$ and with a fixed radius: motion around a circle lying in a plane orthogonal to the vector $\boldsymbol{\omega}(\mathbf{x_0})$. To see this, choose the unit vectors $\mathbf{f}$ and $\mathbf{j}$ such that $\mathbf{e} = \frac{\omega(\mathbf{x_0})}{|\omega(\mathbf{x_0})|}$, $\mathbf{f}$ and $\mathbf{j}$ form an orthonormal basis for $\mathbb{R}^3$. Let

$$\mathbf{h}(0) = d\,\mathbf{e} + L\,\cos(\varphi)\mathbf{f} + L\,\sin(\varphi)\mathbf{j}$$

[5]In this analogy with rigid body motion the factor $\frac{1}{2}$ is a flaw but its removal by absorbtion in the definition of the curl symbol would have uncomfortable consequences for vector analysis.

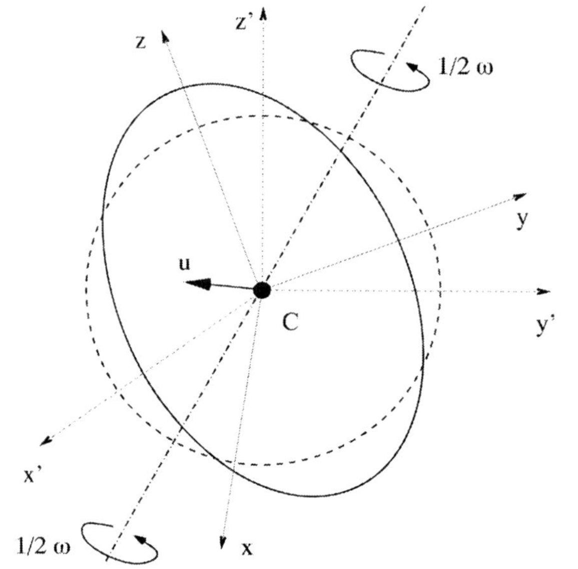

Figure 2.6. *The instantaneous deformation of a small spherical fluid particle can often be decomposed into three constituents:* (i) *uniform translation at the velocity* **u** *of the particle's centre C;* (ii) *rigid rotation with angular velocity* $\frac{1}{2}\,\boldsymbol{\omega}$*, where* $\boldsymbol{\omega}$ *is the vorticity at C;* (iii) *pure rate of strain (rate of change of shape), which in the principal axes* (x,y,z) *of rate of strain (the coordinate axes are represented as* x'*,* y'*,* z'*) takes the form of an elongation in the x and z directions and a foreshortening in the y direction.*

with φ being the angular coordinate in the $(\mathbf{f},\mathbf{j})$-plane, and observe that

$$\mathbf{h}(t) = d\,\mathbf{e} + L\,\cos(\varphi + \theta\,t)\mathbf{f} + L\,\sin(\varphi + \theta\,t)\mathbf{j}$$

satisfies

$$\mathbf{h}'(t) = -L\theta\,\sin(\varphi + \theta\,t)\mathbf{f} + L\theta\cos(\varphi + \theta\,t)\mathbf{j} = \frac{1}{2}\,\boldsymbol{\omega}(\mathbf{x_0}) \wedge \mathbf{h}(t).$$

By the uniqueness theorem for the linear differential equation (2.9) this must be the solution with initial data $\mathbf{h}(0)$. Thus the flow (2.9) is a rotation with constant angular velocity, the axis of rotation being given by $\boldsymbol{\omega}(\mathbf{x_0})$.

Splitting the vector field on the right side of (2.8) into these three components, we have described the elements that make up the local flow behavior at the leading order (see Figure 2.6). It is tempting to conclude that the flow (2.8) can therefore be viewed as a combination of a translation, a deformation, and a rigid rotation; this is true to some extent.

Recall that if A is a 3×3 matrix and $\mathbf{b_0} \in \mathbb{R}^3$ is a vector, the unique solution of the initial value problem

$$\begin{cases} \mathbf{z}'(t) = A\,\mathbf{z}(t) + \mathbf{b_0}, \\ \mathbf{z}(0) = \mathbf{z_0} \in \mathbb{R}^3 \end{cases}$$

is given by

$$\mathbf{z}(t;\mathbf{z_0}) = e^{At}\mathbf{z_0} + \Big(\int_0^t e^{A(t-s)}\,ds\Big)\,\mathbf{b_0}, \qquad t \in \mathbb{R},$$

where $e^{At} = \sum_{k=1}^{\infty} \frac{A^k t^k}{k!}$, the convergence being uniform on compact intervals of $\mathbb{R}$. Given another 3×3 matrix B, it is known that $e^{(A+B)t} = e^{At}e^{Bt}$ for all $t \in \mathbb{R}$ if and only if A and B commute. The solution to (2.8) is therefore

$$\mathbf{h}(t) = e^{(D_S(\mathbf{x_0})+D_A(\mathbf{x_0}))t}\,\mathbf{h}(0) + \Big(\int_0^t e^{\big(D_S(\mathbf{x_0})+D_A(\mathbf{x_0})\big)(t-s)}\,ds\Big)\,\mathbf{u}(\mathbf{x_0}),$$

and earlier we established that the maps $\mathbf{h} \mapsto e^{D_S(\mathbf{x_0})t}\mathbf{h}$ and $\mathbf{h} \mapsto e^{D_A(\mathbf{x_0})t}\mathbf{h}$ represent a deformation along three fixed axes and a rotation about a fixed axis, respectively. Therefore, if the matrices $D_S(\mathbf{x_0})$ and $D_A(\mathbf{x_0})$ commute, then the flow (2.8) represents a rotation with a fixed axis followed by a deformation along three fixed axes (and vice versa, by commutativity),

$$\mathbf{h}(0) \to e^{D_S(\mathbf{x_0})t}\Big(e^{D_A(\mathbf{x_0})t}\,\mathbf{h}(0)\Big),$$

and followed by a translation by the vector

$$\Big(\int_0^t e^{(D_S(\mathbf{x_0})+D_A(\mathbf{x_0}))(t-s)}\Big)\,\mathbf{u}(\mathbf{x_0}).$$

Unfortunately, an easy calculation shows that the matrices $D_S(\mathbf{x_0})$ and $D_A(\mathbf{x_0})$ commute if and only if $D\mathbf{u}(\mathbf{x_0})$ commutes with its transpose, and this is generally not true even if (2.1) holds. Therefore, it is not true that, locally,[6] the flow can always be decomposed into a rotation with a fixed axis, a deformation along three fixed axes, and a translation. However, performing n pairs of such rotations combined with deformations, all being succeeded by a translation, gives a reasonable approximation of the flow if n is large enough, in view of Trotter's product formula

$$e^{(A+B)t} = \lim_{n\to\infty}\Big(e^{At/n}\,e^{Bt/n}\Big)^n \quad \text{uniformly on compact intervals,}$$

which is valid even for matrices that do not commute; cf. [43].

We would like to emphasize the **local nature** of the rotation measured by the vorticity; from the knowledge of vorticity nothing can be concluded with respect to a global rotation of the water.

Example 2.1. In $\mathbb{R}^3$, for the two-dimensional flow with velocity

$$\mathbf{u} = (y, 0, 0)$$

the fluid is not rotating in any sense despite having vorticity $\boldsymbol{\omega} = (0,0,-1)$. The path of a particle located at (x_0, y_0, z_0) at time $t = 0$ is

$$t \mapsto (x_0 + t\,y_0,\; y_0,\; z_0)$$

[6]We emphasize that these considerations have a physical meaning only locally in space and time since they depend on linearizing.

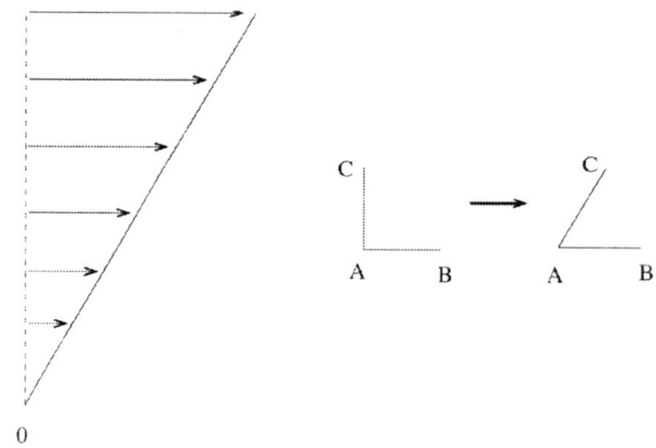

Figure 2.7. *The nonuniform translational flow with vorticity in Example* 2.1 *does not involve a global rotation of the water.*

so that all particles move horizontally in straight lines. Clearly the fluid does not rotate in any sense, but the relative positions of three particles A, B, C (with the two momentarily perpendicular fluid line elements AB and AC oriented as shown in Figure 2.7 in a horizontal plane $z = z_0$) change in time, indicating the presence of some local spin.

Computing $(D\mathbf{u})(\mathbf{x_0})$ explicitly at an arbitrary but fixed point $\mathbf{x_0} \in \mathbb{R}^3$, one can easily check that $(D\mathbf{u})(\mathbf{x_0})$ does not commute with its transpose. Consequently, in view of the above considerations, we do not expect a simple decomposition of the flow into a rotation with angular speed $\frac{1}{2}$ around the negative z-axis, a deformation along the axes

$$\mathbf{e_1} = (0,0,1), \quad \mathbf{e_2} = \frac{1}{\sqrt{2}}(1,1,0), \quad \mathbf{e_3} = \frac{1}{\sqrt{2}}(1,-1,0),$$

given by the eigenvectors of the matrix

$$D_S(\mathbf{x_0}) = \begin{pmatrix} 0 & \frac{1}{2} & 0 \\ \frac{1}{2} & 0 & 0 \\ 0 & 0 & 0 \end{pmatrix}$$

corresponding to the eigenvalues

$$\lambda_1 = 0, \quad \lambda_2 = \frac{1}{2}, \quad \lambda_3 = -\frac{1}{2},$$

followed by a translation in the direction of the x-axis. Indeed, the flow pattern is such that there is no global rotation about the z-axis. This situation illustrates the previous comments: it is not always easy to visualize the local rotational effect of vorticity. ∎

Example 2.2. In the annulus

$$\{(x,y) \in \mathbb{R}^2 : \ 1 < x^2 + y^2 < 4\},$$

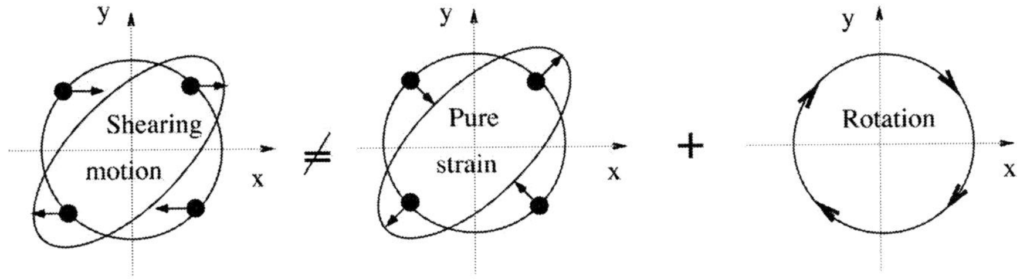

Figure 2.8. *In Example* 2.1 *the pure strain in the* (x,y)*-plane near the origin consists of an elongation in the direction of the main diagonal and a foreshortening in the orthogonal direction, while the rotation is clockwise. However, the shearing motion near the origin (where the translational component vanishes) is not a simple combination of these two constituents.*

for the two-dimensional flow with velocity

$$\mathbf{u}(x,y,t) = \left(\frac{-y}{x^2+y^2}, \frac{x}{x^2+y^2}\right)$$

each particle rotates with angular speed r^{-2} depending on the distance

$$r = \sqrt{x^2+y^2}$$

from the center of the annulus. However, the flow has zero vorticity.

Notice that the variable angular speed of the particles makes it possible for the flow to be irrotational. For example, a particle on the outer boundary $r = 2$ returns to its initial position in time $t = 8\pi$, whereas one on the inner boundary $r = 1$ needs time $t = 2\pi$ to do so. Thus, while each particle describes a circle, the variable angular velocity accounts for the fact that there is no local spin. ∎

Example 2.3. If, in contrast to Example 2.2, we consider the velocity field

$$\mathbf{u}(x,y,t) = (-y,x)$$

in the same annular region $\{(x,y) \in \mathbb{R}^2 : \ 1 < x^2+y^2 < 4\}$. Passing to polar coordinates, one can see that now

$$r'(t) = 0, \qquad \theta'(t) = 1,$$

so that the particle paths are

$$t \mapsto (x(t),\, y(t)) = (r\,\cos(t+t_0), r\,\sin(t+t_0)),$$

with t_0 determined by the location at $t = 0$. The new flow is not irrotational: its vorticity can be computed as $\boldsymbol{\omega} = (0,0,2)$. ∎

We now derive a fundamental equation that describes how the vorticity evolves in a flow. Using the identity

$$(\mathbf{u}\cdot\nabla)\mathbf{u} = \nabla\left(\frac{1}{2}\mathbf{u}\cdot\mathbf{u}\right) + (\nabla\wedge\mathbf{u})\wedge\mathbf{u}$$

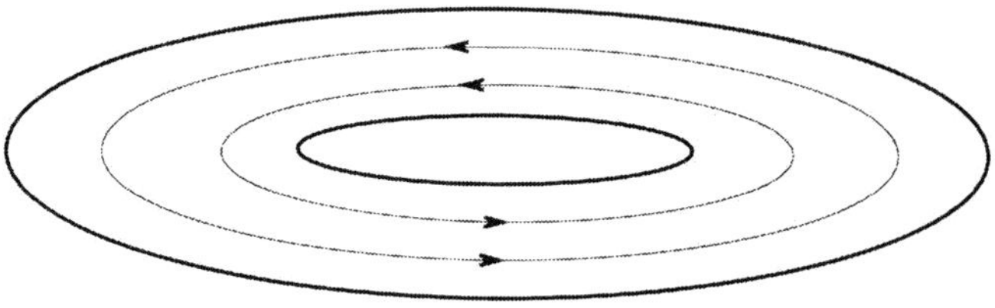

Figure 2.9. *Even if each particle describes a circle, the global rotation is not decisive for the presence of vorticity: a suitable variable angular velocity could prevent the local spin of the particles (e.g., the flow in Example 2.2 is irrotational), while a constant angular velocity (as in Example 2.3) induces vorticity.*

in Euler's equation (2.3), we obtain

$$\mathbf{u}_t + (\nabla \wedge \mathbf{u}) \wedge \mathbf{u} = -\nabla\Big(\frac{1}{2}\mathbf{u}\cdot\mathbf{u} + P + gz\Big).$$

We take the curl to get

$$\frac{\partial \boldsymbol{\omega}}{\partial t} + \nabla \wedge (\boldsymbol{\omega} \wedge \mathbf{u}) = \mathbf{0},$$

in view of the identity

$$\nabla \wedge (\nabla f) = \mathbf{0}, \qquad f \in C^2(\mathbb{R}^3, \mathbb{R}).$$

Using the vector identity

$$\nabla \wedge (\boldsymbol{F} \wedge \boldsymbol{G}) = (\boldsymbol{G}\cdot\nabla)\boldsymbol{F} - (\boldsymbol{F}\cdot\nabla)\boldsymbol{G} + \boldsymbol{F}(\nabla\cdot\boldsymbol{G}) - \boldsymbol{G}(\nabla\cdot\boldsymbol{F})$$

for all $\boldsymbol{F}, \boldsymbol{G} \in C^2(\mathbb{R}^3, \mathbb{R}^3)$, this becomes

$$\frac{\partial \boldsymbol{\omega}}{\partial t} + (\mathbf{u}\cdot\nabla)\boldsymbol{\omega} - (\boldsymbol{\omega}\cdot\nabla)\mathbf{u} + \boldsymbol{\omega}(\nabla\cdot\mathbf{u}) - \mathbf{u}(\nabla\cdot\boldsymbol{\omega}) = \mathbf{0}.$$

In the previous equality the fourth term vanishes because water is assumed to be incompressible, while the fifth term vanishes because of the vector identity

$$\nabla\cdot(\nabla\wedge\boldsymbol{F}) = 0,$$

valid for all $\boldsymbol{F} \in C^2(\mathbb{R}^3, \mathbb{R}^3)$. We therefore obtain

$$\frac{\partial \boldsymbol{\omega}}{\partial t} + (\mathbf{u}\cdot\nabla)\boldsymbol{\omega} = (\boldsymbol{\omega}\cdot\nabla)\mathbf{u},$$

or, alternatively,

$$\frac{D\boldsymbol{\omega}}{Dt} = (\boldsymbol{\omega}\cdot\nabla)\mathbf{u}. \tag{2.10}$$

This is the **vorticity equation**. In analyzing water flows with vorticity, this equation is extremely valuable, in particular because the hydrodynamical pressure has been eliminated: equation (2.10) involves only $\mathbf{u}$ and $\boldsymbol{\omega}$ (which are, of course, related by $\boldsymbol{\omega} = \nabla \wedge \mathbf{u}$).

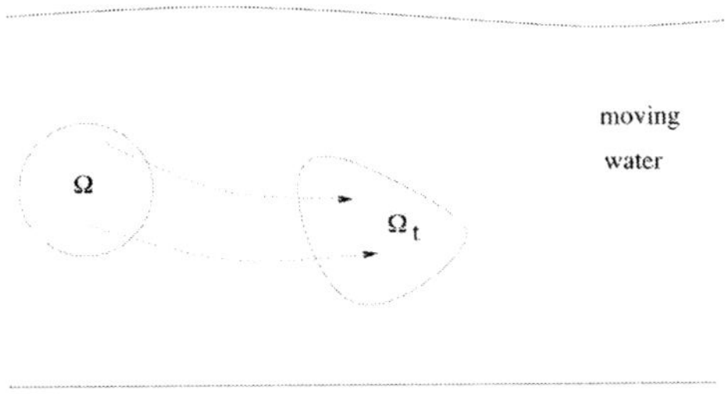

Figure 2.10. Ω_t *is the image of* Ω *as water particles in* Ω *flow for time* t.

In particular, for a two-dimensional water flow (not dependent on the y-coordinate, say), we have $\omega_1 = \omega_3 = 0$ for $\boldsymbol{\omega} = (\omega_1, \omega_2, \omega_3)$, so that

$$(\boldsymbol{\omega} \cdot \nabla)\mathbf{u} = \omega_2 \frac{\partial \mathbf{u}}{\partial y} = 0$$

and consequently

$$\frac{D\boldsymbol{\omega}}{Dt} = 0;$$

that is, *in a two-dimensional flow the vorticity of each individual water particle is constant as the particle moves about.*

For three-dimensional water flows with vorticity, the rather complicated motions that are observed are usually due to the lack of conservation of vorticity; this proves to be a major obstacle to the understanding of crucial properties of the flow. To investigate the vorticity for three-dimensional water flows, it is useful to introduce the **flow map** $\mathbf{x} \mapsto \boldsymbol{\varphi}(t, \mathbf{x})$: with fixed t, this map advances each particle in the water region from its position $\mathbf{x}$ at time $t = 0$ to its position at time t. For fixed t, $\boldsymbol{\varphi} = (\varphi_1, \varphi_2, \varphi_3)$ is an invertible C^1-mapping—this is the **Lagrangian viewpoint.**[7] From the vorticity equation we infer that

$$\boldsymbol{\omega}(t, \boldsymbol{\varphi}(t, \mathbf{x})) = J(t, \mathbf{x})\boldsymbol{\omega}(0, \mathbf{x}), \tag{2.11}$$

where

$$J(t, \mathbf{x}) = \left(\frac{\partial \varphi_i}{\partial x_j}\right)_{\{1 \le i, j \le 3\}}$$

is the Jacobian matrix of the flow map. Indeed, if we define

$$\boldsymbol{F}(t, \mathbf{x}) = \boldsymbol{\omega}(t, \boldsymbol{\varphi}(t, \mathbf{x})), \qquad \boldsymbol{G}(t, \mathbf{x}) = J(t, \mathbf{x})\omega(0, \mathbf{x}),$$

by (2.10) we have

$$\frac{\partial \boldsymbol{F}}{\partial t} = (\boldsymbol{F} \cdot \nabla)\mathbf{u}.$$

[7] There are two standard coordinate systems used in classical fluid dynamics. In Lagrangian coordinates, one describes the fluid by identifying with any one of the particles of the fluid; the observer follows the fluid by picking out any particular particle and keeping track of where it goes, that is, one follows the fluid particles. In **Eulerian coordinates** one describes the fluid from the viewpoint of a fixed observer watching the flow go by; there exists a precise Eulerian state corresponding to a Lagrangian state and vice versa.

On the other hand, with $\nabla \mathbf{u} = \left(\frac{\partial u_i}{\partial x_j}\right)_{\{1\le i,j\le 3\}}$ for $\mathbf{u} = (u_1, u_2, u_3)$, we have

$$\begin{aligned}\frac{\partial \boldsymbol{G}}{\partial t} &= \nabla[\partial_t \boldsymbol{\varphi}(t,\mathbf{x})]\,\boldsymbol{\omega}(0,\mathbf{x}) = \nabla[\mathbf{u}(t,\boldsymbol{\varphi}(t,\mathbf{x}))]\,\boldsymbol{\omega}(0,\mathbf{x}) \\ &= (\nabla \mathbf{u})\,J(t,\mathbf{x})\,\boldsymbol{\omega}(0,\mathbf{x}) = (\boldsymbol{G}\cdot\nabla)\mathbf{u},\end{aligned}$$

so that $\boldsymbol{F}$ and $\boldsymbol{G}$ satisfy the same linear ordinary differential equation (with t as the only independent variable) with the same initial data (and throughout this argument $\mathbf{u}$ is presumed to be known). Thus, by unicity, they are equal and therefore (2.11) holds. An immediate consequence of (2.11) is that *in three-dimensional flows a particle which has no vorticity never acquires it, and, conversely, a particle which is moving rotationally will continue so to move.*

The above result says that fluid particles which were originally vorticity free will continue to be so. An important special case is concerned with the development of water motions which start from rest (for example, motions of initially still water generated by swell from a distant storm traveling into undisturbed water). Thus $\text{curl}\,\mathbf{u} \equiv 0$ initially and then (2.11) ensures that such motions are irrotational at all later times. However, in general, part of the water body may be moving irrotationally and other parts rotationally. As the water moves about, the irrotational part may occupy different regions of space; $\boldsymbol{\omega} \equiv 0$ is a property of those parts of the fluid which are moving irrotationally and not of the regions of space they may temporarily occupy. Thus, if a fluid flow has regions that are irrotational and others where vorticity is present, the spatial region originally occupied by fluid in irrotational motion may pick up vorticity at later times, so, in general, the irrotational region will move around with the flow.

2.3.1 Irrotational flows

Since the fluid domains, D, that we consider are simply connected, curl-free vector fields $\mathbf{u}: D \to \mathbb{R}^3$ are gradients of some function $\phi: D \to \mathbb{R}$: for any velocity field $\mathbf{u}$ with zero vorticity there exists a **velocity potential** ϕ, prescribed up to an additive function of time by the line integral

$$\phi(\mathbf{x}) = \int_{\mathbf{x}_0}^{\mathbf{x}} u_1\,dx + u_2\,dy + u_3\,dz$$

(for which the independence upon the path of integration, joining the fixed basis point $\mathbf{x}_0 \in D$ to the arbitrary point $\mathbf{x} \in D$, is ensured by the vanishing of the vorticity $\boldsymbol{\omega}$ in view of Stokes's theorem), such that

$$\mathbf{u} = \nabla\phi.$$

Notice that the equation for mass conservation (2.1) can be restated as $\Delta\phi = 0$: harmonic functions, and analytic functions in the two-dimensional setting, are relevant to irrotational incompressible flows.

While for simply connected domains $D \subset \mathbb{R}^2$ there is no velocity potential for two-dimensional water flows $\mathbf{u} = (u_1, u_2): D \to \mathbb{R}^2$ with nonzero vorticity, one can nevertheless find a **stream function** $\psi: D \to \mathbb{R}$ such that

$$u_1 = \psi_y, \qquad u_2 = -\psi_x.$$

Then up to a function of time

$$\psi(x,y)=\int_{(x_0,y_0)}^{(x,y)} u_1\,dy-u_2\,dx,$$

with the path of integration being any curve in D joining the fixed point $(x_0,y_0)\in D$ to the arbitrary point $(x,y)\in D$, the independence of the above line integral on the path being ensured by the incompressibility condition (2.1).

2.3.2 Currents

By a **current** we understand a water flow with a flat surface (typically pre-existing before the arrival of the waves). For two-dimensional steady periodic currents moving at constant speed $c>0$, the vorticity and the mass flux specify the entire flow. Indeed, consider the two-dimensional flow

$$\Big(u_1(x-ct,z),0,u_3(x-ct,z)\Big)$$

in a layer of water with the free surface $z=0$ and the flat bed $z=-d$, where $d>0$ is the constant depth. Introducing the **stream function** $\psi(x-ct,z)$ with

$$\begin{cases}\psi_x=-u_3,\\ \psi_z=u_1-c,\end{cases}$$

the **(relative) mass flux** m is defined as

$$m=\int_{-d}^{0}[u_1(x-ct,z)-c]\,dz.$$

Notice that

$$m=\psi(x-ct,0)-\psi(x-ct,-d)$$

is a constant since the kinematic boundary conditions (2.4) and (2.5), in this setting, simply state that $u_3=0$ on $z=0$ and on $z=-d$, so that $\psi(x-ct,0)$ and $\psi(x-ct,-d)$ are constants. Since ψ is uniquely determined up to an additive constant, and it is constant on both boundaries, we may set $\psi=0$ on $z=0$ and $\psi=-m$ on $z=-d$. The vorticity $\boldsymbol{\omega}$ has at most the second component nonzero, so we write $\boldsymbol{\omega}=(0,\omega(x-ct,y),0)$ and by abuse of notation we identify the vorticity $\boldsymbol{\omega}$ with the scalar ω. Consequently, we can express the vorticity as $\Delta\psi=\omega$, thus obtaining the Dirichlet boundary problem

$$\begin{cases}\psi(x,0)=0 \quad\text{and}\quad \psi(x,-d)=-m & \text{for} \quad x\in\mathbb{R},\\ \Delta\psi(x,y)=\omega(x,y) \quad \text{for} & -d<z<0,\ x\in\mathbb{R}\end{cases}$$

with a periodicity condition in the x-variable. If $\omega\in C^1$ is 2π-periodic in the x-variable throughout the strip $\mathbb{R}\times[-d,0]$, the above problem has a unique solution[8] $\psi\in C^2$.

[8]Uniqueness is ensured by maximum principles and the existence of a solution follows from Schauder estimates; for both aspects we refer the reader to the Appendix to Chapter 3.

It is of interest to provide the relatively explicit formula for the solution: for $\omega(x,z) = \sum_{n=0}^{\infty} \omega_n(z) e^{inx}$ we seek

$$\psi(x,z) = \sum_{n=0}^{\infty} \psi_n(z) e^{inx}.$$

For $n = 0$ we have

$$\psi_0''(z) = \omega_0(z), \qquad -d < z < 0,$$

with $\psi_0(0) = 0$ and $\psi_0(-d) = -m$, so that

$$\psi_0(z) = z\Big(\frac{m}{d} + \frac{1}{d}\int_{-d}^{0} s\omega_0(s)\,ds + \int_{-d}^{z} \omega_0(s)\,ds\Big) + \int_{z}^{0} s\omega_0(s)\,ds,$$

and for every fixed $n \geq 1$ we have to solve the boundary value problem

$$\begin{cases} \psi_n''(z) - n^2\,\psi(z) = \omega_n(z), & -d < z < 0, \\ \psi_n(0) = \psi_n(-d) = 0. \end{cases} \tag{2.12}$$

Writing the second-order differential equation in (2.12) as a first-order system for the vector $\Psi_n = \begin{pmatrix} \psi_n \\ \frac{1}{n}\psi_n' \end{pmatrix}$, by the variation of constants formula we have

$$\Psi_n(z) = e^{Az}\Psi_n(0) + \int_0^z e^{A(z-s)} \begin{pmatrix} 0 \\ \frac{1}{n}\,\omega_n(s) \end{pmatrix} ds, \qquad z \in [-d,0],$$

where

$$A = \begin{pmatrix} 0 & n \\ n & 0 \end{pmatrix}, \qquad e^{Az} = \sum_{k=0}^{\infty} \frac{A^k z^k}{n!} = \begin{pmatrix} \cosh(nz) & \sinh(nz) \\ \sinh(nz) & \cosh(nz) \end{pmatrix}.$$

We now impose the two boundary conditions on the general solution and obtain (for $n \geq 1$)

$$\psi_n(z) = -\frac{1}{n}\frac{\sinh(nz)}{\sinh(nd)}\int_{-d}^{0} \sinh\Big(n(s+d)\Big)\,\omega_n(s)\,ds + \frac{1}{n}\int_0^z \sinh\Big(n(z-s)\Big)\,\omega_n(s)\,ds.$$

Knowledge of ψ provides us with the flow velocity; in addition, the pressure is easily determined since its value on $z = 0$ is P_{atm} (by the dynamic boundary condition) and its gradient in the fluid is found using the Euler equation.

Vorticity is thus adequate for the specification of a current, and the previous Fourier series representation shows that, if in a two-dimensional steady current ω is independent of x (or, equivalently, is time independent), so will be ψ and thus u_1, while $u_3 \equiv 0$. Irrotational flows (zero vorticity) model still water (in which case there is no current) as well as currents which are uniform with depth, whereas the simplest rotational setting is that of linearly sheared currents of constant nonzero vorticity. Constant nonzero vorticity is generaly regarded as giving a good description of tidal currents; cf. [250]. These are the most regular and predictable currents; indeed, on areas of the continental shelf and in many coastal inlets they are the most significant currents; cf. [167].

When swell, originating from a distant disturbance of the sea, propagates through a region with an underlying strong current, the wave-current interaction should not be overlooked. Spectacular examples are the wave-current interactions at the Columbia River entrance (where appreciable tidal currents make it one of the most hazardous regions for navigation in the world, since here the wave height can easily be doubled in just a few hours) and those found off the eastern coast of South Africa (where $6m$ high sea waves from southwest meeting the Agulhas current can be tripled in height over a few wavelengths, which has lead to many oil-tanker wreckages); cf. [167].

As pointed out above, constant vorticity gives a good description of tidal currents; in the absence of waves, these flows above the flat bed $z = -d$ have a flat free surface $z = 0$, and the fluid velocity is of the form

$$\mathbf{u}(x,y,z,t) = \Big(u_0 + \omega z, 0, 0\Big)$$

for some constants u_0 and ω with $\omega \neq 0$. The vorticity in this case can be easily computed as $(0,\omega,0)$, being identified with the scalar ω. The positive vorticity case $\omega > 0$, in which the velocity at the surface exceeds that on the bed, is appropriate for the ebb current; negative vorticity (that is, $\omega < 0$) is appropriate for the flood current; cf. [319]. Notice that tides refer to the vertical motion of water (the rise and fall of the tide) caused by the gravitational forces due to the relative motions of Moon, Sun, and Earth, whereas tidal currents refer to the alternating, horizontal movement of water associated with the rise and fall of the tide: the current associated with a rising tide is called the *flood*, and the current associated with a falling tide is called the *ebb*.

On the other hand, the prime source of some ocean currents are long-duration winds; constant vorticity does not give a good description of these currents, a depth-dependent vorticity being more adequate. Also, at the mouth of an estuary the out-flowing current generally exhibits a nonuniform vorticity distribution [294].

2.4 Appendix for Chapter 2

2.4.1 Surface tension

The outermost layer of water molecules has a different environment from the molecules in the interior of the fluid. The water molecules interact with their close neighbors through the short-range attractive internal forces which have their origin in the electrical structure of the molecules (van der Waal forces). A molecule is an aggregation of atomic nuclei and electrons that is sufficiently stable to possess observable properties—and few molecules are more difficult to decompose than H_2O. In water, each hydrogen nucleus is bound to the central oxygen atom by a pair of electrons that is shared between them and only two of the six outer-shell electrons of oxygen are used for this purpose, leaving four electrons which are organized into two nonbonding pairs. The four electron pairs surrounding the oxygen tend to arrange themselves as far from each other as possible in order to minimize repulsions between these clouds of negative charge. This would ordinarily result in a regular tetrahedral geometry (the regular tetrahedron being the only way one can dispose four points in space so that the distance between any two of them is the same) in which the angle between the two segments joining the center and any two vertices should be about 109.5°. However, because

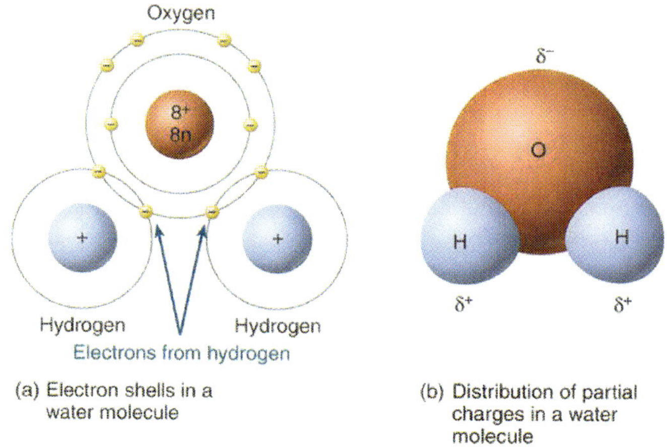

Figure 2.11. *The water molecule.*
[from the NOAA website]

of the presence of the very negative lone electron pairs, the two hydrogen atoms are pushed closer together as the two lone pairs try to get as far away from each other as possible. The result is a distorted tetrahedral arrangement in which the $H - O - H$ angle is 104.5°. Since the positively charged hydrogen atoms are not on opposite sides of the negatively charged oxygen atom, we see that, although the water molecule as a whole is neutral, its opposite sides have opposite charges; the entire molecule has an electrical polarity—like a very weak battery. Despite the fact that these electrical charges are weak, they make water molecules attract each other, forming bonds between the positively charged side of one water molecule and the negatively charged side of another water molecule (opposites attract), the electrostatic force bonding the molecules together being the van der Waal force. For a water molecule in the bulk of the fluid, these forces cancel each other due to the surrounding molecules because they are, on average, directionally uniform. Not so for a molecule adjacent to the surface, where these forces add up to a resultant macroscopic surface force acting in tangential directions; only the tangential components matter since the surface molecules experience practically no force from the air region above the surface.[9] Further, the inward surface direction of the force is already accounted for if the resultant attractive force on inner molecules is taken to be zero. For our purposes, it is appropriate to adopt the view that there acts within the water surface a **surface tension**, that is, a force per unit length (the water body is, as it were, covered by a membrane of negligible thickness—the molecules at the water's free surface cling together, behaving like a weak membrane) satisfying the following conditions:

(a) the surface tension is perpendicular to any line drawn in the surface and has the same magnitude for all directions of the line;

(b) the surface tension has the same value at all points on the surface.

We will consider surface waves over water regions with a rigid impermeable bottom and no lateral boundaries. Naturally, the phenomenon previously reported holds also at the innermost layer where the water is in contact with the rigid bottom. The cohesive forces at the bottom are negligible compared with the internal forces (pressure) acting in response to the weight of the water body,

[9]The air has a very low density compared with water, $12 \cdot 10^{-4}\, g/cm^3$ compared with $1\, g/cm^3$ at $20°\, C$, so that the attraction between surface water molecules and atmospheric molecules is small.

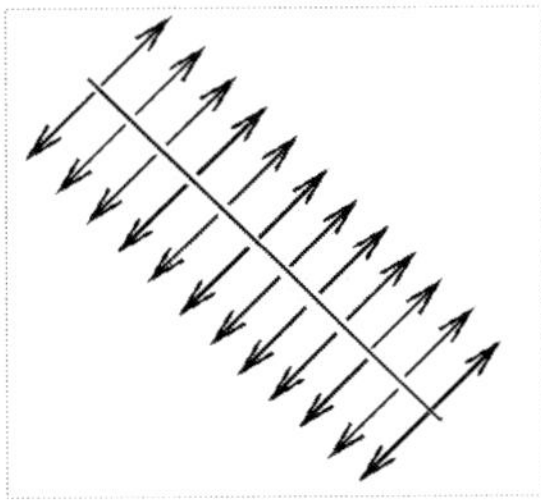

Figure 2.12. *Surface tension is perpendicular to any line drawn in the surface.*

so that, for all practical considerations, the surface tension phenomenon is restricted to the free surface. Water has the greatest surface tension coefficient $\sigma = 73 \cdot 10^{-3}\, N/m$, where the Newton unit is $1\,N = 1\,kg \cdot m/s^2$, of all common liquids except mercury (the only metal that is liquid at normal temperatures). The surface "membrane" formed over any accumulation of water by the cohesive tendency of water molecules at the surface makes it possible for objects which are much denser than water to float. For example, a steel needle (sewing needle) carefully laid on the water surface will float although it is five times as dense as water. A closer examination of the floating needle reveals that it rests in a small depression in the surface of the water, enabling the tension of the surface to act in a direction inclined to the horizontal and thus having a vertical component which can support the weight of the needle.[10] This is indicative of the fact that the effects of surface tension are important on nonflat surfaces where it changes direction from point to point.

Field data and experimental evidence [251, 269] confirm that surface tension produces only waves of small amplitude (not in excess of a few mm). A quantitative estimate for the regime when surface tension can be neglected is provided by the parameter $\hat{\kappa} = \frac{4\pi^2\sigma}{\rho g \lambda^2}$ which measures the relative importance of surface tension and gravity for traveling water waves with wavelength λ, so that small $\hat{\kappa}$ means that gravity is dominant. This parameter arises as follows: the propagation speed c of periodic traveling gravity-capillary surface waves of small amplitude, in water without currents and of average depth d, is given by

$$c = \sqrt{g(1+\hat{\kappa})\frac{\tanh(kd)}{k}},$$

where $k = \frac{2\pi}{\lambda}$ is the wavenumber [308]. Setting $g = 9.8\,m/s^2$, we infer that $\hat{\kappa} < \frac{1}{n^2}$ for $\frac{\lambda}{n} > 1.7\,cm$, so that for wavelengths in excess of several cm we can neglect surface tension.

2.4.2 Dissipation of energy

A crucial assumption in the derivation of the governing equations is that the water flow is frictionless (inviscid): we assumed that the surrounding water exerts a stress (i.e., force per unit area of contact)

[10]We do not discuss the geometry of water drops lying on a solid material, nor do we dwell on the capillarity phenomena occurring in thin glass tubes filled partially with water, where the surface tension at the contact surface with the solid plays a crucial role. Notice that without surface tension, water would spread out over a surface in a very thin film instead of forming droplets. Also, when water drips from a tap it does not immediately fall but lies suspended until it reaches a certain weight, as though it is covered with a stretched skin. A further demonstration of the apparent surface tension in the water surface is the fact that many insects skate across the surface while others hang bat-like from beneath it.

on a lump of water in the form of pressure acting in a direction normal to the boundary of the lump, while in reality (internal) friction also accounts for a tangential component of the stress.

While the incompressibility condition (2.1) remains valid, the inclusion of friction requires that the equation of motion, (2.3), is replaced by the **Navier–Stokes equation**

$$\frac{D\mathbf{u}}{Dt} = -\frac{1}{\rho}\nabla P + (0,0,-g) + \nu\,\Delta\mathbf{u},$$

where ν is the *kinematic viscosity* of the water,[11] the **viscosity** of a fluid being a physical property used to measure the magnitude of the frictional force due to microscopic interactions between the fluid molecules, which dissipates the energy of the fluid flow by converting it partly into heat. For simplicity, in discussing the appropriate boundary conditions, if friction is included in the modelling, we deal only with two-dimensional patterns (independent of the y-variable). The kinematic boundary condition on the free surface, (2.4), remains unchanged, but that on the flat bed, (2.5), has to be changed to the **no-slip condition**

$$\mathbf{u} = 0 \quad \text{on} \quad z = -d.$$

While casual observation seems to indicate that (2.5) is accurate, experimental evidence shows that there is a very thin layer, called a **boundary layer**, across which the tangential component of velocity undergoes a smooth but rapid adjustment from its mainstream value to precisely zero at the solid bottom (this assertion is of a physical character founded on experiment). Finally, the dynamic boundary condition, (2.6), also changes drastically since, in contrast to inviscid theory where the stress exerted at the air-water interface is solely in the form of pressure acting in a normal direction, there are frictional stresses in normal and tangential directions, the normal stress condition being

$$P - P_{atm} = \frac{2\rho\nu}{\sqrt{1+h_x^2}}\Big[\frac{\partial u_1}{\partial x}h_x^2 + \frac{\partial u_3}{\partial z} - \Big(\frac{\partial u_1}{\partial z} + \frac{\partial u_3}{\partial x}\Big)h_x\Big] \quad \text{on} \quad z = h(x,t),$$

while the tangential stress condition is

$$h_x\Big(\frac{\partial u_1}{\partial z} + \frac{\partial u_3}{\partial x}\Big)(h_x^2 - 1) + 2\Big(\frac{\partial u_1}{\partial x} - \frac{\partial u_3}{\partial z}\Big)h_x^2 = 0 \quad \text{on} \quad z = h(x,t)$$

(we omit any tangential component generated by the action of the wind). Notice that we recover (2.6) by formally letting $\nu \to 0$ in the normal stress condition, but the tangential stress condition has no correspondent in inviscid theory.

The governing equations, with friction effects taken into consideration, are currently intractable. Even for a fluid occupying the entire space, to show that smooth solutions of the Euler equations can be obtained from solutions of the Navier–Stokes equation in the vanishing viscosity limit $\nu \to 0$ is an open problem. Adding a solid boundary adds to the intricacy of the problem and dealing with a free boundary makes it even more difficult.

However, it is experimentally well established that all large-scale features of water waves are well predicted by the inviscid governing equations presented in Section 2.2 since water waves suffer only slight attenuation through the three main processes of energy dissipation: bottom friction, surface dissipation, and internal dissipation (see [206]).

- The effects of bottom friction and surface dissipation are confined to very thin boundary layers. The thickness of these boundary layers is roughly $5\sqrt{\frac{\nu}{ß}}$, where ß is the wave frequency [206], in the sense that 97% of the adjustment due to the friction effects related to the influence

[11] Generally, ν varies with temperature so that, for the sake of comparison, we provide its values for different fluids at 15°: $0.0178\,cm^2/s$ for water, $0.022\,cm^2/s$ for alcohol, $0.15\,cm^2/s$ for air, $1.0\,cm^2/s$ for olive oil.

of the boundaries occurs within this region. This formula is especially useful since field data and experiments show that for wave trains that propagate in water of locally uniform but nevertheless changing depth, the frequency ß $= \frac{2\pi c}{\lambda}$, where λ is the wavelength and c the speed, is practically constant [206] so that the number of crests per unit time is always the same—with the change of depth an adjustment of the speed takes place but the frequency is an invariant of the motion. The frequency is usually measured in $hertz = s^{-1}$. An illustrative example is the following: the typical frequency of ocean swell in the North Pacific toward the American coast of California, in a region where the ocean depth averages $2800\,m$, lies mostly in the range from $\frac{1}{4}\,hertz$ to $\frac{1}{9}\,hertz$ according to the data provided by the US National Oceanic and Atmospheric Administration (NOAA), and with $\nu = 1\,mm^2/s$ we deduce that the corresponding boundary layer thickness is smaller than $15\,mm$. Another illustrative example is that of waves with wavelength less than $1\,m$ propagating in a channel of depth $d = 2\,m$ (e.g., the "Canal du Midi" in southwest France): the typical propagation speed is $\sqrt{gd}$ (cf. [177]) so that the boundary layers have, in such a setting, a thickness of about $1\,mm$.

- The internal dissipation is caused by the presence of viscosity in the Navier–Stokes equation (the viscous term $\nu\,\Delta\mathbf{u}$ is missing in the Euler equations) but leads to a very weak damping: the reduction of the wave amplitude, $a(t)$, is according to $a(t) \approx e^{-\frac{8\pi^2 \nu t}{\lambda^2}}\,a(0)$; cf. [233]. With $\nu = 1\,mm^2/s$, the time needed for a $10\,\%$ loss of amplitude is about $T \approx \frac{\lambda^2}{80\pi^2 \nu} \approx \frac{\lambda^2}{8}\,s$ when λ is measured in cm. Thus when $\lambda \leq 1.7\,cm$ this time is less than $0.4\,s$, while if $\lambda \geq 50\,cm$, this time is about $5\,min$. This means that capillary and capillary-gravity waves are suppressed by viscosity almost immediately while gravity waves are affected very little.[12] Swell usually has wavelengths counted in tens of m (cf. [177]) so that days would be needed for a $10\,\%$ loss of amplitude. It is therefore not surprising (cf. [177]) that swell from winter storms in high latitudes, both north and south, commonly reaches the equatorial region as it travels with speeds above $50\,km/h$: swell generated by storms just south of Newfoundland Banks has been reported close to Sierra Leone, $4000\,km$ away. Waves traveling over oceanic distances of $1000\,km$ in the time it takes to lose one-tenth of their amplitude were observed (see [233]).

The previous semiempirical considerations justify the view that nonturbulent gravity water waves are hardly affected by friction; over the long distances needed for a noticeable attenuation of the waves due to frictional dissipation, other effects (e.g., interaction with other waves) will prevail as a source of extracting and redistributing energy. For this reason, throughout the modern research literature, the studies on gravity waves are basically restricted to analyzing the governing equations presented in Section 2.2.

2.4.3 Flow-invariant sets

Let $\Omega \subset \mathbb{R}^n$ be a closed C^1-hypersurface in $\mathbb{R}^n$ and let $F : \mathbb{R}^n \to \mathbb{R}^n$ be a C^1-function. We say that the set Ω is flow invariant for F if every solution $X : [t_0, T) \to \mathbb{R}^n$ of the differential equation

$$X'(t) = F(X(t))$$

which meets Ω at time $t = t_0$ (that is, with $X(t_0) \in \Omega$) remains in Ω for $t \in [t_0, T)$. Based on some considerations made in [261], we now present the following result.

[12]In special circumstances, greatly enhanced dissipation can occur within a surface boundary layer, e.g., for water covered by a thin layer of a contaminant such as oil; in which case the dissipative effect is considerably increased: this is the mechanism responsible for the proverbial calming effect of "oil on troubled waters" [206].

Theorem A 2.1 (Bony [26] and Brézis [32]). *If $\nu(Y)\cdot F(Y)=0$ whenever $\nu(Y)$ is normal to Ω at $Y\in\Omega$, then Ω is flow invariant for F.*

Proof. Let $X:[t_0,T)\to\mathbb{R}^n$ be a solution with $X(t_0)\in\Omega$ and assume that we do not have $X(t)\in\Omega$ for all $t\in(t_0,T)$. Since Ω is closed, the set $\{t\in(t_0,T):\; X(t)\notin\Omega\}$ is an open set; thus it can be represented as an at most countable union of disjoint open intervals. By assumption this set is not empty, so we find $(a,b)\subset(t_0,T)$ such that $X(t)\notin\Omega$ for all $t\in(a,b)$, while $X(a)\in\Omega$. Without loss of generality, we may assume that $b<\infty$. We now define the continuous function $\delta:[a,b)\to[0,\infty)$, where

$$\delta(t)=\inf_{Y\in\Omega}\{|X(t)-Y|\},\qquad t\in[a,b),$$

represents the distance from $X(t)$ to the set Ω. Notice that, since Ω is closed, we have by construction that $\delta(a)=0$ while $\delta(t)>0$ for $t\in(a,b)$. Indeed, if for some $t_1\in(a,b)$ we would have $\delta(t_1)=0$, then there is a sequence $Y_n\in\Omega$ with $\lim_{n\to\infty}|Y_n-X(t_1)|=0$, which leads to the contradiction $X(t_1)\in\Omega$. Fix $t\in(a,b)$ and let $Y(t)\in\Omega$ be the (not necessarily unique) nearest point to $X(t)\in\Omega$: if $Y_n\in\Omega$ are such that $\lim_{n\to\infty}|Y_n-X(t)|=\delta(t)$, the sequence $\{Y_n\}_{n\geq 1}$ being obviously bounded, any of its limit points has to belong to Ω and can play the role of $Y(t)$. For any $h\in(0,b-t)$ we clearly have

$$0<\delta(t+h)=|X(t+h)-Y(t+h)|\leq|X(t+h)-Y(t)|$$

and

$$0<|X(t)-Y(t)|=\delta(t),$$

so that

$$\begin{aligned}\delta(t+h)-\delta(t)&\leq|X(t+h)-Y(t)|-|X(t)-Y(t)|\\&=\frac{|X(t+h)-Y(t)|^2-|X(t)-Y(t)|^2}{|X(t+h)-Y(t)|+|X(t)-Y(t)|}.\end{aligned}\tag{2.13}$$

On the other hand, the functions $X,\,Y:[a,b]\to\mathbb{R}^n$ are obviously bounded. Let $M_0>0$ be such that

$$|X(t)|+|Y(t)|\leq M_0,\qquad t\in[a,b].\tag{2.14}$$

For fixed points $X_1,\,X_2\in\mathbb{R}^n$ with $|X_1|+|X_2|\leq 2\,M_0$ we define the C^1-function

$$f_{12}:[0,1]\to\mathbb{R}^n,\quad f_{12}(s)=F(sX_1+(1-s)X_2),\qquad s\in[0,1],$$

and observe that

$$|f_{12}(1)-f_{12}(0)|\leq\sup_{s\in[0,1]}\{|f_{12}'(s)|\}\leq|X_1-X_2|\sup_{|X|\leq 2M_0}\{|F'(X)|\}.$$

Since the left-hand side is precisely $|F(X_1)-F(X_2)|$, denoting

$$M=\sup_{|X|\leq 2M_0}\{|F'(X)|\}\geq 0,$$

we proved that the restriction of the C^1-function $F:\mathbb{R}^n\to\mathbb{R}^n$ to the set $\{X\in\mathbb{R}^n:|X|\leq M_0\}$ is Lipschitz because

$$|F(X_1)-F(X_2)|\leq M\,|X_1-X_2|\quad\text{whenever}\quad|X_1|,|X_2|\leq M_0.\tag{2.15}$$

For a fixed $t \in [a,b)$, the differential equation yields

$$X(t+h) = X(t) + \int_t^{t+h} F(X(s))\,ds, \quad h \in [0, b-t],$$

so that

$$X(t+h) - Y(t) = X(t) - Y(t) + h\,F(X(t)) + \int_t^{t+h} \Big(F(X(s)) - F(X(t))\Big)\,ds\,.$$

Since by (2.14) and (2.15) we have

$$\begin{aligned} |F(X(t)| &\le |F(X(0))| + |F(X(t) - F(X(0))| \\ &\le |F(X(0))| + M\,|X(t) - X(0)| \le |F(X(0))| + 2\,M M_0 \end{aligned}$$

and[13]

$$\begin{aligned} \left|\int_t^{t+h} \Big(F(X(s)) - F(X(t))\Big)\,ds\right| &\le \int_t^{t+h} |F(X(s)) - F(X(t))|\,ds \\ &\le M \int_t^{t+h} |X(s) - X(t)|\,ds = o(h), \end{aligned}$$

taking the inner product of this expression with itself, we infer

$$|X(t+h) - Y(t)|^2 = |X(t) - Y(t)|^2 + 2h\Big(X(t) - Y(t)\Big) \cdot F(X(t)) + o(h).$$

Therefore, dividing (2.13) by $h > 0$ and subsequently letting $h \downarrow 0$, we obtain

$$\limsup_{h \downarrow 0} \frac{\delta(t+h) - \delta(t)}{h} \le \frac{\Big(X(t) - Y(t)\Big) \cdot F(X(t))}{|X(t) - Y(t)|}. \tag{2.16}$$

But by construction, the vector $(X(t) - Y(t))$ is normal to Ω at $Y(t) \in \Omega$. Indeed, the tangent space to Ω at $Y(t)$, $T_{Y(t)}\Omega$, has dimension $(n-1)$, and the normal direction at $Y(t)$ is the unique direction orthogonal to this space, or, equivalently, the direction orthogonal to all tangent vectors $f'(0)$ of C^1-curves $f : (-\varepsilon, \varepsilon) \to \Omega$ with $f(0) = Y(t)$, since $T_{Y(t)}\Omega$ is simply the space of all vectors obtained in this way. But if $s \mapsto f(s)$ is such a curve, then the C^1-function $s \mapsto |X(t) - f(s)|^2$ has a positive local minimum $|X(t) - Y(t)|^2 > 0$ at $s = 0$. Therefore, its derivative vanishes at $s = 0$; that is,

$$2\,f'(0) \cdot \Big(X(t) - Y(t)\Big) = 0$$

since $f(0) = Y(t)$. But this means precisely that the vector $(X(t) - Y(t)) \neq 0$ is orthogonal to the vector $f'(0)$. Consequently, by hypothesis,

$$\Big(X(t) - Y(t)\Big) \cdot F(Y(t)) = 0.$$

Substracting this zero term from the numerator on the right-hand side of (2.16) we get that, for any $t \in (a,b)$,

$$\limsup_{h \downarrow 0} \frac{\delta(t+h) - \delta(t)}{h} \le \frac{\Big(X(t) - Y(t)\Big) \cdot \Big(F(X(t)) - F(Y(t))\Big)}{|X(t) - Y(t)|}.$$

[13]By $f(s) = o(s)$ we mean $\lim_{s \to 0} \frac{|f(s)|}{s} = 0$.

In combination with (2.15) this yields

$$\limsup_{h\downarrow 0} \frac{\delta(t+h)-\delta(t)}{h} \le M\,|X(t)-Y(t)| = M\,\delta(t), \qquad t\in(a,b).$$

Since $\delta(a)=0$ we deduce the integral inequality

$$\delta(t) \le M\int_a^t \delta(s)\,ds, \qquad t\in(a,b). \tag{2.17}$$

Indeed, for $\varepsilon>0$, let us define the continuous function

$$\gamma:[a,b]\to\mathbb{R}, \qquad \gamma(t)=\delta(t)-\varepsilon\,t-M\int_a^t\delta(s)\,ds, \qquad t\in[a,b],$$

so that the previous inequality amounts to

$$\limsup_{h\downarrow 0}\frac{\gamma(t+h)-\gamma(t)}{h}\le -\varepsilon, \qquad t\in(a,b). \tag{2.18}$$

This, we claim, ensures that γ is decreasing on $[a,b]$ so that, in particular,

$$\delta(t)-\varepsilon\,t-M\int_a^t\delta(s)\,ds\le -\varepsilon a, \qquad t\in[a,b],$$

which leads us to (2.17) in the limit $\varepsilon\downarrow 0$. To prove our monotonicity claim, assume the existence of $t_1,t_2\in[a,b]$ with $\gamma(t_1)<\gamma(t_2)$. In view of the continuity of γ we may assume $t_1,t_2\in(a,b)$. Consider now the set

$$\left\{t\in[t_1,t_2]:\ \gamma(t)<\frac{\gamma(t_1)+\gamma(t_2)}{2}\right\}$$

and let $\xi\in[t_1,t_2]$ be the supremum of all elements in this nonempty set (t_1 being an element of the set). Clearly $\xi<t_2$ so that, by the definition of ξ, we must have

$$\gamma(\xi+h)\ge\gamma(\xi)=\frac{\gamma(t_1)+\gamma(t_2)}{2} \quad \text{for all} \quad h\in(0,t_2-\xi).$$

This, however, leads to

$$\limsup_{h\downarrow 0}\frac{\gamma(\xi+h)-\gamma(\xi)}{h}\ge 0,$$

which contradicts (2.18) because $\xi\in(a,b)$.

With (2.17) established, we now choose some $\varepsilon>0$ and define

$$V(t)=\varepsilon+M\int_a^t\delta(s)\,ds, \qquad t\in[a,b].$$

Then $V(t)\ge\varepsilon$ for all $t\in[a,b)$ and so (2.17) implies

$$V'(t)\le M\,V(t), \qquad t\in(a,b).$$

This yields at once

$$V(t)\le V(0)e^{M(t-a)}, \qquad t\in[a,b).$$

In view of (2.17) and the fact that $V(a)=\varepsilon$ we infer that

$$\delta(t)\le\varepsilon\,e^{M(t-a)}, \qquad t\in[a,b).$$

Now letting $\varepsilon \downarrow 0$ leads us to the contradiction that $\delta(t) = 0$ for all $t \in [a,b]$. This completes the proof. $\square$

The connection of Theorem A 2.1 to the kinematic boundary condition $\frac{DS}{Dt} = 0$ is established as follows. Take the closed C^1-hypersurface

$$\Omega = \{(x,y,x,t) \in \mathbb{R}^4 : \; z - h(x,y,t) = 0\}$$

and notice that we can write the differential equation

$$\mathbf{x}'(t) = \mathbf{u}(\mathbf{x}(t),t)$$

in the equivalent form

$$X'(t) = F(X(t))$$

with

$$X = (x,y,z,t), \qquad F(X) = \Big(u_1(x,y,z,t),\, u_2(x,y,z,t),\, u_3(x,y,z,t),\, 1\Big).$$

Written this way, the condition $\nu(X)\cdot F(X) = 0$ is precisely $\frac{DS}{Dt} = 0$. An important fact should be pointed out. Guided by the case of a rigid boundary, intuition might suggest that, at any fixed time t, if $z = h(x,y,t)$, then the velocity field $\mathbf{u}(x,y,z,t)$ should be tangent to the free surface at (x,y,z). Expressing this in the form

$$u_3(x,y,z,t) = u_1(x,y,z,t)h_x(x,y,t) + u_2(x,y,z,t)h_y(x,y,t),$$

we see that if this were so, we should have $h_t(x,y,t) = 0$ all along the free surface, which means $h_t \equiv 0$ and thus the free surface would be time independent, but observation confirms that this is not the case unless the water's surface is flat.

2.4.4 The compression rate of a fluid

Here we show that a liquid flowing with velocity $\mathbf{u}$ is compressed at the rate $\nabla\cdot\mathbf{u}$.

Indeed, the motion of a fluid is described in Lagrangian coordinates by $\mathbf{y} = \mathbf{y}(t;\mathbf{x})$, representing, at time $t \geq t_0$, the position of the particle that at time $t = t_0$ was located at $\mathbf{x} = (x_1,x_2,x_3)$: $\mathbf{y}(t_0;\mathbf{x}) = \mathbf{x}$. The partial derivative of $\mathbf{y} = (y_1,y_2,y_3)$ with respect to time, for $\mathbf{x}$ fixed, is the velocity of the flow. Let V be a volume in the fluid and $V(t)$ the volume occupied by the particles $\mathbf{y}(t;\mathbf{x})$ with $\mathbf{x} \in V$; then we know from multivariable calculus that

$$\int_{V(t)} 1\, d\mathbf{x} = \int_V \det J(t)\, d\mathbf{x},$$

where $J(t)$ is the Jacobian matrix with entries $J_{ij}(t) = \frac{\partial y_i}{\partial x_j}$ for $i,j = 1,2,3$. This shows that the ratio of a volume of fluid to its (small) initial volume is $\det J(t)$ and therefore the rate at which fluid is compressed is $\frac{\partial}{\partial t}\det J(t)$. Observe that $J(t_0) = I_3$, the 3×3 identity matrix. From continuity we infer that $J(t)$ is invertible for all $t \in [t_0,t_1]$ for some $t_1 > t_0$. If we denote $A(t) = J_t(t)[J(t)]^{-1}$, we clearly have $J_t(t) = A(t)J(t)$ so that

$$J(t+s) = J(t) + s\,A(t)J(t) + o(s) = [I_3 + s\,A(t)]\;J(t) + o(s), \qquad t_0 < t \leq t+s < t_1,$$

and the multiplication rule for the determinants yields

$$\det J(t+s) = \det[I_3 + s\,A(t)]\det J(t) + o(s).$$

On the other hand, expanding the determinant as a polynomial of degree 3 in the variable s, we obtain

$$\det[I_3 + s\,A(t)] = 1 + s\,[\operatorname{trace} A(t)] + O(s^2),$$

where $[\operatorname{trace} A(t)]$ is the sum of the diagonal entries of the 3×3 matrix $A(t)$. Combining these, we get

$$\det J(t+s) = (1 + s\,[\operatorname{trace} A(t)])\det J(t) + o(s),$$

or

$$\frac{\det J(t+s) - \det J(t)}{s} = [\operatorname{trace} A(t)]\det J(t) + \frac{o(s)}{s}.$$

Passing to the limit $s \to 0$, we obtain

$$\frac{\partial}{\partial t}\det J(t) = [\operatorname{trace} A(t)]\det J(t),$$

and since $\det J(t_0) = 1$, we find that

$$\det J(t) = e^{\int_{t_0}^{t}[\operatorname{trace} A(s)]ds}, \qquad t \in [t_0, t_1].$$

We now observe that

$$\frac{\partial}{\partial t} J_{ij} = \frac{\partial}{\partial t}\frac{\partial y_i(t;\mathbf{x})}{\partial x_j} = \frac{\partial}{\partial x_j}\frac{\partial y_i(t,\mathbf{x})}{\partial t} = \frac{\partial}{\partial x_j} u_i(t,\mathbf{x})$$

yields

$$\operatorname{trace} A(t_0) = \operatorname{trace} J_t(t_0) = \sum_{i=1}^{3}\frac{\partial u_i}{\partial x_i}(\mathbf{x}, t_0).$$

In conclusion $\frac{\partial}{\partial t}\det J(t)|_{t=t_0} = \sum_{i=1}^{3}\frac{\partial u_i}{\partial x_i}(\mathbf{x}, t_0)$. Since the chosen instant $t = t_0$ and the location $\mathbf{x}$ were arbitrary, we deduce that a fluid that is flowing with velocity $\mathbf{u}$ is being compressed at the rate $\nabla \cdot \mathbf{u}$, and for this reason the velocity field of an incompressible fluid is divergence free. Such vector fields are often called *solenoidal*.

2.4.5 Sobolev spaces

In the study of partial differential equations it is convenient to incorporate the L^p norms of the derivative of a function into a Banach space norm. The scale of Sobolev spaces is especially useful in the analysis of elliptic partial differential equations, where these spaces are better suited than the scale of C^k-functions to express the regularity of solutions (see the Appendix to Chapter 3).

Given a domain $R \subset \mathbb{R}^n$ whose boundary ∂R is of class C^1, the **Sobolev spaces** $W^{k,s}(R)$ with $k \geq 0$ integer and $1 < s < \infty$ are useful in the study of partial differential equations in R (the cases $s = 1$ and $s = \infty$ are also sometimes used but the lack of reflexivity of these spaces leads to additional complications[14]). They consist of locally integrable functions $h : R \to \mathbb{R}$ such that

[14]Nevertheless, the case $s = \infty$ is particularly interesting since a function $f : R \to \mathbb{R}$ is locally Lipschitz if and only if $f \in W^{1,\infty}_{loc}(R)$, in the sense that $f \in W^{1,\infty}(R_0)$ for every domain R_0 whose closure $\overline{R_0}$ lies in R.

for integers $i_1,\dots,i_n \geq 0$ with $i_1+\cdots+i_n \leq k$ the distributional derivative $\partial_{x_1}^{i_1}\cdots\partial_{x_n}^{i_n} h$ belongs to $L^s(R)$. The spaces $W^{k,s}(R)$ are Banach spaces—respectively Hilbert spaces for $s=2$, in which case they are denoted $H^k(R)$—if endowed with the norm

$$\|h\|_{W^{k,s}(R)} = \Big(\sum_{0\leq i_1+\cdots+i_n\leq k} \|\partial_{x_1}^{i_1}\cdots\partial_{x_n}^{i_n} h\|_{L^s(R)}^s \Big)^{1/s},$$

where

$$\|h\|_{L^s(R)} = \Big(\int_R |h(x_1,\dots,x_n)|^s\,dx_1\cdots dx_n \Big)^{1/s}.$$

While the definition for a Sobolev function requires only that it belong to $L^s(R)$, and so is, strictly speaking, an equivalence class of functions whose values can be assigned arbitrarily in sets of measure zero, functions in $W^{1,s}(R)$ with $s>n$ have a representative that is continuous, and functions in $W^{1,s}(R)$ with $1<s<n$ have representatives that are absolutely continuous on almost all lines. The borderline case $s=n$ is somewhat delicate: one can exhibit nowhere-continuous functions in $W^{1,n}(\mathbb{R}^n)$.

We can incorporate boundary conditions in the space $W^{k,s}(R)$ if $k\geq 1$: if $n=1$, we simply have (choosing a representative) that $W^{1,s}(R)\subset C(\overline{R})$; while for $n\geq 2$, functions in $W^{1,s}(R)$ have a trace on ∂R that belongs to $L^s(\partial R)$. If we denote by $W_0^{k,s}(R)$ the closure in $W^{k,s}(R)$ of the test functions, that is, of the C^∞-functions of compact support in R, then $h\in W^{1,s}(R)$ is in $W_0^{1,s}(R)$ if and only if its trace is zero.[15] Notice that ∂R has n-dimensional Lebesgue measure zero so that restricting an integrable function in R to the boundary has no direct meaning, the difficulty being resolved by the notion of the trace operator, which associates to $h\in C(\overline{R})$ its continuous restriction to ∂R and extends to a bounded linear operator from $W^{1,s}(R)$ to $L^s(\partial R)$. On the other hand, we can extend functions in $W^{1,s}(R)$ to functions in $W^{1,s}(\mathbb{R}^n)$ with compact support.

A further important fact is that, in Sobolev spaces, an L^s bound on a higher derivative implies bounds on lower derivatives: for $1<s<n$ there exists a constant $c_s>0$ such that

$$\Big(\int_{\mathbb{R}^n} |f(x_1,\dots,x_n|^{\frac{ns}{n-s}}\,dx_1\cdots dx_n \Big)^{\frac{1}{s}-\frac{1}{n}} \leq c_s \Big(\int_{\mathbb{R}^n} \sum_{j=1}^n |\partial_{x_j} f(x_1,\dots,x_n)|^s\,dx_1\cdots dx_n \Big)^{\frac{1}{s}}$$

for all $f\in W^{1,s}(\mathbb{R}^n)$. Also, if $f,g\in W^{1,s}(R)\cap L^\infty(R)$, then $fg\in W^{1,s}(R)\cap L^\infty(R)$ and the product rule for derivatives

$$\frac{\partial(fg)}{\partial x_i} = \frac{\partial f}{\partial x_i} g + f\frac{\partial g}{\partial x_i}, \qquad i=1,\dots,n,$$

holds a.e. in R. Moreover, if $f\in W^{1,s}(R)$ and $F\in C^1(\mathbb{R})$ with $F'\in L^\infty(\mathbb{R})$ and $F(0)=0$, then $F(f)\in W^{1,s}(R)$ and the chain rule

$$\frac{\partial F(f)}{\partial x_i} = F'(f)\frac{\partial f}{\partial x_i}, \qquad i=1,\dots,n,$$

holds a.e. in R. Let us also note the following compactness property: for $1<s<n$, a bounded sequence $\{f_j\}_{j\geq 1}$ in $W^{1,s}(R)$, with R bounded, contains a subsequence that converges in $L^r(R)$ for each $1\leq r<\frac{ns}{n-s}$. Good modern sources for Sobolev spaces are [111] and [112].

[15] In general, there is a description of the Sobolev space $W^{1,s}(R)$ as the closure of functions in $C^\infty(R)\cap W^{1,s}(R)$ under the norm $\|\cdot\|_{W^{k,s}(R)}$.

2.5 Notes for Chapter 2

The restriction to two-dimensional flows is very advantageous for reasons of analytic tractability. Few rigorous results on regular three-dimensional wave patterns (for which the free surface is genuinly two-dimensional) exist and all the research literature is restricted to irrotational flows with surface waves of small amplitude. For the existence of doubly periodic irrotational waves (diamond waves) generated by the nonlinear interaction of two wave trains with the same small amplitude that propagate in directions at an oblique angle (neither parallel nor orthogonal), see the discussion in [156]. As for irrotational solitary waves, it turns out that there are no well-localized steady solutions of the governing equations that are of positive elevation with a genuinly two-dimensional free surface [81]. For three-dimensional flows with vorticity, which model wave-current interactions, even numerical simulations are scarce.

Chapter 3

Wave-current interactions

"… the progress of physics will to a large extent depend on the progress of nonlinear mathematics, of methods to solve nonlinear equations."

W. Heisenberg

The full complexity of wave-current interactions is far beyond our ability to grasp, but, as pointed out in Chapter 2, these water flows exhibit regularities which can be studied in detail. The present chapter is devoted to an in-depth study of two-dimensional symmetric periodic traveling waves propagating at the free surface of water with a flat bed in a flow without stagnation points.

Given $c > 0$, we are looking for two-dimensional periodic steady waves traveling at speed c; that is, the space-time dependence of the free surface, of the pressure, and of the velocity field has the form $(x - ct)$ and is periodic with period $L > 0$. Let the y-axis point vertically upward, with the origin lying on the mean water level,

$$\int_0^L \eta(x)\,dx = 0,$$

so that the wave profile $y = \eta(x - ct)$ oscillates about the flat surface $y = 0$. The flat bed is given by $y = -d$, with $d > 0$ representing the mean water depth. Let (u, v) be the velocity field. We define, up to an additive constant, the **stream function** $\psi(x, y)$ by

$$\psi_x = -v, \quad \psi_y = u - c,$$

and we let

$$\omega = u_y - v_x = \Delta\psi$$

be the vorticity of the flow, modelling the presence of a current beneath the surface wave, as discussed in the previous chapter.

The free boundary problem for steady periodic water waves can be simplified by eliminating time by the change of frame $(x - ct, y) \mapsto (x, y)$. In the new reference frame, in which the origin moves in the direction of propagation of the wave with wave speed c, the wave is stationary and the flow is steady. The mass flux across $x = x_0$ at time t_0 is

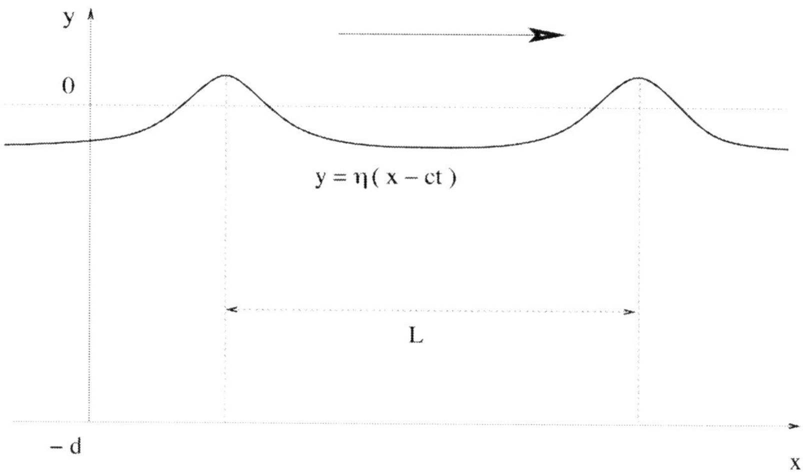

Figure 3.1. *A periodic traveling water wave.*

$\int_{-d}^{\eta(x_0-ct_0)} u(x_0 - ct_0, y)\,dy$, and we define the **relative mass flux** (relative to the uniform flow at speed c) as

$$p_0 = \int_{-d}^{\eta(x)} \Big(u(x,y) - c\Big)\,dy,$$

which is independent of x by the kinematic boundary conditions (2.4) and (2.5). The boundary conditions (2.4) and (2.5) also ensure that $\psi(x,y)$ is constant on the free surface $y = \eta(x)$ and on the flat bed $y = 0$. We normalize ψ by choosing $\psi = 0$ on the free surface. Then, as a consequence of our definition of p_0, the flat bed is the streamline $\psi = -p_0$. From the equation of mass conservation (2.1) we deduce that $\int_0^L v(\xi, y)\,d\xi$ is independent of y and (2.5) shows that this constant is zero, so that ψ is L-periodic in x.

The waves we investigate are such that **the wave speed c is larger than the horizontal fluid velocity u**, that is,

$$u < c \quad \text{throughout the closed fluid domain,} \tag{3.1}$$

so that $p_0 < 0$. The traditional criterion for wave breaking requires that the horizontal fluid velocity near the wave crest exceeds the wave speed (cf. [18]), and for irrotational flows u attains its maximum at the wave crest[16] (see Chapter 4). Consequently, for irrotational flows (3.1) arises from the requirement that the wave is not near the breaking state. Field evidence indicates that the assumption (3.1) is also appropriate for rotational waves that are not near the breaking state and for whom the flow is without stagnation points[17]; cf. [294, 296]. We will show that the validity of (3.1) guarantees the existence of a function γ, called the **vorticity function**, such that $\omega = \gamma(\psi)$ throughout the fluid. Thus

$$\Delta\psi = \gamma(\psi).$$

[16]The presence of a strong nonuniform underlying current might alter this feature.

[17]A point where $(u - c, v) = 0$ is called an (instantaneous) stagnation point. Notice that (3.1) prevents the existence of stagnation points in the flow.

In conclusion, the assumption (3.1) is physically reasonable and mathematically advantageous. Throughout this chapter we assume its validity.

Let

$$\Gamma(p) = \int_0^p \gamma(-s)\,ds, \qquad p_0 \le p \le 0,$$

have maximum value Γ_{max} for $p \in [p_0, 0]$. From the Euler equation (2.3) we obtain **Bernoulli's law**, which states that

$$E = \frac{(c-u)^2 + v^2}{2} + gy + P + \Gamma(-\psi)$$

is constant throughout the flow. Therefore, the dynamic boundary condition (2.6) is equivalent to stating that

$$\frac{\psi_x^2 + \psi_y^2}{2} + g(y+d) = Q \quad \text{on} \quad y = \eta(x),$$

where $Q = E - P_{atm} + gd$. The hydraulic head E, and hence also Q, has for any flow a constant value that will be regarded as a parameter for the family of solutions that we will construct. In our analysis the mean water depth d is not fixed but will depend on the particular solution.

Summarizing the above considerations, we can reformulate the governing equations from Section 2.2 as the free boundary problem

$$\begin{cases} \Delta\psi = \gamma(\psi) \quad \text{in} \quad -d < y < \eta(x), \\ |\nabla\psi|^2 + 2g(y+d) = Q \quad \text{on} \quad y = \eta(x), \\ \psi = 0 \quad \text{on} \quad y = \eta(x), \\ \psi = -p_0 \quad \text{on} \quad y = -d, \end{cases} \tag{3.2}$$

which is to be solved for functions that are of period L in the x-variable. For simplicity we set $L = 2\pi$. The approach we present follows that developed in [71], with minor changes; some aspects are simplified and certain minor inaccuracies are removed (the main one being the fact that the vorticity therein had the opposite sign to the physical vorticity). The main difficulties associated with the problem (3.2) are its nonlinear character and the fact that the free surface $y = \eta(x)$ is unknown. The latter difficulty can be overcome by introducing a coordinate transform devised by Dubreil-Jacotin in 1934 (see [99]). Since ψ is constant both on the free surface and on the flat bed, with $y \mapsto \psi(x,y)$ strictly decreasing in view of (3.1), for every fixed x the height

$$h = y + d$$

above the flat bed is a single-valued function of ψ. This leads us to the change of variables

$$q = x, \quad p = -\psi$$

that transforms the (unknown) fluid domain

$$D_\eta = \{(x,y) :\ x \in (-\pi,\pi),\ -d < y < \eta(x)\}$$

of one wavelength into the (known) rectangular domain

$$R = (-\pi,\pi) \times (p_0, 0).$$

We choose the location of the wave crest at $(0, \eta(0))$.

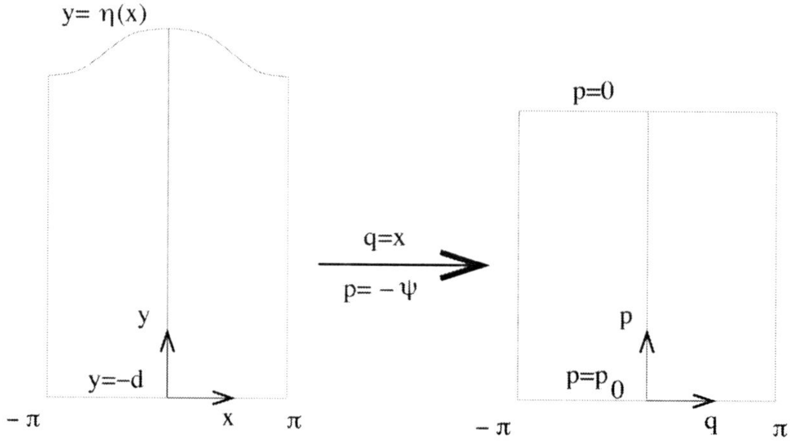

Figure 3.2. *The hodograph transform.*

Since

$$h_q = \frac{v}{u-c}, \qquad h_p = \frac{1}{c-u}, \tag{3.3}$$

we have

$$v = -\frac{h_q}{h_p}, \qquad u = c - \frac{1}{h_p}, \qquad \partial_x = \partial_q - \frac{h_q}{h_p}\,\partial_p, \qquad \partial_y = \frac{1}{h_p}\,\partial_p.$$

Consequently,

$$\partial_q\,\omega = \Big(\partial_x + \frac{h_q}{h_p}\,\partial_p\Big)\,\omega = \Big(\partial_x - \frac{v}{c-u}\,\partial_y\Big)\,\omega.$$

Taking the curl of the Euler equation (2.3) we get $(c-u)\omega_x - v\omega_y = 0$. Consequently, $\omega_q = 0$ and ω is a function of p throughout the rectangle R, $\omega = \gamma(-p)$, with

$$\gamma(-p) = \gamma(\psi) = \omega = \partial_y u - \partial_x v = \frac{1}{h_p}\,\partial_p\Big(c - \frac{1}{h_p}\Big) - \Big(\partial_q - \frac{h_q}{h_p}\,\partial_p\Big)\Big(-\frac{h_q}{h_p}\Big),$$

and (3.2) becomes

$$\begin{cases} (1+h_q^2)h_{pp} - 2h_q h_p h_{qp} + h_p^2 h_{qq} = \gamma(-p)h_p^3 & \text{in} \quad p_0 < p < 0, \\ 1 + h_q^2 + (2gh - Q)h_p^2 = 0 \quad \text{on} \quad p = 0, \\ h = 0 \quad \text{on} \quad p = p_0, \end{cases} \tag{3.4}$$

with h of period 2π in the q-variable satisfying

$$\frac{1}{2\pi}\int_{-\pi}^{\pi} h(q,0)\,dq = d, \tag{3.5}$$

and with (3.1) replaced by

$$h_p > 0 \quad \text{throughout the closed rectangle} \quad \overline{R}. \tag{3.6}$$

The boundary problems (3.2), under the assumption (3.1), and (3.4), under the assumption (3.6), are equivalent. To see this one has to recover ψ, the free surface η, and the mean depth d, assuming that $h \in C^k_{per}(\overline{R})$ with $k \geq 2$ solves (3.4) and satisfies (3.6), where the subscript *per* indicates 2π-periodicity in the q-variable. The free surface is easily identified as $\eta(x) = h(x,0) - d$, where $d = \frac{1}{2\pi}\int_{-\pi}^{\pi} h(q,0)\,dq$. To recover ψ we define the $C^{k-1}_{per}(\overline{R})$-function

$$F(q,p) = \frac{1}{h_p(q,p)}.$$

For a fixed $x_0 \in \mathbb{R}$ we solve the ordinary differential equation

$$\psi_y(x_0,y) = -F(x_0, -\psi(x_0,y)) \tag{3.7}$$

with initial data $\psi(x_0,\eta(x_0)) = 0$. Since $F \in C^1(\overline{R})$ there is a unique local solution. Moreover, (3.6) ensures $F \geq \delta > 0$ throughout $\overline{R}$ for some $\delta > 0$, so that $y \mapsto \psi(x_0,y)$ is strictly increasing as y decreases and the solution can be continued uniquely until it reaches the value $-p_0$ at some $y(x_0) < \eta(x_0)$. This shows that for any $x \in \mathbb{R}$ we can define $\psi(x,y)$ on some interval $[y(x),\eta(x)]$ with $y(x) < \eta(x)$. By uniqueness for (3.7), the fact that F is 2π-periodic in q ensures the 2π-periodicity of ψ in the x-variable. For $x \in \mathbb{R}$ fixed, differentiating the expression with respect to $y \in [y(x),\eta(x)]$ and using (3.7), we see that

$$y + d - h(x, -\psi(x,y))$$

is independent of $y \in [y(x),\eta(x)]$. Since at $y = \eta(x)$ the expression is zero, we infer that

$$y + d - h(x,-\psi(x,y)) = 0, \qquad y \in [y(x),\eta(x)]. \tag{3.8}$$

As $\psi(x,y(x)) = -p_0$ yields

$$h\Big(x, -\psi(x,y(x))\Big) = 0, \qquad x \in \mathbb{R},$$

we deduce that

$$y(x) = -d, \qquad x \in \mathbb{R}.$$

The fact that ψ is of class C^k follows by the dependence on data for the solutions to (3.7). To show that the constructed ψ satisfies (3.2), observe that we already know that $\psi = -p_0$ on $y = -d$, and $\psi = 0$ on $y = \eta(x)$. Differentiating (3.8) with respect to x, we obtain

$$\psi_x(x,y) = \frac{h_q(x,-\psi(x,y))}{h_p(x,-\psi(x,y))},$$

while (3.7) and the definition of F yield

$$\psi_y(x,y) = \frac{-1}{h_p(x,-\psi(x,y))}.$$

The nonlinear boundary condition in (3.2) follows at once. As for the partial differential equation, differentiating (3.7) with respect to y, we get

$$\begin{aligned}\psi_{yy}(x,y) = \psi_y(x,y)\,F_p(x,-\psi(x,y)) &= -\,F(x,-\psi(x,y))\,F_p(x,-\psi(x,y))\\ &= \frac{h_{pp}}{h_p^3}(x,-\psi(x,y)),\end{aligned}$$

while differentiating the identity

$$\psi_x(x,y)h_p(x,-\psi(x,y)) = h_q(x,-\psi(x,y))$$

with respect to x yields

$$\psi_{xx}(x,y) = \Big(\frac{h_{qq}}{h_p} - 2h_{qp}\frac{h_q}{h_p^2} + h_{pp}\frac{h_q^2}{h_p^3}\Big)(x,-\psi(x,y)).$$

Now

$$\psi_{xx} + \psi_{yy} = \gamma(\psi)$$

follows from the quasi-linear partial differential equation in (3.4). Defining $v = -\psi_x$ and $u = \psi_y + c$, we obtain a solution of the original problem, where $c > 0$ specifies the moving frame in which the wave is stationary.

For further considerations we notice that if the solution to (3.4) is more regular, $h \in C^{3,\alpha}_{per}(\overline{R})$ for some $\alpha \in (0,1)$, then $\gamma \in C^{1,\alpha}([0,|p_0|])$ and the physical solution

$$(u,v,\eta) \in C^{2,\alpha}_{per}(\overline{D_\eta}) \times C^{2,\alpha}_{per}(\overline{D_\eta}) \times C^{3,\alpha}_{per}(\mathbb{R}),$$

where $\overline{D_\eta}$ is the closure of the fluid domain

$$D_\eta = \{(x,y) : \; x \in \mathbb{R}, \; -d < y < \eta(x)\}.$$

3.1 Steady periodic waves of small amplitude

Given $p_0 < 0$, $\alpha \in (0,1)$, and the vorticity function $\gamma \in C^{1,\alpha}(\overline{R})$, we seek solutions $h \in C^{3,\alpha}_{per}(\overline{R})$ to the problem (3.4), subject to the condition (3.6). Such solutions correspond to solutions

$$\eta \in C^{3,\alpha}_{per}(\overline{D_\eta}), \quad u,v \in C^{2,\alpha}_{per}(\overline{D_\eta}), \quad P \in C^{1,\alpha}_{per}(\overline{D_\eta})$$

of the governing equations, with specified vorticity γ. Notice that a solution h that depends on q corresponds to an undulating free surface, while q-independent solutions h correspond to **laminar flows**—parallel shear flows with a flat free surface and such that each particle moves horizontally with a speed that depends only on the height above the flat bed—representing pure currents.

Theorem 3.1. *Let*

$$\gamma_\infty = \|\gamma\|_{C[p_0,0]}, \quad p_1 = \max\{p \in [p_0,0] : \Gamma(p) = \Gamma_{max}\}$$

and assume that

$$g > \frac{\sqrt{2}}{3}\gamma_\infty^{3/2}\,|p_1|^{1/2} + \frac{2\sqrt{2}}{5}\gamma_\infty^{1/2}\,|p_1|^{3/2}. \tag{3.9}$$

Then there is a C^1-curve $\mathfrak{C}_{loc}$ of solutions $h \in C^{3,\alpha}_{per}(\overline{R})$. Moreover, the solution curve $\mathfrak{C}_{loc}$ contains precisely one function that is independent of q.

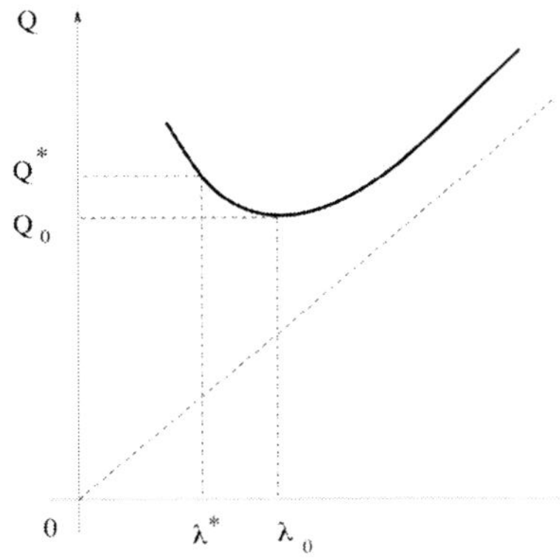

Figure 3.3. *Functional dependence for laminar flows.*

The proof of this result relies on applying the Crandall–Rabinowitz theorem (see the Appendix) and for this we have to identify the bifurcating parameter. Notice that the laminar flow solutions to (3.4) are given explicitly by

$$H(p,\lambda) = \int_0^p \frac{ds}{\sqrt{\lambda - 2\Gamma(s)}} + \frac{Q-\lambda}{2g} = \int_{p_0}^p \frac{ds}{\sqrt{\lambda - 2\Gamma(s)}}, \qquad p_0 \le p \le 0, \tag{3.10}$$

the parameters λ and Q being related by

$$0 < \int_{p_0}^0 \frac{ds}{\sqrt{\lambda - 2\Gamma(s)}} = \frac{Q-\lambda}{2g} \tag{3.11}$$

with $0 \le 2\Gamma_{max} < \lambda < Q$.

The bifurcation parameter

$$\lambda = \frac{1}{H_p^2(0,\lambda)}$$

is the square of the current velocity at the surface in the moving frame, $[c - u(0,0)]^2$. We point out that λ is not a single-valued function of Q, as the function $\lambda \mapsto Q(\lambda)$ is strictly convex for $\lambda > 0$, its minimum Q_0 on $(0,\infty)$ being attained at the unique point $\lambda_0 > 0$ where

$$\int_{p_0}^0 \Big(\lambda - 2\Gamma(s)\Big)^{-3/2} ds = \frac{1}{g}.$$

For every $Q > Q_0$ there is exactly one $\lambda > \lambda_0$ satisfying (3.11), and generally only for certain $Q > Q_0$ there is another solution $\lambda \in (2\Gamma_{max}, \lambda_0)$. We will see that the bifurcation point λ^* is located to the left of λ_0.

3.1.1 The linearization

In order to find waves of small amplitude, we first linearize the problem (3.4) about a laminar solution H. We seek solutions $h \in C^{3,\alpha}_{per}(\overline{R})$, even in the q-variable and zero on $p = p_0$, of the form

$$h = H + \varepsilon m.$$

The symmetry of the waves is expressed by requiring m to be even. Denoting

$$a(p,\lambda) = \sqrt{\lambda - 2\Gamma(p)},$$

at order ε we obtain, for $m \in C^{3,\alpha}_{per}(\overline{R})$ even in the q-variable, the boundary problem

$$\begin{cases} (a^3 m_p)_p + a\, m_{qq} = 0 \quad \text{in} \quad R, \\ a^3 m_p = gm \quad \text{on} \quad p = 0, \\ m = 0 \quad \text{on} \quad p = p_0. \end{cases} \tag{3.12}$$

We claim that $m \in C^{3,\alpha}_{per}(\overline{R})$, even in the q-variable, has the Fourier series representation

$$m(q,p) = \sum_{k=0}^{\infty} m_k(p)\cos(kq) \quad \text{in} \quad C^2_{per}(\overline{R}) \tag{3.13}$$

with $C^{3,\alpha}[p_0,0]$ coefficients

$$m_0(p) = \frac{1}{2\pi}\int_{-\pi}^{\pi} m(q,p)\,dq, \qquad m_k(p) = \frac{1}{\pi}\int_{-\pi}^{\pi} m(q,p)\cos(kq)\,dq, \qquad k \geq 1.$$

The other cases being similar, it suffices to prove that $\sum_{k=0}^{\infty} m_k''(p)\cos(kq)$ converges in $C_{per}(\overline{R})$. Notice that

$$|m_0''(p)| = \left|\frac{1}{2\pi}\int_{-\pi}^{\pi} m_{pp}(q,p)\,dq\right| \leq \|m\|_{C^2_{per}(\overline{R})},$$

$$m_k''(p) = \frac{1}{\pi}\int_{-\pi}^{\pi} m_{pp}(q,p)\cos(kq)\,dq = -\frac{1}{k\pi}\int_{-\pi}^{\pi} m_{ppq}(q,p)\sin(kq)\,dq, \qquad k \geq 1,$$

by using integration by parts in the last step. Thus for $N \geq n \geq 1$ we have

$$\begin{aligned} \Big|\sum_{k=n}^{N} m_k''(p)\cos(kq)\Big|^2 &\leq \Big(\sum_{k=n}^{N} |m_k''(p)|\Big)^2 = \Big(\sum_{k=n}^{N} \frac{1}{k\pi}\Big|\int_{-\pi}^{\pi} m_{ppq}(q,p)\sin(kq)\,dq\Big|\Big)^2 \\ &\leq \Big\{\sum_{k=n}^{N}\frac{1}{k^2}\Big\}\Big\{\sum_{k=n}^{N}\Big(\frac{1}{\pi}\int_{-\pi}^{\pi} m_{ppq}(q,p)\sin(kq)\,dq\Big)^2\Big\} \\ &\leq \frac{\pi^2}{6}\Big\{\frac{1}{\pi}\int_{-\pi}^{\pi} m^2_{ppq}(q,p)\,dq\Big\} \leq \frac{\pi}{3}\,\|m\|^2_{C^3_{per}(\overline{R})}. \end{aligned}$$

Therefore, $\sum_{k=0}^{\infty} m_k''(p)\cos(kq)$ converges in $C_{per}(\overline{R})$. Similarly $\sum_{k=0}^{\infty} m_k(p)\cos(kq)$ converges in $C_{per}(\overline{R})$, and the convergence of the right-hand side of (3.13) to $m(\cdot,p)$ in $L^2[-\pi,\pi]$ for every fixed $p \in [p_0,0]$ yields $m(q,p) = \sum_{k=0}^{\infty} m_k(p)\cos(kq)$ in $C_{per}(\overline{R})$. Then $\sum_{k=0}^{\infty} m_k''(p)\cos(kq)$ converges to $m_{pp}(q,p)$ in the sense of distributions, and this permits us to identify the limit of $\sum_{k=0}^{\infty} m_k''(p)\cos(kq)$ in $C_{per}(\overline{R})$ as being precisely $m_{pp}(q,p)$. Repeating this procedure we establish the validity of (3.13).

From (3.13) we deduce that m is a solution to (3.12) if and only if each m_k solves the Sturm–Liouville problem

$$\begin{cases} (a^3 M_p)_p = k^2 a M & \text{in} \quad (p_0, 0), \\ a^3 M_p = gM & \text{on} \quad p = 0, \\ M = 0 & \text{on} \quad p = p_0, \end{cases} \tag{3.14}$$

with m being q-dependent amounting to $m_k \not\equiv 0$ for some $k \geq 1$. We seek solutions of period 2π so that we investigate (3.14) for $k = 1$.

The variational approach

We associate to (3.14), with $k = 1$, the minimization problem

$$\mu(\lambda) = \inf_{\varphi \in H^1(p_0,0),\ \varphi(p_0)=0,\ \varphi \not\equiv 0} \{\mathbb{F}(\varphi, \lambda)\} \tag{3.15}$$

with

$$\mathbb{F}(\varphi,\lambda) = \frac{-g\,\varphi^2(0) + \displaystyle\int_{p_0}^0 a^3 \varphi_p^2\,dp}{\displaystyle\int_{p_0}^0 a\varphi^2\,dp}.$$

This is suggested by the fact that for a solution M of (3.14) we have $\mathbb{F}(M,\lambda) = -k^2$, while the choice of the function space is motivated by the quest for the largest possible Hilbert space for which $\mathbb{F}$ is well defined, with the boundary condition on $p = p_0$ captured while that on $p = 0$ is encoded in the form of $\mathbb{F}(\cdot,\lambda)$. For each $\lambda > 2\Gamma_{max}$ the existence of $\mu(\lambda) \in \mathbb{R}$ is ensured since if

$$\varepsilon(\lambda) = \inf_{p \in [p_0, 0]} \{a(p,\lambda)\} > 0,$$

then

$$\begin{aligned} \int_{p_0}^0 a^3 \varphi_p^2\,dp + \frac{4g^2}{\varepsilon^4(\lambda)} \int_{p_0}^0 a\varphi^2\,dp &\geq \varepsilon^3(\lambda) \int_{p_0}^0 \varphi_p^2\,dp + \frac{4g^2}{\varepsilon^3(\lambda)} \int_{p_0}^0 \varphi^2\,dp \\ &\geq 4g \int_{p_0}^0 \varphi\varphi_p\,dp = 2g\varphi^2(0) \end{aligned} \tag{3.16}$$

whenever $\varphi \in H^1(p_0,0)$ satisfies $\varphi(p_0) = 0$. Thus

$$\mu(\lambda) \geq -\frac{4g^2}{\varepsilon^4(\lambda)}.$$

We now claim that the infimum in (3.15) is attained by a function $M \in C^{3,\alpha}[p_0,0]$. Choose a minimizing sequence $\varphi_n \in H^1(p_0,0)$ with $\varphi_n(p_0) = 0$ and such that $\mathbb{F}(\varphi_n,\lambda) \to \mu(\lambda)$. Since $\mathbb{F}(\theta\varphi,\lambda) = \mathbb{F}(\varphi,\lambda)$ for any number $\theta \neq 0$, we can normalize the sequence $\{\varphi_n\}_{n\geq 1}$ by setting $\int_{p_0}^0 a\,\varphi_n^2\,dp = 1$ so that, using (3.16), we infer that

$$\mathbb{F}(\varphi_n,\lambda) = -g\,\varphi_n^2(0) + \int_{p_0}^0 a^3\,(\partial_p\varphi_n)^2\,dp \geq \frac{1}{2}\int_{p_0}^0 a^3\,(\partial_p\varphi_n)^2\,dp - \frac{2g^2}{\varepsilon^4(\lambda)}$$

$$\geq \frac{\varepsilon^3(\lambda)}{2}\int_{p_0}^0 (\partial_p\varphi_n)^2\,dp - \frac{2g^2}{\varepsilon^4(\lambda)}.$$

Since $\mathbb{F}(\varphi_n,\lambda) \to \mu(\lambda)$ we deduce that the sequence $\{\int_{p_0}^0 (\partial_p\varphi_n)^2\,dp\}_{n\geq 1}$ is bounded. As

$$\frac{1}{\varepsilon(\lambda)} = \frac{1}{\varepsilon(\lambda)}\int_{p_0}^0 a\,\varphi_n^2\,dp \geq \int_{p_0}^0 \varphi_n^2\,dp, \qquad n \geq 1,$$

we have that $\{\varphi_n\}_{n\geq 1}$ is bounded in the Hilbert space $H^1(p_0,0)$ and consequently (cf. [111]) has a weakly convergent subsequence $\{\varphi_{n_k}\}$ with limit $M \in H^1(p_0,0)$. Notice that $\partial_p\,\varphi_{n_k} \rightharpoonup M_p$ weakly in $L^2(p_0,0)$ and $\varphi_{n_k}(p_0) = 0$ yield

$$\varphi_{n_k}(p) = \int_{p_0}^p \partial_p\,\varphi_{n_k}(s)\,ds \to \int_{p_0}^p M_p(s)\,ds = M(p) \quad \text{at every} \quad p \in [p_0,0]. \tag{3.17}$$

From $\partial_p\,\varphi_{n_k} \rightharpoonup M_p$ we cannot generally infer the a.e. convergence of some subsequence as wild oscillations are possible, and nonlinear operations with weakly convergent sequences are generally prohibited (cf. [111]). Fortunately, the sequence $\{\varphi_{n_k}\}$ is minimizing, and the functional $\mathbb{F}(\cdot,\lambda)$ has suitable structural properties:

$$\int_{p_0}^0 a^3\,M_p^2\,dp \leq \liminf_{n_k\to\infty}\int_{p_0}^0 a^3\,(\partial_p\,\varphi_{n_k})^2\,dp \tag{3.18}$$

since

$$\int_{p_0}^0 a^3\,(\partial_p\varphi_{n_k})^2\,dp - \int_{p_0}^0 a^3\,M_p^2\,dp = \int_{p_0}^0 a^3(\partial_p\varphi_{n_k} - M_p)^2\,dp$$

$$+2\int_{p_0}^0 a^3(\partial_p\varphi_{n_k})M_p\,dp - 2\int_{p_0}^0 a^3\,M_p^2\,dp,$$

and $a^3 M_p \in L^2(p_0,0)$ together with $\partial_p\varphi_{n_k} \rightharpoonup M_p$ weakly in $L^2(p_0,0)$ ensure that the sum of the last two terms converges toward zero as $n_k \to \infty$. From (3.17) and (3.18) we infer that

$$-g\,M^2(0) + \int_{p_0}^0 a^3\,M_p^2\,dp \leq \liminf_{n_k\to\infty}\Big\{-g\,\varphi_{n_k}^2(0) + \int_{p_0}^0 a^3\,(\partial_p\varphi_{n_k})^2\,dp\Big\},$$

and since the sequence $\{\varphi_{n_k}\}$ is minimizing for $\mathbb{F}(\cdot,\lambda)$, the infimum is a minimum attained at $M \in H^1(p_0,0)$. Actually, $M \in C^{3,\alpha}[p_0,0]$. Indeed, as a minimum, M satisfies the Euler–Lagrange equation associated to the variational problem,

$$0 = \frac{d}{d\varepsilon}\,\mathbb{F}(M+\varepsilon\varphi,\lambda)\Big|_{\varepsilon=0} \tag{3.19}$$

for every $\varphi \in H^1(p_0,0)$ with $\varphi(p_0) = 0$. Observe that $\mathbb{F}(M,\lambda) = \mu(\lambda)$ and $\int_{p_0}^0 a\,M^2\,dp = 1$ by the dominated convergence theorem in view of (3.17), the normalization of $\{\varphi_{n_k}\}$, and the fact that the boundedness of $\{\varphi_{n_k}\}$ in $H^1(p_0,0)$ ensures an $L^\infty[p_0,0]$ uniform bound for $\{\varphi_{n_k}\}$. This allows us to explicitate (3.19) as

$$-gM(0)\varphi(0)+\int_{p_0}^{0} a^3\varphi_p M_p\,dp=\mu(\lambda)\int_{p_0}^{0} aM\varphi\,dp \tag{3.20}$$

for every $\varphi\in H^1(p_0,0)$ with $\varphi(p_0)=0$. Choosing φ smooth and compactly supported in $(p_0,0)$, we infer that

$$(a^3 M_p)_p=-\mu\, aM \quad \text{in} \quad H^{-1}(p_0,0). \tag{3.21}$$

Since $a\in C^{2,\alpha}[p_0,0]\subset H^2(p_0,0)$ and $M\in H^1(p_0,0)$, we have $aM\in H^1(p_0,0)$ so that $a^3M_p\in H^2(p_0,0)$ and $M_p\in H^2(p_0,0)$, that is, $M\in H^3(p_0,0)\subset C^2[p_0,0]$. Consequently, (3.21) holds classically and $a\in C^{2,\alpha}[p_0,0]$ yields $M\in C^{3,\alpha}[p_0,0]$. Multiplying (3.21) by some $\varphi\in H^1(p_0,0)$ with $\varphi(p_0)=0$ and integrating yields

$$-a^3(0)M_p(0)\varphi(0)+\int_{p_0}^{0} a^3\varphi_p M_p\,dp=\mu(\lambda)\int_{p_0}^{0} aM\varphi\,dp.$$

Choosing above and in (3.20) $\varphi(p)=p-p_0$, we obtain the missing boundary condition at $p=0$ so that the minimizer $M\in C^{3,\alpha}[p_0,0]$ is a classical solution of the (weighted) Sturm–Liouville problem

$$\begin{cases} (a^3 M_p)_p=-\mu\, a\, M & \text{in} \quad (p_0,0),\\ a^3 M_p=gM & \text{on} \quad p=0,\\ M=0 & \text{on} \quad p=p_0. \end{cases} \tag{3.22}$$

Remark. For the existence of linear waves it is necessary that $\mu(\lambda)=-1$ for some $\lambda>2\Gamma_{max}$, and we will see that this also suffices for the existence of nonlinear waves of small amplitude. For this we first prove that $\mu(\lambda)$ depends smoothly on $\lambda>2\Gamma_{max}$. Since for $\lambda>g+2\Gamma_{max}$ we have $a(\lambda,p)=\sqrt{\lambda-2\Gamma(p)}>\sqrt{g}$ for $p\in[p_0,0]$ and thus

$$\int_{p_0}^{0}(a^3\varphi_p^2+a\varphi^2)\,dp>\sqrt{g}\int_{p_0}^{0}(g\varphi_p^2+\varphi^2)\,dp\ge 2g\int_{p_0}^{0}\varphi\varphi_p\,dp=g\varphi^2(0)$$

whenever $\varphi\in H^1(p_0,0)$ is such that $\varphi(p_0)=0$, we deduce from (3.15) that $\mu(\lambda)>-1$ for $\lambda>g+2\Gamma_{max}$. Consequently, the existence of some $\lambda>2\Gamma_{max}$ with $\mu(\lambda)=-1$ is ensured if $\mu(\lambda)\le-1$ for some $\lambda>2\Gamma_{max}$. In this context notice that, cf. Example 3.2 below, given $p_0<0$, there are constant vorticities $\gamma<0$ such that $\mu(\lambda)>-1$ for all $\lambda>2\Gamma_{max}$, which explains the need for some assumption of type (3.9).

Example 3.2 (large constant negative vorticity). If $\gamma<0$ constant satisfies

$$|\gamma|>\frac{1}{2}+\frac{g^2}{2|p_0|^3}\Big(\frac{3}{2}+|p_0|\Big)^2,$$

then

$$(|2\gamma|)^{3/2}>\sqrt{|2\gamma|}>\frac{g\,(\frac{3}{2}+|p_0|)}{|p_0|^{3/2}},$$

while $\Gamma(p) = \gamma p$ on $[p_0, 0]$ with $\Gamma_{max} = \gamma p_0$, so that for $\lambda > 2\Gamma_{max}$ we have

$$a(p,\lambda) = \sqrt{\lambda - 2\Gamma(p)} > \sqrt{2\Gamma(p_0) - 2\Gamma(p)} = \sqrt{2|\gamma|(p - p_0)}, \qquad p \in [p_0, 0].$$

Therefore, if $\varphi \in H^1(p_0, 0)$ with $\varphi(p_0) = 0$, $\varphi \not\equiv 0$, then we have

$$\begin{aligned}
\int_{p_0}^{0} (a^3 \varphi_p^2 + a\varphi^2)\,dp &> (|2\gamma|)^{3/2} \int_{p_0}^{0} \varphi_p^2 (p - p_0)^{3/2}\,dp + \sqrt{|2\gamma|} \int_{p_0}^{0} \varphi^2 \sqrt{p - p_0}\,dp \\
&\geq \frac{g}{|p_0|^{3/2}} \int_{p_0}^{0} \varphi_p^2 (p - p_0)^{3/2}\,dp + \frac{g\,(\frac{3}{2} + |p_0|)}{|p_0|^{3/2}} \int_{p_0}^{0} \varphi^2 \sqrt{p - p_0}\,dp \\
&= \frac{g}{|p_0|^{3/2}} \left\{ \int_{p_0}^{0} \varphi_p^2 (p - p_0)^{3/2}\,dp + |p_0| \int_{p_0}^{0} \varphi^2 \sqrt{p - p_0}\,dp + \frac{3}{2} \int_{p_0}^{0} \varphi^2 \sqrt{p - p_0}\,dp \right\} \\
&\geq \frac{g}{|p_0|^{3/2}} \left\{ \int_{p_0}^{0} \varphi_p^2 (p - p_0)^{3/2}\,dp + \int_{p_0}^{0} \varphi^2 (p - p_0)^{3/2}\,dp + \frac{3}{2} \int_{p_0}^{0} \varphi^2 \sqrt{p - p_0}\,dp \right\} \\
&\geq \frac{g}{|p_0|^{3/2}} \left\{ 2 \int_{p_0}^{0} \varphi \varphi_p (p - p_0)^{3/2}\,dp + \frac{3}{2} \int_{p_0}^{0} \varphi^2 \sqrt{p - p_0}\,dp \right\} = g\,\varphi^2(0),
\end{aligned}$$

and the characterization (3.15) yields $\mu(\lambda) > -1$ for $\lambda > 2\Gamma_{max}$. ∎

Monotonicity of the ground state

In this subsection we will prove that $\lambda \mapsto \mu(\lambda)$ is real-analytic[18] for $\lambda > 2\Gamma_{max}$, with $\mu(\lambda)$ depending monotonically on $\lambda > 2\Gamma_{max}$ whenever $\mu(\lambda) < 0$.

Let $M(p,\lambda)$ be the $C^{3,\alpha}[p_0, 0]$-eigenfunction of (3.22) corresponding to the eigenvalue $\mu(\lambda)$, normalized[19] by requiring $M(0,\lambda) = 1$. Notice that

$$M(p,\lambda) = \varphi(p, \lambda, \mu(\lambda)),$$

where $\varphi(p,\lambda,\mu)$ is the unique solution of the linear differential equation

$$(a^3 \varphi_p)_p = -\mu a \varphi \quad \text{in} \quad (p_0, 0) \tag{3.23}$$

with initial data

$$\begin{cases} \varphi(0) = 1, \\ \varphi'(0) = \dfrac{g}{a^3(0)}, \end{cases} \tag{3.24}$$

depending analytically on (λ, μ) by the dependence of solutions on parameters. The variational approach that provided the existence of $\mu(\lambda)$ yields

$$\mu \int_{p_0}^{0} a M^2\,dp + g - \int_{p_0}^{0} a^3 M_p^2\,dp = 0, \tag{3.25}$$

[18]In the sense that near any point it may be represented by a convergent power series on some interval of positive length centered at that point.

[19]For $\mu < 0$, $M(0,\lambda) = 0$ would imply $M \equiv 0$. This can be easily seen by multiplying the differential equation for M by M, and integrating by parts on $[p_0, 0]$.

which we regard as a functional relation between $\mu(\lambda)$ and λ. Differentiating (3.23) with respect to μ, multiplying by $\varphi = M$, and integrating on $[p_0, 0]$ yields, in view of (3.24) and the fact that $M(p_0, \lambda, \mu) = 0$, that

$$\mu \int_{p_0}^{0} aMM_\mu \, dp - \int_{p_0}^{0} a^3 M_p M_{p\mu} \, dp = -\int_{p_0}^{0} aM^2 \, dp.$$

Therefore, the partial derivative with respect to μ of the function of (μ, λ) on the left side of (3.25) equals

$$\int_{p_0}^{0} aM^2 \, dp + 2\mu \int_{p_0}^{0} aMM_\mu \, dp - 2\int_{p_0}^{0} a^3 M_p M_{p\mu} \, dp = -\int_{p_0}^{0} aM^2 \, dp < 0.$$

By the implicit function theorem (cf. [38]) we deduce the real-analytic dependence of $\mu(\lambda)$ on $\lambda > 2\Gamma_{max} \geq 0$.

Denoting $\dot{a} = \frac{\partial a}{\partial \lambda}$ and so on, we compute

$$\dot{a} = \frac{1}{2a}, \qquad \dot{a}_p = -\frac{a_p}{2a^2}.$$

From (3.22) we obtain that $\dot{M}$ satisfies

$$\begin{cases} (a^3 \dot{M}_p)_p + \dfrac{3}{2}(aM_p)_p = -\dot{\mu} aM - \dfrac{1}{2a}\mu M - \mu a \dot{M} & \text{in} \quad (p_0, 0), \\ \dfrac{3}{2} aM_p + a^3 \dot{M}_p = g\dot{M} \quad \text{at} \quad p = 0, \\ \dot{M} = 0 \quad \text{at} \quad p = p_0. \end{cases}$$

Multiplying the above differential equation by M and the differential equation in (3.22) by $\dot{M}$, integrating on $(p_0, 0)$, and substracting the outcomes, we obtain

$$\dot{\mu} \int_{p_0}^{0} aM^2 \, dp = -\frac{1}{2}\mu \int_{p_0}^{0} a^{-1} M^2 \, dp + \frac{3}{2}\int_{p_0}^{0} aM_p^2 \, dp. \tag{3.26}$$

Consequently, we have the following proposition.

Proposition 3.3. *$\lambda \mapsto \mu(\lambda)$ is increasing in any interval where it is negative and the solution λ^* to $\mu(\lambda) = -1$, if it exists, is unique. Moreover, a solution exists if and only if* $\lim_{\lambda \downarrow 2\Gamma_{max}} \mu(\lambda) < -1$.

3.1.2 Existence of nonlinear waves of small amplitude

Existence of solutions for the linearization

Assuming the validity of (3.9) we prove the existence of nontrivial solutions to the linearized problem (3.12). The previous developments show that this amounts to proving that for some $\lambda > 2\Gamma_{max}$ we have $\mu(\lambda) \leq -1$. Let

$$p_1 = \max\{p \in [p_0, 0] : \Gamma(p) = \Gamma_{max}\}$$

and define for $k > \frac{1}{2}$ and $n \geq 2$ the function

$$\varphi_n(p) = \begin{cases} 0, & p_0 \leq p \leq p_n, \\ (p-p_n)^k, & p_n \leq p \leq 0, \end{cases}$$

where

$$p_n = \left(1 - \frac{1}{n}\right) p_1 + \frac{1}{n}\, p_0 < 0.$$

Clearly $\varphi_n \in H^1(p_0, 0)$ is such that $\varphi_n(0) = 0$ and $\varphi_n \not\equiv 0$; the reason we introduced $p_n < 0$ instead of simply setting $p_n = p_1$ was to prevent $\varphi_n \equiv 0$ in the special case when $p_1 = 0$. We have

$$a(p, 2\Gamma_{max}) = \sqrt{2\Gamma_{max} - 2\Gamma(p)} = \sqrt{2\Gamma(p_1) - 2\Gamma(p)} \leq \sqrt{2\gamma_\infty |p_1 - p|}, \quad p \in [p_0, 0],$$

so that

$$\begin{aligned}
&\int_{p_0}^0 a^3(p, 2\Gamma_{max}) \Big(\partial_p \varphi_n(p)\Big)^2 dp + \int_{p_0}^0 a(p, 2\Gamma_{max}) \varphi_n^2(p) dp \\
&\leq (2\gamma_\infty)^{3/2} k^2 \int_{p_n}^0 |p - p_1|^{3/2} (p - p_n)^{2k-2}\, dp \\
&\quad + (2\gamma_\infty)^{1/2} \int_{p_n}^0 |p - p_1|^{1/2} (p - p_n)^{2k}\, dp \\
&= (2\gamma_\infty)^{3/2} k^2 \Big\{ \int_{p_n}^{p_1} |p - p_1|^{3/2} (p - p_n)^{2k-2}\, dp + \int_{p_1}^0 |p - p_1|^{3/2} (p - p_n)^{2k-2}\, dp \Big\} \\
&\quad + (2\gamma_\infty)^{1/2} \Big\{ \int_{p_n}^{p_1} |p - p_1|^{1/2} (p - p_n)^{2k}\, dp + \int_{p_1}^0 |p - p_1|^{1/2} (p - p_n)^{2k}\, dp \Big\} \\
&\leq (2\gamma_\infty)^{3/2} k^2 \Big\{ |p_1 - p_n|^{3/2} \int_{p_n}^{p_1} (p - p_n)^{2k-2}\, dp + \int_{p_1}^0 (p - p_n)^{2k-1/2}\, dp \Big\} \\
&\quad + (2\gamma_\infty)^{1/2} \Big\{ |p_1 - p_n|^{1/2} \int_{p_n}^{p_1} (p - p_n)^{2k}\, dp + \int_{p_1}^0 (p - p_n)^{2k+1/2}\, dp \Big\}.
\end{aligned}$$

The last expression can be computed explicitly as

$$\begin{aligned}
&(2\gamma_\infty)^{3/2} k^2 \Big\{ \frac{|p_n|^{2k+1/2}}{2k+1/2} + \frac{3(p_1 - p_n)^{2k+1/2}}{(2k-1)(4k+1)} \Big\} \\
&\quad + (2\gamma_\infty)^{1/2} \Big\{ \frac{|p_n|^{2k+3/2}}{2k+3/2} + \frac{(p_1 - p_n)^{2k+3/2}}{(2k+1)(4k+3)} \Big\} \\
&= \varphi_n^2(0) \Big\{ (2\gamma_\infty)^{3/2} k^2 \frac{|p_n|^{1/2}}{2k+1/2} + (2\gamma_\infty)^{1/2} \frac{|p_n|^{3/2}}{2k+3/2} \Big\} \\
&\quad + \varphi_n^2(0) \Big\{ (2\gamma_\infty)^{3/2} k^2 \frac{3(p_1 - p_n)^{2k+1/2}}{|p_n|^{2k}(2k-1)(4k+1)} \\
&\quad + (2\gamma_\infty)^{1/2} \frac{(p_1 - p_n)^{2k+3/2}}{|p_n|^{2k}(2k+1)(4k+3)} \Big\}
\end{aligned}$$

since $\varphi_n^2(0) = |p_n|^{2k}$. On the other hand, by construction we have

$$\frac{|p_1 - p_n|}{|p_n|} \le 1,$$

while $\lim_{n\to\infty} |p_1 - p_n| = 0$. Using (3.9), we can find $k > \frac{1}{2}$ sufficiently small and some integer $N \ge 2$ such that for some $\varepsilon > 0$ we have

$$(2\gamma_\infty)^{3/2} k^2 \frac{|p_n|^{1/2}}{2k+1/2} + (2\gamma_\infty)^{1/2} \frac{|p_n|^{3/2}}{2k+3/2} < g - \varepsilon$$

whenever $n \ge N$. With this specified value of k we can now choose $n \ge N$ large enough to ensure

$$(2\gamma_\infty)^{3/2} k^2 \frac{3(p_1 - p_n)^{2k+1/2}}{|p_n|^{2k}(2k-1)(4k+1)} + (2\gamma_\infty)^{1/2} \frac{(p_1 - p_n)^{2k+3/2}}{|p_n|^{2k}(2k+1)(4k+3)} < \varepsilon.$$

This provides us with φ_n satisfying

$$\int_{p_0}^0 a^3(p, 2\Gamma_{max}) \Big(\partial_p \varphi_n(p)\Big)^2 dp + \int_{p_0}^0 a(p, 2\Gamma_{max}) \varphi_n^2(p) dp < g\, \varphi_n^2(0).$$

Since a depends continuously on λ, the previous inequality ensures that $\mathbb{F}(\varphi_n, \lambda) < -1$ for some $\lambda > 2\Gamma_{max}$. At this specific λ we have $\mu(\lambda) < -1$.

Example 3.4 (Nonnegative vorticity). If $\gamma \ge 0$, then $\Gamma_{max} = 0$, $p_1 = 0$, and the linearized problem has solutions as (3.9) holds true. In Example 3.2 we saw that this cannot be expected for negative constant vorticities with $|\gamma|$ sufficiently large! ■

The appropriate setting for local bifurcation

To check that we can apply the Crandall–Rabinowitz theorem in the present setting, we let

$$T = \{(q,p) :\ q \in [-\pi, \pi],\ p = 0\}, \qquad B = \{(q,p) :\ q \in [-\pi, \pi],\ p = p_0\}$$

be the top, respectively, the bottom, of the closed rectangle $\overline{R}$, and we define the Banach spaces

$$X = \{w \in C^{3,\alpha}_{per}(\overline{R}) :\ w = 0 \quad \text{on} \quad B\}, \qquad Y = C^{1,\alpha}_{per}(\overline{R}) \times C^{2,\alpha}_{per}(T),$$

where the subscript "*per*" means 2π-periodicity and evenness in the q-variable. If $H(p,\lambda)$ are laminar flows, set

$$h(q,p) = H(p,\lambda) + w(q,p) \quad \text{with} \quad w \in X$$

and write for $\lambda > 2\Gamma_{max}$ the system (3.4) in operator form:

$$\mathcal{F}(\lambda, w) = 0 \quad \text{with} \quad w \in X,$$

where $\mathcal{F} : (2\Gamma_{max}, \infty) \times X \to Y$ is given by $\mathcal{F} = (\mathcal{F}_1, \mathcal{F}_2)$ with

$$\begin{aligned}\mathcal{F}_1(\lambda, w) &= (1 + w_q^2)(H_{pp} + w_{pp}) - 2w_q(H_p + w_p)w_{pq} \\ &\quad + (H_p + w_p)^2 w_{qq} - \gamma(-p)(H_p + w_p)^3\end{aligned} \tag{3.27}$$

and

$$\mathcal{F}_2(\lambda, w) = \Big(1 + w_q^2 + [2g(H + w) - Q](H_p + w_p)^2\Big)\Big|_T. \tag{3.28}$$

By the equation satisfied by H we have

$$\mathcal{F}(\lambda, 0) = 0 \quad \text{for all} \quad \lambda > 2\Gamma_{max}.$$

The linearized operator $\mathcal{F}_w = (\mathcal{F}_{1w}, \mathcal{F}_{2w})$ at $w = 0$ is given by

$$\begin{cases} \mathcal{F}_{1w}(\lambda, 0) = \partial_p^2 + H_p^2 \partial_q^2 - 3\gamma(-p) H_p^2 \partial_p \quad \text{in} \quad R, \\ \mathcal{F}_{2w}(\lambda, 0) = 2(\lambda^{-1} g - \lambda^{1/2} \partial_p)\Big|_T. \end{cases} \tag{3.29}$$

The linear eigenvalue problem (3.22) with $\mu = -1$, expresses the fact that the function $M(p)\cos(q)$ belongs to the null space of $\mathcal{F}_w(\lambda, 0)$.

The null space

Assuming (3.9), we know that there is a unique $\lambda^* > 2\Gamma_{max}$ with $\mu(\lambda^*) = -1$ so that the null space $ker\{\mathcal{F}_w(\lambda^*, 0)\}$ contains at least one element $w(q, p) = M(p)\cos(q)$, where $M \in C^{3,\alpha}[p_0, 0]$ is the unique eigenfunction of (3.22) corresponding to the eigenvalue $\mu(\lambda^*) = -1$, normalized by $M(0) = 1$. The null space is one-dimensional. Indeed, if $m \in C^{3,\alpha}_{per}(\overline{R})$ belongs to the null space, then we proved before that its Fourier coefficients m_k satisfy (3.14) so that $m_1(p)$ is a constant multiple of $M(p)$ while if $m_k \not\equiv 0$ for some $k \geq 2$, we would have

$$\frac{-g\, m_k^2(0) + \displaystyle\int_{p_0}^0 a^3 (\partial_p m_k)^2 \, dp}{\displaystyle\int_{p_0}^0 a\, m_k^2 \, dp} = -k^2 < -1,$$

contradicting the minimizing property of $\mu(\lambda^*) = -1$. As for m_0, from the differential equation and from the boundary condition at $p = p_0$ in (3.14) with $k = 0$ we get

$$m_0(p) = A_0 \int_{p_0}^p a^{-3}(s, \lambda^*)\, ds, \qquad p \in [p_0, 0],$$

for some $A_0 \in \mathbb{R}$, and the boundary condition at $p = 0$ yields $A_0 = 0$ unless

$$\frac{1}{g} = \int_{p_0}^0 a^{-3}(p, \lambda^*)\, dp,$$

which is impossible. Indeed, the last relation is the defining property of λ_0 as the point where, in the context of laminar flows, the strictly convex function $\lambda \mapsto Q(\lambda)$ attains its minimum. However,

$$\lambda^* < \lambda_0, \tag{3.30}$$

by the monotonicity property of $\mu(\lambda)$ and the fact that $\mu(\lambda^*) = -1$ while $\mu(\lambda_0) = 0$. To see that $\mu(\lambda_0) = 0$, note that $\mathbb{F}(\int_{p_0}^{p} a^{-3}(s,\lambda_0)ds,\, \lambda_0) = 0$ yields $\mu(\lambda_0) \le 0$. On the other hand, $\mu(\lambda_0) \ge 0$ since $\mathbb{F}(\varphi,\lambda_0) \ge 0$ for any $\varphi \in H^1(p_0,0)$ such that $\varphi(p_0) = 0$, as

$$g\varphi^2(0) = g\Big(\int_{p_0}^{0} \varphi_p(p)dp\Big)^2 = g\Big(\int_{p_0}^{0} a^{3/2}(p,\lambda_0)\varphi_p(p)a^{-3/2}(p,\lambda_0)dp\Big)^2$$
$$\le g\Big(\int_{p_0}^{0} a^3(p,\lambda_0)\varphi_p^2(p)dp\Big)\Big(\int_{p_0}^{0} a^{-3}(p,\lambda_0)dp\Big) = \int_{p_0}^{0} a^3(p,\lambda_0)\varphi_p^2(p)dp.$$

The range

The fact that the operator $\mathcal{F}_w(\lambda^*,0) : X \to Y$ has a closed range of codimension one is ensured if we prove that the pair $(\mathcal{A},\, \mathcal{B}) \in Y$ belongs to the range if and only if it satisfies the orthogonality condition

$$\int\!\!\int_R \mathcal{A}(q,p)a^3(p,\lambda^*)\varphi^*(q,p)dqdp + \frac{1}{2}\int_T \mathcal{B}(q)a^2(0,\lambda^*)\varphi^*(q,0)dq = 0, \tag{3.31}$$

where

$$\varphi^*(q,p) = M(p)\cos(q) \in X$$

generates the null space $ker\,\{\mathcal{F}_w(\lambda^*,0)\}$. Indeed, assuming the validity of the characterization (3.31), the range $\mathcal{R}(\mathcal{F}_w(\lambda^*,0))$ is clearly closed in Y. Notice that $\lambda^* > 0$ as $\lambda^* > 2\Gamma_{max}$ and $\Gamma(0) = 0$ ensures $2\Gamma_{max} \ge 0$. On the other hand, $a(p,\lambda^*) > 0$ for $p \in [p_0,0]$ and $\mathbb{F}(M,\lambda^*) = -1$ ensure $M(0) \ne 0$. Since $a^2(0,\lambda^*) = \lambda^*$, we deduce from (3.31) that $(0,\cos(q)) \notin \mathcal{R}(\mathcal{F}_w(\lambda^*,0))$. Notice that if $(\mathcal{A}_1,\, \mathcal{B}_1),\, (\mathcal{A}_2,\, \mathcal{B}_2) \in Y \setminus \mathcal{R}(\mathcal{F}_w(\lambda^*,0))$, then

$$(\mathcal{A}_1,\, \mathcal{B}_1) - c(\mathcal{A}_2,\, \mathcal{B}_2) \in \mathcal{R}(\mathcal{F}_w(\lambda^*,0))$$

for

$$c = \frac{\displaystyle\int\!\!\int_R \mathcal{A}_1 a^3\varphi^*\,dqdp + \frac{1}{2}\int_T \mathcal{B}_1 a^2\varphi^*\,dq}{\displaystyle\int\!\!\int_R \mathcal{A}_2 a^3\varphi^*\,dqdp + \frac{1}{2}\int_T \mathcal{B}_2 a^2\varphi^*\,dq},$$

with a nonzero denominator as $(\mathcal{A}_2,\, \mathcal{B}_2) \notin \mathcal{R}(\mathcal{F}_w(\lambda^*,0))$. Therefore, the Banach space $Y/\mathcal{R}(\mathcal{F}_w(\lambda^*,0))$ is one-dimensional, so $\mathcal{R}(\mathcal{F}_w(\lambda^*,0))$ has codimension one.

As for the necessity of (3.31), if $\mathcal{F}_w(\lambda^*,0)\varphi = (\mathcal{A},\, \mathcal{B})$, then multiplying the partial differential equation

$$\mathcal{A} = \mathcal{F}_{1w}(\lambda^*,0)\varphi = \varphi_{pp} + H_p^2\varphi_{qq} - 3\gamma(-p)H_p^2\varphi_p$$

by $a^3\varphi^*$, integrating by parts, and using the fact that

$$\mathcal{B} = \mathcal{F}_{2w}(\lambda^*,0)\varphi = 2\Big(\frac{g}{\lambda^*}\varphi - \varphi_p\sqrt{\lambda^*}\Big)\Big|_T$$

in combination with $\mathcal{F}_w(\lambda^*,0)\varphi^* = 0$ and $H_p = a^{-1}$, we get (3.31).

The proof of the sufficiency of (3.31) is technically more intricate. To ensure a coercivity condition (see the remark at the end of this subsection), let us introduce the closed subspaces

$$X_0 = \left\{ \phi \in X : \int_{-\pi}^{\pi} \phi(q,p)dq = 0 \text{ for all } p \in [p_0, 0] \right\} \subset X,$$
$$Y_0 = \left\{ (\mathcal{A}, \mathcal{B}) \in Y : \int_{-\pi}^{\pi} \mathcal{A}(q,p)dq = 0 \text{ for all } p \in [p_0, 0], \int_T \mathcal{B}\, dq = 0 \right\} \subset Y.$$

Given $(\mathcal{A}, \mathcal{B}) \in Y$ such that (3.31) holds, since

$$H_p = a^{-1}, \quad a_p = -\gamma(-p)a^{-1}, \quad a(0) = \sqrt{\lambda^*},$$

using (3.29), we see that $\mathcal{F}_w(0,\lambda^*)\phi = (\mathcal{A}, \mathcal{B})$ for some $\phi \in X$ if and only if

$$\begin{cases} \left(a^3 \partial_p \phi_0\right)_p = a^3 \mathcal{A}_0 \quad \text{in} \quad (p_0, 0), \\ g\phi_0 - a^3 \partial_p \phi_0 = \dfrac{1}{2} a^2 \mathcal{B}_0 \quad \text{at} \quad p = 0, \\ \phi_0 = 0 \quad \text{at} \quad p = p_0 \end{cases} \tag{3.32}$$

and

$$\begin{cases} (a^3 \varphi_p)_p + a\varphi_{qq} = a^3 (\mathcal{A} - \mathcal{A}_0) \quad \text{in} \quad R, \\ g\varphi - a^3 \varphi_p = \dfrac{1}{2} a^2 (\mathcal{B} - \mathcal{B}_0) \quad \text{on} \quad T, \\ \varphi = 0 \quad \text{on} \quad B \end{cases} \tag{3.33}$$

for

$$\varphi = \phi - \phi_0 \in X_0, \quad (\mathcal{A} - \mathcal{A}_0, \mathcal{B} - \mathcal{B}_0) \in Y_0,$$

where $\mathcal{B}_0 \in \mathbb{R}$, $\phi_0 \in C^{3,\alpha}[p_0, 0]$, $\mathcal{A}_0 \in C^{1,\alpha}[p_0, 0]$ are given by

$$\mathcal{B}_0 = \frac{1}{2\pi} \int_T \mathcal{B}\, dq, \quad \phi_0(p) = \frac{1}{2\pi} \int_{-\pi}^{\pi} \phi(q,p)dq, \quad \mathcal{A}_0(p) = \frac{1}{2\pi} \int_{-\pi}^{\pi} \mathcal{A}(q,p)dq.$$

Claim 1. *For any* $(\mathcal{A}_0, \mathcal{B}_0) \in C^{1,\alpha}[p_0, 0] \times \mathbb{R}$ *the problem* (3.32) *has a unique solution* $\phi_0 \in C^{3,\alpha}[p_0, 0]$.

Indeed, from the differential equation in (3.32) we infer that

$$a^3(p)\phi_0'(p) = C + \int_{p_0}^{p} a^3(s,\lambda^*)\mathcal{A}_0(s)ds, \qquad p \in [p_0, 0],$$

for some constant $C \in \mathbb{R}$, so that the boundary condition at $p = p_0$ yields

$$\phi_0(p) = C \int_{p_0}^{p} a^{-3}(s,\lambda^*)ds + \int_{p_0}^{p} a^{-3}(s,\lambda^*)\left(\int_{p_0}^{s} a^3(\tau,\lambda^*)\mathcal{A}_0(\tau)d\tau\right)ds, \; p \in [p_0, 0].$$

The boundary condition at $p = 0$ becomes

$$C\left\{1 - g\int_{p_0}^{0} a^{-3}(p,\lambda^*)dp\right\} = g\int_{p_0}^{0} a^{-3}(s,\lambda^*)\left(\int_{p_0}^{s} a^3(\tau,\lambda^*)\mathcal{A}_0(\tau)d\tau\right)ds$$
$$- \int_{p_0}^{0} a^3(\tau,\lambda^*)\mathcal{A}_0(\tau)d\tau - \frac{1}{2}a^2(0,\lambda^*)\mathcal{B}_0.$$

We already proved (in the analysis of the null space) that

$$\frac{1}{g} = \int_{p_0}^{0} a^{-3}(p,\lambda_0)\,dp < \int_{p_0}^{0} a^{-3}(p,\lambda^*)\,dp.$$

Consequently, the constant C is always uniquely determined and Claim 1 is proved.

The above considerations reduce the proof of the sufficiency of (3.31) to showing that if (3.31) holds, then (3.33) has a solution $\varphi \in X_0$. Notice that both integrals in (3.31) vanish if they are evaluated on $\mathcal{A}_0$, respectively, $\mathcal{B}_0$. Consequently, we have to prove that if $(\mathcal{A}, \mathcal{B}) \in Y_0$ are satisfying (3.31), then (3.33) has a solution $\varphi \in X_0$.

Claim 2. *With $a = a(p,\lambda^*)$ and $(\mathcal{A}, \mathcal{B}) \in Y_0$, we claim that for every $\varepsilon \in (0,1)$ the approximate problem*

$$\begin{cases} -\varepsilon a^3\, v^{(\varepsilon)} + (1+\varepsilon)\left(a^3\, v_p^{(\varepsilon)}\right)_p + (1+\varepsilon)a\, v_{qq}^{(\varepsilon)} = a^3\,\mathcal{A} \quad \text{in} \quad R, \\ g\, v^{(\varepsilon)} - (1+\varepsilon)a^3\, v_p^{(\varepsilon)} = \frac{1}{2}a^2\,\mathcal{B} \quad \text{on} \quad T, \\ v^{(\varepsilon)} = 0 \quad \text{on} \quad B \end{cases} \tag{3.34}$$

has a unique solution $v^{(\varepsilon)} \in X_0$.

To prove the claim we introduce the space

$$\mathbb{H} = \left\{ \varphi \in H^1_{per}(R):\ \varphi \text{ even in } q,\ \int_{-\pi}^{\pi} \varphi(q,p)\,dq = 0 \text{ a.e. in } [p_0,0],\ \varphi = 0 \text{ a.e. on } B \right\}.$$

Notice that $\varphi \mapsto \int_{-\pi}^{\pi} \varphi(q,p)\,dq$ is by Fubini's theorem a bounded linear map from $H^1_{per}(R) \subset L^1(R)$ to $L^1[p_0,0]$, and the trace operator is also bounded from $H^1_{per}(R)$ to $L^2(B)$ (cf. [111]) so that $\mathbb{H}$ is a Hilbert space, being a closed subspace of the Hilbert space $H^1_{per}(R)$. A function $\varphi \in \mathbb{H}$ is a weak solution to (3.34) if

$$\begin{aligned} (1+\varepsilon)\int\!\!\int_R a^3 \varphi_p \phi_p\, dq\,dp + (1+\varepsilon)\int\!\!\int_R a\varphi_q \phi_q\, dq\,dp \\ + \varepsilon \int\!\!\int_R a^3\, \varphi\phi\, dq\,dp - g\int_T \varphi\phi\, dq \\ = -\frac{1}{2}\int_T \mathcal{B}\, a^2 \phi\, dq - \int\!\!\int_R \mathcal{A}\, a^3 \phi\, dq\,dp \end{aligned} \tag{3.35}$$

for all $\phi \in \mathbb{H}$. For $\varphi \in \mathbb{H} \cap C^3_{per}(\overline{R})$ we have (cf. Section 3.1.1)

$$\varphi(q,p) = \sum_{k=1}^{\infty} \varphi_k(p)\cos(kq) \quad \text{in} \quad C^2_{per}(\overline{R}),$$

with $\varphi_k \in C^3[p_0,0]$ given by

$$\varphi_k(p) = \frac{1}{\pi}\int_{-\pi}^{\pi} \varphi(q,p)\cos(kq)\,dq, \qquad p \in [p_0,0], \quad k \geq 1.$$

Clearly $\varphi_k(p_0)=0$ for any $k\geq 1$, and

$$\int\int_R a^3\varphi_p^2\,dq\,dp = \pi\sum_{k=1}^{\infty}\int_{p_0}^{0} a^3\,(\partial_p\varphi_k)^2\,dp,$$

while

$$\int\int_R a\varphi_q^2\,dq\,dp = \pi\sum_{k=1}^{\infty}k^2\int_{p_0}^{0} a\,\varphi_k^2\,dp,\qquad \int_T \varphi^2\,dq = \pi\sum_{k=1}^{\infty}\varphi_k^2(0).$$

Taking into account the mimimization problem (3.15), we deduce that

$$\int\int_R a^3\varphi_p^2\,dq\,dp + \int\int_R a\varphi_q^2\,dq\,dp \geq \pi\sum_{k=1}^{\infty}\int_{p_0}^{0}\Big\{a^3\,(\partial_p\varphi_k)^2 + a\,\varphi_k^2\Big\}\,dp$$

$$\geq \pi\,g\sum_{k=1}^{\infty}\varphi_k^2(0) = g\int_T \varphi^2\,dq.$$

Consequently,

$$\int\int_R a^3\varphi_p^2\,dq\,dp + \int\int_R a\varphi_q^2\,dq\,dp \geq g\int_T \varphi^2\,dq. \tag{3.36}$$

Moreover, since $\mathbb{H}\cap C^3_{per}(\overline{R})$ is dense in $\mathbb{H}$ (cf. [112]), (3.36) holds for all $\varphi\in\mathbb{H}$.

Since $\inf_{p\in[p_0,0]}\{a(p,\lambda^*)\}>0$, using (3.36) we see that the left-hand side of (3.35) defines a bounded and coercive bilinear form in $\mathbb{H}$, while the right-hand side defines a bounded linear functional on $\mathbb{H}$. An application of the Lax–Milgram theorem (see, e.g., [111]) yields the existence and uniqueness of a weak solution $v^{(\varepsilon)}\in\mathbb{H}$ to (3.34). Standard elliptic regularity theory (see [33]) yields $v^{(\varepsilon)}\in X_0$. Moreover, we have the **Schauder estimates** (see [129, Chapter 8])

$$\|v^{(\varepsilon)}\|_{C^{1,\alpha}_{per}(\overline{R})} \leq C\Big(\|\mathcal{A}\|_{C^{0,\alpha}_{per}(\overline{R})} + \|\mathcal{B}\|_{C^{1,\alpha}_{per}(T)} + \|v^{(\varepsilon)}\|_{L^\infty(R)}\Big) \tag{3.37}$$

with the constant C depending only on $\|a\|_{C^{1,\alpha}_{per}(\mathbb{R})}$ and on $\inf_{p\in[p_0,0]}\{a(p,\lambda^*)\}$.

Claim 3. *If $(\mathcal{A},\mathcal{B})\in Y_0$ satisfy (3.31), then for any sequence $\varepsilon_k\downarrow 0$ the sequence $\{v^{(\varepsilon_k)}\}_{k\geq 1}$ is bounded in $C^{1,\alpha}_{per}(\overline{R})$.*

Indeed, the existence of some $\varepsilon_k\downarrow 0$ with $\|v^{(\varepsilon_k)}\|_{C^{1,\alpha}_{per}(\overline{R})}\to\infty$ implies by (3.37) that $\|v^{(\varepsilon_k)}\|_{L^\infty(R)}\to\infty$. But then (3.37) ensures that the normalized functions

$$v_k = \frac{v^{(\varepsilon_k)}}{\|v^{(\varepsilon_k)}\|_{L^\infty(R)}}$$

are bounded in $C^{1,\alpha}_{per}(\overline{R})$. The compact embedding $C^{1,\alpha}_{per}(\overline{R})\subset C^1_{per}(\overline{R})$ enables us to extract a subsequence $\{v_{n_k}\}$ that converges in $C^1_{per}(\overline{R})$ to some v with $\|v\|_{L^\infty(R)}=1$. Consider now

(3.35) with $\varepsilon = \varepsilon_{n_k}$ and $\varphi = v^{(\varepsilon_{n_k})}$, divide by $\|v^{(\varepsilon_{n_k})}\|_{L^\infty(R)}$, and pass to the limit $n_k \to \infty$ to obtain

$$\int\int_R a^3 v_p \phi_p \, dq dp + \int\int_R a v_q \phi_q \, dq dp = g \int_T v\phi \, dq, \qquad \phi \in \mathbb{H}.$$

Thus $v \in C^1_{per}(\overline{R})$ is a weak solution (in $\mathbb{H}$) of the problem

$$\begin{cases} (a^3 v_p)_p + a\, v_{qq} = 0 \quad \text{in} \quad R, \\ g\, v - a^3 v_p = 0 \quad \text{on} \quad T, \\ v = 0 \quad \text{on} \quad B. \end{cases} \tag{3.38}$$

By elliptic regularity $v \in X_0$. Notice that (3.38) is precisely (3.12) so that $v \in ker\,\{\mathcal{F}_w(0,\lambda^*)\}$. Since $ker\,\{\mathcal{F}_w(0,\lambda^*)\}$ is one-dimensional, the function v is a multiple of $\varphi^*(q,p) = M(p)\cos(q)$, where $M \in C^{3,\alpha}[p_0,0]$ is the eigenfunction of (3.22) corresponding to $\mu(\lambda) = -1$, that is,

$$v = \delta\varphi^* \tag{3.39}$$

for some $\delta \in \mathbb{R}$. If $(\mathcal{A},\, \mathcal{B}) \in Y_0$ satisfy (3.31), we can infer more about this limit v. Indeed, performing in (3.35), with $\varepsilon = \varepsilon_{n_k}$, $\varphi = v^{(\varepsilon_{n_k})}$, $\phi = \varphi^*$, an integration by parts, and using the fact that φ^* solves (3.38) and (3.31) holds true, we get

$$\int\int_R a^3\, \varphi^*\, v^{(\varepsilon_{n_k})}\, dq dp + g \int_T \varphi^*\, v^{(\varepsilon_{n_k})}\, dq = 0.$$

Dividing by $\|v^{(\varepsilon_{n_k})}\|_{L^\infty(R)}$ and passing to the limit $n_k \to \infty$, the previous relation yields

$$\int\int_R a^3\, \varphi^*\, v\, dq dp + g \int_T \varphi^*\, v\, dq = 0.$$

But then (3.39) forces $v \equiv 0$, which contradicts $\|v\|_{L^\infty(R)} = 1$ and proves Claim 3.

Claim 3 and the compactness of $C^{1,\alpha}_{per}(\overline{R}) \subset C^1_{per}(\overline{R})$ yield the existence of a subsequence $\{v^{(\varepsilon_{n_k})}\}$ converging to some limit w in $C^1_{per}(\overline{R})$. Passing to the limit $n_k \to \infty$ in (3.35) with $\varepsilon = \varepsilon_{n_k}$, $\varphi = v^{(\varepsilon_{n_k})}$, we obtain that w is a weak solution (in $\mathbb{H}$), and by elliptic regularity a classical solution $w \in X_0$, of

$$\begin{cases} (a^3 w_p)_p + a\, w_{qq} = a^3 \mathcal{A} \quad \text{in} \quad R, \\ g\, w - a^3 w_p = \frac{1}{2} a^2 \mathcal{B} \quad \text{on} \quad T, \\ w = 0 \quad \text{on} \quad B, \end{cases} \tag{3.40}$$

which is precisely (3.33). This completes the proof of the sufficiency of (3.31).

Remark. Let us emphasize a delicate technical point. It was essential to split X into X_0 and its topological complement corresponding to the zero Fourier mode. Indeed, if we try to solve (3.34) directly in X for $(\mathcal{A},\, \mathcal{B}) \in Y$, then the function

$$\varphi_0(p) = \int_{p_0}^{p} a^{-3}(s,\lambda_0)\, ds, \qquad p \in [p_0,0],$$

would be in a corresponding Hilbert space $\mathbb{H}$, in the definition of which we dispense of the condition that the means over $[-\pi,\pi]$ should vanish for a.e. $p \in [p_0,0]$. Evaluated on (φ_0,φ_0) the bilinear form in (3.35) becomes

$$2\pi(1+\varepsilon)\int_{p_0}^{0} a^3(p,\lambda^*)a^{-6}(p,\lambda_0)\,dp + 2\pi\varepsilon\int_{p_0}^{0} a^3(p,\lambda^*)\varphi_0^2(p)\,dp - 2\pi\, g\,\varphi_0^2(0).$$

But $\varphi_0(0) = \frac{1}{g}$ by the defining property of λ_0, and (3.30) yields $a(p,\lambda^*) < a(p,\lambda_0)$ on $[p_0,0]$ so that

$$\int_{p_0}^{0} a^3(p,\lambda^*)a^{-6}(p,\lambda_0)\,dp < \int_{p_0}^{0} a^{-3}(p,\lambda_0)\,dp = \varphi_0(0) = \frac{1}{g}.$$

We see that for $\varepsilon > 0$ small enough, coerciveness is lost!

The transversality condition

To ensure the applicability of the Crandall–Rabinowitz theorem (see the Appendix) it suffices to check the transversality condition

$$[\mathcal{F}_{\lambda w}(\lambda^*,0)](1,\varphi^*) \notin \mathcal{R}(\mathcal{F}_w(\lambda^*,0)).$$

Since $a_p = -\gamma(-p)a^{-1}$, where $a = a(\cdot,\lambda^*)$, from (3.29) we compute

$$\mathcal{F}_{\lambda w}(\lambda^*,0) = -\left(a^{-4}\,\partial_q^2 + 3a_p a^{-3}\,\partial_p,\ \left\{2g\left(\frac{1}{\lambda^*}\right)^2 + \frac{1}{\sqrt{\lambda^*}}\,\partial_p\right\}\Big|_T\right).$$

By the previous characterization of the range $\mathcal{R}(\mathcal{F}_w(\lambda^*,0))$, it suffices to check that

$$\begin{aligned}&\int\!\!\int_R a^3\varphi^*\left\{a^{-4}\,\varphi^*_{qq} + 3a_p a^{-3}\,\varphi^*_p\right\}dq\,dp\\ &+\int_T\left\{\frac{g}{\lambda^*}(\varphi^*)^2 + \frac{1}{2}\sqrt{\lambda^*}\,\varphi^*\varphi^*_p\right\}dq \neq 0.\end{aligned} \tag{3.41}$$

Since

$$\varphi^*(q,p) = M(p)\cos(q)$$

with $M \not\equiv 0$ solving (3.22) with $\mu = -1$, and $a(0) = \sqrt{\lambda^*}$, we have that

$$\begin{cases} a^3\,\varphi^*_{pp} + 3a^2 a_p\,\varphi^*_p - a\,\varphi^* = 0 & \text{in} \quad (p_0,0),\\ (\lambda^*)^{3/2}\varphi^*_p = g\,\varphi^* & \text{at} \quad p = 0,\\ \varphi^* = 0 & \text{at} \quad p = p_0. \end{cases} \tag{3.42}$$

Thus

$$\begin{aligned}\int\!\!\int_R a_p\varphi^*\varphi^*_p\,dq\,dp &= \int_T a\,\varphi^*\varphi^*_p\,dq - \int\!\!\int_R a\,\varphi^*\varphi^*_{pp}\,dq\,dp - \int\!\!\int_R a\,(\varphi^*_p)^2\,dq\,dp\\ &= \int_T a\,\varphi^*\varphi^*_p\,dq + \int\!\!\int_R \varphi^*\left(3a_p\varphi^*_p - a^{-1}\varphi^*\right)dq\,dp - \int\!\!\int_R a\,(\varphi^*_p)^2\,dq\,dp\end{aligned}$$

so that

$$\int\int_R a_p \varphi^* \varphi_p^* dq dp = -\frac{1}{2}\int_T a\varphi^* \varphi_p^* dq + \frac{1}{2}\int\int_R \left\{a^{-1}(\varphi^*)^2 + a(\varphi_p^*)^2\right\} dq dp.$$

Consequently, we can express (3.41) as

$$\int\int_R a^{-1}\varphi^* \varphi_{qq}^* dq dp + \frac{3}{2}\int\int_R a^{-1}(\varphi^*)^2 dq dp + \frac{3}{2}\int\int_R a(\varphi_p^*)^2 dq dp$$
$$+ \int_T \left\{\frac{g}{\lambda^*}(\varphi^*)^2 - \sqrt{\lambda^*}\varphi^*\varphi_p^*\right\} dq \neq 0$$

since $a(0) = \sqrt{\lambda^*}$. But the boundary condition at $p = 0$ in (3.42) yields the vanishing of the boundary integral, and since $\varphi_{qq}^* = -\varphi^*$ we deduce that (3.41) equals

$$\frac{1}{2}\int\int_R a^{-1}(\varphi^*)^2 dq dp + \frac{3}{2}\int\int_R a(\varphi_p^*)^2 dq dp > 0.$$

This completes the proof of the transversality condition and of Theorem 3.1.

Conclusion

The previous considerations show that the existence of $\lambda^* > 2\Gamma_{max}$ with $\mu(\lambda^*) = -1$ ensures the existence of nonlinear waves of small amplitude. The condition (3.9) is sufficient to ensure the existence of λ^*. While the hypotheses of the Crandall–Rabinowitz theorem are generally only sufficient but not necessary for local bifurcation, in our case the condition $\mu(\lambda^*) = -1$ is necessary and sufficient for local bifurcation in the specified spaces. Indeed, the proof of sufficiency is also contained in the previous considerations: the existence of a bifurcating curve implies that the linearized problem (3.12) has a nontrivial solution, and we saw that this is possible only if $\mu(\lambda^*) = -1$ for some $\lambda^* > 2\Gamma_{max}$.

3.1.3 The dispersion relation

Since for the laminar flows

$$H_p^{-1}(0,\lambda) = (c-u)\Big|_{\text{at the flat surface}} = \sqrt{\lambda},$$

we see that waves of small amplitude occur exactly when the velocity of the laminar flows reaches the critical speed $\sqrt{\lambda^*}$ at the flat surface. In some special cases it is possible to compute λ^* explicitly.

Irrotational flow

If $\gamma \equiv 0$, then $a(p,\lambda) \equiv \sqrt{\lambda}$ and the problem (3.22) with $\mu = -1$ becomes

$$\begin{cases} M_{pp} = \lambda^{-1} M & \text{on} \quad (p_0, 0), \\ M_p = g\,\lambda^{-3/2} M & \text{at} \quad p = 0, \\ M = 0 & \text{at} \quad p = p_0. \end{cases}$$

The general solution of the differential equation with the boundary condition at $p = p_0$ is

$$M(p) = \delta \sinh\Big(\frac{p-p_0}{\sqrt{\lambda}}\Big), \qquad p \in [p_0, 0],$$

with $\delta \in \mathbb{R}$. For $\delta \neq 0$ the boundary condition at $p = 0$ is equivalent to the implicit equation

$$\lambda - g \tanh\Big(\frac{|p_0|}{\sqrt{\lambda}}\Big) = 0. \tag{3.43}$$

As a function of λ, the left-hand side can be easily seen to be a strictly increasing function from $(0,\infty)$ onto $(-g,\infty)$, so that $\lambda^* > 0$ is its unique root. We can compute λ^* explicitly in terms of the depth d of the laminar water flow at which bifurcation occurs. Indeed, by (3.10) we have

$$H(p,\lambda^*) = \frac{p-p_0}{\sqrt{\lambda^*}}, \qquad p \in [p_0, 0],$$

so $H_p^{-2}(p,\lambda^*) \equiv \lambda^*$ and in the physical variables we get the uniform current

$$u(x,y) - c \equiv -\sqrt{\lambda^*}, \quad v(x,y) \equiv 0, \qquad -d \le y \le 0.$$

But the definition

$$p_0 = \int_{-d}^{0} \Big(u(x,y) - c\Big)\, dy$$

of the relative mass flux yields $d\sqrt{\lambda^*} = |p_0|$ so that (3.43) becomes **the dispersion relation for 2π-periodic irrotational gravity water waves**

$$(c-u)\Big|_{\text{at the flat surface}} = \sqrt{g \tanh(d)}. \tag{3.44}$$

To elucidate the meaning of "dispersion," notice that in our analysis we assumed that the wave period is precisely 2π. We have to write the value of the bifurcation parameter for periodic traveling waves with wavelength L. For this, given the wavelength $L > 0$, let

$$\kappa = \frac{2\pi}{L}$$

be the associated **wavenumber**, representing the number of cycles of this periodic wave that appear in a spatial window of length 2π in the direction of wave propagation. Consider a solution of period L to the governing equations for periodic traveling water waves

$$\begin{cases} (u-c)u_x + vu_y = -P_x \quad \text{in} \quad -d < y < \eta(x-ct), \\ (u-c)v_x + vv_y = -P_y - g \quad \text{in} \quad -d < y < \eta(x-ct), \\ u_x + v_y = 0 \quad \text{in} \quad -d < y < \eta(x-ct), \\ u_y - v_x = \omega \quad \text{in} \quad -d < y < \eta(x-ct), \\ v = (u-c)\eta' \quad \text{on} \quad y = \eta(x-ct), \\ P = P_{atm} \quad \text{on} \quad y = \eta(x-ct), \\ v = 0 \quad \text{on} \quad y = -d, \end{cases} \tag{3.45}$$

and perform the change of variables (scaling)

$$\begin{cases} \tilde{\eta} = \kappa\eta, & \tilde{c} = c, & \tilde{u} = u, & \tilde{v} = v, & \tilde{P} = P, \\ \tilde{x} = \kappa x, & \tilde{y} = \kappa y, & \tilde{t} = \kappa t, & \tilde{g} = \kappa^{-1} g, & \tilde{\omega} = \kappa^{-1}\omega. \end{cases} \tag{3.46}$$

The new variables (dependent and independent) satisfy a 2π-periodic system in the $\tilde{x}$-variable of a form almost identical to (3.45), the only difference being that g should be replaced by $\tilde{g}$ and ω by $\tilde{\omega}$ (corresponding to replacing γ by $\tilde{\gamma} = \kappa^{-1}\gamma$). In view of (3.44), we obtain the **dispersion relation for periodic irrotational gravity water waves**

$$(c - u^*) = \sqrt{\frac{g}{\kappa}\,\tanh(\kappa d)}, \tag{3.47}$$

where

$$(c - u^*) = (c - u)\Big|_{\text{at the flat surface}}.$$

Let us now point out some important aspects:

- The function on the right-hand side of (3.47) being strictly decreasing in κ, we deduce that the speed $(c - u^*)$ exhibits a monotonically increasing dependence on the wavelength $L = 2\pi\kappa^{-1}$. This is the **dispersive effect**: within the linear framework (where the superposition principle applies) waves of different lengths travel at different speeds. A group of waves of different wavelengths starting together would spread out, so that after a while the larger waves are at the front.

- **Shallow-water waves** are encountered in the limit $\delta \to 0$, where $\delta = \frac{d}{L}$ is the "shallowness parameter." Writing (3.47) in the form

  $$(c - u^*) = \sqrt{gd\,\frac{\tanh(2\pi\delta)}{2\pi\delta}},$$

  since $\lim_{\delta\to 0} \frac{\tanh(2\pi\delta)}{2\pi\delta} = 1$, for shallow-water waves we have

  $$(c - u^*) \approx \sqrt{gd}. \tag{3.48}$$

  Numerically $\sqrt{\frac{\tanh(s)}{s}}$ lies between 0.97 and 1 for $s < 0.44$ so that (3.48) falls short of (3.47) by at most 3% provided that $\delta < 0.07$, which for practical purposes is the appropriate range for shallow-water waves; cf. [206]. The importance of (3.48) lies in that according to it, all waves, if sufficiently long compared with the average water depth, travel at the same speed. This explains why a ride in a speed boat on a relatively calm sea is smoother at higher speeds: if the boat has a speed inferior to the **critical speed** $\sqrt{gd}$, the waves created by the boat's displacement will overtake it and consequently will create a disturbance in front of the boat, hindering the displacement. To overcome this, the boat must be capable of a sudden burst of power that carries it beyond the critical speed before the wave ahead of it has had the time to form; this is similar to the difficulty in making an aircraft break the sound barrier, that is, travel faster than waves in the air. Setting $g = 9.8\,m/s^2$, it is easy to work out the critical speed. For example, if the depth of the water is $6m$, then the critical speed is $7.8\,m/s = 28\,km/h$.

- Another important regime for water waves is that of **deep water waves**, obtained in the limit $\delta \to \infty$. Writing (3.47) in the form

$$(c-u^*) = \sqrt{gd\,\frac{\tanh(2\pi\delta)}{2\pi\delta}},$$

since $\lim_{\delta\to\infty}\tanh(2\pi\delta)=1$, for deep water waves we have

$$(c-u^*) \approx \sqrt{\frac{gL}{2\pi}}. \tag{3.49}$$

Numerically $\sqrt{\tanh(s)}$ lies between 0.97 and 1 for $s > 1.75$ so that (3.49) falls short of (3.47) by at most 3% provided that $\delta > 0.28$, so $d > 0.28\,L$ characterizes for practical purposes the deep water regime (cf. [206]). The most common cause for waves at sea is due to the wind blowing over the surface of the sea, so (3.49) tells us that waves produced in a storm at sea will travel away from the storm region at speeds proportional to the square root of their wavelengths. In the North Atlantic Ocean, where the average water depth is $3.3\,km$, under normal conditions waves range from those having lengths of $50\,m$ to those having lengths of $100\,m$. Formula (3.49) tells us that the shorter waves travel at $30\,km/h$ and the longer ones travel faster at $40\,km/h$ to a rough approximation that is, however, not far from reality (cf. [177]). In the South Pacific, where the average depth is in excess of $4\,km$, wavelengths around $300\,m$ can be encountered, and such waves travel at speeds around $70\,km/h$. Notice, however, that even in the deep sea we can encounter shallow-water waves: **tsunami waves** often have wavelengths in excess of hundreds of km, and the deepest point in the ocean is in the Marianas Trench in the Western Pacific Ocean at approximately $11021\,m$.

- The **time period** of the wave, $\hbar$ (the ratio between wavelength and wave speed), is much easier to measure than wavelength. For waves of small amplitude we obtain from (3.47) the approximate formula

$$\hbar = \frac{2\pi}{\sqrt{g\kappa\tanh(\kappa d)}},$$

so that, for a fixed undisturbed depth d, the function $\kappa \mapsto \hbar(\kappa)$ is decreasing. Since $\kappa = 2\pi/L$ we infer that the time period of the wave is increasing as a function of wavelength L. The numerical value of the period for typical surface gravity waves varies roughly from $0.1\,s$ to $30\,s$ (cf. [206]).

- Let us investigate the dependence of $(c-u^*)$ on the average depth d for fixed **frequency** $\beta = \kappa(c-u^*)$, representing the number of cycles of the wave that pass by any fixed point during a time interval of length 2π. From (3.47) we get

$$\beta^2 = g\kappa\tanh(\kappa d) = g\,\frac{\beta}{c-u^*}\tanh\Big(\frac{\beta d}{c-u^*}\Big). \tag{3.50}$$

This shows that as the average depth d varies gradually becoming smaller and smaller (at a fixed frequency β), the speed $(c-u^*)$ has to decrease. The interest of this consideration lies in the fact that as ocean waves approach the coast, passing through water of gradually less and less depth, the frequency ß is practically constant (so that the number of crests approaching the shore per unit time is equal to the number far from the shore, as a glance at these waves will confirm). Let us consider, for example, a wave with time period $\hbar = 8s$ having wavelength $L = 100m$ in water of large uniform depth. By (3.50) we have

$$c - u^* = \frac{g}{ß}\tanh\Big(\frac{ßd}{c-u^*}\Big) < \frac{g}{ß},$$

which yields $\frac{ßd}{c-u^*} > ßd\,\frac{ß}{g} = \frac{ß^2}{g}d$ so that

$$c - u^* = \frac{g}{ß}\tanh\Big(\frac{ßd}{c-u^*}\Big) > \frac{g}{ß}\tanh\Big(\frac{ß^2}{g}d\Big).$$

The two previously displayed relations show that for ß fixed we have

$$(c-u^*) \approx \frac{g}{ß} \approx 45\,km/h \quad \text{for large} \quad d.$$

Numerically, since $\tanh(s) > 0.95$ for $s > 1.5$, and $\frac{ß^2}{g}d \approx \frac{d}{16m}$, the above estimate has an accuracy within 5% for $d \geq 24m$. Both the wave speed and the wavelength are reduced with passage into gradually shallower water: at a depth of $1\,m$, using (3.48), the speed of this wave with time period $\hbar = 8s$ (as the frequency ß is preserved, $\hbar = 2\pi/ß$ also remains constant) is

$$(c-u^*) \approx \sqrt{gd} \approx 11\,km/h,$$

and, since $L = \frac{2\pi(c-u^*)}{ß}$, the wavelength is reduced to about $25\,m$—a fourfold reduction!

Flows with constant vorticity

If $\gamma \neq 0$ is a constant, the substitution

$$M(p) = \frac{1}{\sqrt{\lambda - 2\gamma p}}\,M_0\Big(\frac{\sqrt{\lambda-2\gamma p}}{\gamma}\Big)$$

transforms the differential equation in (3.22) with $\mu = -1$ into $M_0'' = M_0$. Since $M_0(p_0) = 0$, we deduce that up to a multiplicative constant

$$M(p) = \frac{1}{\sqrt{\lambda-2\gamma p}}\sinh\Big(\frac{\sqrt{\lambda-2\gamma p}-\sqrt{\lambda-2\gamma p_0}}{\gamma}\Big), \quad p \in [p_0, 0].$$

The boundary condition at $p = 0$ in (3.22) is then equivalent to $\lambda^* > 0$ being a solution of the equation

$$\tanh\Big(\frac{\sqrt{\lambda}-\sqrt{\lambda-2\gamma p_0}}{\gamma}\Big) = \frac{\lambda}{\gamma\sqrt{\lambda}-g}. \tag{3.51}$$

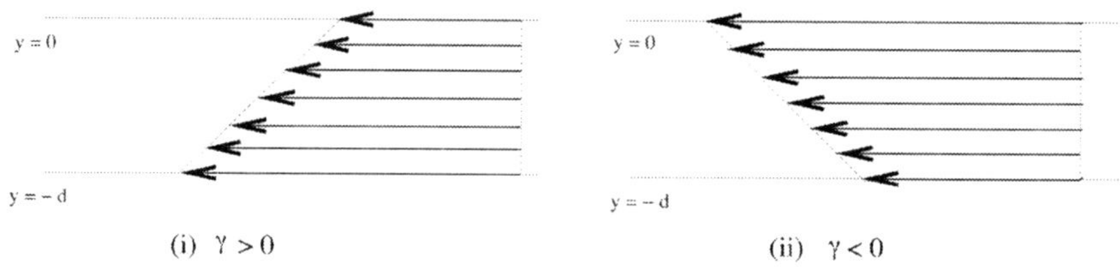

Figure 3.4. *The current in the moving frame.*

From (3.10) we obtain that the bifurcating laminar flow is given by

$$H(p,\lambda^*) = \frac{\sqrt{\lambda^* - 2\gamma p_0} - \sqrt{\lambda^* - 2\gamma p}}{\gamma}, \qquad p \in [p_0, 0].$$

For the corresponding velocity field in physical coordinates we have $v = 0$, $u_y = \gamma$, and $(c - u^*) = (c-u)\big|_{\text{at the flat surface}} = \frac{1}{H_p(0,\lambda^*)}$, so that

$$(u - c, v) = (-\sqrt{\lambda^*} + \gamma y, 0), \qquad -d \le y \le 0. \tag{3.52}$$

From the definition

$$p_0 = \int_{-d}^{0} \Big(u(x,y) - c\Big)\, dy$$

of the relative mass flux we get $|p_0| = d\sqrt{\lambda^*} + \frac{\gamma}{2} d^2$, and thus

$$d = \frac{\sqrt{\lambda^* - 2\gamma p_0} - \sqrt{\lambda^*}}{\gamma} > 0, \tag{3.53}$$

the other root being negative. The previous formula enables us to express (3.51) as

$$\tanh(d) = \frac{\lambda^*}{g - \gamma\sqrt{\lambda^*}}.$$

Solving for $\sqrt{\lambda^*} = (c - u^*)$, we obtain the dispersion relation for 2π-periodic rotational gravity water waves with constant vorticity

$$(c - u^*) = -\frac{\gamma}{2}\tanh(d) + \frac{1}{2}\sqrt{\gamma^2 \tanh^2(d) + 4g\tanh(d)}.$$

As with irrotational waves, replacing g by $\kappa^{-1} g$, d by κd, and γ by $\kappa^{-1}\gamma$, we obtain the **dispersion relation for rotational periodic gravity water waves with constant vorticity**

$$(c - u^*) = -\frac{\gamma}{2\kappa}\tanh(\kappa d) + \frac{1}{2\kappa}\sqrt{\gamma^2 \tanh^2(\kappa d) + 4g\kappa\tanh(\kappa d)}. \tag{3.54}$$

For $\gamma = 0$ we see that (3.54) particularizes to (3.47). Since the right-hand side of (3.54) is strictly decreasing in γ, a negative vorticity enhances the intrinsic wave speed $(c - u^*)$ with respect to the case of an irrotational flow, whereas a positive vorticity diminishes it.

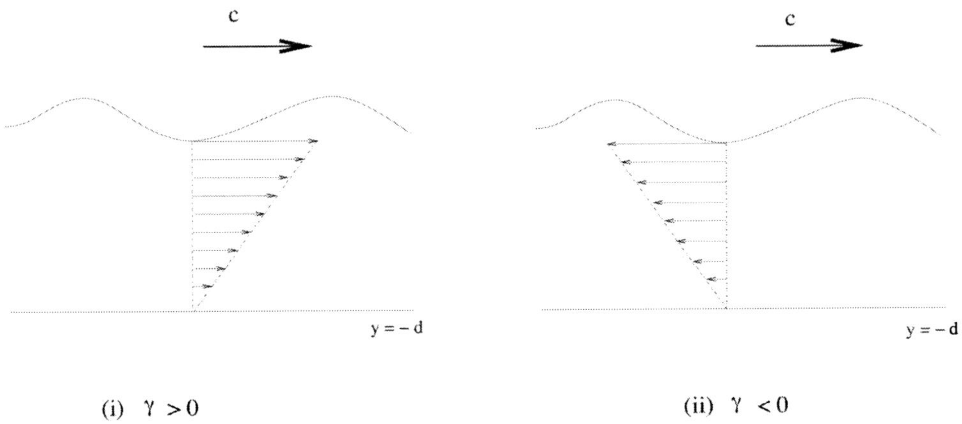

Figure 3.5. *Wave-current interaction:* (i) *favorable current;* (ii) *adverse current.*

A smaller speed facilitates the appearance of waves and this suggests that a current with $\gamma > 0$ is a **favorable current** while $\gamma < 0$ corresponds to an **adverse current**.

To further clarify this issue, notice that the bifurcating laminar flow with velocity field (3.52) is such that

$$(u-c)\Big|_{y=0} = -\sqrt{\lambda^*} < 0$$

along the flat surface and, using (3.53), on the flat bed

$$(u-c)\Big|_{y=-d} = -\sqrt{\lambda^* - 2\gamma p_0} < 0.$$

This current is positively sheared for $\gamma > 0$ and negatively sheared for $\gamma < 0$.

In the physical variables the waves of small amplitude whose existence is ensured by the local bifurcation result arise as genuine nonlinear solutions representing small perturbations of the laminar flow

$$(c - \sqrt{\lambda^*} + \gamma y,\, 0), \qquad -d \le y \le 0. \tag{3.55}$$

Given $p_0 < 0$ and the normalized wavelength $L = 2\pi$, we found solutions in the moving frame, specifying thus v, P, the wave profile η, and the horizontal velocity $(u-c)$. This raises **the issue of how to determine the wave speed**—a nontrivial matter even in the context of irrotational flows, where the issue was settled by Stokes in 1847 (see [287]). In the physical variables the velocity field (u,v) beneath the wave can be decomposed into a current of vorticity γ that is steady and aligned in the plane of the wave motion, of the form $(\gamma(y+d),0)$, and a wave-induced velocity field

$$(U_0(x-ct,y),\, V_0(x-ct,y)) = \Big(u(x-ct,y) - \gamma(y+d),\, v(x-ct,y)\Big)$$

whose horizontal average beneath the trough level is a uniform current of strength $\mathfrak{S} = \frac{1}{2\pi}\int_{-\pi}^{\pi} u(x,-d)\,dx$. Indeed, in the moving frame, for each fixed $y \in [-d, \eta(-\pi)]$, we have

$$\int_{-\pi}^{\pi} v(x,y)\,dx = 0$$

since v is odd in the x-variable as h is even in the q-variable, while

$$\int_{-\pi}^{\pi}\Big(u(x,y)-\gamma(y+d)\Big)dx-\int_{-\pi}^{\pi}u(x,-d)dx=\int_{-\pi}^{\pi}\int_{-d}^{y}\Big(u_y(x,y)-\gamma\Big)dydx$$
$$=\int_{-d}^{y}\int_{-\pi}^{\pi}v_x(x,y)dxdy=0$$

since v is 2π-periodic in the x-variable. Consequently, at any horizontal level beneath the trough level $y=\eta(-\pi)$ the average of (U_0,V_0) over one wavelength is precisely $(\mathfrak{S},0)$. We find c by requiring $\mathfrak{S}=0$: the wave-induced motion is a pure wave motion without an underlying current.

In the case of the laminar flow (3.55), we have

$$u(x-ct,y)=c-\sqrt{\lambda^*}+\gamma y,\quad v(x-ct,y)=0,\qquad -d\le y\le 0,$$

so that

$$c=\sqrt{\lambda^*}+\gamma d=\sqrt{\lambda^*-2\gamma p_0}$$

in view of (3.53). This formula is a good approximation of the speed of the waves of small amplitude.

We proved in the context of discussing the possibility $\mu(\lambda)=-1$ that for functions $\gamma\ge 0$ there always exists a solution $\lambda^*>2\Gamma_{max}=0$. Notice that a solution $\lambda^*>2\Gamma_{max}$ of $\mu(\lambda)=-1$, if it exists, is unique since $\lambda\mapsto\mu(\lambda)$ is strictly increasing wherever $\mu(\lambda)<0$. We also saw that $\gamma<0$ constant with $|\gamma|$ sufficiently large might prevent local bifurcation. It is of interest to find the necessary and sufficient condition for local bifurcation in the case of constant vorticity $\gamma<0$. We discuss the general case in which we do not fix the wavelength $L=2\pi$. Defining the wavenumber $\kappa=2\pi/L$, the necessary and sufficient condition for local bifurcation is the existence of a solution $\lambda>2\Gamma_{max}$ of (3.51) in which we replace g by $\kappa^{-1}g$, γ by $\kappa^{-1}\gamma$ and p_0 by κp_0: one can check that $\tilde{\psi}(\tilde{x},\tilde{y})=\kappa\,\psi(x,y)$, so that $\tilde{p}_0=\kappa p_0$. That is, we seek roots $\lambda>2\Gamma_{max}$ of

$$f(\lambda)=\tanh\left(\frac{2p_0\kappa}{\sqrt{\lambda}+\sqrt{\lambda-2\gamma p_0}}\right)-\frac{\lambda\kappa}{\gamma\sqrt{\lambda}-g}.\tag{3.56}$$

(i) For $\gamma=0$ we have $\Gamma\equiv 0$. The smooth function $f:(0,\infty)\to\mathbb{R}$ is such that $\lim_{\lambda\to\infty}f(\lambda)=\infty$ while $\lim_{\lambda\downarrow 0}f(\lambda)=-1$ so that there exists a root $\lambda^*>0$.

(ii) For constant $\gamma>0$ we have $\Gamma(p)=\gamma p$ so that $\Gamma_{max}=0$. The smooth function $f:(0,g^2/\gamma^2)\to\mathbb{R}$ has a root since $\lim_{\lambda\uparrow g^2/\gamma^2}f(\lambda)=\infty$ while $\lim_{\lambda\downarrow 0}f(\lambda)=-\tanh\left(\kappa\sqrt{\frac{2|p_0|}{\gamma}}\right)<0$.

(iii) For constant $\gamma<0$ we have $\Gamma(p)=\gamma p$ so that $\Gamma_{max}=\gamma p_0$ and local bifurcation occurs if and only if there is a solution $\lambda>2\gamma p_0$ of (3.56). The function $f:(2\gamma p_0,\infty)\to\mathbb{R}$ is smooth and strictly increasing, as the sum of two strictly increasing functions. Clearly $\lim_{\lambda\to\infty}f(\lambda)=\infty$ so that **for waves with wavelength L in a flow of constant vorticity**

$\gamma < 0$, the necessary and sufficient condition for local bifurcation is

$$\tanh\left(\kappa\sqrt{\frac{2p_0}{\gamma}}\right) > \frac{2\gamma p_0 \kappa}{g - \gamma\sqrt{2\gamma p_0}}, \tag{3.57}$$

expressing $\lim_{\lambda \downarrow 2\gamma p_0} f(\lambda) < 0$. Passing to the limit $\kappa \to 0$ and $\kappa \to \infty$, we see that the wavelength L must be sufficiently large for local bifurcation to occur (for very short wavelengths there are no genuine waves).

Flows with nonconstant vorticity

The only quantitative result for general vorticities is the **Burns condition**, obtained in the shallow-water limit (cf. the discussion in [40, 162]): the intrinsic wave speed

$$\sqrt{\lambda^*} = (c-u)\Big|_{\text{at the flat surface}}$$

is required to solve

$$\int_{-d}^{0} \frac{dy}{[U(y) - \sqrt{\lambda^*}]^2} = \frac{1}{g}, \tag{3.58}$$

where $(U(y), 0)$ is the unique current with the given vorticity and zero speed at the flat free surface $y = 0$. With λ^* determined from (3.58), the laminar flow that approximates these shallow water waves of small amplitude is given in the moving frame by the current

$$\left(U(y) - \sqrt{\lambda^*}, 0\right), \qquad -d \le y \le 0.$$

The absence of stagnation points, condition (3.1), requires that $\sqrt{\lambda^*} > U(y)$ throughout $[-d,0]$, and this means that the integral in (3.58) is not a singular integral.

It is instructive to compare the outcome of the Burns condition with the exact results presented before for constant vorticity. If γ is constant, then $U(y) = \gamma y$ on $[-d,0]$ and (3.58) takes the form

$$\frac{1}{\gamma}\left\{\frac{1}{\sqrt{\lambda^*}} - \frac{1}{\sqrt{\lambda^*} + \gamma d}\right\} = \frac{1}{g},$$

from which we infer that

$$\sqrt{\lambda^*} = \frac{-\gamma d \pm \sqrt{\gamma^2 d^2 + 4gd}}{2}.$$

Since the absence of stagnation points requires $\sqrt{\lambda^*} > -\gamma d$, we obtain the unique solution

$$\sqrt{\lambda^*} = \frac{-\gamma d + \sqrt{\gamma^2 d^2 + 4gd}}{2}.$$

This is precisely the shallow-water limit of the exact dispersion relation (3.54). Indeed, since the shallowness parameter $\delta = \frac{\kappa d}{2\pi}$, we can write (3.54) in the form

$$(c-u^*) = -\frac{\gamma d}{2}\frac{\tanh(2\pi\delta)}{2\pi\delta} + \frac{1}{2}\sqrt{\gamma^2 d^2\left(\frac{\tanh(2\pi\delta)}{2\pi\delta}\right)^2 + 4gd\,\frac{\tanh(2\pi\delta)}{2\pi\delta}}$$

and $\lim_{s\to 0}\frac{\tanh(s)}{s} = 1$.

3.2 Rotational steady waves of large amplitude

While λ was the appropriate parameter for the local bifurcation analysis, to extend the obtained curve of solutions to a global continuum of solutions (a maximal connected set in the function space X) it is more convenient to use the parameter Q. Checking the hypotheses of the Crandall–Rabinowitz local bifurcation theorem in terms of the parameter Q is very intricate: even the dependence of the trivial (laminar flow) solutions $H(p,\lambda)$ on Q is implicit. However, recall that for waves of small amplitude we reformulated the governing equations as the nonlinear boundary value problem (3.4) for h, and we subsequently set $h(q,p) = H(p) + w(q,p)$ to obtain the formulation (3.27)–(3.28), in which

$$Q = \lambda + 2g\int_{p_0}^{0}\frac{ds}{\sqrt{\lambda - 2\Gamma(s)}}. \tag{3.59}$$

Throughout the local bifurcation analysis it was understood that Q is given by (3.59) in terms of λ. As pointed out in the beginning of Section 3.1, near the bifurcation point $\lambda^* < \lambda_0$, the map $\lambda \mapsto Q(\lambda)$ given by (3.59) is a bijection. Therefore, restricting if necessary the local bifurcation curve, we can regard (3.27)–(3.28) as a functional equation of the type $F(Q,w) = 0$ near $Q^* = Q(\lambda^*) > Q_0$. The results of Section 3.1 ensure that near Q^* the solution set consists of the trivial (laminar flow) solution curve and a local bifurcation curve $\mathcal{C}_{loc} = \{(Q(s), w(s)) : |s| < \varepsilon_0\}$. Notice that (3.59) ensures that $Q'(\tau) = 0$ if and only if $\lambda'(\tau) = 0$, in view of (3.30), the strict convexity of $\lambda \mapsto Q(\lambda)$, and the defining property of λ_0.

3.2.1 The global continuum

Define $\mathcal{G} = (\mathcal{G}_1, \mathcal{G}_2) : \mathbb{R}\times X \to Y = Y_1 \times Y_2$ by

$$\mathcal{G}_1(Q,h) = (1+h_q^2)h_{pp} - 2h_q h_p h_{qp} + h_p^2 h_{qq} - \gamma(-p)h_p^3, \tag{3.60}$$

$$\mathcal{G}_2(Q,h) = [1 + h_q^2 + (2gh - Q)h_p^2]\Big|_{p=0}, \tag{3.61}$$

where

$$X = \{h \in C^{3,\alpha}_{per}(\overline{R}) : h = 0 \text{ on } B\}, \quad Y_1 = C^{1,\alpha}_{per}(\overline{R}), \quad Y_2 = C^{2,\alpha}_{per}(T),$$

with the subscript "*per*" meaning periodicity and evenness in the q-variable.[20] For $\delta > 0$, we seek solutions in the open set $\mathcal{O}_\delta \subset \mathbb{R} \times X$ given by

$$\mathcal{O}_\delta = \left\{ (Q,h) \in \mathbb{R} \times X : \; h_p > \delta \text{ in } \overline{R}, \; h < \frac{Q-\delta}{2g} \text{ on } T \right\}.$$

The definition of $\mathcal{O}_\delta$ ensures that the nonlinear operator $h \mapsto \mathcal{G}_1(Q,h)$ is uniformly elliptic and that the boundary condition $\mathcal{G}_2(Q,h)$ is uniformly oblique, in the sense that it is bounded away from being tangential on T. Notice that the Crandall–Rabinowitz local bifurcation theorem ensures that for some $\varepsilon_0 > 0$ we have a local bifurcation curve $\mathcal{C}_{loc} = \{(Q(s), h^s) \in \mathbb{R} \times X : \; |s| < \varepsilon_0\}$, with $(Q(0), h^0)$ being the local bifurcation solution (Q^*, H^*). Define[21]

$$\begin{aligned} \mathcal{C}_{loc}^+ &= \{(Q(s), h^s) \in \mathbb{R} \times X : \; 0 \le s < \varepsilon_0\}, \\ \mathcal{C}_{loc}^- &= \{(Q(s), h^s) \in \mathbb{R} \times X : \; -\varepsilon_0 < s \le 0\}. \end{aligned} \tag{3.62}$$

Let $\mathcal{S}_\delta^0$ be the connected component of $\overline{\{(Q,h) \in \mathcal{O}_\delta : \; \mathcal{G}(Q,h) = 0, \, h_q \not\equiv 0\}} \subset \mathbb{R} \times X$ that contains (Q^*, H^*). Thus, for $\delta > 0$ sufficiently small, $\mathcal{S}_\delta^0$ contains the local bifurcation curve $\mathcal{C}_{loc}$. The structure of the set $\mathcal{S}_\delta^0$ is as follows.

Theorem 3.5 (global bifurcation). *Let $\delta > 0$. Then one of the following alternatives holds.*

(i) *$\mathcal{S}_\delta^0$ is unbounded in $\mathbb{R} \times X$.*

(ii) *$\mathcal{S}_\delta^0$ contains another trivial point $(Q(\lambda), H(\lambda))$ with $\lambda \neq \lambda^*$.*

(iii) *$\mathcal{S}_\delta^0$ contains a point $(Q,h) \in \partial\mathcal{O}_\delta$.*

Moreover, there exists a continuous curve $\mathcal{K}_\delta$ in $\mathbb{R} \times X$ for which $\mathcal{C}_{loc}^+ \subset \mathcal{K}_\delta \subset \mathcal{S}_\delta^0$ and at each point $\mathcal{K}_\delta$ has a locally analytic reparametrization. Either $\mathcal{K}_\delta$ is unbounded in $\mathbb{R} \times X$ or it contains a point $(Q,h) \in \partial\mathcal{O}_\delta$, or else $\mathcal{K}_\delta$ is a closed loop.

The third alternative means that $\mathcal{S}_\delta^0$ contains a solution h for which either $h_p = \delta$ at some point in $\overline{R}$ or $2gh = Q - \delta$ at some point on T. As for $\mathcal{K}_\delta = \{(Q(s), h(s))\}_{s \ge 0}$ with $s \mapsto (Q(s), h(s))$ continuous from $[0,\infty)$ to $\mathbb{R} \times X$, for every $s_0 > 0$ we can find some $\varepsilon > 0$ and a real-analytic bijection $\rho : (-1,1) \to (s_0 - \varepsilon, s_0 + \varepsilon)$ such that $t \mapsto (Q(\rho(t)), h(\rho(t)))$ is real-analytic for $t \in (-1,1)$, while for $s \in [0, \varepsilon_0)$ we have $\mathcal{C}_{loc}^+$. The possible loop means that $\mathcal{K}_\delta$ has a continuous parametrization $\{(Q(s), h(s))\}_{s \ge 0}$ with $(Q(0), h(0)) = (Q^*, H^*)$, and such that for some $s_0 > 0$ we have $(Q(s_0), h(s_0)) = (Q^*, H^*)$ with $(Q(s+s_0), h(s+s_0)) = (Q(s), h(s))$ for all $s \ge 0$. We remark that the analyticity is not in the independent variables (q,p) but is in function space. Consequently, no additional regularity is required for the solutions; in particular, the vorticity function γ does not have to be real-analytic.

---

[20] For simplicity, we do not change the notation when referring to h_q with $h \in X$, since it will be clear from the context if we require evenness or oddness in the q-variable.

[21] These two curves correspond to first setting the origin in the moving frame such that for the solutions with a nonflat surface the wave profile is strictly monotone for $x \in (0,\pi)$, and subsequently choosing between the location of the wave crest or trough at $x = 0$; see Lemma 3.9 for a more precise statement.

Theorem 3.5 follows by combining the global bifurcation theory of Rabinowitz with real-analytic global bifurcation theory (see the Appendix to Chapter 3). To see this we have to verify certain structural properties.

Lemma 3.6 (proper map). *If $K \subset Y$ is compact and D is a closed, bounded set in $\overline{\mathcal{O}_\delta}$, then $\mathcal{G}^{-1}(K) \cap D$ is compact in $\mathbb{R} \times X$.*

Proof. Let $\{(f_j, g_j)\}$ be a convergent sequence in $Y = Y_1 \times Y_2$. If $(Q_j, h_j) \in \overline{\mathcal{O}_\delta}$ are such that $\mathcal{G}(Q_j, h_j) = (f_j, g_j)$ and the sequence $\{h_j\}$ is bounded in X, while $\{Q_j\}$ is bounded in $\mathbb{R}$, we wish to prove that there exists a subsequence of $\{(Q_j, h_j)\}$ that converges in $\mathbb{R} \times X$.

The function $\theta_j = \partial_q f_j$ satisfies

$$\begin{aligned} &[1+(\partial_q h_j)^2]\,\partial_p^2\theta_j - 2(\partial_q h_j)(\partial_p h_j)\,\partial_{qp}^2\theta_j + (\partial_p h_j)^2\,\partial_q^2\theta_j \\ &\quad = F(\partial_q h_j, \partial_p h_j, \partial_q^2 h_j, \partial_{qp}^2 h_j, \partial_p^2 h_j) + \partial_q f_j \quad \text{in} \quad R \end{aligned} \tag{3.63}$$

and

$$(\partial_q h_j)\partial_q\theta_j + (2gh_j - Q)(\partial_p h_j)\partial_p\theta_j = G(\partial_q h_j, \partial_p h_j) + \frac{1}{2}\partial_q g_j \text{ on } p = 0, \tag{3.64}$$

$$\theta_j = 0 \quad \text{on} \quad p = p_0, \tag{3.65}$$

where F and G are cubic polynomial expressions. As the index j runs over the positive integers, the right side of (3.63) is the sum of a term that is uniformly bounded in $C^{1,\alpha}_{per}(\overline{R})$ and a term belonging to a convergent sequence in $C^{0,\alpha}_{per}(\overline{R})$. Hence it is compact in $C^{0,\alpha}_{per}(\overline{R})$. Similarly, the right side of (3.64) is compact in $C^{1,\alpha}_{per}(T)$. By passing to subsequences and relabelling, we may assume that these sequences are convergent in $C^{0,\alpha}_{per}(\overline{R})$ and $C^{1,\alpha}_{per}(T)$, respectively. Taking differences, we obtain from (3.63)–(3.65) the equations

$$\begin{aligned} &[1+(\partial_q h_j)^2]\,\partial_p^2(\theta_j - \theta_k) - 2(\partial_q h_j)(\partial_p h_j)\,\partial_{qp}^2(\theta_j-\theta_k) + (\partial_p h_j)^2\,\partial_q^2(\theta_j-\theta_k) \\ &\qquad\qquad = F_{jk} \quad \text{in} \quad R, \\ &(\partial_q h_j)\partial_q(\theta_j-\theta_k) + (2gh_j - Q)(\partial_p h_j)\partial_p(\theta_j-\theta_k) = G_{jk} \text{ on } p = 0, \\ &\theta_j - \theta_k = 0 \quad \text{on} \quad p = p_0, \end{aligned}$$

where $\|F_{jk}\|_{C^{0,\alpha}_{per}(\overline{R})} \to 0$ and $\|G_{jk}\|_{C^{1,\alpha}_{per}(T)} \to 0$ as $j,k \to \infty$. Applying now the Schauder estimates[22] to $(\theta_j - \theta_k)$, we deduce that

$$\|\theta_j - \theta_k\|_{C^{2,\alpha}_{per}(\overline{R})} \le C\left(\|\theta_j - \theta_k\|_{C^{0}_{per}(\overline{R})} + \|F_{jk}\|_{C^{0,\alpha}_{per}(\overline{R})} + \|G_{jk}\|_{C^{1,\alpha}_{per}(T)}\right) \tag{3.66}$$

with $C > 0$ independent of j, k. We claim that (3.66) yields

$$\|\theta_j - \theta_k\|_{C^{2,\alpha}_{per}(\overline{R})} \to 0 \quad \text{as} \quad j,k \to \infty. \tag{3.67}$$

[22] The periodicity in the q-variable allows us to ignore the apparent nonsmoothness of the boundary of R. Since the lateral boundaries of R are irrelevant by periodicity, and on the top boundary we have an oblique boundary condition, while on the bottom boundary we have a homogeneous Dirichlet boundary condition, we combine the estimates in [129, Theorem 6.30], with those in [129, Theorem 6.6]. This is allowed as both estimates are local and the different types of boundary conditions occur on separate parts of the boundary.

Indeed, otherwise we can find a constant $\varepsilon > 0$ and a sequence $k_n \to \infty$ such that $\|\theta_{k_n} - \theta_{k_m}\|_{C^{2,\alpha}_{per}(\overline{R})} \geq \varepsilon$. The normalized functions $w_{nm} = \frac{\theta_{k_n} - \theta_{k_m}}{\|\theta_{k_n} - \theta_{k_m}\|_{C^0_{per}(\overline{R})}} \in C^{2,\alpha}_{per}(\overline{R})$ would be such that $\sup_{(q,p)\in\overline{R}}\{|w_{nm}(q,p)|\} = 1$ and

$$\|w_{nm}\|_{C^{2,\alpha}_{per}(\overline{R})} \leq C\Big(1 + \frac{1}{\varepsilon}\|F_{k_n k_m}\|_{C^{0,\alpha}_{per}(\overline{R})} + \frac{1}{\varepsilon}\|G_{k_n k_m}\|_{C^{1,\alpha}_{per}(T)}\Big)$$

in view of (3.66). Due to the compact embedding $C^{2,\alpha}_{per}(\overline{R}) \subset C^2_{per}(\overline{R})$, we find a subsequence of $\{w_{nm}\}$ that converges in $C^2_{per}(\overline{R})$ to some function w satisfying $\sup_{(q,p)\in\overline{R}}\{|w(q,p)|\} = 1$. Passing if necessary to subsequences, we can find corresponding to this subsequence of $\{w_{nm}\}$ a subsequence of the bounded sequence $\{(Q_j, h_j\} \in \mathcal{O}_\delta$ that converges in $\mathbb{R} \times C^3_{per}(\overline{R})$ to some limit (Q,h). Thus

$$\begin{cases} (1+h_q^2)w_{pp} - 2h_q h_p w_{qp} + h_p^2 w_{qq} = 0 & \text{in} \quad R, \\ h_q w_q + (2gh - Q)h_p w_p = 0 & \text{on} \quad p = 0, \\ w = 0 \quad \text{on} \quad p = p_0. \end{cases}$$

By the weak maximum principle (see Section 3.5.5), the maximum and the minimum of w in $\overline{R}$ are attained on $p = 0$ or on $p = p_0$. Since $w \not\equiv 0$, one of these values, say, the maximum, is nonzero and therefore attained at some point $(q_0, 0) \in T$. Clearly $w_q(q_0, 0) = 0$, even if $q_0 = \pm\pi$, due to periodicity, while Hopf's maximum principle (see Section 3.5.5) yields $w_p(q_0, 0) > 0$. We obtain a contradiction from the boundary condition on $p = 0$, since $h_p(q_0, 0) \geq \delta > 0$ and $Q - 2gh(q_0, 0) \geq \delta > 0$ as $(Q,h) \in \overline{\mathcal{O}_\delta}$. Thus (3.67) must hold true.

By (3.67) we know that all third-order derivatives of h_j, except for $\partial^3_{ppp}h_j$, are Cauchy in $C^\alpha_{per}(\overline{R})$. Using the partial differential equation $\mathcal{G}_1(Q_j, h_j) = f_j$ to express $\partial^2_{pp}h_j$ in terms of other partial derivatives of h_j of order at most 2, we infer that $\{\partial^3_{ppp}h_j\}$ is also a Cauchy sequence in $C^\alpha_{per}(\overline{R})$. Therefore, the original sequence $\{h_j\}$ has a subsequence that converges in X. The corresponding subsequence of $\{Q_j\}$ must converge in $\mathbb{R}$ since $\mathcal{G}_2(Q_j, h_j) = g_j$. □

Lemma 3.7 (Fredholm map). *For each $(Q,h) \in \mathcal{O}_\delta$, the linearized operator $\mathcal{G}_h(Q,h)$ is a Fredholm operator of index 0 from X to Y.*

Proof. Differentiating (3.60) and (3.61), we obtain

$$\begin{aligned} \mathcal{G}_{1h}(Q,h) = (1+h_q^2)\partial_p^2 - 2h_p h_q\, \partial^2_{qp} + h_p^2 \partial_q^2 + 2h_{pp}h_q\,\partial_q \\ -2h_{qp}h_p\,\partial_q - 2h_{qp}h_q\,\partial_p + 2h_{qq}h_p\,\partial_p - 3\gamma(-p)h_p^2\,\partial_p, \end{aligned} \tag{3.68}$$

$$\mathcal{G}_{2h}(Q,h) = 2h_q\,\partial_q + 2(2gh - Q)h_p\,\partial_p + 2gh_p^2. \tag{3.69}$$

The linear operators $\mathcal{G}_{1h}(Q,h)$ are uniformly elliptic, in $(q,p) \in \overline{R}$ as well as in $(Q,h) \in \mathcal{O}_\delta$, because $h_p^2 > \delta^2$. The linear operators $\mathcal{G}_{2h}(Q,h)$ are uniformly oblique because $(Q - 2h)h_p > \delta^2$.

Fix $(Q,h) \in \mathcal{O}_\delta$ and denote by $\mathcal{L}$ the operator $\mathcal{G}_h(Q,h) : X_0 \to Y_0$, where

$$X_0 = \{h \in C^{2,\alpha}_{per}(\overline{R}) :\ h = 0 \text{ on } B\}, \quad Y_0 = C^{0,\alpha}_{per}(\overline{R}) \times C^{1,\alpha}_{per}(T).$$

Furthermore, choose

$$\lambda > 2g \sup_{q\in[-\pi,\pi]} \{h_p^2(q,0)\}$$

and denote by $\mathcal{L}_\lambda : X_0 \to Y_0$ the operator $w \mapsto \mathcal{L}w - \lambda\big(w, w\big|_T\big)$. Then, for any $F = (f,g) \in Y_0$, the oblique derivative problem $\mathcal{L}_\lambda \theta = F$ has a unique solution $w_F \in X_0$ (cf. [129, Theorem 6.31]), and the Schauder estimate

$$\|\theta\|_{C^{2,\alpha}_{per}(\overline{R})} \le C_0\left(\|\theta\|_{C^0_{per}(\overline{R})} + \|\mathcal{G}_{1h}(Q,h)\theta\|_{C^{0,\alpha}_{per}(\overline{R})} + \|\mathcal{G}_{2h}(Q,h)\theta\|_{C^{1,\alpha}_{per}(T)}\right) \tag{3.70}$$

holds for all $\theta \in C^{2,\alpha}_{per}(\overline{R})$, for some constant $C_0 > 0$ that depends only on $\|h\|_{C^{2,\alpha}_{per}(\overline{R})}$ (cf. [129, Theorem 6.30]). Consequently, the inverse mapping $(\mathcal{L}_\lambda)^{-1} : F \mapsto w_F$ is bounded from Y_0 to X_0, and the compact embeddings $C^{2,\alpha}_{per}(\overline{R}) \subset C^{0,\alpha}_{per}(\overline{R})$, $C^{2,\alpha}_{per}(T) \subset C^{1,\alpha}_{per}(T)$, allow us to regard it as a compact map from Y_0 to Y_0. Notice that for $F \in Y_0$, $w \in X_0$ solves $\mathcal{L}w = F$ if and only if

$$\frac{1}{\lambda} w + (\mathcal{L}_\lambda)^{-1} w = \frac{1}{\lambda} (\mathcal{L}_\lambda)^{-1} F. \tag{3.71}$$

Moreover, the operator $w \mapsto \frac{1}{\lambda} w + (\mathcal{L}_\lambda)^{-1} w$ from Y_0 to Y_0 is Fredholm, as a compact perturbation of the Fredholm operator $w \mapsto \frac{1}{\lambda} w$ (cf. the discussion in Section 3.5.1). Consequently, $ker(\mathcal{L}) \subset X_0$ is finite-dimensional. Also, $range(\mathcal{L}) \subset Y_0$ is closed. Indeed, if $\mathcal{L}w_n = F_n \to F$, using the reformulation (3.71) in terms of a Fredholm operator (having a closed range), we infer the existence of $w \in Y_0$ satisfying $w + \lambda(\mathcal{L}_\lambda)^{-1} w = (\mathcal{L}_\lambda)^{-1} F$. Thus $w \in X_0$ and $\mathcal{L}w = F$. These considerations show that $\mathcal{G}_h(Q,h) : X_0 \to Y_0$ is a Fredholm operator, as the codimension of its closed range cannot be infinite-dimensional. Were it so, then we would be able to find a sequence $\{F_n\}_{n\ge 1}$ of linearly independent elements of Y_0 that are not in the range of $\mathcal{L}$. But then $\frac{1}{\lambda}(\mathcal{L}_\lambda)^{-1} F_n$ are linearly independent elements of Y_0 that do not belong to the range of the Fredholm operator $\frac{1}{\lambda}\mathcal{L}_\lambda^{-1}\mathcal{L}$ on the left side of (3.71), and this is impossible.

To show that $\mathcal{G}_h(Q,h) : X \to Y$ is a Fredholm operator, let $\theta \in X$. Taking the q-derivatives of $\mathcal{G}_{1h}(Q,h)\theta$ and $\mathcal{G}_{2h}(Q,h)\theta$, classical Schauder estimates (cf. [129, Theorem 6.30]) provide the existence of a constant $C_1 > 0$, independent of θ, such that

$$\|\theta_q\|_{C^{2,\alpha}_{per}(\overline{R})} \le C_1\left(\|\theta_q\|_{C^0_{per}(\overline{R})} + \|\partial_q \mathcal{G}_{1h}(Q,h)\theta\|_{C^{0,\alpha}_{per}(\overline{R})} + \|\partial_q \mathcal{G}_{2h}(Q,h)\theta\|_{C^{1,\alpha}_{per}(T)}\right).$$

Since

$$\theta_{pp} = \frac{\mathcal{G}_{1h}(Q,h)\theta + F(\theta_q, \theta_p, \theta_{qp}, \theta_{qq})}{1 + h_q^2}$$

for some cubic polynomial expression F, differentiating this identity with respect to p, and using the previous estimate, we get

$$\|\theta_{ppp}\|_{C^{0,\alpha}_{per}(\overline{R})} \le C_2\left(\|\theta\|_{C^1_{per}(\overline{R})} + \|\mathcal{G}_{1h}(Q,h)\theta\|_{C^{1,\alpha}_{per}(\overline{R})} + \|\mathcal{G}_{2h}(Q,h)\theta\|_{C^{2,\alpha}_{per}(T)}\right)$$

for some constant $C_2 > 0$, independent of θ. Hence

$$\|\theta\|_{C^{3,\alpha}_{per}(\overline{R})} \le C\left(\|\theta\|_{C^1_{per}(\overline{R})} + \|\mathcal{G}_{1h}(Q,h)\theta\|_{Y_1} + \|\mathcal{G}_{2h}(Q,h)\theta\|_{Y_2}\right) \tag{3.72}$$

for all $\theta \in X$, for a constant $C > 0$ that depends only on $\|h\|_X$. The fact that $\mathcal{G}_h(Q,h) : X \to Y$ is Fredholm is now plain. Indeed, its kernel is finite-dimensional since this was the case for $\mathcal{G}_h(Q,h) : X_0 \to Y_0$. The fact that its range is closed can be seen as follows. If $\theta_n \in X$ are such that $\mathcal{G}_h(Q,h)\theta_n = F_n \to F$ in Y, then (3.70) ensures that $\{\theta_n\}$ is a Cauchy sequence in X_0, and subsequently (3.72) ensures that $\{\theta_n\}$ is a Cauchy sequence in X. Its limit $\theta \in X$ clearly satisfies $\mathcal{G}_h(Q,h)\theta = F$. As for the fact that the codimension of the range is finite-dimensional, if this were not so, then we would be able to find infinitely many linearly independent elements in Y that do not belong to the range. As $Y \subset Y_0$, this contradicts the fact that $\mathcal{G}_h(Q,h) : X_0 \to Y_0$ is Fredholm.

Finally, notice that at the local bifurcation point, $\mathcal{G}_h(Q^*, H^*)$ has index 0 since its null space has dimension 1 and its range codimension 1. The index being a continuous function of (Q,h) in the open and connected set $\mathcal{O}_\delta$, we deduce that the index of $\mathcal{G}_h(Q,h)$ vanishes for all $(Q,h) \in \mathcal{O}_\delta$. □

Lemma 3.8 (spectral properties).

(i) *For every $M > 0$ there exist constants $c_1, c_2 > 0$ such that for all $(Q,h) \in \mathcal{O}_\delta$ with $|Q| + \|h\|_X \leq M$, we have*

$$c_1 \|\psi\|_X \leq \mu^{\frac{\alpha}{2}} \|(A-\mu)\psi\|_{Y_1} + \mu^{\frac{1+\alpha}{2}} \|B\psi\|_{Y_2}$$

for all $\psi \in X$ and for all real $\mu \geq c_2$, where $A = A(Q,h) = \mathcal{G}_{1h}(Q,h)$ and $B = B(Q,h) = \mathcal{G}_{2h}(Q,h)$.

(ii) *Define the spectrum*

$$\Sigma(Q,h) = \{\mu \in \mathbb{C} : \ A-\mu \text{ not isomorphic from } X_0 = \{\psi \in X : \ B\psi = 0\}, \\ \text{endowed with the norm } \|\psi\|_{Y_1} + \|A\psi\|_{Y_1}, \text{ onto } Y_1\}.$$

Then $\Sigma(Q,h)$ consists entirely of eigenvalues of finite multiplicity with no finite accumulation points. Furthermore, there is a neighborhood $\mathcal{N}$ of $[0,\infty)$ in the complex plane such that $\Sigma(Q,h) \cap \mathcal{N}$ is a finite set.

(iii) *For all $(Q,h) \in \mathcal{O}_\delta$, the boundary operator $\mathcal{G}_{2h}(Q,h) : X \to Y_2$ is onto.*

Proof. For (i), we follow the standard argument of Agmon [2] (see also [140]) by introducing an additional independent variable t, thus augmenting $(q,p) \in \mathbb{R}^2$ to $(q,p,t) \in \mathbb{R}^3$, and considering the elliptic operator $A + \partial_t^2$ over the cylindrical domain $\Omega = R \times (-2,2) \subset \mathbb{R}^3$, with a periodicity condition in the q-variable, and with a homogeneous Dirichlet boundary condition on the boundary of Ω corresponding to $t = \pm 2$. The new operator is clearly uniformly elliptic and satisfies the complementing condition on the entire boundary of Ω: the two faces of $\partial\Omega$ where q is fixed do not matter by periodicity, on the three faces of the rectangular prism $\partial\Omega$ where $p = p_0$ and $t = \pm 2$, we have homogeneous Dirichlet boundary conditions, and on the remaining face, $p = 0$, we have an oblique boundary condition.[23] This ensures (cf. [3, 4]) the validity of Schauder estimates of type (3.70). We now apply

[23]The theory presented in Section 3.5.4 on $\mathbb{R}^2$ extends to $\mathbb{R}^3$: in the principal part of the boundary operator we replace formally the tangential derivatives by multiplications with a symbol ξ_j, and the normal derivative by $(-i\partial_y)$, with analogous operations performed in the principal part of the differential operator. By direct computation one can check that Dirichlet and oblique boundary conditions are of complementing type (as was the case in $\mathbb{R}^2$), whereas tangential boundary conditions are not (in contrast to the case in $\mathbb{R}^2$).

these estimates to the function $w(q,p,t) = e(t)\varphi(t)\psi(q,p)$, where $\psi \in X$, $\varphi : \mathbb{R} \to \mathbb{R}$ is a C^∞-smooth function with support in $(-2,2)$ and such that $\varphi(t) = 1$ for $t \in [-1,1]$, while $e(t) = e^{i\sqrt{\mu}t}$ with $\mu > 1$: for some constant $C > 0$, independent of μ and ψ,

$$\|w\|_{C^{2,\alpha}_{per}(\overline{\Omega})} \le C\Big(\|w\|_{C^{0,\alpha}_{per}(\overline{\Omega})} + \|(A+\partial_t^2)w\|_{C^{0,\alpha}_{per}(\overline{\Omega})} + \|Bw\|_{C^{1,\alpha}_{per}(T\times[-2,2])}\Big), \tag{3.73}$$

with the index "*per*" indicating periodicity in the q-variable. Notice that

$$(A+\partial_t^2)w = e(t)\varphi(t)(A-\mu)\psi + [2i\mu^{1/2}e(t)\varphi'(t) + e(t)\varphi''(t)]\,\psi.$$

Straightforward calculations show the existence of constants $C_j > 0$, $j = 0,\dots,4$, that are independent of $\mu > 1$ and of $\psi \in X$, such that

$$\|e(t)\varphi(t)\psi\|_{C^{2,\alpha}_{per}(\overline{\Omega})} \ge C_0\Big(\|\psi\|_{C^{2,\alpha}_{per}(\overline{R})} + \mu\|\psi\|_{C^{0,\alpha}_{per}(\overline{R})}\Big),$$

$$\|e(t)\varphi(t)\psi\|_{C^{0,\alpha}_{per}(\overline{\Omega})} \le C_1\,\mu^{\alpha/2}\|\psi\|_{C^{0,\alpha}_{per}(\overline{\Omega})},$$

$$\|[2i\mu^{1/2}e(t)\varphi'(t) + e(t)\varphi''(t)]\,\psi\|_{C^{0,\alpha}_{per}(\overline{R})} \le C_2\,\mu^{(\alpha+1)/2}\|\psi\|_{C^{0,\alpha}_{per}(\overline{\Omega})},$$

$$\|(A-\mu)e(t)\varphi(t)\psi\|_{C^{0,\alpha}_{per}(\overline{R})} \le C_3\,\mu^{\alpha/2}\,\|(A-\mu)\psi\|_{C^{0,\alpha}_{per}(\overline{R})},$$

$$\|B\,e(t)\varphi(t)\psi\|_{C^{1,\alpha}_{per}(T\times[-2,2])} \le C_4\,\mu^{(\alpha+1)/2}\|B\psi\|_{C^{1,\alpha}_{per}(T)}.$$

These inequalities in combination with (3.73) yield

$$\frac{C_0}{C(C_3+C_4)}\|\psi\|_{C^{2,\alpha}_{per}(\overline{R})} \le \mu^{\alpha/2}\|(A-\mu)\psi\|_{C^{0,\alpha}_{per}(\overline{R})} + \mu^{(1+\alpha)/2}\|B\psi\|_{C^{1,\alpha}_{per}(T)}, \quad \psi \in X,$$

provided $C_0\,\mu^{(1-\alpha)/2} \ge C\,(C_1+C_2)$. A similar type of inequality can be derived for $\partial_q\psi$. Subsequently, expressing $\partial_p^2\psi$ in terms of $(A+\partial_t^2)\psi$ and the other derivatives of ψ of order less than 2, the assertion in (i) follows for $\mu > 1$ sufficiently large.

As for (ii), it follows as in the proof of Lemma 3.7 that the operator $(A-\mu, B)$ is Fredholm of index 0 from X to Y. By (i) it has a trivial null space, and hence it is one-to-one and onto if μ is sufficiently large. Now (i) ensures that X_0 is a Banach space and $A : X_0 \to Y_1$ is a bounded linear operator. Moreover, if $\mu \ge c_2$, then $(A-\mu) : X_0 \to Y_1$ is bijective, with a compact inverse due to the compact embedding $X \subset Y_1$. This implies (ii) by the Riesz–Schauder theory; cf. [171]. To prove (iii), notice that the existence of a value of μ such that $(A-\mu, B)$ maps X onto Y ensures that B maps X onto Y_2. □

The statement in Theorem 3.5 concerning the set $\mathcal{S}_\delta^0$ follows now by applying the global bifurcation method of Rabinowitz using the generalized Leray–Schauder degree, while the statement concerning the curve $\mathcal{K}_\delta$ is a consequence of the global analytic bifurcation approach[24] (see the discussion in the Appendix, Section 3.5.3).

[24] In both cases we do not actually use the Crandall–Rabinowitz local bifurcation theorem per se since we did not show its applicability in terms of (Q,h); cf. the discussion at the beginning of this subsection. However, all relevant properties of the local bifurcation curve can be expressed in terms of (Q,h), as pointed out before. In both global bifurcation approaches, it is only the conclusion of the Crandall–Rabinowitz local bifurcation theorem that is needed, as one can see by going through the steps presented in Section 3.5.3.

3.2.2 Nodal pattern and bounds

In this subsection we prove that along the global continuum $\mathcal{S}_\delta^0$ and along the curve $\mathcal{K}_\delta$ constructed in the previous subsection, the nodal configuration inherited from the eigenfunction of the linearized problem at the bifurcation point (Q^*, H^*), is preserved. This will allow us to eliminate the alternative (ii) in Theorem 3.5 and to rule out the possibility that $\mathcal{K}_\delta$ is a closed loop. We also simplify the alternative concerning the unboundedness, reducing it, by means of bounds along the continuum, to the statement that the solutions approach a flow with a weak stagnation point (that is, a point where $u = c$).

Nodal pattern

Let Ω be the open rectangle $(0,\pi) \times (p_0, 0)$, which has half the size of R, and denote its sides (excluding the corners) by

$$\partial\Omega_t = \{(q,0) : \; q \in (0,\pi)\}, \qquad \partial\Omega_b = \{(q, p_0) : \; q \in (0,\pi)\},$$
$$\partial\Omega_l = \{(0,p) : \; p \in (p_0,0)\}, \qquad \partial\Omega_r = \{(p,\pi) : \; p \in (p_0,0)\}.$$

We will show that all $(Q,h) \in \mathcal{S}_\delta^0$ with $h_q \not\equiv 0$ satisfy either

$$\begin{aligned} &h_q < 0 \text{ in } \Omega \cup \partial\Omega_t, \quad &&h_{qp} < 0 \text{ on } \partial\Omega_b, \quad &&h_{qq} < 0 \text{ on } \partial\Omega_l, \quad &&h_{qq} > 0 \text{ on } \partial\Omega_r, \\ &h_{qqp}(0,p_0) < 0, \quad &&h_{qqp}(\pi,p_0) > 0, \quad &&h_{qq}(0,0) < 0, \quad &&h_{qq}(\pi,0) > 0, \end{aligned} \tag{3.74}$$

or the exact opposite inequalities. Note that $h \in X$ ensures that $h = 0$ on $\partial\Omega_b$, with $h(q,p)$ even and 2π-periodic in q, so that $h_q = 0$ on $\partial\Omega_l \cup \partial\Omega_r$. Recall the notation (3.62) of the two components of the local bifurcation curve $\mathcal{C}_{loc}$.

Lemma 3.9. *The inequalities* (3.74) *hold in a small neighborhood of* (Q^*, H^*) *in* $\mathbb{R} \times X$, *along the curve* $\mathcal{C}_{loc}^+ \setminus \{(Q^*, H^*)\}$ *coming out of* (Q^*, H^*).

Proof. As pointed out in the beginning of the section, along $\mathcal{C}_{loc}$ it is easier to work with (λ, w) instead of (Q,h). Since $w = h - H(\lambda, p)$ with $H(\lambda, p)$ independent of q, it suffices to check the validity of (3.74) for w along $\mathcal{C}_{loc}$.

From the Crandall–Rabinowitz local bifurcation theorem we know that, for $\varepsilon > 0$ small enough, along $\mathcal{C}_{loc}$ we have

$$w^\varepsilon(q,p) = \varepsilon\, M(p) \cos q + o(\varepsilon) \quad \text{in} \quad C^{3,\alpha}(\overline{\Omega}), \tag{3.75}$$

where $M \in C^{3,\alpha}[p_0, 0]$ is the solution to

$$\begin{cases} (a^3 M_p)_p = aM & \text{on} \quad (p_0, 0), \\ a^3 M_p = gM & \text{at} \quad p = 0, \\ M = 0 & \text{at} \quad p = p_0, \end{cases} \tag{3.76}$$

with $a(p,\lambda) = \sqrt{\lambda - 2\Gamma(p)}$, normalized by $M(0) = 1$. Since M was a ground state, we expect that $M(p) \neq 0$ for $p \in (p_0, 0)$, as is the case for Sturm–Liouville problems. Indeed, the existence of some $p^* \in (p_0,0)$ with $M(p^*)=0$ would yield, after multiplying by M the differential equation for M, and integrating on $[p^*,0]$,

$$g\,M^2(0) - \int_{p^*}^0 a^3 M_p^2\,dp = \int_{p^*}^0 a M^2\,dp.$$

Therefore, the function $\varphi \in H^1(p_0,0)$, given by

$$\varphi(p) = \begin{cases} M(p), & p \in [p^*,0], \\ 0, & p \in [p_0,p^*], \end{cases}$$

would solve the minimization problem $\mathbb{F}(\varphi,\lambda^*) = -1$, formulated in Section 3.1.1. The considerations in Section 3.1.1 ensure the uniqueness of M. Indeed, the difference M of two presumed ground states would satisfy (3.76), with the additional condition $M(0)=0$, in which case $M \equiv 0$ is the solution of a linear second-order differential equation with initial data $M(0) = M_p(0) = 0$. This uniqueness property yields $M = \varphi$ on $[p_0,0]$. But then $M(p_0) = M'(p_0) = 0$ so that $M \equiv 0$ since M solves a linear second-order ordinary differential equation. The obtained contradiction proves that

$$M(p) > 0, \qquad p \in (p_0,0). \tag{3.77}$$

Clearly $M'(p_0) > 0$, as $M'(p_0) = 0$ would force $M \equiv 0$ on $[p_0,0]$. Notice that $M(0) = 1$, $M'(p_0) > 0$, and (3.77) ensure, by means of differentiating (3.75), the validity of (3.74) if $\varepsilon > 0$ is small enough. □

Simarly we can show that along the curve $\mathcal{C}_{loc}^- \setminus \{(Q^*,H^*)\}$ the precise opposite inequalities to (3.74) hold. Recall from Section 3.2.1 that along $\mathcal{C}_{loc}^+$ the wave crest is at $x=0$, while along $\mathcal{C}_{loc}^-$ its location is at $x = \pm\pi$.

Lemma 3.10. *For any $(Q,h) \in \mathcal{S}_\delta^0$ with $h_q \not\equiv 0$, either the inequalities* (3.74) *or their exact opposites hold.*

Proof. The set $\mathcal{S}_\delta^0$ is by construction connected in $\mathbb{R} \times X$. Furthermore, the set of all $(Q,h) \in \overline{\mathcal{O}_\delta}$ that satisfy the inequalities (3.74) is an open set in $\mathbb{R} \times X$, nonempty by the preceding lemma. Therefore, if the statement were false, there would exist $(Q,h) \in \mathcal{S}_\delta^0$ with $h_q \not\equiv 0$, for which one of the inequalities in (3.74) is an equality although it is a limit of elements for which all of the inequalities in (3.74) are strict, or it is a limit of elements for which all of the opposite inequalities to (3.74) are strict. We consider the case when along this sequence all of the inequalities in (3.74) are strict, the case for the exact opposite of (3.74) being entirely analogous. The following reasoning concerns this element (Q,h).

Taking the derivative with respect to q of both the partial differential equation (3.60) and the boundary condition (3.61), $\varphi = \partial_q h$ satisfies

$$\begin{cases} (1+h_q^2)\varphi_{pp} - 2h_q h_p \varphi_{qp} + h_p^2 \varphi_{qq} + 2(h_q h_{pp} - h_p h_{qp})\varphi_q \\ \qquad\qquad + [2h_p h_{qq} - 2h_q h_{qp} - 3\gamma(-p)h_p^2]\varphi_p = 0 \quad \text{in} \quad \Omega, \\ 2h_q \varphi_q + 2gh_p^2 \varphi + 2h_p(2gh - Q)\varphi_p = 0 \quad \text{on} \quad p = 0, \\ \varphi = 0 \quad \text{on} \quad p = p_0. \end{cases} \tag{3.78}$$

By the weak maximum principle, we have $0 \geq \varphi$ since h is the limit in $C^{3,\alpha}_{per}(\overline{\Omega})$ of functions h_n satisfying $\partial_q h_n < 0$ in Ω (in view of maximum principles due to the validity of (3.74) for each h_n). Since $h_q \not\equiv 0$, the strong maximum principle yields $0 > h_q = \varphi$ in Ω, and we have the usual Hopf inequalities on $\partial\Omega_b \cup \partial\Omega_l \cup \partial\Omega_r$, where $\varphi = 0$ attains its maximum. To check the validity of (3.74), it remains to prove that $\varphi < 0$ on $\partial\Omega_t$, $\varphi_q > 0$ at the top right corner, $\varphi_q < 0$ at the top left corner, $\varphi_{qp} > 0$ at the bottom right corner, and $\varphi_{qp} < 0$ at the bottom left corner.

If $\varphi = 0$ at some point $(q_0, 0)$ with $q_0 \in (0,\pi)$, then $\varphi_q(q_0,0) = 0$. By the boundary condition on $p = 0$ in (3.78) we have $h_p(2gh - Q)\varphi_p = 0$ at $(q_0, 0)$, whence $\varphi_p(q_0,0) = 0$ as $h \in \overline{\mathcal{O}_\delta}$ ensures $h_p(q_0,0) \geq \delta$ and $Q - 2g\,h(q_0,0) \geq \delta$. This contradicts the Hopf condition $\varphi_p(q_0,0) > 0$. Therefore, $\varphi < 0$ on $\partial\Omega_t$.

At the top right corner $(\pi,0)$ we will use Serrin's edge point lemma. By oddness we have $h_q = h_{qqq} = h_{qp} = h_{qpp} = 0$ at $(\pi,0)$. Since we proved above that $h_{qq}(\pi,p) = \varphi_q(\pi,p) > 0$ for $p \in (p_0,0)$, we have by continuity that $h_{qq}(\pi,0) = \varphi_q(\pi,0) \geq 0$. Let us suppose that $h_{qq}(\pi,0) = \varphi_q(\pi,0) = 0$ and seek a contradiction. Differentiating the boundary condition on $p = 0$ in (3.78) with respect to q, we get $2h_p(2gh - Q)\varphi_{qp} = 0$ at $(\pi,0)$, whence $\varphi_{qp}(\pi,0) = 0$. Hence all first and second derivatives of $\varphi = h_q$ vanish at $(\pi,0)$, which contradicts Serrin's edge point lemma. Therefore, $h_{qq}(\pi,0) > 0$. The top left corner $(0,0)$ is treated in the same way.

At the bottom right corner (π, p_0) we will again use Serrin's edge point lemma. By oddness we know that $h_q = h_{qqq} = h_{qp} = h_{qpp} = 0$ at (π, p_0). We already know that $h_{qp} < 0$ on $\partial\Omega_b$, and $h_{qp}(\pi, p_0) = 0$, so that $h_{qqp}(\pi, p_0) \geq 0$. Moreover, differentiating the boundary condition on $p = p_0$ in (3.78) with respect to q, we get $\varphi_q = h_{qq} = 0$ along $p = p_0$. In particular, $\varphi_q = h_{qq} = 0$ at (π, p_0). Now if $h_{qqp}(\pi, p_0) = 0$, then we would have all the first and second derivatives of $\varphi = h_q$ vanishing at (π, p_0). This contradicts Serrin's edge point lemma. Therefore, $h_{qqp}(\pi, p_0) > 0$. The bottom left corner $(0, p_0)$ is treated in the same way. This concludes the proof. ☐

At this stage, we have ensured that the nodal pattern (3.74) or its exact opposite holds along $\mathcal{S}^0_\delta$, unless there is a laminar flow solution $(Q(\lambda), H(\cdot,\lambda))$. We now show that the only laminar flow solution is the local bifurcation point (Q^*, H^*).

Lemma 3.11. *The only solution in $\mathcal{S}^0_\delta$ that is independent of q is the bifurcation point (Q^*, H^*).*

Proof. If, say, $(Q_n, h_n) \in \mathcal{S}^0_\delta$, with $\partial_q h_n \not\equiv 0$ satisfying (3.74), converge to $(Q(\lambda), H(\cdot,\lambda)) \in \mathcal{S}^0_\delta$ in $\mathbb{R} \times X$, define

$$\varphi_n = \frac{\partial_q h_n}{\|\partial_q h_n\|_{C^{2,\alpha}_{per}(\overline{R})}}$$

of unit norm in $C^{2,\alpha}_{per}(\overline{R})$. Take a subsequence $\{\varphi_{n_k}\}$ converging to some $\varphi \in C^2_{per}(\overline{R})$ and use the Schauder estimate of type (3.70) for $\partial_q h_{n_k}$, solving a uniformly oblique elliptic boundary value problem of type (3.78), to deduce that $\{\varphi_{n_k}\}$ converges in $C^{2,\alpha}_{per}(\overline{R})$. Since each φ_{n_k} has mean zero in the q-variable at each fixed $p \in [p_0, 0]$, the limit φ is of the form $\partial_q m$, with $m \in C^{2,\alpha}_{per}(\overline{R})$ such that $\partial_q m \in C^{2,\alpha}_{per}(\overline{R})$ and $\|\partial_q m\|_{C^{2,\alpha}_{per}(\overline{R})} = 1$. Passing to the limit $n_k \to \infty$ in the uniformly oblique elliptic boundary value problem of type (3.78) for

φ_{n_k}, we get

$$\begin{cases} \varphi_{pp} + H_p^2 \varphi_{qq} - 3\gamma(-p) H_p^2 \varphi_p = 0 \quad \text{in} \quad R, \\ 2gH_p^2 \varphi + 2H_p(2gH - Q)\varphi_p = 0 \quad \text{on} \quad p = 0, \\ \varphi = 0 \quad \text{on} \quad p = p_0. \end{cases} \tag{3.79}$$

Expand the function φ, which is odd in the q-variable for each fixed $p \in [p_0, 0]$, in a sine series for fixed $p \in [p_0, 0]$,

$$\varphi(q,p) = \sum_{k=1}^{\infty} \sin(kq) f_k(p),$$

where $f_k \in C^2[p_0, 0]$ is given by

$$f_k(p) = \frac{2}{\pi} \int_0^{\pi} \sin(kq)\varphi(q,p)\,dq, \qquad k \geq 1. \tag{3.80}$$

Notice that $H_p(0) = \frac{1}{\sqrt{\lambda}}$ and $\lambda = Q - 2gH(0)$, in accordance with the calculations in the beginning of Section 3.1. Therefore, multiplying (3.79) by $\sin(q)$ and integrating over $[-\pi, \pi]$, we infer that $f_1 \in C^2[p_0, 0]$ satisfies

$$\begin{cases} (a^3 \partial_p f_1)_p = a f_1 \quad \text{in} \quad (p_0, 0), \\ f_1'(0) = g\lambda^{-3/2} f_1(0), \\ f_1(p_0) = 0, \end{cases} \tag{3.81}$$

with a computed at the laminar flow $(Q(\lambda), H(\cdot,\lambda))$. Since $\partial_q h_{n_k}(q,p) < 0$ for $(q,p) \in \Omega$ and $\partial_q h_{n_k} = 0$ on $p = p_0$, we deduce that $\varphi \leq 0$ in Ω and $\varphi = 0$ on $p = p_0$. Since $\varphi \not\equiv 0$ throughout Ω, the strong maximum principle ensures $\varphi < 0$ in Ω. Consequently, we have

$$f_1(p) < 0, \qquad p \in (p_0, 0). \tag{3.82}$$

Now by (3.81) f_1 is an eigenfunction of (3.22) with eigenvalue $\mu = -1$. The nonvanishing of f_1 shows that it must be the ground state; that is, $\mu = -1$ is the lowest eigenvalue. Indeed, if there is an eigenvalue $\mu_0 < -1$, the considerations in Lemma 3.9 show that we can normalize the corresponding eigenfunction $M \in C^2[p_0, 0]$ by setting $M(0) = 1$, which ensures $M(p) > 0$ for $p \in (p_0, 0]$. Using integration by parts and the boundary value problems satisfied by f_1 and M, we would obtain the contradiction

$$\begin{aligned} 0 &= a^3 M_p f_1 \Big|_{p=0} - a^3 M \partial_p(f_1) \Big|_{p=0} \\ &= \int_{p_0}^{0} (a^3 M_p f_1)_p\,dp - \int_{p_0}^{0} (a^3 M\, \partial_p f_1)_p\,dp \\ &= \int_{p_0}^{0} (a^3 M_p)_p f_1\,dp - \int_{p_0}^{0} (a^3 \partial_p f_1)_p M\,dp = -(1+\mu_0) \int_{p_0}^{0} a M f_1\,dp. \end{aligned}$$

The contradiction is due to the fact that $a > 0$, $M > 0$, and $f_1 < 0$ on $(p_0, 0)$. Therefore, $\mu = -1$ is the lowest eigenvalue and the results in Proposition 3.3 on the location of the unique root of $\mu(\lambda) = -1$ yield $\lambda = \lambda^*$. □

We now summarize the outcome of the considerations made in this section. The previous considerations eliminate the alternative (ii) in the global bifurcation (Theorem 3.5), and also the possibility that $\mathcal{K}_\delta$ is a closed loop.

Proposition 3.12. *The connected set* $\mathcal{S}_\delta^0$ *either is unbounded in* $\mathbb{R} \times X$ *or contains a point* $(Q,h) \in \partial\mathcal{O}_\delta$. *The same is true for the curve* $\mathcal{K}_\delta$. *Moreover, any solution different from the local bifurcation solution* (Q^*, H^*), *belonging to one of these sets, satisfies the properties specified in* (3.74).

Bounds along the continuum

We now simplify the first alternative of the global bifurcation theorem.

Lemma 3.13. *If* $\sup_{h\in\mathcal{S}_\delta^0}\{\|h\|_X\} = \infty$, *then* $\sup_{h\in\mathcal{S}_\delta^0}\{Q + \|h_p\|_{L^\infty(R)}\} = \infty$. *Similarly, if* $\sup_{h\in\mathcal{K}_\delta}\{\|h\|_X\} = \infty$, *then* $\sup_{h\in\mathcal{K}_\delta}\{Q + \|h_p\|_{L^\infty(R)}\} = \infty$.

Proof. The proofs being similar, we show only that if $\sup_{h\in\mathcal{K}_\delta}\{Q + \|h_p\|_{L^\infty(R)}\}$ is bounded, then $\sup_{h\in\mathcal{K}_\delta}\{\|h\|_X\} < \infty$. We have

$$0 \le h(q,p) = y + d \le \eta(0) + d \le \frac{|p_0|}{\inf_{\overline{D_\eta}}\{c-u\}} = |p_0| \sup_{\overline{R}}\{h_p\}$$

by the definition of the relative mass flux p_0. Thus h is bounded along $\mathcal{K}_\delta$. Consider now $\varphi = h_q$, which satisfies the elliptic boundary value problem (3.78) and is, for fixed $p \in [p_0, 0]$, odd in the q-variable. By the strong maximum principle there is no interior maximum unless $\varphi \equiv 0$. But $\varphi = 0$ on $p = p_0$, while on $p = 0$ we have in view of (3.4) that

$$h_q^2 = h_p^2(Q - 2gh) - 1 \le Q h_p^2 - 1 \quad \text{on} \quad p = 0.$$

Thus $\sup_{h\in\mathcal{K}}\{\|h_q\|_{L^\infty(R)}\} < \infty$, so that h is bounded in $C^1_{per}(\overline{R})$ along $\mathcal{K}_\delta$. We now apply the Schauder-type estimates due to Lieberman and Trudinger [205]—see the discussion of the boundary value problem (3.134)—with

$$F(\Theta,h,\xi,r) = (1+\xi_1^2)r_{22} - 2\xi_1\xi_2 r_{12} + \xi_2^2 r_{11} - \gamma(-p)\xi_2^3,$$

$$G(\Theta,h,\xi) = 1 + \xi_1^2 + (2gh - Q)\xi_2^2.$$

By using cut-off functions, the result in [205] is applicable in a subset of $C^2_{per}(\overline{R})$, where $Q - 2gh > \delta$ on $p = 0$, and $h_p > \delta$ in $\overline{R}$. These conditions together with the fact that $F_{rr} = 0$ ensure the applicability of the estimate (3.135), since by direct calculation we check first that

$$\frac{\Lambda_2}{\Lambda_1} \le 1 + \frac{1+h_q^2}{h_p^2}, \qquad \chi = 2\xi_2(Q - 2gh) > 2\delta^2,$$

and subsequently also the validity of all required inequalities (cf. the discussion at the end of Section 3.5.4). We infer that all along $\mathcal{K}_\delta$ the $C^{2,\alpha}_{per}(\overline{R})$ norm of the solution $h \in X$ is uniformly bounded. To prove $C^{3,\alpha}_{per}(\overline{R})$ a priori bounds for $h \in \mathcal{K}_\delta$, notice that $\varphi = \partial_q h$

satisfies (3.78). The Schauder estimates for the regular oblique derivative problem (cf. [129, Theorem 6.30]) and the $C^{2,\alpha}_{per}(\overline{R})$ a priori bounds for $h \in \mathcal{K}_\delta$ yield the uniform boundedness of the $C^{2,\alpha}_{per}(\overline{R})$ norm of h_q all along $\mathcal{K}_\delta$. Thus, to obtain the uniform boundedness of $\|h\|_X$ along $\mathcal{K}_\delta$, we have only to prove uniform $C^{2,\alpha}_{per}(\overline{R})$ estimates for h_p along $\mathcal{K}_\delta$. We already have uniform estimates on all the third derivatives of h except h_{ppp}. In order to get these, we express h_{pp} from the partial differential equation in (3.4) in terms of the other derivatives of h of order less than or equal to 2. Thus $\|h_{pp}\|_{C^{1,\alpha}_{per}(\overline{R})}$ is uniformly bounded along $\mathcal{K}_\delta$. This is the missing ingredient to show that $\|h\|_X$ is bounded along $\mathcal{K}_\delta$. □

Let now $\mathcal{S}^0 = \bigcup_{\delta>0} \mathcal{S}^0_\delta$. By definition, the continua $\mathcal{S}^0_\delta$ increase as $\delta > 0$ decreases. The previous considerations show that for any $\delta > 0$, one of the following four alternatives holds:

(1) there exists a sequence $(Q_n, h_n) \in \mathcal{S}^0_\delta$ with $\lim_{n\to\infty} Q_n = \infty$;

(2) there exists a sequence $(Q_n, h_n) \in \mathcal{S}^0_\delta$ with $\lim_{n\to\infty}[\max_{\overline{R}} \partial_p h_n] = \infty$;

(3) there exists a $(Q, h) \in \mathcal{S}^0_\delta$ with $\partial_p h = \delta$ somewhere in $\overline{R}$;

(4) there exists a $(Q, h) \in \mathcal{S}^0_\delta$ with $Q - 2gh = \delta$ somewhere on T;

and the same alternatives are valid if we replace $\mathcal{S}^0_\delta$ by $\mathcal{K}_\delta$. By returning to the original form of the governing equations, we will now show that these alternatives mean that the continuum $\mathcal{S}^0$ contains waves that approach a flow with a weak stagnation point, that is, a point in the fluid where $u = c$. This follows from some uniform lower bounds on the flow speed at the wave crest and on the pressure throughout the fluid.

Lemma 3.14. *We have either* $\sup_{\overline{D_{\eta n}}}\{u_n\} \to c$ *or* $\inf_{\overline{D_{\eta n}}}\{u_n\} \to -\infty$ *along some sequence in* $\mathcal{S}^0$.

Proof. Note that from the definition of the relative mass flux p_0,

$$\inf_{-d\le y\le \eta(x)} [c - u(x,y)] \le \frac{|p_0|}{\eta(x)+d} \le \sup_{-d\le y\le \eta(x)} [c - u(x,y)].$$

Therefore,

$$Q = [c - u(0,\eta(0))]^2 + 2g[\eta(0)+d] \le \sup_{\overline{D_\eta}} [c-u]^2 + \frac{2g|p_0|}{\inf_{\overline{D_\eta}}[c-u]}. \tag{3.83}$$

If for some $\delta > 0$ the alternative (1) holds, then (3.83) shows that $\sup_{\overline{D_{\eta n}}}\{u_n\} \to c$ or $\inf_{\overline{D_{\eta n}}}\{u_n\} \to -\infty$, where (u_n, v_n, η_n) is the solution of the water-wave problem corresponding to h_n. If for some $\delta > 0$ the alternative (2) holds, then $\sup_{\overline{D_{\eta n}}}\{u_n\} \to c$ as $\partial_p h_n = \frac{1}{c-u_n}$. If for a sequence $\delta_n \downarrow 0$ the alternative (3) holds, then there exists a sequence $(Q_n, h_n) \in \mathcal{S}^0$ such that $\inf_{\overline{R}}\{\partial_p h_n\} \to 0$. Therefore $\inf_{\overline{D_{\eta n}}}\{u_n\} \to -\infty$. Finally, if for a sequence $\delta_n \to 0$ the alternative (4) holds, then we find $(Q_n, h_n) \in \mathcal{S}^0$ such that $\inf_T [Q_n - 2gh_n] \to 0$. Using the nonlinear boundary condition in (3.4), we get

$$\frac{1}{(\partial_p h_n)^2} \leq \frac{1+(\partial_q h_n)^2}{(\partial_p h_n)^2} = Q_n - 2gh_n \quad \text{on} \quad T.$$

Therefore $\sup_{\overline{R}} \{\partial_p h_n\} \to \infty$. Hence $\sup_{\overline{D_{\eta n}}} \{u_n\} \to c$. □

Lemma 3.15 (speed at the crest). *For any nonlaminar flow we have*

$$u(0,\eta(0)) > c - \sqrt{\lambda_0}, \tag{3.84}$$

where $\lambda_0 > 2\Gamma_{max} \geq 0$ is the unique minimum of the strictly convex function $\lambda \mapsto Q(\lambda)$ defined by (3.11) *for $\lambda > 2\Gamma_{max}$.*

Proof. To prove this claim, notice that we can write the partial differential equation in (3.4) as

$$\gamma(-p) + \left(\frac{1+h_q^2}{2h_p^2}\right)_p = \frac{h_{qq}}{h_p} - \frac{h_q}{h_p^2} h_{pq}.$$

We evaluate this beneath the crest, that is, on $q = 0$. The nodal pattern (3.74) and the fact that $h_q = 0$ on $q = 0$ yield that the right side is negative. Integrating along $q = 0$ from $p < 0$ to 0, we get

$$-\Gamma(p) + \frac{1}{2h_p^2(0,0)} - \frac{1}{2h_p^2(0,p)} < 0.$$

Therefore, either $[c - u(0,\eta(0))]^2 \leq 2\Gamma_{max}$, or

$$h_p^2(p,0) < \frac{1}{[c-u(0,\eta(0))]^2 - 2\Gamma(p)}, \qquad p \in [p_0, 0],$$

so

$$\eta(0) + d = h(0,0) = \int_{p_0}^{p} h_p(0,p)dp < \int_{p_0}^{p} \frac{1}{\sqrt{[c-u(0,\eta(0))]^2 - 2\Gamma(p)}}. \tag{3.85}$$

A similar procedure beneath the wave trough, along $q = \pi$, yields

$$\eta(\pi) + d > \int_{p_0}^{p} \frac{1}{\sqrt{[c-u(\pi,\eta(\pi))]^2 - 2\Gamma(p)}}. \tag{3.86}$$

On the other hand, evaluating the nonlinear boundary condition in (3.4) at $q = 0$ and at $q = \pi$, where $v = 0$, we deduce that

$$Q = [c - u(0,\eta(0))]^2 + 2g[\eta(0)+d] = [c - u(\pi,\eta(\pi))]^2 + 2g[\eta(\pi)+d].$$

Since for a nonlaminar flow $\eta(0) + d > \eta(\pi) + d$, we have

$$[c - u(0,\eta(0))]^2 < [c - u(\pi,\eta(\pi))]^2,$$

while (3.85) and (3.86) combined yield

$$Q\Big([c-u(0,\eta(0))]^2\Big) > Q\Big([c-u(\pi,\eta(\pi))]^2\Big).$$

The function $\lambda \mapsto Q(\lambda)$ defined by (3.11) being strictly convex for $\lambda > 2\Gamma_{max}$, with a minimum at λ_0, the previous two inequalities are possible only if

$$[c-u(0,\eta(0))]^2 < \lambda_0 < [c-u(\pi,\eta(\pi))]^2.$$

The bound (3.84) follows at once. □

Lemma 3.16 (pressure). *The pressure throughout the fluid has the following lower bound:*

$$P \geq P_{atm} - \frac{1}{2}\|\gamma\|^2_{L^\infty[p_0,0]}[\eta(0)+d]^2. \tag{3.87}$$

Proof. From the Euler equation and the definition of the stream function we compute

$$\Delta P = 2(\psi_{xx}\psi_{yy}-\psi_{xy}^2) = (\psi_{xx}+\psi_{yy})^2 - \psi_{xx}^2 - \psi_{yy}^2 - 2\psi_{xy}^2 \leq (\Delta\psi)^2 = \gamma^2(\psi).$$

With $M = \|\gamma\|^2_{L^\infty[p_0,0]}$, we get

$$\Delta\Big(P - \frac{M}{2}(y+d)^2\Big) \leq \gamma^2(\psi) - M \leq 0$$

throughout the fluid. By the weak maximum principle, the minimum of $P - \frac{M}{2}(y+d)^2$ occurs on the free surface or on the flat bed. But on $y=-d$ we have $\psi_x = \psi_{xx} = 0$ as ψ is constant there, so that the Euler equation yields $P_y = -g < 0$ on $y=-d$. Using Hopf's form of the maximum principle, the minimum of the function must occur on $y = \eta(x)$, where $P = P_{atm}$ and $y+d \leq \eta(0)+d$. The inequality (3.87) follows. □

We are now in a position to prove the main result on the existence of periodic traveling waves of large amplitude.

Theorem 3.17. *Let the wave speed $c > 0$, the wavelength $L = 2\pi$, and the relative mass flux $p_0 < 0$ be given. For a constant $\alpha \in (0,1)$, let the vorticity function $\gamma \in C^{1,\alpha}[0,|p_0|]$ satisfy (3.9). Then there exists a continuum $\mathcal{S}^0$ of solutions (u,v,η) with regularity $C^{2,\alpha}_{per}(\overline{D_\eta}) \times C^{2,\alpha}_{per}(\overline{D_\eta}) \times C^{3,\alpha}_{per}(\mathbb{R})$ and satisfying the following properties:*

(i) *u, v, and η have period 2π in x;*

(ii) *within each period a nonflat wave profile η has a single maximum (crest) and a single minimum (trough);*

(iii) *if $x=0$ is the location of the wave crest, then u and η are symmetric while v is antisymmetric around the line $x=0$, with $v(x,y)>0$ for $y>-d$ and $x\in(0,\pi)$;*

(iv) *the wave profile is strictly decreasing from crest to trough;*

(v) *the flow beneath each wave has no stagnation points; that is, $u < c$ throughout the fluid.*

Moreover, $\mathcal{S}^0$ contains a single laminar flow (with a flat surface $\eta = 0$ and all streamlines parallel to the flat bed), and there is a sequence (u_n, v_n, η_n) for which $\max_{\overline{D_{\eta_n}}}[u_n] \uparrow c$.

While we cannot provide a quantitative estimate of the meaning of "large amplitude," the existence of waves such that the flow beneath the wave approaches a weak stagnation point shows that the solutions we construct are not small perturbations of the flat free surface. Indeed, at the local bifurcation the corresponding laminar flow is without weak stagnation points.

Proof. The existence of a continuum $\mathcal{S}^0$ of solutions with the stated regularity properties was shown in the considerations from the previous subsection. All that is left to prove is that at the boundary of the continuum of solutions we approach stagnation. Recall Bernoulli's law from the beginning of Chapter 3: throughout the fluid

$$E = \frac{(c-u)^2 + v^2}{2} + gy + P + \Gamma(-\psi)$$

is constant (although the value of the constant E may vary along $\mathcal{S}^0$). Evaluating E at the wave crest, we get

$$E = \frac{[c - u(0,\eta(0))]^2}{2} + g\eta(0) + P_{atm}.$$

The last two displayed relations yield

$$P - P_{atm} + \frac{(c-u)^2 - [c - u(0,\eta(0))]^2}{2} + \frac{v^2}{2} = g[\eta(0) - y] - \Gamma(-\psi).$$

From Lemmas 3.15 and 3.16 we obtain that

$$\frac{1}{2}(c-u)^2 \le g[\eta(0)+d] + \frac{1}{2}\|\gamma\|^2_{L^\infty[0,|p_0|]}[\eta(0)+d]^2 + \frac{\lambda_0}{2} + \|\Gamma\|_{L^\infty[0,|p_0|]} \tag{3.88}$$

throughout the fluid. If $\inf_{\overline{D_{\eta_n}}}\{u_n\} \to -\infty$ along some sequence in $\mathcal{S}^0$, then the left side of (3.88) would become unbounded, and consequently $\eta_n(0) + d_n \to \infty$. But

$$|p_0| = \int_{-d_n}^{\eta_n(0)} [c - u_n(0,y)]\,dy$$

is fixed. Consequently, there exists a sequence y_n such that $u_n(0, y_n) \to c$. $\square$

3.2.3 Numerical simulation of waves of large amplitude

The form (3.4) of the governing equations provides a tractable formulation for numerical simulations. The computational domain $\overline{R}$ is partitioned into equal rectangular cells where the number of cells is variable and is determined by convergence criteria: m refers to the

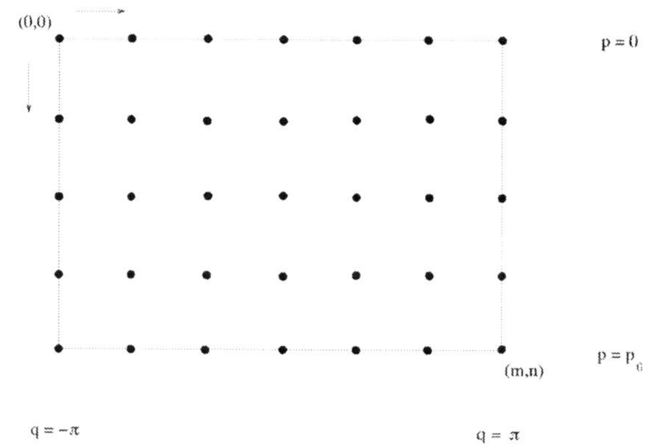

Figure 3.6. *Indexing of q and p by i and j, respectively, on the rectangular domain $\overline{R}=[-\pi,\pi]\times[p_0,0]$, for the discretization of the water-wave problem.*

number of grid points along the q-axis and n to that along the p-axis. Let i index the q-variable and j the p-variable, as shown in Figure 3.6.

The equations in (3.4) are discretized by finite differences to yield a system of nonlinear algebraic equations involving the numerical value $h_{i,j}$ of h at each gridpoint (i,j) and, for the equations governing the top boundary, the value of Q. For the terms h_q, h_p, h_{pp}, and h_{qq} in R (that is, for $1\le i\le m-1$ and $1\le j\le n-1$), one can use central differences, that is,

$$h_q(i,j)=\frac{h_{i+1,j}-h_{i-1,j}}{2\Delta q},\quad h_{qq}(i,j)=\frac{h_{i+1,j}-2h_{i,j}+h_{i-1,j}}{(\Delta q)^2},$$
$$h_p(i,j)=\frac{h_{i,j+1}-h_{i,j-1}}{2\Delta p},\quad h_{pp}(i,j)=\frac{h_{i,j+1}-2h_{i,j}+h_{i,j-1}}{(\Delta p)^2},$$

while the mixed derivative h_{qp} can be approximated as

$$h_{qp}(i,j)=\frac{h_{i+1,j-1}-h_{i-1,j+1}-h_{i-1,j-1}+h_{i-1,j+1}}{4\Delta p\Delta q}.$$

On $p=0$ one approximates $h_p(i,0)$ by the forward difference $\frac{h_{i,0}-h_{i,1}}{\Delta p}$ since we are at the edge of the domain and cannot use a centered difference (there is no $h_{i,-1}$ to use). Periodicity is expressed by $h_{0,j}=h_{m,j}$, and the boundary condition on $p=p_0$ by $h_{i,n}=0$. The value $\gamma(i,j)$ is a given number γ_j, while g and Q are constants. One writes the discretized problem in the form $F(\Xi_0)=0$, where F is a polynomial map from $\mathbb{R}^{mn}$ to $\mathbb{R}^{mn}$ and Ξ is the mn-vector whose elements are all the $h_{i,j}$'s with[25] $0\le i\le m-1$ and $0\le j\le n-1$, in row-major order, say. More precisely, for $j\ge 1$ we use the discretized partial differential equation to obtain an algebraic equation at (i,j), with the provision that $h_{-1,j}$ is replaced by $h_{m-1,j}$ as an expression of periodicity, while for $j=0$ we use the m algebraic equations obtained by imposing at $(i,0)$ with $0\le i\le m-1$ the discretized boundary condition on $p=0$. For fixed Q, the system $F(\Xi)=0$ can be solved by a Newton iteration: starting

[25]The indexes $i=m$ and $j=n$ are taken care of by periodicity and by the boundary condition on $p=p_0$, respectively.

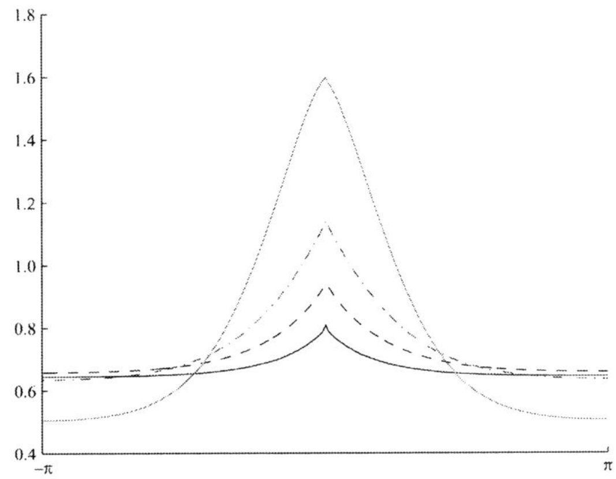

Figure 3.7. *Numerical simulation of waves near stagnation with* $p_0 = -2$ *for various constant vorticities: as* γ *runs through the successive values* 4, 2, 0, −2.95, *the amplitude of the wave near stagnation decreases. The vorticity effect on the shape of the wave profile is clearly visible. [From J. Ko and W. Strauss, J. Fluid Mech.* 608 (2008), *pp.* 197–215. *Reproduced with permission from Cambridge University Press]*

with some Ξ_0, we define

$$\Xi_{n+1} = \Xi_n - [J_F(\Xi_n)]^{-1} F(\Xi_n), \qquad n \geq 0,$$

where we denoted by $J_F(\Xi)$ the Jacobian matrix of F at Ξ, and the sequence $\{\Xi_n\}$ should converge to a root of F. This convergence is not always guaranteed; e.g., if the initial guess Ξ_0 is quite far from the location of the root of F, convergence might fail. However, for constant vorticity the results discussed in Section 3.1.3 provide us with a good approximation of the local bifurcation solution (Q^*, H^*). To effectively calculate waves along the curve $\mathcal{K}_\delta$, one can use (cf. [180, 181]) the efficient numerical continuation library TRILINOS written in C^{++} at Sandia National Laboratories. Once (Q^*, H^*) is found, one gets onto $\mathcal{C}_{loc}$ by providing the program with an approximate solution by shooting in the direction of the eigenvector associated to the linearization. Once on the local bifurcation curve $\mathcal{C}_{loc}$, the continuation proceeds. For further details of the implementation of this numerical simulation we refer the reader to the discussion in [180].

Let us point out some general observations that can be drawn from the simulations in [180, 181].

(i) The amplitude increases along $\mathcal{K}_\delta$, with maximum amplitude near stagnation.

(ii) For constant nonzero vorticity the weak stagnation point (where u approaches c) is either at the wave crest or on the bottom directly below the crest. It is therefore a stagnation point as $v = 0$ at both locations. For nonconstant vorticity the weak stagnation point is always located on the vertical line beneath the crest, and is therefore a stagnation point. It might happen that stagnation occurs at an interior point of the fluid domain.

(iii) A nonzero vorticity distribution can have several main effects on the gross appearance of a wave: the interaction of a wave train with a current of large constant negative vorticity produces a surface profile that has reduced overall steepness, but increased crest-trough asymmetry (with a broader trough and a sharper crest) than in the irrotational case. For large positive constant vorticity the wave appears to be steeper than in the irrotational case.[26] This conclusion is supported by the experimental evidence provided in [294].

(iv) The depth varies only slightly along $\mathcal{K}_\delta$.

(v) The energy Q increases at first along $\mathcal{K}_\delta$, away from the local bifurcation (Q^*, H^*), but then may decrease.

At present all these numerical observations are open theoretical problems. For some progress in the theoretical confirmation of (ii) see [73, 306].

3.3 Symmetry of rotational steady water waves

Swell entering a region of still water is appropriately modelled as a wave train propagating in irrotational flow—these are the Stokes waves. Garabedian [122] proved the ubiquity of regular wave trains under the assumption of irrotational flow by showing that if every streamline has one minimum and one maximum per period (below the wave trough, respectively, below the wave crest), then the wave is symmetric around the crest. The variational proof given in [122] was later simplified by Toland [299]. Subsequently, Okamoto and Shoji [245] presented a proof of symmetry based on the assumption that the surface wave profile has a positive/negative derivative from trough to crest, respectively, from crest to trough. Since in a frame moving with the speed of the wave the free surface is a streamline, the result in [245] improves upon Garabedian's result.

In the context of wave-current interactions it might appear reasonable to expect that due to a possible strong shear induced by the current, one encounters asymmetric wave trains. The aim of this section is to show that in the absence of stagnation points in the flow, symmetry is not affected by the presence of a current. Since the simultaneous generation of waves and depth-varying currents is experimentally very complicated [294], one can not simply expect to establish experimentally the validity of this fact by creating in the laboratory flows corresponding to all physically interesting classes of vorticity distribution; a theoretical approach is therefore appropriate.

Theorem 3.18. *Steady rotational*[27] *periodic gravity water waves without stagnation points in the underlying flow and with a surface wave profile that is monotone between troughs and crests are symmetric. Moreover, unless the entire free surface is horizontally flat, there can be no horizontal parts of the wave profile.*

Proof. Consider a steady wave with a nonflat free surface that is monotone between crests and troughs. If the wave profile is horizontally flat near the crest/trough, then by the location of the crest/trough we mean the middle point of the corresponding horizontal interval. Let us

[26] Notice that the sign of the vorticity in [180, 181] is the opposite of the one we use.

[27] We assume that the vorticity function $\gamma \in C^{1,\alpha}[p_0, 0]$ and the solution h to (3.4) is in the space $C^3_{per}(\overline{R})$ for some $\alpha \in (0, 1)$. The considerations in the previous two sections ensure the existence of symmetric waves. The present result establishes that monotonicity implies symmetry.

assume that the horizontal position of the wave trough is at $q = -\pi$, while that of the wave crest is located in the interval $[0,\pi)$. For a reflection parameter $\lambda \in (-\pi,0)$ the reflection of q about λ is given by $q^\lambda \equiv 2\lambda - q$, and the associated *reflection function* is

$$w(q,p;\lambda) = h(q,p) - h(2\lambda - q,p), \qquad (q,p) \in [\lambda, 2\lambda+\pi] \times [p_0,0].$$

To apply maximum principles in our setting notice that if $\Omega \subset R$ is a rectangle with the horizontal sides supported on those of the rectangle R and if $h,\tilde{h} \in C^2_{per}(\overline{\Omega})$ are solutions to the water-wave problem (3.4) with $h_p > 0$ in $\overline{R}$, then the uniformly elliptic operator with continuous coefficients

$$\begin{aligned} \mathcal{L} = \left(1+h_q^2\right)\partial_p^2 + h_p^2\partial_q^2 - 2h_p h_q \partial_p\partial_q + \left[\tilde{h}_{pp}\left(h_q+\tilde{h}_q\right) - 2h_p\tilde{h}_{pq}\right]\partial_q \\ + \left[\tilde{h}_{qq}\left(h_p+\tilde{h}_p\right) - 2\tilde{h}_q\tilde{h}_{pq} - \gamma(-p)\left(h_p^2+h_p\tilde{h}_p+\tilde{h}_p^2\right)\right]\partial_p \end{aligned} \tag{3.89}$$

satisfies $\mathcal{L}(h-\tilde{h}) = 0$ in Ω. In particular, for a solution $h \in C^2_{per}(\overline{R})$ with $h_p > 0$ throughout $\overline{R}$ and for any $\lambda \in (-\pi,0]$, the function $\tilde{h}(q,p) = h(2\lambda - q,p)$ is also a solution so that $\mathcal{L}\,w(q,p;\lambda) = 0$. Notice also that since $h_p > 0$ in $\overline{R}$, we have that the coefficient of $\partial_q\partial_p$ in $\mathcal{L}$ vanishes at a point (q,p) if and only if $h_q(q,p) = 0$.

The reflection function $w(\cdot,\cdot;\lambda)$ satisfies the boundary conditions

$$\begin{cases} w(\lambda,p;\lambda) = 0 & \text{for} \quad p \in [p_0,0], \\ w(q,p_0;\lambda) = 0 & \text{for} \quad q \in [\lambda, 2\lambda+\pi]. \end{cases} \tag{3.90}$$

Indeed, the first property is immediate from the definiton of $w(q,p;\lambda)$, and the second follows from the boundary condition $h=0$ on $p=p_0$. Since the surface wave profile is nondecreasing from trough to crest by assumption, the reflection function satisfies $w(q,0;\lambda) \geq 0$ for $\lambda > -\pi$ close enough to $-\pi$. Let

$$\lambda_0 = \sup\{\lambda \in (-\pi,0] : \ w(q,0;\lambda) \geq 0 \quad \text{for all} \quad q \in [\lambda, 2\lambda+\pi]\}.$$

One of the following two cases can occur:

(I) $\lambda_0 = 0$;

(II) $\lambda_0 \in (-\pi,0)$, in which case we cannot have $w(q,0;\lambda_0) > 0$ for all $q \in (\lambda_0, 2\lambda+\pi]$ so that there exists $q_0 \in (\lambda_0, 2\lambda_0+\pi]$ for which $w(q_0,0;\lambda_0) = 0$. Since $w(q,0;\lambda_0) \geq 0$ on $[\lambda_0, 2\lambda_0+\pi]$, at q_0 the graphs of the functions $q \mapsto h(q,0)$ and $q \mapsto h(2\lambda_0 - q,0)$ intersect and are tangent to each other.

Assume first that (I) holds. The periodicity of h and the definition of λ_0 and of the reflection function yield the additional boundary conditions

$$\begin{cases} w(\pi,p;0) = 0 & \text{for} \quad p \in [p_0,0], \\ w(q,0;0) \geq 0 & \text{for} \quad q \in [0,\pi]. \end{cases} \tag{3.91}$$

Consider the rectangle $\Omega = (0,\pi) \times (p_0,0)$. Since $w \in C^2(\overline{\Omega})$, by the strong maximum principle—Lemma 3.27 (i)—there cannot exist an interior point $(q,p) \in \Omega$ where

$w(q,p;0) \le 0$ unless w vanishes everywhere in Ω, in view of the boundary conditions (3.90) and (3.91). If w vanishes throughout Ω, we have symmetry, so suppose on the contrary that

$$w(q,p;0) > 0 \quad \text{in} \quad \Omega.$$

At the trough $(\pi,0)$ both w and w_p vanish by (3.91). Since $h_q(\pm\pi,0) = 0$ we get

$$w_q(\pi,0;0) = 0,$$

while differentiating the boundary condition on $p = 0$ in (3.4) with respect to q yields

$$0 = 2h_q h_{qq} + (2gh_q)h_p^2 + 2(2gh - Q)h_p h_{qp} = 2(2gh - Q)h_p h_{qp} \quad \text{at} \quad (\pi,0),$$

forcing $h_{qp}(\pi,0) = 0$ since $(2gh - Q)h_p = 1$ at $(\pi,0)$ according to the same boundary condition. A similar argument holds for the reflection $h(-q,p)$, whence

$$w_{qp}(\pi,0) = 0.$$

Keeping the definition of $w(q,p;0)$ in mind, we thus see that

$$w = w_q = w_p = w_{qq} = w_{qp} = w_{pp} = 0 \quad \text{at the trough} \quad (\pi,0),$$

where also $h_q = 0$. According to Lemma 3.27(iii) and the observation about the form of $\mathcal{L}$, this contradicts the earlier conclusion that $w(q,p;0) > 0$ in Ω, and we infer that if (I) is the case, then $h(q,p) = h(-q,p)$ for $(q,p) \in \Omega$, meaning that the wave is symmetric about the crest located at $q = 0$.

Let us now consider the possibility (II). We increase the domain of definition of w by the following extension:

$$w(q,p;\lambda_0) = h(q,p) - h(2\lambda_0 + 2\pi - q,p), \quad (q,p) \in [2\lambda_0 + \pi, \lambda_0 + \pi] \times [p_0,0].$$

If we redefine $\Omega \equiv (\lambda_0, \lambda_0 + \pi) \times (p_0,0)$, it turns out that periodicity guarantees $w \in C^2(\overline{\Omega})$. Furthermore, note that as long as $2\lambda + \pi$ lies to the left of the wave crest, $w(q,0;\lambda) \ge 0$ will hold for $\lambda \le q \le 2\lambda + \pi$ by the monotonicity of the surface profile between trough and crest. Therefore, $2\lambda_0 + \pi$ lies to right of—or at least in line with—the wave crest. Consequently (II) implies that $h(q,p)$ is nonincreasing for $q \in (2\lambda_0 + \pi, \pi)$, yielding the useful boundary condition

$$w(q,0;\lambda_0) \ge 0, \qquad q \in [\lambda_0, \lambda_0 + \pi].$$

Summing up the situation, we have

$$\begin{cases} w(\lambda_0,p;\lambda_0) = w(\lambda_0 + \pi, p;\lambda_0) = 0 \quad \text{for} \quad p \in [p_0,0], \\ w(q,p_0;\lambda_0) = 0 \quad \text{and} \quad w(q,0;\lambda_0) \ge 0 \quad \text{for} \quad q \in [\lambda_0, \lambda_0 + \pi]. \end{cases}$$

As in dealing with (I), we may apply Lemma 3.27(i) to conclude that

$$w(q,p;\lambda_0) > 0 \quad \text{in } \Omega \text{ unless } w \text{ vanishes identically.}$$

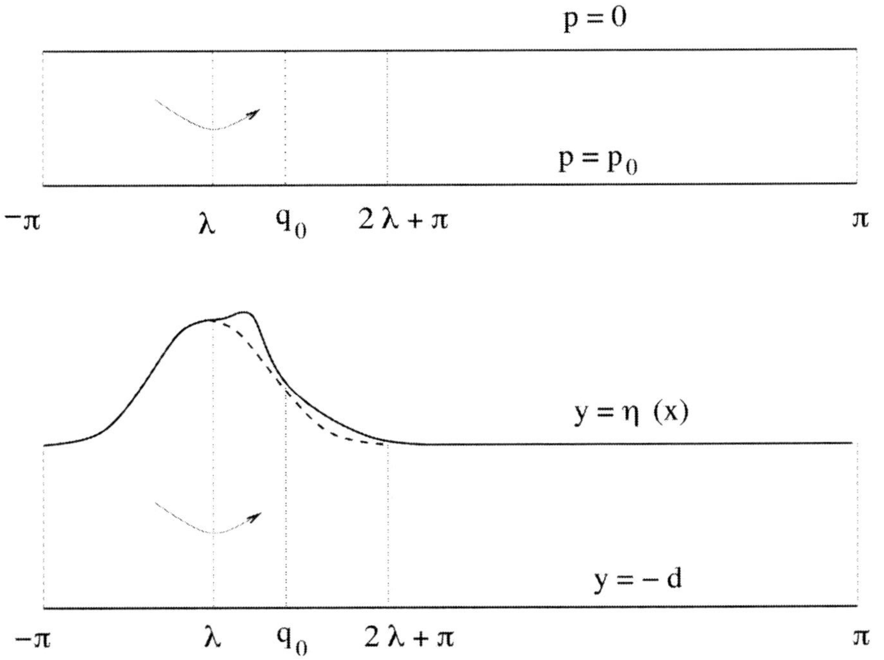

Figure 3.8. *Symmetrization in the (q,p)-variables (top) and in the physical variables (bottom).*

Let us first assume that $w(q,p;\lambda_0)>0$ in Ω. At $(q_0,0)$ the tangency condition ensures $w_q(q_0,0)=0$, meaning that $h_q(q_0,0)=-h_q(2\lambda_0-q_0,0)$. Since $h(q_0,0)=h(2\lambda_0-q_0,0)$, the nonlinear boundary condition in (3.4) forces $h_p(q_0,0)=h_p(2\lambda_0-q_0,0)$. Thus

$$\nabla w(q_0,0;\lambda_0)=(0,0),$$

contradicting Lemma 3.27(ii) and by extension the assumption $w(q,p;\lambda_0)>0$ in Ω. Consequently $w\equiv 0$ in $\overline{\Omega}$. Then $h(q,0)=h(\pi,0)$ whenever $2\lambda_0+\pi\le q\le\pi$ since $w(2\lambda_0+\pi,0;\lambda_0)=0$ yields $h(2\lambda_0+\pi,0)=h(-\pi,0)=h(\pi,0)$ and the map $q\mapsto h(q,0)$ is nonincreasing on that interval as we established that $2\lambda_0+\pi$ lies to the right of the wave crest. We also have that $h(q,p)=h(2\lambda_0-q,p)$ for all $q\in[\lambda_0,2\lambda_0+\pi]$ so that $q=\lambda_0$ must be the location of the wave crest. But by assumption $\lambda_0<0$ and the wave crest is in $[0,\pi)$.

These considerations show that h must be symmetric about the line $q=0$, with the wave crest at $(0,0)$ in the (q,p)-variables. To conclude the proof we must show that there are no flat (horizontal) parts of the free surface profile. Assume there exists $0\le q_1<q_2\le\pi$ with $h(q_1,0)=h(q_2,0)$. The monotonicity of the surface wave between crest and trough yields then $h_q(q,0)=0$ for $q\in[q_1,q_2]$. Now notice that by the symmetry of h about $q=0$ and by its periodicity in the q-variable with period 2π, we must have $h_q(\pm\pi,p)=h_q(0,p)=0$ for all $p\in[p_0,0]$, Since $h_q(q,p_0)=0$ for $q\in[0,\pi]$ by the bottom boundary condition in (3.4), while $h_q(q,0)\le 0$ for $q\in[0,\pi]$ due to the monotonicity of the surface wave profile, we have $h_q\le 0$ all along the boundary of the rectangle $[0,\pi]\times[p_0,0]$. Since

$h_q(\frac{q_1+q_2}{2},0)=0$ and h_q satisfies a second-order elliptic equation with no zero order term[28] in $(0,\pi)\times(p_0,0)$, by Lemma 3.27(ii) we should have $h_{qp}(\frac{q_1+q_2}{2},0)>0$. But $h_p(q,0)$ is constant on $[q_1,q_2]$, in view of the fact that $h(q,0)$ is constant there and the nonlinear boundary condition in (3.4) holds true, so $h_{qp}(\frac{q_1+q_2}{2},0)=0$. The obtained contradiction completes the proof. ☐

Remark. A careful look at the previous proof shows that in the moving frame every streamline is symmetric about $x=0$, and every streamline but the flat bed increases strictly from $x=-\pi$ to $x=\pi$. For this, it suffices to notice that the streamline $\psi=-p$ is precisely the graph of the function $q\mapsto h(q,p)$.

3.4 Regularity of the streamlines

In this section we investigate the regularity properties of the free surface and of the streamlines beneath the surface for the waves without stagnation points in the underlying flow whose existence was established in the previous sections.

Theorem 3.19. *Assume that the vorticity function $\gamma\in C^{1,\alpha}[p_0,0]$ for some $\alpha\in(0,1)$. Then for a steady periodic wave without stagnation points in the underlying flow[29] each streamline beneath the surface is real-analytic while the free surface is smooth. Moreover, if the vorticity function is real-analytic in a neighborhood of $[p_0,0]$, then the free surface is the graph of a real-analytic function.*

Remark. Using the notation introduced earlier, the statement above means that for a given $\gamma\in C^{1,\alpha}[p_0,0]$, if $h\in C^{3,\alpha}_{per}(\overline{R})$ is a solution to (3.4) for which

$$\inf_{(x,y)\in\overline{D_\eta}}\{c-u(x,y)\}>\varepsilon>0, \tag{3.92}$$

then the map $q\mapsto h(q,p)$ must be real-analytic for all $p\in[p_0,0)$ and is smooth for $p=0$. Moreover, if γ is real-analytic, then $q\mapsto h(q,0)$ is also real-analytic.

Proof. Let us first show that we can write the nonlinear boundary condition in (3.4) in the equivalent form

$$h_p=\sqrt{\frac{1+h_q^2}{Q-2gh}}\quad\text{on}\quad p=0. \tag{3.93}$$

Indeed, since $h_p=\frac{1}{c-u}$, by (3.92) we have $h_p<\frac{1}{\varepsilon}$ throughout $\overline{R}$ and the boundary condition on $p=0$ in (3.4) yields

$$Q-2gh=\frac{1+h_q^2}{h_p^2}>\varepsilon^2(1+h_q^2)\geq\varepsilon^2\quad\text{on}\quad p=0, \tag{3.94}$$

[28] Obtained by differentiating the differential equation in (3.4) with respect to q.

[29] We assume the regularity ensured by the previous existence theory; that is, $h\in C^{3,\alpha}_{per}(\overline{R})$ satisfies $h_p>0$ throughout $\overline{R}$.

so that $h_p = \pm\sqrt{\frac{1+h_q^2}{Q-2gh}}$ on $p=0$. The choice of sign is determined by noticing that by (3.94) the expression $\frac{1+h_q^2}{Q-2gh}$ is strictly positive and $h_p(0,0) \geq 0$ as the maximum of $h(q,p) = y+d$ is attained at the wave crest $(0,0)$.

Since for γ real-analytic the maps[30] $(q,h,s) \mapsto \sqrt{\frac{1+s^2}{Q-2gh}}$ and

$$(q,p,h,W,M) \mapsto (1+w_1^2)M_3 - 2w_1w_2M_2 + w_2^2M_1 - \gamma(-p)w_2^3$$

are real-analytic in $\mathbb{R} \times U_0 \times \mathbb{R}$, respectively, in $\mathbb{R}^2 \times U_0 \times U_1 \times \mathbb{R}^3$, where

$$U_0 = \{h \in \mathbb{R} : \ Q - 2gh > \varepsilon\}, \qquad U_1 = \{(w_1, w_2 \in \mathbb{R}^2 : \ 0 < w_2 < \tfrac{1}{\varepsilon}\},$$

the classical regularity theory for elliptic nonlinear boundary problems (see Example 3.26 in the Appendix) ensures that if $\gamma \in C^{k,\alpha}[p_0,0]$ for some integer $k \geq 0$, then the solution $h \in C^{3,\alpha}_{per}(\overline{R})$ to (3.4) belongs to the space $C^{k+2,\alpha}_{per}(\overline{R})$, while if γ is real-analytic, so will be the solution h. Notice now that applying ∂_q to (3.4) we obtain a nonlinear elliptic boundary value problem for $\tilde{h} = h_q$ that enters into the framework of Example 3.26 so that $\partial_q h \in C^{3,\alpha}_{per}(\overline{R})$. Iterating this procedure we see that even if $\gamma \in C^{1,\alpha}[p_0,0] \setminus C^{2,\alpha}[p_0,0]$, we will have $\partial_q^m h \in C^{3,\alpha}_{per}(\overline{R})$ for any $m \geq 0$. In particular, since the streamlines are given by $q \mapsto h(q,p)$ with $p \in [p_0,0]$ fixed, all streamlines are smooth.

It remains to prove that for $\gamma \in C^{1,\alpha}[p_0,0]$ all streamlines beneath the surface are real-analytic. Let D be the strip $\{(q,p) \in \mathbb{R}^2 : \ q \in \mathbb{R}, \ p_0 \leq p \leq 0\}$, define the Banach spaces $X = \{h \in C^{3,\alpha}_{per}(\overline{R}) : \ h(q,p_0) = 0\}$, $Y = C^{1,\alpha}_{per}(\overline{R})$, $Z = C^{2,\alpha}_{per}(\mathbb{R})$, and consider the open set

$$\mathcal{O} = \{h \in X : \ \inf_{(q,p)\in\overline{R}} h_p(q,p) > 0\} \subset X.$$

For $h \in \mathcal{O}$ we define $F(h) \in Y \times Z$ by

$$\Big((1+h_q^2)h_{pp} - 2h_ph_q h_{pq} + h_p^2 h_{qq} - \gamma(-p)h_p^3, \ [1+h_q^2 + (2gh-Q)h_p^2]\Big|_{p=0}\Big).$$

Denoting by $\mathcal{A}$ a real-analytic dependence, we see that

$$F \in \mathcal{A}(\mathcal{O}, Y \times Z). \tag{3.95}$$

Furthermore, if τ_a denotes the translation by the amount ap in the q-variable,

$$\tau_a f(q,p) := f(q+ap,p), \qquad (q,p) \in D,$$

then

$$\partial_q(\tau_a f) = \tau_a(\partial_q f), \qquad \partial_p(\tau_a f) = \tau_a(\partial_p f + a\partial_q f).$$

Using these relations, a direct calculation shows that

$$F(\tau_a h) - \tau_a F(h) = aK(\tau_a h, a), \tag{3.96}$$

[30] Here $W = (w_1, w_2)$ stands for (h_q, h_p) and $M = (M_1, M_2, M_3)$ for (h_{qq}, h_{qp}, h_{pp}).

where the operator $K = (K_1, K_2)$ is given by

$$K_1(h,a) = (2h_{qp} - ah_{qq}) - \gamma h_q[3h_p^2 - 3ah_p h_q + a^2 h_q^2],$$
$$K_2(h,a) = (2gh - Q)(2h_q h_p - ah_q^2)\Big|_{p=0}.$$

Note that (3.96) holds true for any $h \in \mathcal{O}$ and any $a \in \mathbb{R}$ with $|a|$ sufficiently small to ensure that $\tau_a h \in \mathcal{O}$. Note also that

$$K \in \mathcal{A}(\mathcal{O} \times \mathbb{R}, Y \times Z). \tag{3.97}$$

Let now $h^0 \in \mathcal{O}$ be a solution to (3.4) with $\gamma \in C^{1,\alpha}[p_0, 0]$. Then $\Phi(h^0, 0) = F(h^0) = 0$, where

$$\Phi : \mathcal{O} \times \mathbb{R} \to Y \times Z, \quad \Phi(h,a) = F(h) - aK(h,a) + \left(0, \lambda(h_p - h_p^0 - ah_q^0)\Big|_{p=0}\right),$$

with $\lambda > 0$ being a positive number chosen so that (3.99) below holds true.[31] Since $F(h^0) = 0$ as h^0 is a solution to (3.4), and $(\partial_p(\tau_a h^0))(q,0) = h_p^0(q,0) + a\, h_q^0(q,0)$, we deduce in view of (3.96) that

$$\Phi(\tau_a h^0, a) = \tau_a F(h^0) = 0. \tag{3.98}$$

Using the implicit function theorem for real-analytic functions (see [38]), we now show that $a \mapsto \tau_a h^0$ is the unique solution of $\Phi(h,a) = 0$ near $(h^0, 0)$. Indeed, by (3.95) and (3.97), Φ is real-analytic in both variables in a neighborhood of $(h^0, 0)$ in $\mathcal{O} \times \mathbb{R}$. Denoting the Fréchet derivative

$$DF(h_0) =: (L, T) : X \to Y \times Z,$$

we observe that the definition of $\mathcal{O}$ ensures that L is a uniformly elliptic operator. Moreover, L satisfies the weak maximum principle, since it has no zero order term [129]. The boundary operator T is of uniform oblique type and has the form

$$Th = [2h_q^0 h_q + 2h_p^0(2gh^0 - Q)h_p + 2g(h_p^0)^2 h]\Big|_{p=0}.$$

We claim that $D_1\Phi(h^0, 0)$ defined by

$$\left(D_1\Phi(h^0,0)\right)h := \frac{d}{d\varepsilon}\Phi(h^0 + \varepsilon h, 0)\Big|_{\varepsilon=0} = (L, T + \lambda\partial_p)h$$

is an isomorphism from X onto $Y \times Z$, provided $\lambda > 0$ is such that

$$\lambda > \sup_{q \in \mathbb{R}} \left\{2h_p^0(Q - 2gh^0)\Big|_{p=0}\right\}. \tag{3.99}$$

Indeed, in view of the definition of $\mathcal{O}$, for such $\lambda > 0$ the boundary operator $T + \lambda\partial_p$ is uniformly oblique, while L is uniformly elliptic. The classical approach to elliptic boundary

[31] The rationale for introducing the term involving λ is the following. If τ_a^0 is the standard translation in the q-variable, i.e. $\tau_a^0 f(q,p) = f(a+q,p)$, then $F(\tau_a^0 h^0) = \tau_a^0 F(h^0) = 0$. Hence, writing $DF(h^0)$ for the Fréchet derivative of F at h_0, this equivariance implies that $DF(h^0)[h_q^0] = 0$. In particular, the linear operator $DF(h^0)$ has a nontrivial kernel if h^0 is not constant in the q-variable, while the Fréchet derivative of Φ at $(h^0, 0)$ will be an isomorphism.

value problems presented in the Appendix shows that $(L, T + \lambda\partial_p)$ is a Fredholm operator of index zero. Hence it suffices to show that it is injective. Assume in contrast to our claim that there is a nonzero $h \in X$ such that $(L, T + \lambda\partial_p)h = 0$. We may assume that h has a positive maximum (otherwise consider $-h$). By the weak maximum principle [129] and the boundary condition $h(\cdot, p_0) = 0$ we conclude that h takes its positive maximum on $p = 0$, say, at $(q_0, 0)$. Then $h_q(q_0, 0) = 0$, whereas $h_p(q_0, 0) > 0$ by Hopf's maximum principle [116]. This contradicts our assumption $(Th + \lambda h_p)|_{p=0} = 0$ since $h_q(q_0, 0) = 0$ yields that $(Th + \lambda h_p)$ evaluated at $(q_0, 0)$ equals

$$h_p(q_0,0)\Big[\lambda - 2h^0_p(q_0,0)\Big(Q - 2gh^0(q_0,0)\Big)\Big] + 2g\Big[h^0_p(q_0,0)\Big]^2 h(q_0,0) > 0$$

in view of (3.99) as $h(q_0, 0) > 0$. Thus for $\lambda > 0$ satisfying (3.99), we have

$$D_1\Phi(h^0, 0) \in \mathrm{Isom}(X, Y \times Z).$$

By the implicit function theorem for real-analytic maps [38], we now conclude the existence of some $\varepsilon > 0$ and some $\varphi \in \mathcal{A}((-\varepsilon, \varepsilon), \mathcal{O})$ such that in a sufficiently small neighborhood of $(h^0, 0) \in X \times \mathbb{R}$ all solutions of $\Phi(h, a) = 0$ are given by $(h, a) = (\varphi(a), a)$. Taking into account (3.98), by uniqueness we deduce that $\tau_a h^0 = \varphi(a)$ for $a \in (-\varepsilon, \varepsilon)$. In particular, given $(q, p) \in \mathbb{R} \times [p_0, 0)$, we have

$$[a \mapsto h^0(ap + q, p)] \in \mathcal{A}((-\varepsilon, \varepsilon), \mathbb{R}).$$

Real analyticity being a local property, we conclude that $[q \mapsto h^0(q, p)] \in \mathcal{A}(\mathbb{R}, \mathbb{R})$ for any $p \in [p_0, 0)$, and the proof is completed. $\square$

Remark. Notice that h is not real-analytic in R if γ is not real-analytic, but nevertheless all streamlines beneath the surface are real-analytic. It is also worth pointing out that the absence of stagnation points, expressed by the condition (3.92) which implies $h_p > 0$ throughout $\overline{R}$, is not a mere technicality: in the irrotational case $\gamma \equiv 0$ there exist steady periodic waves (the Stokes waves of greatest height; see [298]) with a stagnation point at their crest where the free surface fails to be C^1 and is just Lipschitz continuous.

3.5 Appendix for Chapter 3

3.5.1 Fredholm operators

An operator between two Banach spaces X and Y is said to be *compact* if it maps bounded sets into sets with compact closure.[32] Linear compact operators have finite- dimensional or, in a meaningful sense, almost finite-dimensional range. More precisely, a uniform limit of linear operators of finite rank is compact, and if Y is a separable Hilbert space the converse is also true[33] (see [208, Chapter 1]).

[32]Related to this is the concept of a *completely continuous* operator, the defining property of which is that it maps weakly convergent sequences into norm convergent sequences. Since weakly convergent sequences are bounded, compact operators are completely continuous, but the converse is not always true (see [78, Chapter VI]). However, since a Banach space is reflexive if and only if every bounded sequence contains a weakly convergent subsequence (see [321, Chapter V]), for reflexive Banach spaces X a completely continuous operator to Y will be compact.

[33]Although it fails in full generality, e.g., if $X = Y = \mathcal{B}(H)$, the space of bounded linear operators on an infinite-dimensional Hilbert space H.

We therefore expect that linear compact operators on separable Hilbert spaces are similar to the finite-dimensional case. However, one important concept for finite-dimensional linear operators does not generalize to the setting of compact linear operators: for finite-dimensional vector spaces X and Y the *index* of a linear transformation $T : X \to Y$ is defined by

$$\text{ind}(T) = \dim(ker(T)) - \dim(coker(T)), \tag{3.100}$$

where $coker(T) = Y/range(T)$, one has

$$\text{ind}(T) = \dim(X) - \dim(Y), \tag{3.101}$$

and, if $T^* : Y^* \cong Y \to X^* \cong X$ is the adjoint operator of T, then

$$\dim(ker(T)) = \dim(coker(T^*)), \qquad \dim(ker(T^*)) = \dim(coker(T)). \tag{3.102}$$

In particular, $\text{ind}(T) = -\text{ind}(T^*)$. The relations (3.102) give an indication of the degrees of freedom and the constraints associated to solving equations involving T. Indeed, considering the four equations

$$\text{(i) } Tx = y, \quad \text{(ii) } Tx = 0, \quad \text{(iii) } T^* f = g, \quad \text{(iv) } T^* f = 0, \tag{3.103}$$

we have the following **Fredholm alternative**: the second equation has a solution space $X_0 \subset X$ with $\dim(X_0) \geq 0$, the fourth equation has a solution space $Y_0 \subset Y^*$ with $\dim(Y_0) = \dim(X_0) - \text{ind}(T)$, in which case the first equation is solvable precisely for those $y \in Y$ such that $f(y) = 0$ for all $f \in Y_0$, and the third equation is solvable for precisely those $g \in X^*$ such that $g(x) = 0$ for all $x \in X_0$. In particular, if the second and fourth equations have only the trivial solution, then the first and third equations can be solved uniquely for any right-hand side.

To generalize the index concept to the case when X and Y are infinite-dimensional Banach spaces, instead of trying to make sense of the right-hand side of (3.101) we use (3.100) and introduce the class of *Fredholm operators* as being those bounded linear operators $T : X \to Y$ for which $ker(T)$ is finite-dimensional, $range(T)$ is closed, and $coker(T)$ is finite-dimensional, in which case we can define the index of T by (3.100). Let us provide a brief rationale for these requirements. A normed space X is called a topological direct sum of its two subspaces X_1 and X_2 if the addition map from $X_1 \times X_2$ is a topological isomorphism (that is, bijective and continuous). If this is the case, then X_1 and X_2 have to be closed subspaces with $X_1 \cap X_2 = \{0\}$ and such that X is the direct sum of X_1 and X_2 (that is, each $x \in X$ has a unique representation of the form $x = x_1 + x_2$ with $x_1 \in X_1$ and $x_2 \in X_2$). Moreover, the quotient map is a topological isomorphism from X/X_1 into X_2 (see [214, Chapter 7]). In proving the finite-dimensional relations (3.102) one relies on this type of decomposition for the subspaces $ker(T) \subset X$ and $range(T) \subset Y$, which is always granted in the finite-dimensional setting. However, while the Banach spaces X for which every closed subspace X_1 has a topological complement are precisely the Hilbert spaces, in which case X_2 is the orthogonal complement of X_1 (see [78, Chapter III]), for general Banach spaces X a closed subspace X_1 such that X_1 or X/X_1 is finite-dimensional always has a topological complement (see [214, Chapter 7]). In addition to these considerations, the requirements on $ker(T)$ and $range(T)$ ensure that the so-called *defect numbers* on the right of (3.100) are finite. Notice that there are compact linear operators T for which $range(T)$ is not closed: for a given sequence $a = \{a(k)\}_{k \geq 1} \in l^\infty$, the bounded linear operator $T : l^2 \to l^2$, defined by $(Tx)(k) = a(k)x(k)$ for $k \geq 1$, is compact if and only if $\lim_{k\to\infty} a(k) = 0$ and has closed range if and only if there is a constant $c > 0$ with $|a(k)| \geq c$ if $a(k) \neq 0$ (see [178, Chapter III]).

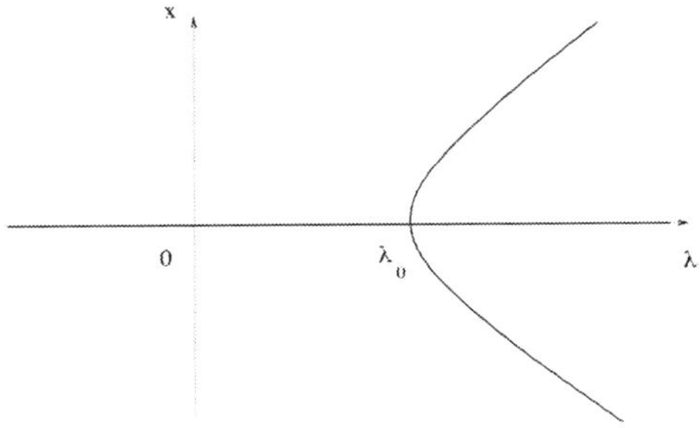

Figure 3.9. *A pitchfork bifurcation.*

Fredholm operators have the following basic properties (see [178, Chapter III]).

- If $T \in \mathcal{F}(X,Y)$, then the adjoint $T^* \in \mathcal{F}(Y^*,X^*)$ and (3.102) holds. In particular, $\text{ind}(T^*) = -\text{ind}(T)$ and the Fredholm alternative holds in the form stated above.[34]
- $T \in \mathcal{B}(X,Y)$ is Fredholm if and only if it is almost invertible, in the sense that there exist $S_1, S_2 \in \mathcal{B}(Y,X)$ as well as finite-rank operators $K_1 \in \mathcal{K}(X,X)$ and $K_2 \in \mathcal{K}(Y,Y)$ such that $S_1T = 1_X + K_1$ and $TS_2 = 1_Y + K_2$, where 1_Z denotes the identity on Z.
- The set $\mathcal{F}(X,Y)$ is open in $\mathcal{B}(X,Y)$ and $\text{ind} : \mathcal{F}(X,Y) \to \mathbb{Z}$ is locally constant[35] with $\text{ind}(T + K) = \text{ind}(T)$ for $T \in \mathcal{F}(X,Y)$, $K \in \mathcal{K}(X,Y)$. Moreover, if $T_1 \in \mathcal{F}(X,Y)$ and $T_2 \in \mathcal{F}(Y,Z)$, then $T_2T_1 \in \mathcal{F}(X,Z)$ with $\text{ind}(T_2T_1) = \text{ind}(T_2) + \text{ind}(T_1)$.

X, Y, Z are Banach spaces with dual spaces X^*, Y^*, Z^*, and we denoted by $\mathcal{B}(X,Y)$ the Banach space of bounded linear operators from X to Y, by $\mathcal{K}(X,Y)$ the closed subspace of $\mathcal{B}(X,Y)$ of all linear compact operators, and by $\mathcal{F}(X,Y)$ the open subset[36] of $\mathcal{B}(X,Y)$ of all Fredholm operators.

3.5.2 Local bifurcation

When describing the structure of the set of solutions of an equation that depends on a parameter, local bifurcation is the appearance of new solutions when the parameter reaches a critical value. If $\mathbb{X}$, $\mathbb{Y}$ are real or complex Banach spaces[37] and $F : \mathbb{R} \times \mathbb{X} \to \mathbb{Y}$ is a map such that

$$F(\lambda, 0) = 0, \qquad \lambda \in \mathbb{R}, \tag{3.104}$$

then local bifurcation addresses the question: for which λ_0 is there a sequence $(\lambda_n, x_n) \in \mathbb{R} \times \mathbb{X}$ with $x_n \neq 0$ of solutions to $F(\lambda_n, x_n) = 0$, converging to $(\lambda_0, 0) \in \mathbb{R} \times \mathbb{X}$? λ_0 is then called a **bifurcation point**.

If $F \in C^1(\mathbb{R} \times \mathbb{X}, \mathbb{Y})$ and if the bounded linear map $\partial_x F(\lambda_0, 0) \in \mathcal{L}(\mathbb{X}, \mathbb{Y})$ is a homeomorphism from $\mathbb{X}$ to $\mathbb{Y}$, by the implicit function theorem (see [22, Chapter 3]) all solutions to $F(\lambda, x) = 0$ in a neighborhood of $(\lambda_0, 0) \in \mathbb{R} \times \mathbb{X}$ lie on a unique curve $\{(\lambda, x) : x = \varphi(\varepsilon)\}$ with $\varphi : (\lambda_0 - \varepsilon, \lambda_0 + \varepsilon) \to \mathbb{X}$

[34] With the *caveat* that if either X or Y is infinite-dimensional, we must specify that the subspaces X_0 and Y_0 are finite-dimensional.

[35] If X and Y are Hilbert spaces the converse also holds: if $T_1, T_2 \in \mathcal{F}(X,Y)$ have the same index, they belong to the same connected component of $\mathcal{F}(X,Y)$ (see [317, Section 12]).

[36] This is a linear subspace only if $\dim(Y)$ is finite, since open linear subspaces coincide with the entire normed space, and if $\dim(Y) = \infty$, then finite-rank operators are not Fredholm.

[37] We mainly discuss the case of real Banach spaces, but all considerations are equally valid in the case of complex scalars.

of class C^1 for some $\varepsilon > 0$. From (3.104) we conclude that $\varphi(\lambda) = 0$ for all $\lambda \in (\lambda_0 - \varepsilon, \lambda_0 + \varepsilon)$. Consequently, a necessary condition for λ_0 to be a local bifurcation point is that $\partial_x F(\lambda_0, 0) : \mathbb{X} \to \mathbb{Y}$ should not be a homeomorphism. Since $\partial_x F(\lambda_0, 0) \in \mathcal{L}(\mathbb{X}, \mathbb{Y})$ as F is C^1, in view of the open mapping theorem (see [262, Chapter III]) this is equivalent to $\partial_x F(\lambda_0, 0)$ not being a bijection. This condition is, however, not sufficient for local bifurcation, as can be seen by regarding $\mathbb{X} = \mathbb{C}$ as a Banach space over $\mathbb{R}$ (we do not identify $\mathbb{X}$ with $\mathbb{R}^2$ since we want to take advantage of multiplication by complex numbers) and consider the compact operator $F \in C^1(\mathbb{R} \times \mathbb{X}, \mathbb{X})$ given by

$$F(\lambda, z) = z - \lambda z - i\,|z|^2 z, \qquad \lambda \in \mathbb{R}, \quad z \in \mathbb{X}. \tag{3.105}$$

Then $\partial_x F(\lambda, 0) \in \mathcal{L}(\mathbb{X}, \mathbb{X})$ is the multiplication operator $(1 - \lambda)$ which fails to be bijective at $\lambda_0 = 1$. But $\lambda_0 = 1$ is not a bifurcation point since a solution $z \neq 0$ of $F(\lambda, z) = 0$ should satisfy $(1 - \lambda)\,|z|^2 = i\,|z|^4$ after multiplication by $\bar{z}$ and this is impossible for $\lambda \in \mathbb{R}$.

The following result is particularly well suited for our purposes.

Theorem [the Crandall–Rabinowitz local bifurcation theorem]. *Let $\mathbb{X}$, $\mathbb{Y}$ be Banach spaces and let $F \in C^k(\mathbb{R} \times \mathbb{X}, \mathbb{Y})$ with $k \geq 2$ satisfy*

(i) $F(\lambda, 0) = 0$ *for all* $\lambda \in \mathbb{R}$*;*

(ii) $L = \partial_x F(\lambda_0, 0) \in \mathcal{L}(\mathbb{X}, \mathbb{Y})$ *is a Fredholm operator of index zero with* $ker(L)$ *one-dimensional,* $ker(L) = \{s\xi_0 :\ s \in \mathbb{R}\}$ *with* $\xi_0 \in \mathbb{X}$, $\xi_0 \neq 0$*;*

(iii) *the transversality condition* $[\partial^2_{\lambda x} F(\lambda_0, 0)](1, \xi_0) \notin range(L)$ *holds, where we set* $\partial^2_{\lambda x} F(\lambda_0, 0) = \partial_\lambda [\partial_x F(\lambda, 0)]\big|_{\lambda=\lambda_0} \in \mathcal{L}(\mathbb{R}, \mathcal{L}(\mathbb{X}, \mathbb{Y})) = \mathcal{L}(\mathbb{R} \times \mathbb{X}, \mathbb{Y})$.

Then λ_0 is a bifurcation point: there exist $\varepsilon_0 > 0$ and a branch of solutions

$$\{(\lambda, x) = (\Lambda(s), s\,\chi(s)) :\ s \in \mathbb{R},\ |s| < \varepsilon_0\} \subset \mathbb{R} \times \mathbb{X}$$

of $F(\lambda, x) = 0$ with $\Lambda(0) = \lambda_0$, $\chi(0) = \xi_0$, and such that $s \mapsto \Lambda(s) \in \mathbb{R}$, $s \mapsto s\chi(s) \in \mathbb{X}$ are of class C^{k-1} on $(-\varepsilon_0, \varepsilon_0)$. Furthermore, there exists an open set $U_0 \subset \mathbb{R} \times \mathbb{X}$ with $(\lambda_0, 0) \in U_0$ and

$$\{(\lambda, x) \in U_0 :\ F(\lambda, x) = 0,\ x \neq 0\} = \{(\Lambda(s), s\,\chi(s)) :\ 0 < |s| < \varepsilon_0\}.$$

Let us briefly discuss the hypotheses of this theorem; for a proof of the result we refer the reader to [86] or to [38, Chapter 8]. Example (3.105) shows the importance of hypothesis (ii). Concerning the transversality condition (iii), for $F(\lambda, x) = x(\lambda^2 + x^2)$ with $(\lambda, x) \in \mathbb{R} \times \mathbb{R}$ we have $L = \partial_x F(0, 0) = 0 \in \mathcal{L}(\mathbb{R}, \mathbb{R})$ with $ker(L) = \mathbb{R}$ and $range(L) = \{0\}$ of codimension 1, but the transversality condition is not satisfied since $\partial^2_{\lambda x} F(0, 0) = 0 \in \mathcal{L}(\mathbb{R} \times \mathbb{R}, \mathbb{R})$. In this case $\lambda_0 = 0$ is not a bifurcation point since $F(\lambda, x) = 0$ is possible only if $x = 0$.

Notice that while the Crandall–Rabinowitz local bifurcation theorem has wide applicability, it does not exhaust all possible bifurcations: $\lambda_0 = 0$ is a bifurcation point for $F(\lambda, x) = x(\lambda^3 - x^3)$ with $(\lambda, x) \in \mathbb{R} \times \mathbb{R}$ even though the transversality condition fails.

3.5.3 Global bifurcation and degree theory

Topological degree theory, which arises as a generalization of the winding number of a curve in the complex plane, is a fundamental concept in algebraic and differential topology and has become an important tool in nonlinear analysis.

The Brouwer degree

By a *loop* in the complex plane $\mathbb{C}$ we understand a continuous curve $\gamma : [a, b] \subset \mathbb{R} \to \mathbb{C}$ that is closed, that is, $\gamma(a) = \gamma(b)$. The loop is called *simple* (or a *Jordan curve*) if γ is one-to-one on $[a, b)$. For a point $z_0 \in \mathbb{C}$ that does not lie on a simple loop γ, the notion of "the number of times the loop γ winds counterclockwise[38] around z_0" is formalized as the *index* (or *winding number* of γ around z_0)

$$\operatorname{ind}_\gamma(z_0) = \frac{1}{2\pi i}\int_\gamma \frac{dz}{z-z_0} = \frac{\phi(b)-\phi(a)}{2\pi i},$$

where ϕ is any continuous[39] logarithm on $[a,b]$ of the complex function $z \mapsto z - z_0$. Notice that if γ is a simple loop, then the complement $\mathbb{C}\setminus J$ has by the Jordan–Schönflies theorem (cf. [39, Chapter IV]) two components, one of which is bounded and homeomorphic to the open unit disk, called the *inside* of J and denoted $\mathcal{I}(J)$. By the argument principle (cf. [39, Chapter IX]) the number of roots of $f(z) = z_0$ in $\mathcal{I}(J)$, of a continuous function $f : \overline{\mathcal{I}(J)} \to \mathbb{C}$ which is analytic in $\mathcal{I}(J)$ and with $f(z) \neq z_0$ on J, counted according to multiplicity, is finite and equals $\operatorname{ind}_{f\circ\gamma}(z_0)$. If f is C^1 on J, this can be expressed by a simple change of variables as $\frac{1}{2\pi i}\int_\gamma \frac{f'(z)}{f(z)-z_0}\,dz$ and can be interpreted as the change in the argument of f as z describes the curve γ. If the function f is very complicated, we can look for a simpler function g such that the curve $f\circ\gamma$ can be continuously deformed into $g\circ\gamma$: if there is a continuous map $H : [0,1]\times[a,b] \to \mathbb{C}\setminus\{z_0\}$ such that $H(0,s) = f(\gamma(s))$ and $H(1,z) = g(\gamma(s))$ for $s \in [a,b]$, then $\operatorname{ind}_{f\circ\gamma}(z_0) = \operatorname{ind}_{g\circ\gamma}(z_0)$ (cf. [39, Chapter IV]). Summarizing these considerations, given the inside U of a Jordan loop γ in the plane, we have defined a degree $\deg(f,U,z_0) = \operatorname{ind}_{f\circ\gamma}(z_0)$ which counts the number of solutions of $f(z) = z_0$ in U whenever $f : \overline{U} \to \mathbb{C}$ is continuous and analytic in U. The invariance of this degree with respect to certain deformations of f facilitates its computation.

In trying to extend this approach outside the realm of analytic functions, one runs into several problems. First, given an open bounded set $\Omega \subset \mathbb{R}^n$ and a mapping $F : \Omega \to \mathbb{R}^n$, the number of solutions to $F(x) = 0$ in Ω is generally not well defined: there might be infinitely many solutions[40] and, even if there are finitely many solutions, their number is unstable with respect to small perturbations. For example, consider $F_\varepsilon : (-1,1) \to \mathbb{R}$ given by $x \mapsto x^2 - \varepsilon$. Then F_ε has no zeros for $\varepsilon < 0$, one zero for $\varepsilon = 0$, two zeros for $\varepsilon \in (0,1)$, and no zeros for $\varepsilon \geq 1$, so that

- the number of zeros can change if a zero reaches the boundary;
- zeros where F_x is invertible are stable under small perturbations[41] but the number of zeros can change if two zeros with opposite signs of $\det F_x$ run into each other.

This suggests that, rather than counting the number of zeros, the degree should perhaps just indicate whether there are zeros. For a given open and bounded set $\Omega \subset \mathbb{R}^n$ we say that a continuous map[42]

---

[38]With orientation taken into account, so that encirclings in opposite directions are counted with opposite signs.

[39]Recall that, in general, to define integrals of continuous functions along a continuous curve γ as Riemann–Stieltjes integrals, the curve should be rectifiable: the supremum of the lengths $\sum_{k=1}^n |\gamma(t_k)-\gamma(t_{k-1})|$ of the broken line determined by the (arbitrary but finitely many) points $a \leq t_0 \leq \cdots \leq t_n \leq b$ should be finite—the least upper bound being the length of γ; a curve is rectifiable if and only if both projections on the axes are functions of bounded variation (cf. [112]). However, if the continuous curve γ, not necessarily closed, lies in the domain of analyticity of the holomorphic function f, then Cauchy's theorem enables one to define $\int_\gamma f(z)dz$ by approximating γ uniformly with piecewise smooth curves (that is, continuous curves $\gamma : [a,b] \to \mathbb{C}$ such that there exist finitely many points $a = t_0 \leq \cdots \leq t_n = b$ with γ smooth on $[t_{k-1},t_k]$ for $k = 1,\ldots,n$), whether γ is rectifiable or not (cf. [39]).

[40]As is the case for $F(x,y) = (0, 1-4(x^2+y^2))$ in the unit disk $\Omega = \{(x,y) \in \mathbb{R}^2 : x^2+y^2 < 1\}$.

[41]If F presents a C^1-dependence on (λ,x), where $\lambda \in \mathbb{R}$ is a parameter, and $\det F_x(\lambda_0,x_0) \neq 0$ at a point (λ_0,x_0) where $F(\lambda_0,x_0) = 0$, the implicit function theorem ensures that near (λ_0,x_0) the solution set of $F(\lambda,x) = 0$ is given by a C^1-curve $(\lambda, x(\lambda))$ with $x(\lambda_0) = x_0$.

[42]It suffices to deal with continuous functions $f : \overline{\Omega} \to \mathbb{R}^n$ that have no zeros on $\partial\Omega$. Indeed, the Tietze extension theorem (cf. [96]) yields the existence of a continuous map $\tilde{f} : \mathbb{R}^n \to \mathbb{R}^n$ such that its restriction to $\overline{\Omega}$ is precisely f. We can define $\deg(f,\Omega) := \deg(\tilde{f},\Omega)$, the definition of $\deg(f,\Omega)$ being independent of the choice of the extension $\tilde{f}$ of f (cf. [188]).

$f : \mathbb{R}^n \to \mathbb{R}^n$ is Ω-*admissible* if $f(x) \neq 0$ for all $x \in \partial\Omega$, and we denote by $\mathcal{A}(\Omega)$ the set of all Ω-admissible functions. The *Brouwer degree* is an integer-valued function $\deg : \mathcal{A}(\Omega) \to \mathbb{Z}$ satisfying the following properties.[43]

(P1) Normalization. *If* $Id : \mathbb{R}^n \to \mathbb{R}^n$ *is the identity map, then* $\deg(Id - x_0, \Omega) = 1$ *if* $x_0 \in \Omega$ *and* $\deg(Id - x_0, \Omega) = 0$ *if* $x_0 \notin \Omega$. ($\deg(Id - x_0, \Omega)$ is not defined for $x_0 \in \partial\Omega$.)

(P2) Additivity. *If* Ω_1, Ω_2 *are two open nonempty disjoint subsets of* Ω *such that* $f^{-1}(0) \cap \Omega \subset \Omega_1 \cup \Omega_2$, *then* $\deg(f, \Omega) = \deg(f, \Omega_1) + \deg(f, \Omega_2)$ *for all* $f \in \mathcal{A}(\Omega)$. (It is not necessary to have $\Omega_1 \cup \Omega_2 = \Omega$; only the location of the zeros matters.)

(P3) Homotopy invariance. *For* $h : [0,1] \times \mathbb{R}^n \to \mathbb{R}^n$ *continuous with* $h(t,x) \neq 0$ *for* $(t,x) \in [0,1] \times \partial\Omega$, $\deg(h(t,\cdot), \Omega)$ *is independent of* t.

(P2) says roughly that $\deg(f, \Omega)$ should yield information about the location of zeros in the sense that if f has no zeros in $\Omega \setminus (\Omega_1 \cup \Omega_2)$, then the number of zeros of f in Ω is the sum of the number of zeros of f in Ω_1 and Ω_2; this is, however, not an exact statement as only for special classes of functions is the degree equal to the number of zeros. Property (P3) reflects the feature that $\deg(f, \Omega)$ is preserved if f is continuously deformed[44] so that at each stage of the deformation we do not have zeros on $\partial\Omega$. This facilitates the calculation of $\deg(f, \Omega)$ for complicated functions. Since the characteristic of tools that are developed by means of algebraic topology is that they do not distinguish between homotopic objects (cf. [35]), it is not surprising that it is possible to construct the Brouwer degree by techniques of algebraic topology. Another approach is to show that (P1)–(P3) uniquely define the Brouwer degree (see [188]). We now briefly sketch the main steps of this approach. First, (P1)–(P3) show that a nontrivial degree is a sufficient condition[45] for $f \in \mathcal{A}(\Omega)$ to have a zero in Ω.

(P4) Existence. *If* $f \in \mathcal{A}(\Omega)$ *is such that* $\deg(f, \Omega) \neq 0$, *then* f *has at least a zero in* Ω.

Proof. Indeed, if $f \in \mathcal{A}(\Omega)$ is such that $f^{-1}(0) \cap \Omega = \emptyset$, let $\Omega_1, \Omega_2, \Omega_3, \Omega_4$ be disjoint nonempty open subsets of Ω. By (P2) we have

$$\begin{aligned} \deg(f, \Omega) &= \deg(f, \Omega_1) + \deg(f, \Omega_2), \quad \deg(f, \Omega) = \deg(f, \Omega_3) + \deg(f, \Omega_4), \\ \deg(f, \Omega) &= \deg(f, \Omega_1 \cup \Omega_2) + \deg(f, \Omega_3 \cup \Omega_4), \\ &= \deg(f, \Omega_1) + \deg(f, \Omega_2) + \deg(f, \Omega_3) + \deg(f, \Omega_4). \end{aligned}$$

These relations lead to $\deg(f, \Omega) = 2\deg(f, \Omega)$, i.e., $\deg(f, \Omega) = 0$. □

Further degree properties that follow directly from (P1)–(P3) are (see [188]) as follows.

(P5) Excision. *For every* $f \in \mathcal{A}(\Omega)$, $\deg(f, \Omega) = \deg(f, \Omega_1)$ *whenever* Ω_1 *is an open subset of* Ω *with* $f^{-1}(0) \cap \Omega \subset \Omega_1$ *and such that* $\Omega \setminus \Omega_1$ *has nonempty interior.*[46]

(P6) Continuity. *If* $f, g \in \mathcal{A}(\Omega)$ *are such that* $\sup_{x \in \overline{\Omega}}\{|f(x) - g(x)|\} < \min_{x \in \partial\Omega}\{|f(x)|\}$, *then* $\deg(f, \Omega) = \deg(g, \Omega)$.

---

[43]Sometimes the degree is defined as $\deg(f, \Omega, y)$ for $y \notin f(\partial\Omega)$ but the required translation invariance $\deg(f, \Omega, y) = \deg(f - y, \Omega)$ enables one to view it as dependent just upon f and Ω.

[44]Two continuous functions $f, g : \mathbb{R}^n \to \mathbb{R}^n$ are *homotopic* if there is a continuous mapping $h : [0,1] \times \mathbb{R}^n \to \mathbb{R}^n$ such that $h(0,\cdot) = f$ while $h(1,\cdot) = g$. This defines an equivalence relation. If we add the condition $h(t,x) \neq 0$ for $(t,x) \in [0,1] \times \partial\Omega$, we obtain an equivalence relation $\equiv_\Omega$ in the class of admissible functions $\mathcal{A}(\Omega)$. Property (P3) means that the degree is constant in each equivalence class of the relation $\equiv_\Omega$.

[45]We shall see that $\deg(f, (-2,2)) = 0$ for $f(x) = x^2 - 1$, so that the condition is not necessary.

[46]This assumption is made to avoid pathological cases like $\Omega_1 = \Omega \setminus \{x_0\}$ with $x_0 \in \Omega$.

The next step is the calculation of the degree for linear maps and for maps satisfying certain regularity conditions:

- If $f : \mathbb{R}^n \to \mathbb{R}^n$ is a linear bijection, then $\deg(f,\Omega) = \operatorname{sign}(\det f)$.
- If $f \in \mathcal{A}(\Omega)$ is of class C^1 and zero is a regular value[47] of f, then[48]

$$\deg(f,\Omega) = \sum_{x \in f^{-1}(0)\cap\Omega} \operatorname{sign}[\det df(x)]. \tag{3.106}$$

It is remarkable that formula (3.106) can be extended to all $f \in \mathcal{A}(\Omega)$. A key element in this extension is the fact that by Sard's theorem (see [188]) the set of critical values of f is of Lebesgue measure zero. In particular, every neighborhood of a critical value contains a point that is a regular value. Some nontrivial considerations show then that if $f \in \mathcal{A}(\Omega)$ and $\varepsilon = \inf_{x\in\partial\Omega}\{|f(x)|\} > 0$, then we can find $\tilde{f} \in C^1(\Omega;\mathbb{R}^n)$ with $|f(x) - \tilde{f}(x)| < \varepsilon$ for all $x \in \Omega$, and such that zero is a regular value of $\tilde{f}$. By (P6) we have $\deg(f,\Omega) = \deg(\tilde{f},\Omega)$, and for the computation of $\deg(\tilde{f},\Omega)$ one can use (3.106). This definition is independent of the particular representative $\tilde{f}$ (see, e.g., [188]).

We point out two further important properties of the Brouwer degree (cf. [188]).

(P7) Product. *Let $\Omega_1 \subset \mathbb{R}^n$, $\Omega_2 \subset \mathbb{R}^m$ be open bounded sets. If $f_1 \in \mathcal{A}(\Omega_1)$ and $f_2 \in \mathcal{A}(\Omega_2)$, then $f = (f_1, f_2) \in \mathcal{A}(\Omega)$ for $\Omega = \Omega_1 \times \Omega_2$ and* $\deg(f,\Omega) = \deg(f_1,\Omega_1)\cdot\deg(f_2,\Omega_2)$.

(P8) Reduction of dimension. *For $n > m$ natural numbers, write $\mathbb{R}^n = \mathbb{R}^m \times \mathbb{R}^{n-m}$ and let $\pi_{n-m} : \mathbb{R}^n \to \mathbb{R}^{n-m}$ be the projection onto the last $n-m$ coordinates. If $\Omega \subset \mathbb{R}^n$ is an open bounded set and $f \in \mathcal{A}(\Omega)$ satisfies $\pi_{n-m} f = \pi_{n-m}$, setting $\Omega_m = \Omega \cap (\mathbb{R}^m \times \{0\})$ and identifying $\mathbb{R}^m \times \{0\}$ with $\mathbb{R}^m$, we have* $\deg(f,\Omega) = \deg(f_m,\Omega_m)$, *where $f_m : \Omega_m \to \mathbb{R}^m$ is the restriction of f to Ω_m.*

The Leray–Schauder degree

There is no general analogue of the Brouwer degree on infinite-dimensional Banach spaces. However, Leray and Schauder [201] found that for compact perturbations of the identity on infinite-dimensional Banach spaces a degree, having all of the properties of the Brouwer degree, exists.

Let $U \subset X$ be a subset of the Banach space X. An operator $f : U \subset X \to Y$, with Y a Banach space, is called *finite-dimensional* if its range is finite-dimensional, and is called *compact* if it is continuous and maps bounded sets into relatively compact ones. In Section 3.5.1 we pointed out that, while the uniform limit of linear operators of finite rank is necessarily a compact linear operator, the converse is generally not true. However, compact (nonlinear) operators defined on bounded sets can be uniformly approximated by compact, finite-dimensional (nonlinear) operators. The reason for this is that although the concept of a normed linear space is quite general, the compact subsets within a normed linear space X are "almost" finite-dimensional: if $K \subset X$ is compact, given $\varepsilon > 0$, there exist a finite-dimensional subspace $X_\varepsilon \subset X$ and a continuous (nonlinear) map $P_\varepsilon : K \to X_\varepsilon$ such that $|P_\varepsilon(x) - x| \le \varepsilon$ for all $x \in K$. Indeed, if the points $\{x_k\}_{1\le k\le n}$ in K are such that the balls centered at these points and of radius ε cover K, we may set $P_\varepsilon(x) = \sum_{k=1}^{n} \phi_k(x) x_i$, where $\{\phi_k\}_{1\le k\le n}$ is a partition of unity for K (see [35, Chapter 3]). Now, if $U \subset X$ is bounded

---

[47]A point $y \in \mathbb{R}^n$ is a *regular value* of a C^1-map $f : \Omega \to \mathbb{R}^n$ if at every $x \in f^{-1}(y)$ we have $\det df(x) \neq 0$. All other values of f are called *critical values*, a point $x \in \overline{\Omega}$ where $\det df(x) = 0$ being a *critical point* of f.

[48]Notice that $f : \overline{\Omega} \to \mathbb{R}^n$ is continuous with $0 \notin f(\partial\Omega)$, so that the compactness of $\partial\Omega$ ensures $\inf_{x\in\partial\Omega} = \varepsilon > 0$. The set $\{x \in \Omega : |f(x)| \le \frac{\varepsilon}{2}\}$ is then a compact subset of Ω that contains the set $f^{-1}(0) \cap \Omega$. Since by the inverse function theorem f is invertible in a small neighborhood of any point $x \in f^{-1}(0) \cap \Omega$, there are at most finitely many points in the set $f^{-1}(0) \cap \Omega$. Thus the sum in (3.106) is well defined, provided we set it at zero if we sum over an empty set.

and $f : U \to X$ is compact, let $K = \overline{f(U)}$ and choose $P_{1/n}$ according to the previously described scheme. Then $f_n = P_{1/n} \circ f : U \to X$ are finite rank operators that approximate f uniformly on U.

If $U \subset X$ is a bounded and open subset of the Banach space X, one can easily check that compact perturbations of the identity, of the form $Id - f$ with $f : \overline{U} \to X$ compact, are *proper* (that is, inverse images of compact sets are compact) and map closed subsets of $\overline{U}$ to closed sets in X. Consequently, if $Id - f \neq 0$ on ∂U, we must have $\inf_{x \in \partial U} \{|x - f(x)|\} = \varepsilon_0 > 0$ since $(Id - f)(\partial U)$ is closed. Choosing now for $\varepsilon \in (0, \varepsilon_0)$ a finite-rank operator f_ε and a finite-dimensional subspace $X_\varepsilon \subset X$ such that $f_\varepsilon(\overline{U}) \subset X_\varepsilon$ and $|f(x) - f_\varepsilon(x)| < \varepsilon$ for all $x \in \overline{U}$, if we set $U_\varepsilon = U \cap X_\varepsilon$ and we denote by Id_ε the identity on X_ε, then $(Id_\varepsilon - f_\varepsilon) \in \mathcal{A}(U_\varepsilon)$ since $Id_\varepsilon - f_\varepsilon$ has no zeros on $\partial U_\varepsilon = \partial U \cap X_\varepsilon$. The *Leray–Schauder degree* $\deg(Id - f, U)$ of $Id - f$ on $\overline{U}$ is defined as the Brouwer degree $\deg(Id_\varepsilon - f_\varepsilon, U_\varepsilon)$. This definition is independent of the choices made: the compact set K containing $f(\overline{U})$, the positive number ε, the finite-dimensional subspace X_ε of X, and the map f_ε (cf. [35, Chapter 10]). We now list the important properties of the Leray–Schauder degree for an open bounded subset U of a Banach space X and a compact map $f : \overline{U} \to X$ with $f(x) \neq x$ for $x \in \partial U$. These are consequences of the corresponding properties of the Brouwer degree (cf. [35, Chapter 11]).

(I) Normalization. *For the inclusion* $i : \overline{U} \to X$ *we have* $\deg(i, U) = 1$ *if* $0 \in U$ *and* $\deg(i, U) = 0$ *if* $0 \notin \overline{U}$, *with* $\deg(i, U)$ *undefined if* $0 \in \partial U$.

(II) Existence. *If* $\deg(Id - f, U) \neq 0$, *then* f *has a fixed point*[49] *in* U.

(III) Additivity. *If* U_1, U_2 *are open subsets of* U *such that* $f(x) \neq x$ *for all* $x \in U \setminus (U_1 \cup U_2)$, *then* $\deg(Id - f, U) = \deg(Id - f_1, U_1) + \deg(Id - f_2, U_2)$, *where* f_k *denotes the restriction of* f *to* U_k. (In particular, $\deg(Id - f, \emptyset) = 0$ and the **excision property** holds: if $W \subset U$ is open and contains all the fixed points of f in U, then $\deg(Id - f, U) = \deg(Id - f_W, W)$, where f_W denotes the restriction of f to $\overline{W}$.)

(IV) Continuity. *If* $g : \overline{U} \to X$ *is compact without fixed points on* ∂U, *and* $\sup_{x \in \overline{U}} \{|f(x) - g(x)|\} < \min_{x \in \partial U} \{|f(x) - x|\}$, *then* $\deg(Id - f, U) = \deg(Id - g, U)$.

(V) Product. *If* V *is an open bounded subset of the Banach space* Y *and* $g : \overline{V} \to Y$ *is compact without fixed points on* ∂V, *then* $F = (f, g) : \overline{U} \times \overline{V} \to X \times Y$ *is compact without fixed points on* $\partial(U \times V)$ *and* $\deg(Id_{X \times Y} - F, U \times V) = \deg(Id_X - f, U) \cdot \deg(Id_Y - g, V)$.

(VI) Factor. *If* $X = X_1 \times X_2$ *is a product of two Banach spaces and* $f : X \to X$ *is continuous with* $f : \overline{U} \to X_1 \times \{0_2\}$ *compact without fixed points on the boundary of the open and bounded set* $U \subset X$, *then* $\deg(Id - f, U) = \deg(Id_1 - f_1, U_1)$, *where* Id_1 *is the identity map of* X_1, $U_1 = \{x_1 \in X_1 : (x_1, 0_2) \in U\}$ *is an open and bounded set in* X_1, *and the restriction of* f *to* X_1 *gives rise to the map* $f_1 : X_1 \to X_1$ *by setting* $f(x_1, 0_2) = (f_1(x_1), 0_2)$.

The last property expresses the feature that the Leray–Schauder degree $\deg(Id - f, U)$ is constant on connected components of both arguments.

(VII) Generalized homotopy. *For* $\lambda_2 > \lambda_1$ *real constants and* $W \subset [\lambda_1, \lambda_2] \times X$ *open and bounded, consider the open sets* $W_\lambda = \{x \in X : (\lambda, x) \in W\} \subset X$. *If* $h : \overline{W} \to X$ *is compact such that* $h_\lambda(x) \neq x$ *for all* $x \in \partial W_\lambda$ *and all* $\lambda \in (\lambda_1, \lambda_2)$, *then* $\deg(Id_X - h_\lambda, W_\lambda)$ *is independent of* λ, *where* $h_\lambda : \overline{W_\lambda} \to X$ *is defined by* $h_\lambda(x) = h(\lambda, x)$.

For the calculation of the Leray–Schauder degree, let us note the following key results.

- Recall (cf. [262]) that the elements $\lambda \neq 0$ of the spectrum $\sigma(T) \subset \mathbb{R}$ of a compact linear operator $T : X \to X$ (on a real Banach space X) are all eigenvalues of finite multiplicity, with zero as the only possible accumulation point. If we set $B_R = \{x \in X : |x| < R\}$ for

[49] Zeros of $Id - f$ correspond to fixed points of f. Notice that the example provided in the discussion of the Brouwer degree shows that the presence of fixed points of f in U does not necessarily mean that $\deg(Id - f, U) \neq 0$.

$R > 0$, for $\mu \neq 0$ such that $\frac{1}{\mu} \notin \sigma(T)$, the operator $Id - \mu T$ has no fixed point on ∂B_R and it turns out (cf. [35, Chapter 16]) that

$$\deg(Id - \mu T, B_R) = (-1)^{\beta(\mu)}, \qquad R > 0, \tag{3.107}$$

where $\beta(\mu)$ is the number of eigenvalues (counted by multiplicity) of T that lie in the set $\{\lambda \in \sigma(T) : \lambda\mu > 0 \text{ and } |\lambda| > \frac{1}{|\mu|}\}$.

- To extend the above to nonlinear compact operators, we first notice that if $f : X \to X$ is compact and differentiable at $0 \in X$, then its Fréchet derivative $f'(0)$ is a compact linear operator[50] $T_0 : X \to X$ (cf. [35, Chapter 17]). Then, if $f(0) = 0$ and if $1 \notin \sigma(T_0)$, then for $R > 0$ sufficiently small we have

$$\deg(Id - f, B_R) = \deg(Id - T_0, B_R). \tag{3.108}$$

Global continuum

To illustrate the usefulness of degree theory we sketch its applicability to the extension of the local bifurcation curve C_{loc} provided by the Crandall–Rabinowitz theorem to a maximal connected set of solutions. We place ourselves in the setting of the discussion of local bifurcation (Section 3.5.2), where we assume that $X = Y$ and $F \in C^k(\mathbb{R} \times X, X)$ is of the form $F = Id_X - f$ with $f : \mathbb{R} \times X \to X$ compact such that[51] $f_x(\lambda, 0) = \lambda T$ for some compact linear operator $T : X \to X$. Under these assumptions let S_0 be the maximal connected component of the solution set

$$S = \{(\lambda, x) \in \mathbb{R} \times X : \ F(\lambda, x) = 0\}$$

that contains the bifurcation point $(\lambda_0, 0)$. The closed set $S_0 \subset \mathbb{R} \times X$ contains the local bifurcation curve C_{loc}, and the global bifurcation theorem of Rabinowitz [257] ensures that one of the following alternatives holds (cf. Figure 3.10):

1. S_0 is unbounded in $\mathbb{R} \times X$;
2. S_0 contains another point $(\lambda, 0)$ with $\lambda \neq \lambda_0$.

Proof. To see why this is so, assume the contrary. Then S_0 is bounded in $\mathbb{R} \times X$ and meets $\mathbb{R} \times \{0\}$ only at $(\lambda_0, 0)$. We first claim that S_0 is compact. Indeed, by hypothesis S_0 is bounded so that there exists an interval $[\lambda_1, \lambda_2] \subset \mathbb{R}$ such that $(\lambda, x) \in S_0$ implies $\lambda \in [\lambda_1, \lambda_2]$. Since f is compact, $\overline{f(S_0)}$ must be compact in X. But this set is just the projection of S_0 onto X so that S_0 is a closed subset of the compact set $[\lambda_1, \lambda_2] \times \overline{f(S_0)}$, and thus compact. Now by local bifurcation there is a small neighborhood U_0 of $(\lambda_0, 0)$ in $\mathbb{R} \times X$ in which S intersects $\mathbb{R} \times \{0\}$ only at $(\lambda_0, 0)$. Together with the fact that S_0 intersects $\mathbb{R} \times \{0\}$ only at $(\lambda_0, 0)$, this enables us to find $\varepsilon > 0$ such that the closed set $(\mathbb{R} \setminus (\lambda_0 - \varepsilon, \lambda_0 + \varepsilon)) \times \{0\}$ is disjoint from the compact set S_0, and the only $\lambda \in (\lambda_0 - \varepsilon, \lambda_0 + \varepsilon)$ with $\frac{1}{\lambda} \in \sigma(T)$ is $\lambda = \lambda_0$. The distance between these sets is then a positive number. Let $\delta_0 > 0$ be half that distance and denote the open δ_0-neighborhood of S_0 in $\mathbb{R} \times X$ by U_{δ_0}. If we set $K = \overline{U_{\delta_0}} \cap S$, then K is compact by the same argument we used for S_0. Clearly $S_0 \cap (\partial U_{\delta_0} \cap S) = \emptyset$ but by a topological separation theorem[52] we may write $K = K_1 \cup K_2$, where K_1 and K_2 are disjoint closed

[50] The example $f(x) = x^2$ for $x \in L^\infty[-1, 1]$ shows that the converse does not hold.

[51] In particular, we must have $\lambda_0 \neq 0$ since otherwise no bifurcation can occur at $(\lambda_0, 0)$.

[52] Stating that if P and Q are disjoint closed subsets of a compact space K such that there is no closed connected subset of K that intersects both P and Q, then there exist disjoint closed subsets C_P and C_Q such that $P \subset C_P$, $Q \subset C_Q$, and $C_P \cup C_Q = K$ (cf. [35, Chapter 14]). We can apply this result with $P = S_0$, $Q = (\partial U_{\delta_0} \cap S)$, and $K = \overline{U_{\delta_0}} \cap S$, noticing that the existence of a closed connected subset S_K of K intersecting both P and Q would enable us to extend S_0 to $S_0 \cup S_K$ (the union of two connected sets with nonempty intersection being connected), contradicting the maximality of S_0.

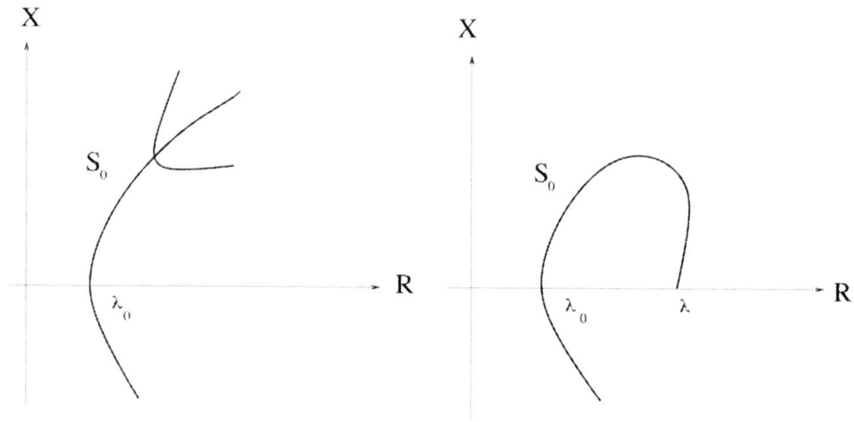

Figure 3.10. *A sketch of the two alternatives for the global continuum. Despite the drawing, it is generally not known that the continuum is a union of curves—except near the bifurcation point* $(\lambda_0, 0)$.

subsets of K with $S_0 \subset K_1$ and $\partial U_{\delta_0} \cap S \subset K_2$. If $\partial U_{\delta_0} \cap S \neq \emptyset$, then the distance $\delta \leq 2\delta_0$ between the compact subsets K_1 and K_2 of $\mathbb{R} \times X$ is positive, while if $\partial U_{\delta_0} \cap S = \emptyset$, we simply set $\delta = \delta_0/2$. Consider the open bounded set $O \subset \mathbb{R} \times X$ given by the connected component of the intersection of U_{δ_0} with the $\delta/2$-neighborhood of K_1 in $\mathbb{R} \times X$ that contains S_0. Notice that if $(\lambda, 0) \in O$, then $|\lambda - \lambda_0| < \varepsilon$, while $\partial O \cap S = \emptyset$ since $\partial O \cap S \subset K$. Moreover, since $(\lambda_0, 0) \in O \cap (\mathbb{R} \times \{0\})$, the intersection $O \cap (\mathbb{R} \times \{0\})$ is a connected subset of the real line and therefore (cf. [232]) it consists of an open interval containing λ_0. See the top left corner in Figure 3.11.

For each real number λ, define the open bounded sets $O_\lambda = \{x \in X : (\lambda, x) \in O\}$ and notice that since O is bounded, $O_\lambda = \emptyset$ for λ sufficiently far from λ_0. If $(\lambda^*, 0) \notin \overline{O}$, we can choose $\rho(\lambda^*) > 0$ such that $|(\lambda^*, 0) - (\lambda, x)| > \rho(\lambda^*)$ for all $(\lambda, x) \in \overline{O}$, while for $(\lambda^*, 0) \in \overline{O}$ with $\lambda^* \neq \lambda_0$, since $(\lambda^*, 0) \notin S$, we can find $\rho(\lambda^*) > 0$ such that $|(\lambda^*, 0) - (\lambda, x)| > \rho(\lambda^*)$ for all $(\lambda, x) \in S$. By diminishing $\rho(\lambda) > 0$ we can ensure in view of (3.108) that

$$\deg(Id_X - f_\lambda, B_{\rho(\lambda)}) = \deg(Id_X - \lambda T, B_{\rho(\lambda)}) \quad \text{for} \quad \lambda \neq \lambda_0 \quad \text{with} \quad (\lambda, 0) \in O, \tag{3.109}$$

where $f_\lambda : X \to X$ is defined by $f_\lambda(x) = f(\lambda, x)$ and B_R stands for the open unit ball in X centered at 0 and of radius R. Furthermore, we can choose $\rho(\lambda)$ so that $\rho : \mathbb{R} \setminus \{\lambda_0\} \to (0, \infty)$ is continuous. For each $\lambda \neq \lambda_0$ we consider the open and bounded set $O_\lambda \setminus \overline{B_{\rho(\lambda)}} \subset X$. The restriction of f_λ to it is compact and has no fixed points on the boundary, so that $\deg(Id_X - f_\lambda, O_\lambda \setminus \overline{B_{\rho(\lambda)}})$ is well defined. We claim that

$$\deg(Id_X - f_\lambda, O_\lambda \setminus \overline{B_{\rho(\lambda)}}) = 0, \qquad \lambda \neq \lambda_0. \tag{3.110}$$

By the empty set property (III), this claim is true for λ sufficiently far from λ_0 since then $O_\lambda = \emptyset$. Now let $\lambda_1 > \lambda_0$ be such that $(\lambda_1, 0) \notin O$ and choose $\lambda_2 > \lambda_1$ such that $O(\lambda_2) = \emptyset$ (see the top right corner of Figure 3.11). If $\Lambda = [\lambda_1, \lambda_2]$, let $W = O \cap (\Lambda \times X)$. Since $O_\lambda \setminus \overline{B_{\rho(\lambda)}} = O_\lambda$ for $\lambda \in \Lambda$ as $(\lambda, 0) \notin O$ for $\lambda > \lambda_1$, the generalized homotopy property (VII) ensures that $\deg(Id_X - f_\lambda, O_\lambda \setminus \overline{B_{\rho(\lambda)}})$ is independent of the choice of $\lambda \in \Lambda$, and its evaluation at λ_2 proves (3.110) for $\lambda = \lambda_1$. For $\lambda_1 < \lambda_0$ such that $(\lambda, 0) \notin O$ we choose now $\lambda_2 < \lambda_1$ sufficiently far and we repeat the procedure, proving thus (3.110) for all λ such that $(\lambda, 0) \notin O$.

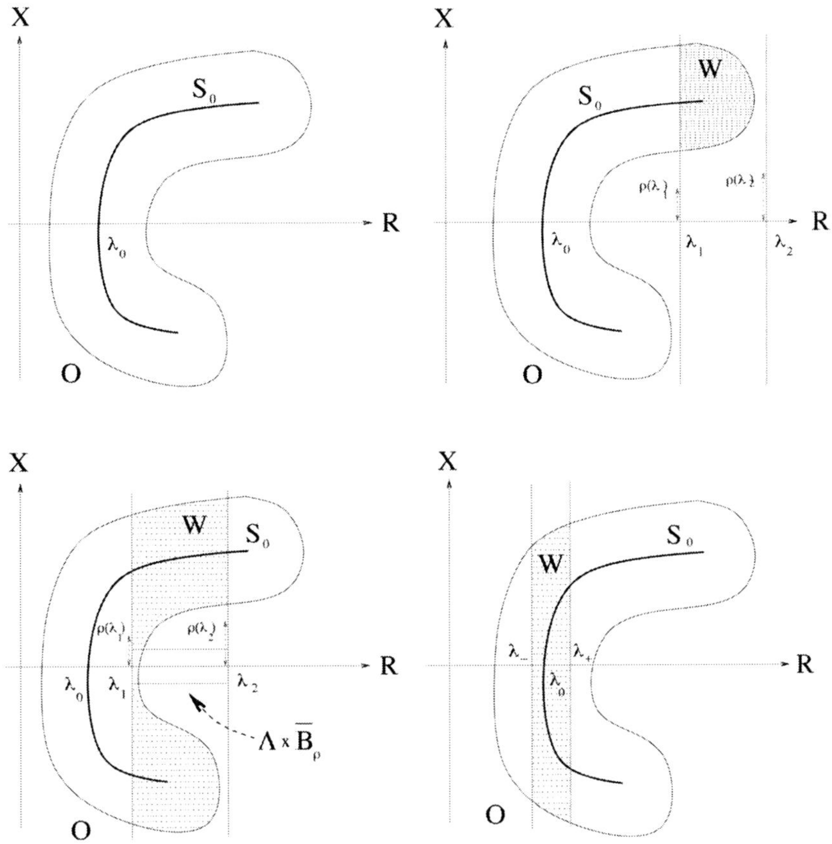

Figure 3.11. *Illustration of the global bifurcation approach by means of degree theory.*

If $\lambda_1 \neq \lambda_0$ with $(\lambda_1, 0) \in \overline{O}$, choose λ_2 such that λ_0 does not belong to the closed interval Λ with endpoints λ_1 and λ_2, with $(\lambda_2, 0) \notin O$ and such that no bifurcation point lies on Λ. Then there is $\rho \in (0, \frac{1}{2} \inf_{\lambda \in \Lambda}\{\rho(\lambda)\})$ such that $(\Lambda \times \overline{B_\rho}) \cap S = \emptyset$. Using the generalized homotopy property (VII) with $W = O \cap (\Lambda \times (X \setminus \overline{B_\rho}))$ we infer that $\deg(Id_X - f_{\lambda_1}, O_{\lambda_1} \setminus \overline{B_\rho}) = 0$ since, as above, $\deg(Id_X - f_{\lambda_2}, O_{\lambda_2} \setminus \overline{B_\rho}) = 0$; see the bottom left corner of Figure 3.11. Since $\rho(\lambda) > \rho$ for $\lambda \in \Lambda$ and there are no zeros of $(Id_X - f_{\lambda_1})$ in $\overline{B_{\rho(\lambda_1)}} \setminus B_\rho \subset \overline{B_{\rho(\lambda_1)}}$, the excision property (III) yields (3.110) for $\lambda = \lambda_1$. This completes the proof of (3.110).

Choose now $\lambda_- < \lambda_0 < \lambda_+$ close enough to λ_0 so that $0 \notin \Lambda = [\lambda_-, \lambda_+]$, $(\lambda, 0) \in O$ for all $\lambda \in \Lambda$ (so that $\lambda = \lambda_0$ is the only $\lambda \in \Lambda$ such that $\frac{1}{\lambda}$ is in the spectrum of T); see the bottom right of Figure 3.11. Using the generalized homotopy property (VII) with $W = O \cap (\Lambda \times X)$ we deduce that

$$\deg(Id_X - f_\lambda, O_\lambda) \quad \text{is independent of} \quad \lambda \in \Lambda.$$

Using now the additive property (III), we get for $\lambda \in [\lambda_-, \lambda_0)$ that

$$\deg(Id_X - f_\lambda, O_\lambda) = \deg(Id_X - f_\lambda, O_\lambda \setminus \overline{B_{\rho(\lambda)}}) + \deg(Id_X - f_\lambda, B_{\rho(\lambda)}) = \deg(Id_X - f_\lambda, B_{\rho(\lambda)})$$

in view of (3.110), and similarly

$$\deg(Id_X - f_\lambda, O_\lambda) = \deg(Id_X - f_\lambda, B_{\rho(\lambda)})$$

for $\lambda \in (\lambda_0, \lambda_+]$. The last three displayed relations yield that $\deg(Id_X - f_\lambda, B_{\rho(\lambda)})$ is independent of $\lambda \in \Lambda \setminus \{\lambda_0\}$. However, by (3.108) and (3.107) we have

$$\deg(Id_X - f_\lambda, B_{\rho(\lambda)}) = \deg(Id_X - \lambda T, B_{\rho(\lambda)}) = (-1)^{\beta(\lambda)}$$

for $\lambda \in \Lambda \setminus \{\lambda_0\}$. This is impossible since $\beta(\lambda)$ is constant on $[\lambda_-, \lambda_0)$ and on $(\lambda_0, \lambda_+]$, but its value jumps by one at $\lambda = \lambda_0$. The obtained contradiction proves that one of the two alternatives must hold. □

Generalizations of the Leray–Schauder degree

Unfortunately, most problems for water waves with vorticity do not appear to be equivalent to finding zeros of a compact perturbation $F(\lambda, x) = x - f(\lambda, x)$ of the identity Id, with $F(\lambda, 0) = 0$ and linear part $Id - \lambda T$ at $x = 0$ for some compact operator T. It is thus of interest to extend Rabinowitz's result to a larger class of mappings. Kielhöfer [174] succeeded in defining a degree for proper Fredholm operators of index zero, which permits the investigation of a more general class of nonlinear equations with linear boundary conditions than those accessible by the classical Leray–Schauder degree. Healey and Simpson [140] subsequently improved this approach to accomodate nonlinear boundary conditions. We briefly outline their methodology. Let $\mathcal{X}$, Y_1, and Y_2 be real Banach spaces with $\mathcal{X} \subset Y_1$ continuously embedded (that is, the inclusion map is continuous). A bounded linear map $(A, B) : \mathcal{X} \to Y = Y_1 \times Y_2$ is said to be *admissible* if

(i) it is a Fredholm operator of index zero with $B : \mathcal{X} \to Y_2$ onto;

(ii) there exist constants $c_1, c_2, \alpha > 0$ such that

$$c_1 \|u\|_{\mathcal{X}} \leq \mu^{\alpha/2} \|(A - \mu)u\|_{Y_1} + \mu^{(1+\alpha)/2} \|Bu\|_{Y_2} \tag{3.111}$$

for all $u \in \mathcal{X}$ and $\mu \in \mathbb{R}$ with $\mu \geq c_2$;

(iii) there exists an open neighborhood $\mathcal{N}$ of $[0, \infty)$ in the complex plane such that $\sigma(A, B) \cap \mathcal{N}$ consists of finitely many eigenvalues, each of finite algebraic multiplicity,[53] where the spectrum $\sigma(A, B)$ is defined as the set of $\mu \in \mathbb{C}$ such that $(A - \mu)$ is not an ismorphism from the Banach space $X_0 = \{u \in \mathcal{X} : \ Bu = 0\}$, equipped with the norm $\|u\|_{Y_1} + \|Au\|_{Y_1}$, onto Y_1. The relation (3.111) ensures that X_0 is a Banach space and $A : X_0 \to Y_1$ is a bounded linear operator. Here $\mathcal{X}$, Y_1 and Y_2 are complexified in the natural way.[54]

Notice that the assumption (iii) ensures that (A, B) has a finite number of positive eigenvalues (counted with algebraic multiplicity), a spectral property reminiscent of that for the compact perturbations of the identity leading to formula (3.107). Also, to compare with the considerations in [139], notice that if we define $\mathcal{N}$ such that $a^{\alpha/2}|b| < \frac{c_1}{k}$ for $(a + ib) \in \mathcal{N}$ with $a \geq c_2$, where $k > 0$ is such that $\|x\|_{Y_1} \leq k\|x\|_{\mathcal{X}}$ for all $x \in \mathcal{X}$, then $a < c_2$ whenever $(a + ib) \in \sigma(A, B) \cap \mathcal{N}$. Indeed, if $u_1 + iu_2$ is the eigenvector corresponding to the eigenvalue $(a + ib) \in \mathcal{N}$ with $a \geq c_2$, choosing $\mu = a$ in (3.111) evaluated for $u = u_i$ $(i = 1, 2)$ leads to a contradiction.

If $\mathcal{U} \subset \mathcal{X}$ is open, we say that the nonlinear map $F : \overline{\mathcal{U}} \to Y$ is *admissible* if $F \in C^2(\mathcal{U}, Y) \cap C^0(\overline{\mathcal{U}}, Y)$, and for each $u \in \mathcal{U}$ the linear map F_u is admissible in the sense defined above, and F is proper (that is, if $K \subset Y$ is compact, then $F^{-1}(K)$ is compact). We say that $y \in Y \setminus F(\partial\mathcal{U})$ is a *regular value* of F if the linear operator F_u is onto (and hence an isomorphism) for all $u \in F^{-1}(y) \cap \mathcal{U}$.

[53] An eigenvalue μ has finite algebraic multiplicity if $\dim[ker(A - \mu Id)^m] = \dim[ker(A - \mu Id)^{m+1}] < \infty$ for some integer $m \geq 1$. The smallest $m \geq 1$ with this property is called the algebraic multiplicity of μ.

[54] If X is a real Banach space, the set $X_{\mathbb{C}}$ of all ordered pairs $z = (x, y)$ of X together with the operations given by $(x_1, y_1) + (x_2, y_2) = (x_1 + x_2, y_1 + y_2)$ and $(a + ib)(x, y) = (ax - by, ay + bx)$ for $x_1, x_2, y_1, y_2 \in X$ and $a, b \in \mathbb{R}$, is the complexification of X.

By properness $F^{-1}(y) \cap \mathcal{U} = F^{-1}(y) \cap \overline{\mathcal{U}}$ is compact and by the inverse function theorem every solution of $F(u) = y$ is isolated so that $F^{-1}(y) \cap \mathcal{U}$ is a finite set, allowing us to define the degree by

$$\deg(F, \mathcal{U}, y) = \sum_{u \in F^{-1}(y) \cap \mathcal{U}} (-1)^{\beta(u)},$$

where $\beta(u)$ is the finite number of positive real eigenvalues (counted with multiplicity) of F_u; if $F^{-1}(y) \cap \mathcal{U} = \emptyset$, we set $\deg(F, \mathcal{U}, y) = 0$. If $y \in Y \setminus F(\partial\mathcal{U})$ is a critical value, by the Smale–Quinn–Sard theorem [278, 256] we may choose a regular value y_* with $\|y - y_*\|_Y < \varepsilon$ suficiently small, and define $\deg(F, \mathcal{U}, y) = \deg(F, \mathcal{U}, y_*)$. Notice that by properness $y \in Y \setminus F(\partial\mathcal{U})$ is at a positive distance $\varepsilon_0 > 0$ from $F(\partial\mathcal{U})$ so that $\varepsilon \in (0, \varepsilon_0)$ ensures $y_* \notin F(\partial\mathcal{U})$. One can show (see [140]) that this definition is independent of the particular choice of the regular value y_*. The generalized degree thus constructed enjoys the properties (II) and (III) of the Leray–Schauder degree, as well as the following strengthened version of homotopy invariance, valid on open, noncylindrical domains $\mathcal{W} \subset [0,1] \times \mathcal{U}$, which is similar to (VII): if we set $\mathcal{W}_\lambda = \{u \in \mathcal{U} :\ (\lambda, u) \in \mathcal{W}\}$ and $\partial\mathcal{W}_\lambda = \{u \in \mathcal{U} :\ (\lambda, u) \in \partial\mathcal{W}\}$, then

$$\deg(H(0,\cdot),\ \mathcal{W}_0,\ y) = \deg(H(1,\cdot),\ \mathcal{W}_1,\ y)$$

for all continuous proper mappings $H : \overline{\mathcal{W}} \to Y$ such that $H \in C^2(\mathcal{W}, Y)$ and $H(\lambda, \cdot)$ admissible on $\mathcal{W}_\lambda$ for each $\lambda \in [0,1]$, with $y \notin H(\lambda, \partial\mathcal{W}_\lambda)$ for each $\lambda \in [0,1]$. For details and proofs we refer the reader to [140].

The generalized degree defined above is instrumental in extending local bifurcation curves to a global continuum, in the vein of Rabinowitz's approach using the Leray–Schauder degree. However, in many cases (e.g., in the context of water waves with vorticity) the previous assumptions (i)–(iii) needed to define the generalized degree depend on an ellipticity property of the underlying partial differential equation that cannot be imposed a priori on all classical solutions. Accordingly, in addition to the two Rabinowitz alternatives obtained in the previous subsection using the Leray–Schauder degree, we must add a third alternative that the continuum terminates due to loss of ellipticity. To be more precise, let X, Y_1, Y_2 be real Banach spaces and consider the problem of extending the local curve of solutions $\mathcal{C}_{loc}$ passing through the local bifurcation point $(\lambda_0, 0)$ for $F : \mathbb{R} \times X \to Y = Y_1 \times Y_2$ to a global continuum $S_0 \subset \mathcal{X} = \mathbb{R} \times X$. If the map F is admissible for an open set $\mathcal{O} \subset \mathcal{X}$, not necessarily bounded, and $\mathcal{C}_{loc} \subset \mathcal{O}$, let S_0 be the closure in $\mathcal{X}$ of the component of the solution set $\{(\lambda, x) \in \mathcal{X} :\ F(\lambda, x) = 0\}$ that contains $(\lambda_0, 0)$. Thus S_0 contains the curve $\mathcal{C}_{loc}$. Then

- S_0 is unbounded in $\mathcal{X}$,
- S_0 contains another point $(\lambda, 0)$ with $\lambda \neq \lambda_0$, or
- S_0 contains a point $(\lambda, x) \in \partial\mathcal{O}$.

The proof is identical to the proof of Rabinowitz's alternatives, except that we now use the generalized degree in place of the Leray–Schauder degree: if neither alternative holds, we argue by contradiction (and $S_0 \subset \mathcal{O}$ ensures the existence of the generalized degree enjoying all the properties needed for the argument pursued for the Leray–Schauder degree, so that the same approach works). Notice that in this setting, in contrast to the case of the Leray–Schauder degree, the fact that $\lambda \in \mathbb{R}$ does not seem to be that decisive. This is indicative of possibilities of considering $\lambda \in \mathbb{R}^n$ with $n \geq 1$ or even with values in a Banach space, but deeper topological tools are required to establish the existence of solution continua in that setting (cf. [140]). In our water-wave context λ will always be a real parameter.

Analytic global bifurcation

The outcome of the global bifurcation approach by degree-theoretical methods was the existence of a global continuum (a maximal connected set, but not necessarily path-connected) of solutions that extends the local bifurcation curve. Dancer [88] noticed that in a real-analytic setting it is possible to strengthen Rabinowitz's result: either there exists an unbounded curve of solutions extending the local bifurcation curve or there exists a curve of solutions joining the local bifurcation point $(\lambda_0,0)$ to some $(\lambda_*,0)$, where $\lambda_* \notin \{0,\lambda_0\}$ is such that $\frac{1}{\lambda_*}$ is an eigenvalue of T of odd algebraic multiplicity. Dancer's approach dealt with bifurcation from eigenvalues of possibly high multiplicity; in the case of bifurcation from a simple eigenvalue (as presented in Section 3.5.2) Buffoni and Toland [38] were able to refine the result. The present account follows to a large extent [38]. The core idea is that rather than using degree-theoretical methods, one combines local bifurcation theory, the implicit function theorem, and the structure theorem for real-analytic varieties to piece together a global continuum.

Let X, Y be real Banach spaces, $U \subset \mathbb{R} \times X$ be an open set, and $F : U \to Y$ be a real-analytic function.[55] Assume that

(G1) $\{(\lambda,0) : \lambda \in \mathbb{R}\} \subset U$ and $F(\lambda,0) = 0$ for all $\lambda \in \mathbb{R}$;

(G2) $F_x(\lambda,x) : X \to Y$ is a Fredholm operator of index zero on the solution set

$$S = \{(\lambda,x) \in U : \ F(\lambda,x) = 0\};$$

(G3) the hypotheses of the Crandall–Rabinowitz local bifurcation theorem (see Section 3.5.2) hold at some $(\lambda_0,0)$;

(G4) bounded closed subsets of S are compact in $\mathbb{R} \times X$.

Then the local bifurcation curve (with $x(s) = s\chi(s)$, using the notation of Section 3.5.2)

$$C_{loc} = \{(\Lambda(s), x(s)) : \ |s| < \varepsilon_0\} \subset U$$

has an extension to a continuous curve $C = \{(\Lambda(s), x(s)) : \ s \in [0,\infty)\} \subset U$ such that the following hold.

(C1) $s \mapsto (\Lambda(s), x(s))$ is continuous from $[0,\infty)$ to $\mathbb{R} \times X$ and $C \subset S$.

(C2) At each point $(\Lambda(s_0), x(s_0))$ with $s_0 \in [0,\infty)$ the curve C has a local analytic reparametrization in the following sense. In a right neighborhood of $s_0 = 0$, C and C_{loc} coincide. For each $s_0 > 0$ there exists $\rho_0 : (-1,1) \to \mathbb{R}$ continuous and injective, with $\rho_0(0) = s_0$ and $t \mapsto (\Lambda(\rho_0(t)), x(\rho_0(t))$ is real-analytic for $t \in (-1,1)$ and injective[56] on $(-1,0)$ and on $(0,1)$.

(C3) One of the following occurs:

(i) $\|(\Lambda(s), x(s))\| \to \infty$ as $s \to \infty$.

(ii) $(\Lambda(s), x(s)) \to \partial U$ as $s \to \infty$.

(iii) C is a closed loop: for some $T > 0$ we have $C = \{(\Lambda(s), x(s)) : \ 0 \leq s \leq T\}$ and $(\Lambda(T), x(T)) = (\lambda_0, 0)$ with $(\Lambda(s+T), x(s+T)) = (\Lambda(s), x(s))$ for all $s \geq 0$.

We present the main features of the proof of this result in a series of steps.

[55] This means (cf. [38, Chapter 4]) that for every $\xi_0 \in U$ there exists $\varepsilon > 0$ such that for $\xi \in U$ with $\|\xi - \xi_0\| < \varepsilon$ we have a power series expansion $F(\xi) = \sum_{k=0}^{\infty} m_k(\xi - \xi_0)^k$, where $m_0 \in Y$ and m_k is for $k \geq 1$ a bounded symmetric multilinear operator from $(\mathbb{R} \times X) \times \cdots \times (\mathbb{R} \times X)$ to Y such that $\sum_{k=0}^{\infty} \varepsilon^k \|m_k\| < \infty$, with $\|m_k\| = \sup_{\|\xi_1\|,\ldots,\|\xi_k\| \leq 1} \|m_k(\xi_1,\ldots,\xi_k)\|$.

[56] The parametrization is locally injective but need not be globally injective; see, e.g., part (C3)(iii). Also, the existence of a local real-analytic parametrization does not mean that the curve C is smooth: the curve $y^2 = x^3$ has the point $(0,0)$ as a "pinch point" (bending so that it becomes tangent to itself) but admits nevertheless the real-analytic parametrization $\{(t^2,t^3) : \ t \in \mathbb{R}\}$.

The first step consists in using the Lyapunov–Schmidt procedure to reduce the investigation of solutions in a neighborhood of a fixed solution $(\lambda_*, x_*) \in S$ to an equivalent problem involving an equation in finite dimensions. Let

$$\mathcal{N} = \{(\lambda, x) \in S : \ ker(F_x(\lambda, x)) = \{0\}\}$$

be the set of nonsingular solutions: by the real-analytic implicit function theorem (cf. [38, Chapter 4]) near such a point the solution set is given by a curve $\{(s, \phi(s)) : \ |s - \lambda| < \varepsilon\}$ for some $\varepsilon > 0$, where $s \mapsto \phi(s)$ is real-analytic from $(\lambda - \varepsilon, \lambda + \varepsilon) \subset \mathbb{R}$ to X with $\phi(0) = x$. Maximal connected subsets of $\mathcal{N}$ are called *distinguished arcs* since there exist a (possibly infinite) open interval $I \subset \mathbb{R}$ and a real-analytic function $\phi : I \to X$ such that the set is precisely $\{(\lambda, \phi(\lambda)) : \ \lambda \in I\}$. With the structure of S near points in $\mathcal{N}$ thus clarified, let us now study the structure of S in a neighborhood of a point $(\lambda_*, x_*) \in S \setminus \mathcal{N}$. The Fredholm property of F can now be exploited to implement the Lyapunov–Schmidt reduction (cf. [38, Chapter 8]). This yields the existence of an open neighborhood V of $(\lambda_*, 0)$ in $\mathbb{R} \times ker(F_x(\lambda_*, x_*)) \cong \mathbb{R}^{q+1}$ ($q \geq 1$ being the finite dimension of the kernel) and of real-analytic maps $\psi : V \to X$ and $h : V \to \mathbb{R}^q$ such that the following hold.

(L1) $(\lambda, \psi(\lambda, \xi)) \in U$ for $(\lambda, \xi) \in V$, with $\psi(\lambda_*, 0) = x_*$.

(L2) For $(\lambda, \xi) \in V$ we have $h(\lambda, \xi) = 0$ if and only if $F(\lambda, \psi(\lambda, \xi)) = 0$. (That is, the infinite-dimensional problem $F(\lambda, x) = 0$ is "reduced" to the finite-dimensional one "find $(\lambda, \xi) \in V$ such that $h(\lambda, \xi) = 0$.")

(L3) If $F(\lambda, x) = 0$ for some $(\lambda, x) \in U$ sufficiently close to (λ_*, x_*), then there exists $\xi \in ker(F_x(\lambda_*, x_*))$ such that $(\lambda, \xi) \in V$ and $x = \psi(\lambda, \xi)$.

(L4) $\dim[ker(F_x(\lambda, \psi(\lambda, \xi))] = \dim[ker(h_\xi(\lambda, \xi))]$ for $(\lambda, \xi) \in V$.

(L5) If $(\lambda_*, x_*) = (\lambda_0, 0)$, then for s with $0 < |s| < \varepsilon_0$ we have $\dim[ker(F_x(\Lambda(s), x(s))] = 1$ if $\Lambda'(s) = 0$, while otherwise $ker(F_x(\Lambda(s), x(s)) = \{0\}$.

In contrast to the case of smooth (C^∞) functions $f : \mathbb{R}^n \to \mathbb{R}$, in which case any closed subset of $\mathbb{R}^n$ could be the set of zeros (cf. [241]), the set of zeros of real-analytic functions $f_k : \mathbb{R}^n \to \mathbb{R}$ with $1 \leq k \leq m$ (called a *real-analytic variety*), has very nice structural properties: it can be stratified into submanifolds of dimensions $0, 1, \ldots, N$ for some $N \geq 0$, with the zero-dimensional manifold being a discrete set of points (see [209]). These structural properties hold since the Weierstrass preparation theorem[57] allows one to establish properties of real-analytic varieties by first transforming the problem into a problem for algebraic varieties (zero sets of polynomials) and subsequently inducting on dimension. We will actually need the following results.[58]

(R1) ("Structure theorem of semianalytic sets") If $\Omega \subset \mathbb{R}^n$ is open, we say that $\mathcal{U} \subset \Omega$ is a *semianalytic set* if it is of the form

$$\mathcal{U} = \{x \in \Omega : \ f_j(x) = 0, j = 1, \ldots, m\} \cap \{x \in \Omega : \ g_l(x) > 0, j = 1, \ldots, r\}$$

with $f_j, g_l : \Omega \to \mathbb{R}$ real-analytic. Every semianalytic set $\mathcal{U}$ is a disjoint union of a finite number of semialgebraic sets, each of which is homeomorphic (by means of a continuous bijection) to an open rectangle $(0,1)^d$, with $(0,1)^0$ understood as a point. The dimension of $\mathcal{U}$ is the maximum of the numbers d_i.

[57]Stating (cf. [186, Chapter 6]) that real-analytic functions $f : \mathbb{R}^n \to \mathbb{R}$, $f \not\equiv 0$, are near a zero $\xi = (\xi_1, \ldots, \xi_n) \in \mathbb{R}^n$, up to multiplication by a function not zero at ξ, a polynomial in one fixed variable, say, ξ_n, with leading coefficient one, and whose coefficients are real-analytic functions in the remaining variables $\xi_1, \ldots, \xi_{n-1}$, and having a zero at ξ.

[58]For the proofs see [230] for algebraic varieties, and [143] for the extension to the real-analytic context (cf. [25]). Alternatively, see the discussion of real-analytic varieties in [38, Chapter 7] and the discussion of real-algebraic varieties in [24, Chapters 2, 3, and 9].

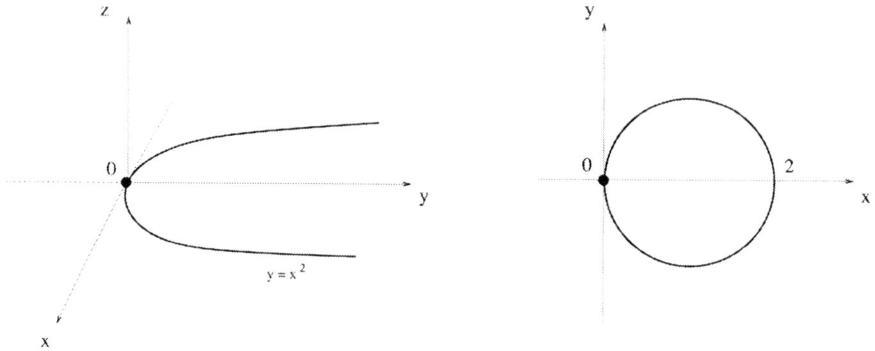

Figure 3.12. *On the left, for the real-analytic variety $\mathcal{A} = \{z = 0\} \subset \mathbb{R}^3$ and the semianalytic set $\mathcal{U} = \{y = x^2, x > 0\} \subset \mathbb{R}^3$, we have $\mathcal{A}_0 = \{z = 0,\ y = x^2\}$; only one of the two branches of $\mathcal{A}_0$ at $(0,0,0)$ lies in the set $\mathcal{A} \cap \mathcal{U}$. On the right, in the open neighborhood $\{x^2 + y^2 < 4\cos^2(\frac{\varepsilon}{2})\}$ of $(0,0)$, the real-analytic variety $\mathcal{A}_0 = \{(x-1)^2 + y^2 = 1\} \subset \mathbb{R}^2$ is the image of $p_1 : (-\pi + \varepsilon, \pi - \varepsilon) \to \mathbb{R}^2$, $p_1(t) = (1 - \cos t, \sin t)$, with $\varepsilon > 0$ small preventing the closures of the two branches to meet again at $(0,2)$.*

(R2) ("Curve selection lemma") For an open set $\Omega \subset \mathbb{R}^n$, let $\mathcal{A} \subset \mathbb{R}^n$ be a real-analytic variety and $\mathcal{U} \subset \Omega$ be a semianalytic set. If $\xi_0 \in \Omega$ is an accumulation point of $\mathcal{A} \cap \mathcal{U}$, then there is an open set $\Omega_0 \subset \Omega$ with $\xi_0 \in \Omega_0$, and a one-dimensional real-analytic variety $\mathcal{A}_0 \subset \mathcal{A} \cap \Omega_0$ such that ξ_0 is an accumulation point of the set $\mathcal{U} \cap \mathcal{A}_0$.

(R3) ("Structure theorem for one-dimensional branches") If ξ_0 is an accumulation point of a one-dimensional real-analytic variety $\mathcal{A}_0$ in $\mathbb{R}^n$, then for some $\delta > 0$ the set $\mathcal{A}_0 \setminus \{\xi_0\}$ consists in $\{\xi \in \mathbb{R}^n : |\xi - \xi_0| < \delta\}$ of an finite even number of disjoint open real-analytic branches with closures that intersect one another only at ξ_0. More precisely, there exist numbers $\varepsilon_k > 0$ and functions $p_k : (-\varepsilon_k, \varepsilon_k) \to \mathcal{A}_0$ such that p_k maps each of the intervals $(-\varepsilon_k, 0)$ and $(0, \varepsilon_k)$ bijectively onto branches of $\mathcal{A}_0$, with $p_k(0) = \xi_0$ and $\bigcup_k p_k(-\varepsilon_k, \varepsilon_k) = \mathcal{A}_0 \cap \{\xi \in \mathbb{R}^n : |\xi - \xi_0| < \delta\}$. Moreover, on $(-\varepsilon_k, \varepsilon_k)$ we can express p_k as the convergent power series $p_k(t) = \sum_{j=1}^{\infty} a_{jk} t^j$ with $a_{jk} \in \mathbb{R}^n$ (t being the *local uniformizing parameter*) and, for each fixed k, the greatest common divisor of $\{j \geq 1 : a_{jk} \neq 0\}$ is 1; see Figure 3.12.

Notice that in (R3) each branch is paired, in a unique way, with another branch, so that their union with the point ξ_0 forms a curve.

Let us first assume that there is some $\varepsilon \in (0, \varepsilon_0)$ with $\Lambda'(s) \neq 0$ for $s \in (-\varepsilon, 0) \cup (0, \varepsilon)$. Then (L5) ensures that there exists a distinguished arc having $(\lambda_0, 0)$ as an endpoint. Let us now define a *route of length* $N \in \mathbb{N} \cup \{\infty\}$ to be a set $\{C_n : 0 \leq n < N\}$ of distinguished arcs (that is, curves representing maximal connected subsets of $\mathcal{N}$), and a set of points $E_n = (\lambda_n, x_n) \in S$ with $0 \leq n < N$ such that the following hold.

(Ra) $E_0 = (\lambda_0, 0)$ and $\{(\Lambda(s), x(s)) : s \in (0, \varepsilon)\} \subset C_0$.

(Rb) For $N > 1$ and $0 \leq n < N - 1$ we have $E_{n+1} \in (\partial C_n \cap \partial C_{n+1}) \setminus E_n$ and there exists an injective real-analytic map $\rho : (-1, 1) \to C_n \cup \{E_{n+1}\} \cup C_{n+1}$ with $\rho(0) = E_{n+1}$.

(Rc) The mapping $n \mapsto C_n$ is injective.

This means that we continue each distinguished arc C_n until we reach a point E_{n+1} that does not lie any more in the set $\mathcal{N}$. At E_{n+1} we can use the structure theorem described in (R2)–(R3), with $\mathcal{A} = \{(\lambda, \xi) \in \mathbb{R}^{q+1} : h(\lambda, \xi) = 0\}$ and $\mathcal{U} = \{(\lambda, \xi) \in \mathbb{R}^{q+1} : \det^2(h_\xi(\lambda, \xi)) > 0\}$, to choose a branch

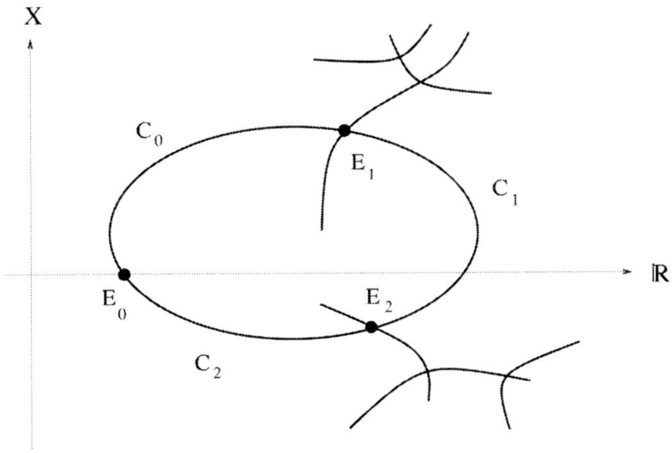

Figure 3.13. *The maximal route is not necessarily a maximal connected subset of S. It is maximal in the sense that we exhausted the possibilities of adding distinguished arcs.*

whose maximal continuation defines C_{n+1}. Indeed, if $(\lambda,x) \in \mathcal{N}$, then the structure of S nearby is given by a real-analytic curve, while at branch points we have an even number of branches out of these points, in accordance with (R3). At $(\lambda,x) \in \overline{\mathcal{N}} \setminus \mathcal{N}$, let B^- be the incoming branch (lying in $\mathcal{N}$) and let B^+ be the outgoing branch, with $t \in (-\varepsilon,\varepsilon)$ the local uniformizing parameter. Since the map $t \to \det(h_\xi(\lambda(t),\xi(t)))$ is a real-analytic function that has no zeros on $(-\varepsilon,0)$, its possible zeros for $t \in (0,\varepsilon)$ do not accumulate at $t=0$. Consequently, we can find some $\varepsilon_1 \in (0,\varepsilon]$ such that for $t \in (0,\varepsilon_1)$ the corresponding portion of the branch B^+ lies in $\mathcal{N}$. This curve is then extended to provide us with an outgoing distinguished arc at (λ,x).

We can think of the above construction as a graph with vertex set $\{E_n\}$ and edges C_n, with C_0 uniquely determined by the local bifurcation theorem and, for a route of length $N > 1$, with each C_{n+1} with $n < N-1$ uniquely determined by C_n and vice versa. Notice that C_n and C_m are disjoint if $n \neq m$. It is also clear that we can choose a continuous reparametrization $\{(\Lambda(s),x(s)) : 0 \le s < N\}$ of a route of length N such that $s \in (n,n+1)$ provides a parametrization of C_n, with the endpoints E_n corresponding to the integers $0 \le n < N$. For a maximal route (see Figure 3.11) we have three alternatives:

(A1) $N = \infty$.

(A2) $N < \infty$ is an integer but $\overline{C_{N-1}}$ is not a compact subset of U.

(A3) $N < \infty$ is an integer and $\overline{C_{N-1}}$ is a compact subset of U.

To prove the validity of the alternatives listed in (C3), suppose that a maximal route $\mathcal{R}$ of length N is bounded in $\mathbb{R}\times X$ and bounded away from the boundary of ∂U, and let us prove that (C3)(iii) must hold. Indeed, since the closure of $\mathcal{R}$ in $\mathbb{R}\times X$ is bounded, by (G4) it is compact in $\mathbb{R}\times X$ so that we can find a sequence $t_n \to N$ with $\{(\Lambda(t_n),x(t_n))\}$ converging to some $(\lambda_*,x_*) \in S$. If (A1) holds, then every neighborhood of (λ_*,x_*) will intersect infinitely many distinguished arcs C_n, which contradicts the structure theorem (R3). If (A2) holds, since $t_n \to N$, we have $(\lambda_*,x_*) \in \partial C_{N-1} \setminus \{E_{N-1}\}$. Using the Lyapunov–Schmidt reduction near (λ_*,x_*) we find that in this neighborhood C_{N-1} corresponds to a branch and we could extend the route further by adding $E_N = (\lambda_*,x_*)$ and continuing along another branch. Since this contradicts the maximality of the route $\mathcal{R}$, (A3) must hold. If (A3) holds, let E_N be the endpoint of C_{N-1} that is different from E_{N-1}. The unique continuation of C_{N-1} as a distinguished arc C_N is ensured by the construction explained before. Since $\mathcal{R}$ is maximal we have $C_N = C_m$ for some $m \in \{0,1,\dots,N-2\}$. If $m \in \{1,\dots,N-3\}$, since C_{m-1} and C_{m+1} are the only continuations of C_m, we must have $C_{N-1} = C_k$ with $k \in \{0,\dots,N-2\}$, which contradicts (Rc). Thus $C_N = C_0$ or $C_N = C_{N-2}$. If $C_N = C_{N-2}$, the unique continuation property forces $N = 1$, with $C_0 \cup C_1 \cup \{E_0\}$ forming a loop. Finally, if $C_N = C_0$, there are two possibilities: $E_N = E_0$ or $E_N = E_1$. The latter one forces $C_{N-1} = C_1$ and contradicts (Rc), so that we must have $C_N = C_0$ with $E_N = E_0$, proving that $\mathcal{R}$ is a loop.

Assume now that the set $\{s \in (-\varepsilon_0,\varepsilon_0): \ \Lambda'(s) = 0\}$ has 0 as an accumulation point. By real-analyticity, $\Lambda(s) = \lambda_0$ for $|s| < \varepsilon_0$. Instead of the subset $\mathcal{N}$ of $S \subset \mathbb{R}\times X$ considered before, let

$$S_1 = \{x \in X: \ (\lambda_0,x) \in U, \ F(\lambda_0,x) = 0\}, \qquad \mathcal{N}_1 = \{x \in S_1: \ \dim[ker(F_x(\lambda_0,x))] = 1\}.$$

In the Lyapunov–Schmidt reduction scheme for (λ_0,x) with $x \in \mathcal{N}_1$, the structure of S_1 near such a point is obtained from that of $\{\xi \in \mathbb{R}: \ h(\lambda_0,\xi) = 0\}$ for a real-analytic function $h: \mathbb{R}^2 \to \mathbb{R}$, and therefore is either an isolated point or an entire interval. In the case of $(\lambda_0,0)$ we have by (L5) that $x(s) \in \mathcal{N}_1$ for all $|s| < \varepsilon_0$. Consequently, the approach presented before can be implemented. Namely, the local bifurcation curve in $\{\lambda_0\}\times X$ is continued analytically. If the curve accumulates at a point (λ_0,x_*) with $x_* \notin \mathcal{N}_1$, we employ Lyapunov–Schmidt near $x_* \in X$, and (R2)–(R3) with $\mathcal{A} = \{\xi \in \mathbb{R}^q: \ h(\lambda_0,\xi) = 0\}$, where $q = \dim[ker(h_\xi(\lambda_0,\xi_*))] \geq 1$, provides us with a branch allowing us to continue the curve further. Indeed, by the rank theorem (cf. [241]) $\dim[ker(h_\xi(\lambda_0,\xi))] = 1$ is equivalent to $\mathrm{rank}[h_\xi(\lambda_0,\xi)] = q-1$ for $h_\xi(\lambda_0,\xi): \mathbb{R}^q \to \mathbb{R}^q$. At x_*, for the incoming branch $B^- \subset \mathcal{N}_1$ we know that h_ξ has rank $(q-1)$, and along the outgoing branch B^+ we know that $\dim[ker(h_\xi(\lambda_0,\xi))] \geq 1$ since otherwise the inverse function theorem would ensure that points $\xi \in B^+$ are isolated. Knowing that $\det[h_\xi(\lambda_0,\xi)] = 0$ on B^+, we can express the condition $\mathrm{rank}[h_\xi(\lambda_0,\xi)] = q-1$ by the request that the sum of the squares of the minors of dimension $(q-1)$ does not vanish. This being a real-analytic function of the local uniformizing parameter in (R3), its zeros cannot accumulate at $t = 0$ since there are no zeros for $t < 0$. Thus, points on the branch B^+ near x_* belong to $\mathcal{N}_1$, the distinguished outgoing arc being obtained by analytic continuation. We obtain the alternatives listed in (C3), with the additional specification that $\lambda = \lambda_0$ along the curve.

We conclude our discussion by pointing out that there are physically relevant equations with solution curves C along which $\lambda = \lambda_0$. For example, if $\mathcal{C}$ is the periodic Hilbert transform, defined as the conjugation operator $\mathcal{C}(e^{int}) = -i\,\mathrm{sgn}(n)e^{int}$ for $n \in \mathbb{Z}$ (with the convention that $\mathrm{sgn}(0) = 0$), then (cf. [38, Chapter 10]) the problem of the existence of periodic steady irrotational waves propagating at the surface of water of infinite depth (modelling deep water waves) can be formulated as seeking zeros of $F(\lambda,w) = \mathcal{C}(w') - \lambda w - \lambda w\mathcal{C}(w') - \lambda\mathcal{C}(ww')$, with $w \in H^1_{per}(\mathbb{R})$ even. Since by Riesz's theorem (see [38, Chapter 10]) the operator $\mathcal{C}: L^2[-\pi,\pi] \to L^2[-\pi,\pi]$ is bounded, one can easily check that with $X = \{w \in H^1_{per}(\mathbb{R}): \ w \text{ even}\}$, $Y = \{w \in L^2_{per}(\mathbb{R}): \ w \text{ even}\}$, the Crandall–Rabinowitz local bifurcation theorem is applicable at the bifurcation point $(0,0)$, the kernel of $L = F_x(0,0)$ being generated by the constant function 1, and $range(L)$ being $\{f \in Y: \ \int_{-\pi}^{\pi} f(s)ds = 0\}$, while $[\partial^2_{\lambda,w}F(0,0)](1,1) = -1$. One can show that all the hypotheses of the global real-analytic bifurcation theorem hold (see [38, Chapters 10 and 11]) with $U = \{(\lambda,w) \in \mathbb{R}\times X: \ 1 - 2\lambda w > 0\}$. The global curve extending the local bifurcation curve is $\{(0,s): \ s \in \mathbb{R}\}$.

3.5.4 Elliptic boundary value problems

This vast subject is central to mathematical physics. Rather than attempting an exhaustive survey we discuss some fundamental results in the illustrative case of second-order equations in smooth planar domains.

Hölder spaces

In our study of periodic water waves we seek classical solutions so that the function $h(q,p)$ should be periodic in the q-variable and at least twice continuously differentiable in both variables. Denoting $R=(-\pi,\pi)\times(p_0,0)$ with $p_0<0$, the $C^k_{per}(\overline{R})$ Banach spaces (here $k\geq 0$ is an integer) of all functions which are periodic in q and k times continuously differentiable in $(q,p)\in\mathbb{R}\times[p_0,0]$, with the norm given by

$$\|h\|_{C^k_{per}(\overline{R})}=\sum_{0\leq i+j\leq k}\ \sup_{(q,p)\in R}|\partial_q^i\partial_p^j f(q,p)|,$$

are not appropriate for elliptic partial differential equations, while the Hölder spaces $C^{k,\alpha}_{per}(\overline{R})$ and the Sobolev spaces $W^{k,s}_{per}(\overline{R})$ are particularly useful.

Hölder continuity is a quantitative measure of continuity (think of it as fractional differentiability), the Hölder spaces $C^{k,\alpha}_{per}(\overline{R})$ with $0\leq\alpha<1$ being designed to fill up the gaps between $C^k_{per}(\overline{R})$ and $C^{k+1}_{per}(\overline{R})$. For $k=0$ and $0\leq\alpha\leq 1$ these spaces are defined as the subspace of functions $h\in C^0_{per}(\overline{R})$ which obey the Hölder continuity bound

$$|f(\Theta)-f(\Theta_1)|\leq C\,|\Theta-\Theta_1|^\alpha$$

for some constant $C>0$ and all $\Theta,\Theta_1\in R$, and endowed with the norm

$$\|f\|_{C^{0,\alpha}_{per}(\overline{R})}=\sup_{\Theta\in R}\{|f(\Theta)|\}+\sup_{\Theta,\Theta_1\in R,\ \Theta\neq\Theta_1}\left\{\frac{|f(\Theta)-f(\Theta_1)|}{|\Theta-\Theta_1|^\alpha}\right\}$$

they are Banach spaces. The restriction to $\alpha\leq 1$ is due to the fact that $\alpha>1$ would imply that all partial derivatives exist and are zero so that the function would be constant. For $\alpha=1$ the space $C^{0,1}_{per}(\overline{R})$ is the space of Lipschitz functions—a space slightly larger than $C^1_{per}(\overline{R})$ since, e.g., the function $\Theta=(q,p)\mapsto|\Theta|=\sqrt{q^2+p^2}$ belongs to $C^{0,1}_{per}(\overline{R})$ but is not continuously differentiable, so that the scale $C^{0,\alpha}_{per}(\overline{R})$ with $0\leq\alpha\leq 1$ provides a near-continuum of spaces between $C^0_{per}(\overline{R})$ and $C^1_{per}(\overline{R})$. For $k\geq 1$ we define $C^{k,\alpha}_{per}(\overline{R})$ as the subspace of functions $h\in C^k_{per}(\overline{R})$ whose norm

$$\|h\|_{C^{k,\alpha}_{per}(\overline{R})}=\|h\|_{C^k_{per}(\overline{R})}+\sum_{i+j=k}\ \sup_{(q,p)\in R}\|\partial_q^i\partial_p^j h(q,p)\|_{C^{0,\alpha}(\overline{R})}$$

is finite. Equipped with these norms, $C^{k,\alpha}_{per}(\overline{R})$ are Banach spaces, and if $f_1\in C^{k_1,\alpha_1}_{per}(\overline{R})$, $f_2\in C^{k_2,\alpha_2}_{per}(\overline{R})$, then the product $f_1f_2\in C^{k,\alpha}_{per}(\overline{R})$, where $k+\alpha=\min\{k_1+\alpha_1,k_2+\alpha_2\}$. Moreover, the Arzela–Ascoli theorem ensures that the inclusion map $C^{k_2,\alpha_2}_{per}(\overline{R})\hookrightarrow C^{k_1,\alpha_1}_{per}(\overline{R})$ is compact if $k_2+\alpha_2>k_1+\alpha_1$. A good reference for Hölder spaces is [129].

To explain why it is necessary to introduce Hölder and Sobolev spaces into elliptic theory (as opposed to the more intuitive $C^k_{per}(\overline{R})$ spaces), for a given function $f:R\to\mathbb{R}$ that is periodic in the q-variable, let us consider the Poisson equation

$$\Delta h=f\quad\text{in}\quad R,\tag{3.112}$$

under the requirement of 2π-periodicity in the horizontal q-direction and with homogeneous Dirichlet boundary conditions[59]

$$h=0\quad\text{on}\quad\partial R=\{(q,0):q\in[-\pi,\pi]\}\cup\{(q,p_0):q\in[-\pi,\pi]\}.\tag{3.113}$$

[59]Boundary conditions have to be prescribed throughout the boundary but the periodicity condition enables us to dispense of discussing the lateral boundaries.

Then for any $f \in L^s(R)$ with $1 < s < \infty$ there is a unique strong solution[60] $h \in W^{2,s}_{per}(R) \cap W^{1,s}_{0,per}(R)$ of (3.112)–(3.113), and

$$\|h\|_{W^{2,s}_{per}(R)} \le \tilde{c}_s\,(\|h\|_{L^s(R)} + \|f\|_{L^s(R)}), \tag{3.114}$$

where $\tilde{c}_s$ depends only on s. Also, for $f \in C^{0,\alpha}_{per}(\overline{R})$ with $\alpha \in (0,1)$ there is a unique classical solution $h \in C^{2,\alpha}_{per}(\overline{R})$ and

$$\|h\|_{C^{2,\alpha}_{per}(\overline{R})} \le c_\alpha\,(\|h\|_{C^0_{per}(\overline{R})} + \|f\|_{C^{0,\alpha}_{per}(\overline{R})}), \tag{3.115}$$

where c_α depends only on α (see [129, Chapter 9], respectively [129, Chapter 6]). The **Schauder estimates**[61] (3.115) and (3.114) assert, roughly speaking, that if Δh belongs to $L^s(\overline{R})$ or has $C^{0,\alpha}$-regularity, then all second derivatives of h belong to $L^s(R)$ or have $C^{0,\alpha}$-regularity as well. This remarkable phenomenon—that control of a special linear combination of derivatives of h at some order implies control of all derivatives of h at that order—is known as **elliptic regularity**.

Example 3.20. The failure of the C^k-spaces to express elliptic regularity can be seen by considering the homogeneous Dirichlet problem for the Poisson equation (3.112) in the open set $B^* = \{(q,p) \in \mathbb{R}^2 : q^2+p^2 < \frac{1}{4}\}$ with

$$f(q,p) = -\frac{qp}{q^2+p^2}\left(2\,f_0^{-1/2} + \frac{1}{4}\,f_0^{-3/2}\right) \in C^0(\overline{B^*}),$$

where $f_0(q,p) = -\frac{1}{2}\ln(q^2+p^2)$. Indeed, a direct calculation shows that

$$h^*(q,p) = qp\left(f_0^{1/2} - \sqrt{\ln 2}\right)$$

belongs to $C^1(\overline{B^*})$ with $h^*_{qq}, h^*_{pp} \in C^0(\overline{B^*})$ and satisfies $\Delta h^* = f$ in B^* with $h^* = 0$ on ∂B^*, while

$$h^*_{qp} = f_0^{1/2} - \sqrt{\ln 2} - \frac{1}{2}\,f_0^{-1/2} + \frac{q^2p^2}{(q^2+p^2)^2}\left(f_0^{-1/2} - \frac{1}{4}\,f_0^{-3/2}\right) \to \infty \quad \text{for} \quad q,p \to 0.$$

This proves that a Schauder estimate of type (3.115) on the C^k-scale, asserting that

$$\|h\|_{C^2(\overline{B^*})} \le c_0\,(\|h\|_{C^0(B^*)} + \|f\|_{C^0(\overline{B^*})}),$$

simply does not exist. Moreover, in this case there is no C^2 solution in $\overline{B^*}$. Indeed, were $h \in C^2(\overline{B^*})$ a solution, the difference $h - h^*$ would be a C^1-function in B^* such that $\partial_q^2(h-h^*)$ and $\partial_p^2(h-h^*)$ exist and their sum vanishes in the open set B^*. Then (see [39, Section IX.7]) the function $h - h^*$ is C^2 (and harmonic) in B^*. But then h^* would be C^2 at the origin and this is not the case. ∎

The previous considerations explain why the scales of Hölder and Sobolev spaces are suitable to express elliptic regularity, in marked contrast to the C^k-spaces. Since we deal with classical solutions and in the degree-theoretic considerations showing the existence of waves of large amplitude as well as in proving the symmetry of the waves we rely upon **maximum principles**, the most suitable setting for our purposes is that of Hölder spaces—the maximum principles lend themselves well to the suprema that appear in the definition of the norms.

[60] For these solutions the derivatives up to second order exist as functions in $L^s(R)$ and (3.112) is to hold a.e. in R. Here $W^{1,s}_{0,per}(R)$ is the class of functions in $W^{1,s}_{per}(R)$ with zero trace on $p=0$ and on $p=p_0$.

[61] Sometimes (3.114) is referred to as a **Calderón–Zygmund inequality**.

Existence theory for linear equations

The introduction of the concept of weak solutions presents the possibility of splitting the investigation of the existence of classical solutions into two steps:

- the existence and uniqueness of weak solutions,
- the regularity of weak solutions,

the first step being generally granted by ellipticity and the second by Schauder estimates. Recall that (cf. [129]) the second-order operator in divergence form

$$\mathcal{L}h = \partial_q(a_{20}h_q + a_{11}h_p + b_1 h) + \partial_p(a_{11}h_q + a_{02}h_p + b_2 h) + c_1 h_q + c_2 h_p + dh$$

with coefficients $a_{ij}, b_i \in C^{1,\alpha}_{per}(\overline{R})$ and $c_i, d \in C^{0,\alpha}_{per}(\overline{R})$ for some $\alpha \in (0,1)$, is called *uniformly elliptic* if there exists a constant $\lambda > 0$ such that

$$\sum_{i+j=2} a_{ij}(q,p)\xi_i\xi_j \geq \lambda(\xi_1^2 + \xi_2^2), \qquad (q,p) \in \overline{R}, \quad \xi_1, \xi_2 \in \mathbb{R}. \tag{3.116}$$

For such operators with homogeneous boundary conditions on $p = 0$ and $p = p_0$ one can show (see [129, Sections 6.2 and 9.5]) that the Schauder estimates (3.115) and (3.114) hold a priori,[62] with c_α depending on α, λ and the $C^{0,\alpha}_{per}(\overline{R})$-norms of the coefficients.

Example 3.21. Consider the simple but nontrivial example

$$\Delta h + dh = f \quad \text{in} \quad R, \tag{3.117}$$

with homogeneous Dirichlet boundary conditions (3.113) and the periodicity requirement in the q-variable. In (3.117) the functions $d, f \in C^{0,\alpha}_{per}(\overline{R})$ are given, with $d \leq 0$ in $\overline{R}$. If $\mathbb{H}$ is the Hilbert space of functions $h \in H^1_{per}(R)$ with $h = 0$ on $p = 0$ and on $p = p_0$ (in the sense of traces on $L^2(\mathbb{R})$; cf. the Appendix to Chapter 2) endowed with the inner product

$$(h,\varphi) = \int\int_R (h_q\varphi_q + h_p\varphi_p + h\varphi)\,dq\,dp, \qquad h, \varphi \in \mathbb{H},$$

a *weak solution* is defined as a function $h \in \mathbb{H}$ such that[63]

$$\int\int_R (h_q\varphi_q + h_p\varphi_p - dh\varphi)\,dq\,dp = -\int\int_R f\varphi\,dq\,dp, \qquad \varphi \in \mathbb{H}. \tag{3.118}$$

Notice that by the Riesz representation theorem there exists a bounded linear operator $\mathcal{L} : \mathbb{H} \to L^2(R)$ such that the left side of (3.118) is precisely $\int\int_R (\mathcal{L}u)\varphi\,dq\,dp$ and consequently (3.118) can be recast as

$$\mathcal{L}h = -f \quad \text{in} \quad L^2(R). \tag{3.119}$$

---

[62] In the sense that they are valid for all possible classical solutions in $C^{2,\alpha}_{per}(\overline{R})$, respectively for all strong solutions in $W^{2,s}_{per}(R)$, even if the existence of such solutions is not guaranteed: on the right-hand side we replace f by $\mathcal{L}h$ and the estimates (3.115) and (3.114) hold for any $h \in C^{2,\alpha}_{per}(\overline{R})$, respectively, any $h \in W^{2,s}_{per}(\overline{R})$; cf. [129, Chapter 6].

[63] A classical solution $h \in C^2_{per}(\overline{R})$ satisfies (3.118), as one can see by multiplying the equation by φ and integrating by parts.

To prove the existence of a unique weak solution we use the Lax–Milgram theorem[64] (see [111, Chapter 6]) which states that on a real Hilbert space H, for any bounded linear functional $F : H \to \mathbb{R}$ and any bilinear bounded and coercive map[65] $B : H \times H \to \mathbb{R}$ there exists a unique $u_F \in H$ with $B(u_F, v) = F(v)$ for all $v \in H$. In our setting, notice that by Poincaré's inequality (see [111, Chapter 5]) there exists a constant $c > 0$ with

$$c \int\int_R h^2\,dqdp \leq \int\int_R (h_q^2 + h_p^2)\,dqdp, \qquad u \in \mathbb{H}, \tag{3.120}$$

so that the inner product

$$[u,v] = \int\int_R (u_q v_q + u_p v_p)\,dqdp, \qquad u,\, v \in \mathbb{H},$$

induces a norm that is equivalent to the original one on $\mathbb{H}$. The inner product $[\cdot,\cdot]$ makes

$$B(u,v) = \int\int_R (u_q\, v_q + u_p\, v_p)\,dqdp$$

coercive, its boundedness and that of $F(\varphi) = -\int\int_R f\varphi\,dqdp$ being also clear.[66] From the Lax–Milgram theorem we infer that for every $f \in L^2(R)$ the equation $(\mathcal{L}+d)h = -f$ has a unique solution h_d^f and the map $f \mapsto h_d^f$ is an isomorphism from $L^2(R)$ to $\mathbb{H}$. Since (see [111, Chapter 5]) $h \mapsto dh$ is compact from $\mathbb{H}$ to $L^2(R)$, we deduce (see Section 3.5.1) that $\mathcal{L} : \mathbb{H} \to L^2(R)$ is a Fredholm operator with $\text{ind}(\mathcal{L}) = \text{ind}(\mathcal{L}+d) = 0$. Since $d \leq 0$ in $\overline{R}$, the weak maximum principle for weak solutions $h \in \mathbb{H}$ (cf. [129, Section 8.1]) ensures that $ker(\mathcal{L}) = \{0\}$. Consequently, Consequently, (3.119) has a unique solution $h_f \in \mathbb{H}$. As for regularity, if $d \in C_{per}^{\infty}(\overline{R})$, then repeatedly using (3.118) with various choices of functions φ (in particular, difference quotients—see [33, Section IX.6] for a detailed exposition) one can show that for every integer $s \geq 0$ there exists a constant $k_s > 0$ such that to each $f \in H_{per}^s(R)$ there corresponds a unique solution $h_f \in H_{per}^{s+2}(R)$ to (3.118) and $\|h_f\|_{H_{per}^{s+2}(R)} \leq k_s\, \|f\|_{H_{per}^s(R)}$. It is now possible to infer existence results in Hölder spaces. Indeed, if $f \in C_{per}^{0,\alpha}(\overline{R})$, we approximate it in $C_{per}^{0,\alpha}(\overline{R})$ by a sequence $f_j \in C_{per}^{2,\alpha}(\overline{R})$; cf. [129, Section 7.2]. Using (see [111, Chapter 5]) the fact that $H_{per}^4(R) \subset C_{per}^{2,\alpha}(\overline{R})$, the previous considerations yield unique weak solutions $h_j \in C_{per}^{2,\alpha}(\overline{R})$ corresponding to f_j. Since the Schauder estimates (3.115) are true a priori, we have

$$\|h_j\|_{C_{per}^{2,\alpha}(\overline{R})} \leq c_\alpha \left(\|h_j\|_{C^0(\overline{R})} + \|f_j\|_{C^{0,\alpha}(\overline{R})}\right), \qquad j \geq 1. \tag{3.121}$$

Since $C_{per}^0(\overline{R}) \subset H_{per}^2(R)$ and

$$\|h_j\|_{H_{per}^2(R)} \leq k_0\, \|f_j\|_{L^2(R)} \leq k_0\, \sqrt{2\pi\,|p_0|}\; \|f_j\|_{C^0(\overline{R})}, \qquad j \geq 1,$$

we deduce first that the real sequence $\{\|h_j\|_{C^0(\overline{R})}\}$ is bounded and then, using (3.121), that $\{h_j\}$ is bounded in $C_{per}^{2,\alpha}(\overline{R})$. The space $C_{per}^2(\overline{R})$ being compactly embedded in $C_{per}^{2,\alpha}(\overline{R})$, we can find a subsequence $\{h_{j_k}\}$ that converges in $C_{per}^2(\overline{R})$ to a classical solution $\hat{h} \in C_{per}^2(\overline{R})$ corresponding to f.

[64] An extension of the Riesz representation theorem for bilinear forms: the particular case when B is the inner product of H shows how closely related this is to the Riesz representation theorem. The Lax–Milgram theorem is a very useful tool in establishing the existence of weak solutions to elliptic partial differential equations.

[65] That is, B is bilinear and there exist constants $b_1, b_2 > 0$ with $|B(u,v)| \leq b_2\, \|u\|_H \|v\|_H$ for all $u,\, v \in H$ and $B(u,u) \geq b_1\, \|u\|_H^2$ for all $u \in H$.

[66] B is actually the new inner product.

By uniqueness of the weak solution $\hat{h} = h_f$. Thus $h_f \in C^2_{per}(\overline{R})$. Since along the subsequence that converges to $\hat{h}$ in $C^2_{per}(\overline{R})$ the right side of (3.121) converges to $\hat{c} = c_\alpha \left(\|\hat{h}\|_{C^0(\overline{R})} + \|f\|_{C^{0,\alpha}(\overline{R})} \right)$, we deduce that $\hat{h} \in C^{2,\alpha}_{per}(\overline{R})$ with $\|\hat{h}\|_{C^{2,\alpha}_{per}(\overline{R})} \le \hat{c}$. For example, given $\varepsilon > 0$, for j_k suficiently large we have by (3.121) that

$$|\partial_q^2 h_{j_k}(\Theta) - \partial_q^2 h_{j_k}(\Theta_1)| \le (\hat{c} + \varepsilon)\,|\Theta - \Theta_1|^\alpha$$

for all $\Theta, \Theta_1 \in R$, so $|\partial_q^2 \hat{h}(\Theta) - \partial_q^2 \hat{h}(\Theta_1)| \le (\hat{c} + \varepsilon)\,|\Theta - \Theta_1|^\alpha$. This yields the desired regularity statement: for $f \in C^{0,\alpha}_{per}(\overline{R})$ the weak solution h_f is a classical solution $h_f \in C^{2,\alpha}_{per}(\overline{R})$ and (3.115) holds. Using again difference quotients one can show (see [129, Chapter 6]) that if $d \in C^\infty_{per}(\overline{R})$, then for $f \in C^{m,\alpha}_{per}(\overline{R})$ we have $h_f \in C^{m+2,\alpha}_{per}(\overline{R})$. ∎

The Schauder estimates (3.115) and (3.114) are not only important for proving regularity but also, by means of the *continuity method* (see [129]), for establishing the solvability of more general elliptic operators with the homogeneous boundary conditions (3.113) and subject to periodicity in the q-variable.

Example 3.22. We will illustrate this approach by considering instead of the Poisson equation (3.112) the variable coefficient uniformly elliptic equation

$$\partial_q\,(a_{20}\,h_q + a_{11}\,h_p) + \partial_p\,(a_{11}\,h_q + a_{02}\,h_p) = f \quad \text{in} \quad R$$

with $a_{ij} \in C^{1,\alpha}_{per}(\overline{R})$, $f \in C^{0,\alpha}_{per}(\overline{R})$. Let us write the boundary value problem in the abstract form $\mathcal{L}h = f$ with $\mathcal{L} : X \to Y$ a bounded linear operator (incorporating the boundary values) from the Banach space

$$X = \{h \in C^{2,\alpha}_{per}(\overline{R}) : \; h(q,0) = h(q,p_0) = 0 \text{ for all } q \in [-\pi,\pi]\}$$

to the Banach space $Y = C^{0,\alpha}_{per}(\overline{R})$. The continuity method consists in introducing a parameter $t \in [0,1]$ connecting the problem $\mathcal{L}h = f$ to the simpler Poisson equation, written in the form $\mathcal{L}_0 h = f$ with $\mathcal{L}_0 : X \to Y$, by means of the operators

$$\mathcal{L}_t = t\mathcal{L} + (1-t)\mathcal{L}_0.$$

For $t = 1$ and $t = 0$ we recover our problem, respectively, the Poisson equation. We know from the previous example that $\mathcal{L}_0 : X \to Y$ is invertible and one can easily check that for all $t \in (0,1)$ we have a uniformly elliptic operator, so that (see [129, Chapters 6 and 9]) the Schauder estimates (3.115) and (3.114) will hold for $t \in [0,1]$; that is,

$$\|h\|_X \le c_\alpha\,(\|h\|_{C^0_{per}(\overline{R})} + \|\mathcal{L}_t h\|_Y), \qquad h \in X, \quad t \in [0,1],$$

$$\|h\|_{H^2_{per}(R)} \le \tilde{c}(\|h\|_{L^2(R)} + \|\mathcal{L}_t h\|_{L^2(R)}), \quad h \in H^2_{per}(R) \cap H^1_{0,per}(R), \quad t \in [0,1],$$

for some constants $c_\alpha, \tilde{c} > 0$. Since $Y \subset L^2(R)$, $X \subset H^2_{per}(R) \cap H^1_{0,per}(R) \subset C^0_{per}(\overline{R})$, the previous two inequalities lead to

$$\|h\|_X \le C\,\|\mathcal{L}_t h\|_Y, \qquad h \in X, \quad t \in [0,1], \tag{3.122}$$

for some constant $C > 0$. Indeed, for $h \in X$ we have

$$\begin{aligned}\|h\|_{L^2(R)}\|\mathcal{L}_t h\|_{L^2(R)} &\ge -\int\!\!\int_R h\,(\mathcal{L}_t h)\,dq\,dp = (1-t)\int\!\!\int_R (h_q^2 + h_p^2)\,dq\,dp \\ &+ t\int\!\!\int_R \left(a_{20} h_q^2 + 2a_{11}\,h_q h_p + a_{02}\,h_p^2\right) dq\,dp \ge [1 - t + \lambda t]\int\!\!\int_R (h_q^2 + h_p^2)\,dq\,dp,\end{aligned}$$

using integration by parts and (3.116). In combination with (3.120) this yields

$$\|\mathcal{L}_t h\|_{L^2(R)} \geq c[1-t+\lambda t]\,\|h\|_{L^2(R)}, \qquad h \in X.$$

We now infer by the continuity method (see [129, Section 5.2]) that the considered problem has a unique classical solution $h \in X$ since $\mathcal{L}_0$ is onto and (3.122) holds. The argument is that we can find a constant $\delta > 0$ such that for every $s \in [0,1]$ for which $\mathcal{L}_s$ in invertible, all operators $\mathcal{L}_t$ with $|t-s| < \delta$ will be invertible (so that, starting at $s=0$ and dividing $[0,1]$ into subintervals of length $\delta/2$, we can reach $t=1$ and all operators $\mathcal{L}_t$ are invertible). Indeed, for $y \in Y$ the equation $\mathcal{L}_t x = y$ is equivalent to

$$\mathcal{L}_s x = y - (\mathcal{L}_t - \mathcal{L}_s)x = y - (t-s)\mathcal{L}x + (t-s)\mathcal{L}_0 x.$$

As the linear operator $\mathcal{L}_s$ is invertible, this is equivalent to $x = \mathcal{L}_s^{-1} y - (t-s)\mathcal{L}_s^{-1}(\mathcal{L}-\mathcal{L}_0)x$. But the map $x \mapsto \mathcal{L}_s^{-1} y - (t-s)\mathcal{L}_s^{-1}(\mathcal{L}-\mathcal{L}_0)x$ is a contraction on X if

$$|t-s|\,\|\mathcal{L}_s^{-1}(\mathcal{L}-\mathcal{L}_0)\|_{\mathcal{B}(X,X)} < 1.$$

Since

$$\begin{aligned} \|\mathcal{L}_s^{-1}(\mathcal{L}-\mathcal{L}_0)\|_{\mathcal{B}(X,X)} &\leq \|\mathcal{L}_s^{-1}\|_{\mathcal{B}(Y,X)}\,\|\mathcal{L}-\mathcal{L}_0\|_{\mathcal{B}(X,Y)} \\ &\leq \|\mathcal{L}_s^{-1}\|_{\mathcal{B}(Y,X)}\,(\|\mathcal{L}\|_{\mathcal{B}(X,Y)} + \|\mathcal{L}_0\|_{\mathcal{B}(X,Y)}) \leq C\,(\|\mathcal{L}\|_{\mathcal{B}(X,Y)} + \|\mathcal{L}_0\|_{\mathcal{B}(X,Y)}) \end{aligned}$$

in view of (3.122), if we set $\delta = \frac{1}{2C\,(\|\mathcal{L}\|_{\mathcal{B}(X,Y)} + \|\mathcal{L}_0\|_{\mathcal{B}(X,Y)})}$, the existence of a unique fixed point (and thus, of a unique solution to $\mathcal{L}_t x = y$) follows from Banach's contraction principle (see [129, Section 5.1]). ∎

The complementing condition

An important issue is whether the previous considerations (a priori estimates, the Fredholm property of the operator corresponding to the boundary value problem, regularity theorems) hold for uniformly elliptic operators[67]

$$\mathcal{L} = \sum_{i+j=2} a_{ij}\,\partial_q^i \partial_p^j + b_1\,\partial_q + b_2\,\partial_p + c \tag{3.123}$$

whose coefficients $a_{ij}, b_i, c \in C^{0,\alpha}_{per}(\overline{R})$ for some $\alpha \in (0,1)$, with linear boundary conditions

$$B_1^+(q)h_q + B_2^+(q)h_p + B_0^+(q)h = B^+(q) \quad \text{on} \quad p=0 \tag{3.124}$$

and

$$B_1^-(q)h_q + B_2^-(q)h_p + B_0^-(q)h = B^-(q) \quad \text{on} \quad p=p_0, \tag{3.125}$$

where $B_1^\pm, B_2^\pm \in C^{1,\alpha}_{per}(\mathbb{R})$ and $B^\pm, B_0^\pm \in C^{2,\alpha}_{per}(\mathbb{R})$ are given functions. A remarkable result (see [3, 4]) reduces this problem to checking the complementing condition.[68]

[67] The uniform ellipticity condition is (3.116) even though the operator is not in divergence form since only the principal part (involving the derivatives of highest order) matters. Notice that if $a_{ij} \in C^{2,\alpha}_{per}(\overline{R})$ and $b_i \in C^{1,\alpha}_{per}(\overline{R})$, then (3.123) can be written in divergence form.

[68] Or Lopatinski–Shapiro condition, formulated as a boundary value problem for an ordinary differential equation (an equivalent algebraic formulation as a "covering condition" exists but is generally more difficult to check). The study of these problems was initiated by a number of mathematicians, with [213, 273] among the pioneering results, and culminated in the very general papers [3, 4]. Results of this type hold more generally for systems of elliptic equations in $\mathbb{R}^n$ with $n \geq 2$ but we are mainly interested in the scalar case (that is, one equation) in $\mathbb{R}^2$.

Definition 3.23. *The complementing condition holds at a boundary point* $\Theta_0^+ = (q_0, 0)$ *or* $\Theta_0^- = (q_0, p_0)$ *if the ordinary differential equation*

$$\xi^2 a_{20}(\Theta_0^\pm) w(y) \mp 2i\xi a_{11}(\Theta_0^\pm) w'(y) - w''(y) a_{02}(\Theta_0^\pm) = 0 \tag{3.126}$$

for $y > 0$, *with the boundary conditions*[69]

$$\begin{aligned} & B_1^\pm(q_0)\xi \mp i B_2^\pm(q_0) w'(0) = 0 \quad \textit{if} \quad \left(B_1^\pm(q_0), B_2^\pm(q_0)\right) \neq (0,0) \\ & \textit{or} \\ & B_0^\pm(q_0) w(0) = 0 \quad \textit{if} \quad B_1^\pm(q_0) = B_2^\pm(q_0) = 0, \end{aligned} \tag{3.127}$$

and

$$|w(y)| \to 0 \quad \textit{for} \quad y \to \infty, \tag{3.128}$$

has only the trivial solution for all $\xi \in \mathbb{R} \setminus \{0\}$.

More precisely, the following three statements are equivalent[70] (see [317, Section 13], and [139, Chapter 3]):

- the complementing condition holds at every boundary point;
- the operator $(\mathcal{L}, \mathbb{B}) : W_{per}^{2,2}(R) \to L^2(R) \times [W_{per}^{3/2-l^+,2}(\mathbb{R}) \times W_{per}^{3/2-l^-,2}(\mathbb{R})]$ is Fredholm, where $l^\pm \in \{0,1\}$ is the order of the boundary operator $\mathbb{B}$;
- the a priori estimate

$$\|h\|_{W_{per}^{2,2}(R)} \le c(\|\mathcal{L}h\|_{L^2(R)} + \|h\|_{W_{per}^{1,2}(R)} + \|\mathbb{B}h\|_{W_{per}^{3/2-l^+,2}(\mathbb{R}) \times W_{per}^{3/2-l^-,2}(\mathbb{R})})$$

  holds for all $h \in W_{per}^{2,2}(R)$.

Pertaining regularity, similar estimates can be obtained for higher-order derivatives (see [317], Section 13): if for an integer $k \ge 0$ we have $a_{ij}, b_i, c \in C_{per}^{k,\alpha}(\overline{R})$, $B_1^\pm, B_2^\pm \in C_{per}^{k+1,\alpha}(\mathbb{R})$, $B_0^\pm$, $B^\pm \in C_{per}^{k+2,\alpha}(\mathbb{R})$, and if the complementing condition holds all along the boundary $p = 0$ and $p = p_0$, then for any $h \in W_{per}^{j+2,2}(R)$ with $0 \le j \le k$, we have[71]

$$\|h\|_{W_{per}^{j+2,2}(R)} \le c_j(\|\mathcal{L}h\|_{W_{per}^{j,2}(R)} + \|h\|_{W_{per}^{j+1,2}(R)} + \|\mathbb{B}h\|_{W_{per}^{j+3/2-l^+,2}(\mathbb{R}) \times W_{per}^{j+3/2-l^-,2}(\mathbb{R})})$$

---

[69] Obtained by writing the principal parts of the differential operator and of the boundary operator formally as a polynomial in ∂_q and ∂_p and subsequently replacing ∂_q by ξ and the derivative in the outward normal direction (that is, ∂_p for Θ_0^+ and $-\partial_p$ for Θ_0^-) by $(-i\,\partial_y)$.

[70] Here the boundary term involves a fractional Sobolev norm induced by the inner product

$$(u,v)_{W_{per}^{s+1/2,2}(\mathbb{R})} = (u,v)_{H_{per}^s(\mathbb{R})} + \int_{-\pi}^{\pi}\int_{-\pi}^{\pi} \frac{\{\partial_q^s u(\Theta) - \partial_q^s u(\Theta_1)\}\{\partial_q^s v(\Theta) - \partial_q^s v(\Theta_1)\}}{|\Theta - \Theta_1|^2} d\Theta d\Theta_1$$
$$+ \int_{-\pi}^{\pi}\int_{-\pi}^{\pi} \frac{\{\partial_p^s u(\Theta) - \partial_p^s u(\Theta_1)\}\{\partial_p^s v(\Theta) - \partial_p^s v(\Theta_1)\}}{|\Theta - \Theta_1|^2} d\Theta d\Theta_1$$

with the integer $s \ge 0$. The Banach spaces $W_{per}^{s+1/2,2}(\mathbb{R})$ are the trace spaces of $H_{per}^{s+1}(R)$, the trace operator from $H_{per}^{s+1}(R)$ to $W_{per}^{s+1/2,2}(\mathbb{R}) \times W_{per}^{s+1/2,2}(\mathbb{R})$ being bounded (see [200, Chapter 15]).

[71] It is often possible to replace the norm $\|h\|_{W_{per}^{j+1,2}(R)}$ on the right side by $\|h\|_{L^2(R)}$. For example, in the case of homogeneous Dirichlet boundary conditions for the Poisson equation (3.112), for any $h \in H_{per}^2(R) \cap H_{0,per}^1(R)$, integration by parts yields $-\int\int_R h\,\Delta h\,dq dp = \int\int_R (h_q^2 + h_p^2) dq dp$, and the left side is at most $\|\Delta h\|_{L^2(R)} \|h\|_{L^2(R)}$ by Schwarz's inequality. This way we recover the estimate (3.114) with $s = 2$.

for a constant c_j independent of h. Moreover, the defect numbers and the index of the Fredholm operator $(\mathcal{L}, \mathbb{B})$ are independent of $j \in \{0,\dots,k\}$. Similar estimates hold in the Hölder space setting (see [236, Section 5.6] and [139, Chapter 3]): if $a_{ij}, b_i, c \in C^{k,\alpha}_{per}(\overline{R})$, $B_1^\pm, B_2^\pm \in C^{k+1,\alpha}_{per}(\mathbb{R})$, $B_0^\pm$, $B^\pm \in C^{k+2,\alpha}_{per}(\mathbb{R})$ for an integer $k \geq 0$ and a constant $\alpha \in (0,1)$, and if the complementing condition holds all along the boundary $p=0$ and $p=p_0$, then

$$\|h\|_{C^{j+2,\alpha}_{per}(\overline{R})} \leq c_j \left(\|\mathcal{L}h\|_{C^{j,\alpha}_{per}(\overline{R})} + \|h\|_{C^{j,\alpha}_{per}(\overline{R})} + \|\mathbb{B}h\|_{C^{j+2-l^+,\alpha}_{per}(\mathbb{R})\times C^{j+2-l^-,\alpha}_{per}(\mathbb{R})} \right)$$

for a constant c_j independent of $h \in C^{j+2,\alpha}_{per}(R)$ with $0 \leq j \leq k$.

Example 3.24. For the Laplacian (3.126) becomes $\xi^2 w - w'' = 0$ for $y > 0$, the general solution of which is $w(y) = d_1 e^{\xi y} + d_2 e^{-\xi y}$ with $d_1, d_2 \in \mathbb{C}$ constants. Without loss of generality, we may assume that $\xi > 0$. To fix ideas, let us first consider the boundary point Θ_0^+. The decay condition (3.128) forces $d_1 = 0$ so that (3.127) becomes $d_2 B_0^+(q_0) = 0$ if $B_1^+(q_0) = B_2^+(q_0) = 0$ and in this case we always have $w \equiv 0$ if and only if $B_0^+(q_0) \neq 0$. On the other hand, if either $B_1^+(q_0) \neq 0$ or $B_2^+(q_0) \neq 0$, (3.127) becomes $B_1^+(q_0) d_2 \xi + i\, B_2^+(q_0) d_2 \xi = 0$ so that $d_2 = 0$. Therefore the complementing condition holds at $\Theta_0^+ = (q_0, 0)$ if and only if at least one of the three values $B_0^+(q_0)$, $B_1^+(q_0)$, $B_2(q_0)^+$ is nonzero.[72] The case of a boundary point Θ_0^- is analogous. ∎

Example 3.25. Consider the operator $\mathcal{L} = \Delta + d$, where $d \in C^{0,\alpha}_{per}(\overline{R})$ for some $\alpha \in (0,1)$, with Dirichlet boundary conditions (that is, $B_1^\pm = B_2^\pm \equiv 0$ and $B_0^\pm \equiv 1$ on $\mathbb{R}$). If $B^\pm \in C^{2,\alpha}_{per}(\mathbb{R})$, let $B \in C^{2,\alpha}_{per}(\overline{R})$ be given by $B(q,p) = \frac{p_0 - p}{p_0} B^+(q) + \frac{p}{p_0} B^-(q)$. Then the inhomogeneous problem $\mathcal{L}h = f$ in R and $h = B$ for $p \in \{0, p_0\}$ is equivalent to the problem $\mathcal{L}w = f - \mathcal{L}B$ in R and $w = 0$ for $p \in \{0, p_0\}$. For any constant $c \geq \max_{(q,p)\in\overline{R}} \{d(q,p)\}$ we saw in Example 3.21 that $\mathcal{L}_c = \mathcal{L} - c : X = C^{2,\alpha}_{0,per}(\overline{R}) \to C^{0,\alpha}_{per}(\overline{R})$ is an invertible bounded linear operator[73]. The considerations in Example 3.24 show that the complementing condition holds for $\mathcal{L}_c$. The associated Schauder estimate ensures that the inverse mapping $\mathcal{L}_c^{-1}$ is a compact mapping from $C^{0,\alpha}_{per}(\overline{R})$ into $C^2_{per}(\overline{R})$, and hence it is also compact as a mapping from $C^{0,\alpha}_{per}(\overline{R})$ into itself. Since the Fredholm index of a compact perturbation of the identity is zero, by the Fredholm alternative applied to the equation $w + c\,\mathcal{L}_c^{-1} w = \mathcal{L}_c^{-1} f$ in $C^{0,\alpha}_{per}(\overline{R})$, with given $f \in C^{0,\alpha}_{per}(\overline{R})$, we always have a solution $w \in C^{0,\alpha}_{per}(\overline{R})$ if the homogeneous equation $w + c\,\mathcal{L}_c^{-1} w = 0$ has only the trivial solution $w = 0$ in $C^{0,\alpha}_{per}(\overline{R})$. Since $\mathcal{L}_c^{-1}$ maps $C^{0,\alpha}_{per}(\overline{R})$ onto X, any solution in $C^{0,\alpha}_{per}(\overline{R})$ to the equation $w + c\,\mathcal{L}_c^{-1} w = \mathcal{L}_c^{-1} f$ must belong to X. Operating on the equation with $\mathcal{L}_c$, we obtain $\mathcal{L}w = f$. Thus uniqueness is a sufficient condition for existence: if the homogeneous problem, $\mathcal{L}w = 0$ in R and $w = 0$ for $p \in \{0, p_0\}$, has only the trivial solution in $C^{2,\alpha}_{per}(\overline{R})$, then the inhomogeneous problem $\mathcal{L}w = f$ in R and $w = B^\pm$ for $p \in \{0, p_0\}$, has a unique solution $w \in C^{2,\alpha}_{per}(\overline{R})$ for all $f \in C^{0,\alpha}_{per}(\overline{R})$ and $B^\pm \in C^{2,\alpha}_{per}(\mathbb{R})$. ∎

Nonlinear elliptic boundary value problems

Finding a central core theory of nonlinear elliptic boundary value problems proved elusive and only a few examples are understood in a really satisfactory way—nonlinearity manifests itself in many ways and this seems to preclude an all-encompassing nonlinear theory, in contrast to the case of linear problems. Moreover, the hydrodynamical context in which the quasi-linear elliptic boundary value problem (3.4), that is of interest to us, arises dictates the search of an as large as possible family

[72] This is an artifact of the two-dimensional character: tangential boundary operators for the Laplacian in dimension $n \geq 3$ are not of complementing type.

[73] Here the lower index 0 means that $w = 0$ on $p = 0$ and on $p = p_0$.

of solutions (since each solution corresponds to a water wave). Within the context of water waves the bifurcation approach pursued in this chapter is more advantageous than the existence approaches towards nonlinear problems presented in [129] since, for example, for waves of small amplitude it provides us with all solutions. However, there are general results that are relevant to our problem. We now briefly present a far-reaching regularity result of this type. Given $p_0 < 0$, for $\Theta = (q,p)$, $h \in \mathbb{R}$, $W \in \mathbb{R}^2$, $M \in \mathbb{R}^3$, let $F(\Theta,h,W,M)$ and $B(q,h,W)$ be 2π-periodic in the q-variable and real analytic[74] in $R_0 \times U_0 \times U_1 \times U_2 \subset \mathbb{R}^8$, respectively, in $\mathbb{R} \times U_0 \times U_1 \subset \mathbb{R}^4$, where $R_0 \subset \mathbb{R}^2$, $U_0 \subset \mathbb{R}$, $U_1 \subset \mathbb{R}^2$, $U_2 \subset \mathbb{R}^3$ are open sets[75] with $\overline{R} \subset R_0$, and consider the nonlinear boundary value problem

$$\begin{cases} F(q,p,h,h_q,h_p,h_{qq},h_{qp},h_{pp}) = 0 \quad \text{in} \quad R, \\ B(q,h,h_q,h_p) = 0 \quad \text{on} \quad p = 0, \\ h = 0 \quad \text{on} \quad p = p_0. \end{cases} \tag{3.129}$$

If for an open set $U \subset C^2_{per}(\overline{R})$ we have

$$\Big(h(q,p), h_q(q,p), h_p(q,p), h_{qq}(q,p), h_{qp}(q,p), h_{pp}(q,p)\Big) \in U_0 \times U_1 \times U_2, \qquad (q,p) \in \overline{R},$$

and the linearization of (3.129) about any element $u \in U$, namely,

$$\begin{cases} \mathcal{L}_u\, v = 0 \quad \text{in} \quad R, \\ \dfrac{d}{dt} B(q,p,u+tv,q_q+tv_q,u_p+tv_p)\Big|_{t=0} = 0 \quad \text{on} \quad p = 0, \\ v = 0 \quad \text{on} \quad p = p_0, \end{cases} \tag{3.130}$$

with

$$\mathcal{L}_u\, v = \frac{d}{dt} F(q,p,u+tv,u_q+tv_q,u_p+tv_p,u_{qq}+tv_{qq},u_{qp}+tv_{qp},u_{pp}+tv_{pp})\Big|_{t=0} = 0,$$

is such that the operator $\mathcal{L}_u$ is second-order uniformly elliptic and the complementing condition holds for (3.130) at every point on $p = 0$ and $p = p_0$, then any solution $h \in U$ of (3.131) is real analytic in R and can be continued across the boundaries $p = 0$ and $p = p_0$ of R as a real-analytic function (see the classical approach in [236, Chapter 6], as well as the discussions in [175, 176], and the recent alternative proof in [240]). A further relevant regularity result holds with lower regularity assumptions (see [3, Chapter 11]): if for some integer $k \geq 0$ and some $\alpha \in (0,1)$ we have that[76] $F \in C^{k,\alpha}(R_0 \times U_0 \times U_1 \times U_2)$, $B \in C^{k+1,\alpha}(R_0 \times U_0 \times U_1)$ are 2π-periodic in the q-variable with $\frac{\partial B}{\partial w_2} \neq 0$ all along[77] $p = 0$, and if the complementing condition holds for the linearization at every point on $p = 0$ and on $p = p_0$, then a $C^{2,\alpha}_{per}(\overline{R})$ solution to (3.129) is of class $C^{k+2,\alpha}_{per}(\overline{R})$.

[74] A function is real analytic in an open set if it is infinitely differentiable and for every point within this set the (infinite) Taylor series converges to the value of the function in a neighborhood of the point. A real-analytic function in $\mathbb{R}^k$ is locally the restriction to $\mathbb{R}^k$ of a complex-analytic function of several variables (see [186, Chapter 4]).

[75] In many cases analyticity is granted only in some open subset, usually characterized by inequalities that are expressed in terms of the function and its first- or second-order derivatives.

[76] In this context, notice that Hölder continuous functions on closures $\overline{O}$ of open sets $O \subset \mathbb{R}^m$ admit extensions to $\mathbb{R}^m$; cf. [112].

[77] This corresponds to an oblique boundary condition for the linearization, that is, $B_2^{\pm} \neq 0$ in (3.124) and (3.125). If this is not the case, in general one has to assume $B \in C^{k+2,\alpha}(R_0 \times U_0 \times U_1)$: for example, if $F = \Delta$ with Dirichlet boundary conditions $h(q,0) = B^+(q)$ and $h(q,p_0) = 0$, then $B^+ \in C^{2,\alpha}_{per}(\mathbb{R})$ is clearly necessary for the solution to belong to $C^{2,\alpha}_{per}(\overline{R})$.

Example 3.26. Let $d(\Theta,h,W)$ and $\{a_{ij}(\Theta,h,W)\}_{i,j\geq 0,i+j=2}$ be 2π-periodic in the q-variable and real analytic (or of class $C^{k,\alpha}$ for some $\alpha\in(0,1)$ and some integer $k\geq 0$) in $R_0\times U_0\times U_1\subset\mathbb{R}^5$, where $R_0\subset\mathbb{R}^2$, $U_0\subset\mathbb{R}$, and $U_1\subset\mathbb{R}^2$ are open sets with $\overline{R}\subset R_0$. Consider the quasi-linear boundary value problem

$$\begin{cases} \mathcal{L}u := \sum\limits_{i+j=2} a_{ij}(q,p,h,h_q,h_p)\,\partial_q^i\partial_p^j h + d(q,p,h,h_q,h_p) = 0 \quad \text{in} \quad R, \\ h_p = b(q,h,h_q) \quad \text{on} \quad p=0, \\ h=0 \quad \text{on} \quad p=p_0, \end{cases} \tag{3.131}$$

where b is 2π-periodic in the q-variable and real-analytic (or of class $C^{k+1,\alpha}$) in $\mathbb{R}\times U_0\times\mathbb{R}$. If for an open set $U\subset C^2_{per}(\overline{R})$ we have

$$h(q,p)\in U_0, \quad \Big(h_q(q,p),h_p(q,p)\Big)\in U_1, \qquad (q,p)\in\overline{R},$$

and[78]

$$(a_{02}\,a_{20}-a_{11}^2)\Big|_{h=u(\Theta),\,W=\nabla u(\Theta)} \geq \varepsilon, \qquad \Theta=(q,p)\in\overline{R}, \tag{3.132}$$

for some $\varepsilon>0$, then any solution $h\in U$ of (3.131) is real-analytic in a neighborhood of $\overline{R}$, or of class $C^{k+2,\alpha}_{per}(\overline{R})$ under the milder $C^{k,\alpha}$ and $C^{k+1,\alpha}$ regularity assumptions on the coefficients of the differential operator, respectively, on the boundary function.

Indeed, the principal part of the linearization of (3.131) about $u\in U$ is precisely

$$\begin{cases} a_{20}\big(q,p,u(q,p),u_q(q,p),u_p(q,p)\big)\,v_{qq} + 2a_{11}\big(q,p,u(q,p),u_q(q,p),u_p(q,p)\big)\,v_{qp} \\ \qquad + a_{02}\big(q,p,u(q,p),u_q(q,p),u_p(q,p)\big)\,v_{pp}=0 \quad \text{in} \quad R, \\ v_p=b_u(q)\,v_q \quad \text{on} \quad p=0, \\ v=0 \quad \text{on} \quad p=p_0, \end{cases} \tag{3.133}$$

with $b_u(q)=\dfrac{\partial b(q,u(q,0),z)}{\partial z}\Big|_{z=u_q(q,0)}$. Clearly (3.132) ensures uniform ellipticity. In this case (3.126) is

$$a_{20}\,\xi^2\,w - 2i\,a_{11}\,\xi\,w' - a_{02}\,w''=0$$

with the general solution

$$w(y)=c_+e^{r_+\xi y}+c_-e^{r_-\xi y},$$

where

$$r_\pm = -i\,\frac{a_{11}}{a_{02}} \pm \frac{\sqrt{a_{20}a_{02}-a_{11}^2}}{a_{02}}$$

and $c_\pm\in\mathbb{C}$ are arbitrary constants. Assuming $\xi>0$ without loss of generality, the decay condition (3.128) yields $c_+=0$ because of (3.132), so that (3.127) becomes

$$b_u(q)\xi\,c_- + i\,r_-\xi\,c_-=0 \quad \text{on} \quad p=0.$$

Since $\Re e(r_-)\neq 0$ we get $c_-=0$ so that $w\equiv 0$ and the complementing condition holds on $p=0$. Checking the complementing condition on $p=p_0$ is similar. ∎

[78]Without loss of generality we assumed that $a_{12}=a_{21}$.

We conclude the discussion by presenting a Schauder-type estimate for nonlinear elliptic equations with nonlinear oblique boundary conditions due to Lieberman and Trudinger [205]. Consider the oblique nonlinear elliptic boundary value problem

$$\begin{cases} F(\Theta,h,Dh,D^2h)=0 \quad \text{in} \quad R, \\ G(q,h,Dh)=0 \quad \text{on} \quad p=0, \\ h=0 \quad \text{on} \quad p=p_0, \end{cases} \tag{3.134}$$

where $F \in C^2(\overline{R}\times\mathbb{R}\times\mathbb{R}^2\times\mathbb{S},\mathbb{R})$ and $G \in C^2(\mathbb{R}\times\mathbb{R}\times\mathbb{R}^2,\mathbb{R})$, $\mathbb{S}$ being the three-dimensional linear space of 2×2 real symmetric matrices, while $Dh=(h_q,h_p)$ and $D^2h=[D_{ij}h]_{1\le i,j\le 2}$ denote the gradient and Hessian of the real-valued function $h(\Theta)$, which is 2π-periodic in the first variable of $\Theta=(q,p)$. By a classical solution of (3.134) we mean a function $h \in C^2(\mathbb{R}\times(p_0,0),\mathbb{R}) \cap C^1(\mathbb{R}\times[p_0,0])$ that is 2π-periodic in the q-variable and satisfies the equations in (3.134) in a pointwise sense. The nonlinear operator F is *elliptic* at $(\Theta,h,\xi,r)\in R\times\mathbb{R}\times\mathbb{R}^2\times\mathbb{S}$ if the matrix $F_r=[\partial F/\partial r_{ij}]_{1\le i,j\le 2}$ is positive definite at (Θ,h,ξ,r), being *uniformly elliptic* if the ratio Λ_2/Λ_1 between the largest and smallest eigenvalue remains bounded. The operator G is *oblique* at $(q,h,\xi)\in\mathbb{R}\times\mathbb{R}\times\mathbb{R}^2$ if $\chi=G_\xi\cdot(0,-1)$ is positive. The uniformly elliptic operator F and the oblique operator G are said to satisfy *natural* structure conditions if there exists a nondecreasing real function μ such that for all $(\Theta,h,\xi,r)\in R\times\mathbb{R}\times\mathbb{R}^2\times\mathbb{S}$,

- $\Lambda_2 \le \mu(|h|)\Lambda_1$;
- $|F(\Theta,h,\xi,0)| \le \Lambda_1\mu(|h|)(1+|\xi|^2)$;
- $(1+|\xi|)|F_\xi|+|F_h|+|F_\Theta| \le \Lambda_1\mu(|h|)(1+|\xi|^2+|r|)$;
- $(1+|r|)(|F_{r\Theta}|+|F_{rh}|+|F_{r\xi}|)+|F_{\xi\xi}|+|F_{\xi\Theta}|+|F_{h\xi}|+|F_{hh}|+|F_{h\Theta}|+|F_{\Theta\Theta}|$ $\le \Lambda_1\mu(|h|+|\xi|)(1+|r|)$;
- $F_{rr}\le 0$ (that is, F is concave with respect to r);
- $|G(q,h,\xi_1,0)| \le \chi\,\mu(|h|)(1+|\xi_1|)$;
- $(1+|\xi|)|G_\xi|+|G_\Theta|+|G_h| \le \chi\,\mu(|h|)(1+|\xi|)$.

Then, if $\alpha\in(0,1)$ and $h\in C^2_{per}(\overline{R})$ is a solution to (3.134), then $h\in C^{2,\alpha}_{per}(\overline{R})$ and there is a positive constant C, depending only on $\|h\|_{C^1_{per}(\overline{R})}$, such that

$$\|h\|_{C^{2,\alpha}_{per}(\overline{R})} \le C. \tag{3.135}$$

3.5.5 Maximum principles

The symmetry approach for periodic traveling waves presented in this chapter was based on the device of moving parallel lines to a critical position and then showing that the solution is symmetric about the limiting line—the moving plane method [128]. Throughout the analysis we employed sharp maximum principles for elliptic partial differential equations, which we present now in a form suitable for our purposes (for a proof we refer the reader to [116]).

Lemma 3.27. *Let $\Omega\subset\mathbb{R}^2$ be a rectangle, $w\in C^2(\overline{\Omega})$, and suppose that $\mathcal{L}w\ge 0$ (≤ 0) in Ω for some uniformly elliptic operator $\mathcal{L}=\sum_{i+j=2}a_{ij}\partial_q^i\partial_p^j+\sum_{i=1}^2 b_i\partial_i$ with continuous coefficients in $\overline{\Omega}$. Then the following hold:*

(i) **The strong maximum principle.** *If w attains its maximum (minimum) in the interior of Ω, then w is a constant in $\overline{\Omega}$.*

(ii) **The Hopf boundary point lemma.** *Let Θ_0 be a point on the smooth part of the boundary $\partial\Omega$ such that $w(\Theta_0) > w(\Theta)$ (or $w(\Theta_0) < w(\Theta)$) for all $\Theta \in \Omega$. Then the outer normal derivative of w at Θ_0 is strictly positive (negative).*

(iii) **The Serrin egde point lemma.** *Let Θ_0 be a corner point of $\partial\Omega$ such that $w(\Theta_0) > w(\Theta)$ (or $w(\Theta_0) < w(\Theta)$) for all $\Theta \in \Omega$. If $a_{11}(\Theta_0) = 0$, and τ is a unit vector outward from Ω at Θ_0, then either the first or the second derivative of w in the direction of τ is strictly positive (negative).*

In the previous result the boundedness of Ω ensures the existence of a maximum and minimum of w (in contrast to this, the unbounded function $w(q,p) = \sin(p)\sinh(q)$ is harmonic in the strip $\mathbb{R} \times [-\pi, 0]$, with $w = 0$ on the boundary $p = 0$ and $p = -\pi$). The lack of a zero-order term in $\mathcal{L}$ is important: the function $w(q,p) = \cos(q/2)\sin(p+\pi)$ satisfies $\Delta w + \frac{5}{4}w = 0$ in $\Omega = (-\pi,\pi) \times (-\pi, 0)$, is zero on $\partial\Omega$, and attains its maximum at the interior point $(0, -\pi/2)$. If a zero-order term $c(q,p) \le 0$ in $\overline{\Omega}$ is added to $\mathcal{L}$ and if the nonnegative maximum of $w \in C^2(\Omega) \cap C^1(\overline{\Omega})$ satisfying $(\mathcal{L}+c)w \ge 0$ in Ω is attained at $\Theta_0 \in \overline{\Omega}$, then the **weak maximum principle** (see [129, Chapter 3]) ensures that, unless w is constant in $\overline{\Omega}$, $\Theta_0 \in \partial\Omega$ and, if Θ_0 lies on the smooth part[79] of $\partial\Omega$, then $\nabla w(\Theta_0) \ne (0,0)$. It is possible to allow $c(q,p) > 0$ in $\overline{\Omega}$ in the following sense (cf. Appendix B2 in [292]). Assume that $w \in C^2(\Omega) \cap C^1(\overline{\Omega})$ satisfies $(\mathcal{L}+c)w \ge 0$ in Ω and $w \le 0$ on $\partial\Omega$, and that there exists $w_0 \in C^2(\Omega) \cap C^0(\overline{\Omega})$ with $w_0(q,p) < 0$ in $\overline{\Omega}$ and $(\mathcal{L}+c)w_0 \ge 0$ in Ω. Then either $w < 0$ in Ω or $w = \beta w_0$ for some constant $\beta \le 0$.

Example 3.28. For $\mathcal{L} = \Delta$ in $\Omega = (-\pi,\pi) \times (p_0, 0)$ we may choose

$$w_0(q,p) = -\cos\left(\frac{q}{2(1+\varepsilon)}\right)\sin\left(\pi\,\frac{p - p_0 + \varepsilon}{|p_0| + 2\varepsilon}\right)$$

to infer that if[80] $c(q,p) < \frac{1}{4} + \frac{\pi^2}{p_0^2}$ in $\overline{\Omega}$, then for $w \in C^2(\overline{\Omega})$ satisfying $(\mathcal{L}+c)w \ge 0$ in Ω and $\max_{\partial\Omega}\{w(q,p)\} = 0$, either $w < 0$ in Ω or $w \equiv 0$ in Ω. Indeed, choose $\varepsilon > 0$ such that

$$c(q,p) \le c_\varepsilon = \frac{1}{4(1+\varepsilon)^2} + \frac{\pi^2}{(|p_0| + 2\varepsilon)^2}, \qquad (q,p) \in \overline{\Omega},$$

and notice that $(\mathcal{L}+c)w_0 = (c - c_\varepsilon)w_0$ in Ω. ∎

3.6 Notes for Chapter 3

In 1847, Stokes [287] initiated the study of nonlinear irrotational periodic traveling water waves. The first rigorous constructions by power series of such waves with small amplitude did not appear until the 1920s in the work of Nekrasov [243], Levi-Civita [202], and Struik [291]. The existence theory of large-amplitude irrotational waves was begun by Krasovskii [187] in 1961 and was considerably refined by Keady and Norbury [172] in 1978, using the methods of global bifurcation theory. Shortly thereafter Toland [297] and McLeod [227] proved that in the closure of the continuum of solutions

[79] It suffices that Ω satisfies an interior sphere condition at $\Theta_0 \in \partial\Omega$; that is, there exists an open ball $B \subset \Omega$ with $\Theta_0 \in \partial B$ (see [129, Chapter 3]).

[80] This is a sharp inequality: the value on the right is the first eigenvalue of $-\Delta$ with homogeneous Dirichlet boundary conditions on $\partial\Omega$. We saw in the above example with $p_0 = -\pi$ that if equality holds, the conclusion might not be valid. These ideas are further developed in [21].

found in [172] there exist waves with stagnation points (points where the vertical component of the fluid velocity field is zero while the horizontal component equals the speed of the wave profile). The existence of such a limiting wave with stagnation and angle of 120° at its crest (called "the wave of greatest height") was already conjectured by Stokes [287] in 1847 and proven by Amick, Fraenkel, and Toland [10] in 1982. Spielvogel [281] proved in 1970 that the surface profile of an irrotational steady wave must always be a graph (the small flaw in his argument was recently corrected by Varvaruca [305, 307]) and, as shown by Lewy [204] in 1952, if such a profile is C^1, then in fact it is real-analytic. This regularity result is sharp since the Stokes wave of greatest height is only Lipschitz continuous at its crest (see the discussion in [298]). Concerning the symmetry of irrotational wave profiles that are monotone between crests and troughs, a variational result of Garabedian [122] in 1965 concluded symmetry if in the moving frame each streamline has one mimimum (below the wave trough) and one maximum (below the wave crest) per period, while Okamoto and Shoji [245] used the moving plane method to prove that the monotoncity of the surface wave profile suffices.

In 1802 Gerstner [125, 126] constructed an explicit example (rediscovered by Rankine [258] in 1863) of a periodic traveling wave in water of infinite depth with a particular nonzero vorticity. In 1934 Dubreil-Jacotin [99] proved for small Hölder continuous vorticity functions γ the existence of steady periodic water waves of small amplitude. Improvements of her result for waves of small amplitude but with a general Hölder continuous vorticity function are due to Goyon [130] in 1958 and Zeidler [322] in 1973. The first existence results for steady rotational water waves that are not small perturbations of a flat free surface were obtained recently by Constantin and Strauss [71] under the assumption that the flow has no stagnation points. The local/global bifurcation approaches presented in Section 3.1/Section 3.2 are versions of the results in [71], sometimes expanded (e.g., in setting up the framework for the local bifurcation, and in the discussion of the dispersion relation) and at some places simplified (e.g., in discussing the nodal properties). The first symmetry results for rotational water waves with vorticity were obtained recently in [53] and were first improved in [152] and subsequently in [50]. The result in Section 3.3 is based on the approach pursued in [50], to which we added the fact that flat horizontal parts cannot occur in the wave profile unless the entire profile is flat (the first partial result in this direction, under the requirement of monotonicity of the vorticity with depth, was obtained in [105]). The regularity result of Section 3.4 is a slight simplification of the approach presented in [56].

In the context of the existence theory for steady periodic waves with vorticity the state of the art is far from that attained for irrotational flows. Moreover, in addition to the questions suggested by the irrotational setting, the presence of vorticity confronts us with new phenomena that have no irrotational counterpart. Among the themes of considerable interest are the following:

- It is reasonable to expect that Q is not constant throughout the continuum $\mathcal{S}_\delta^0$ or along the curve $\mathcal{K}_\delta$ from Section 3.2. Clearly this would be the case if the function $s \mapsto \lambda(s)$ were not constant along the local bifurcation curve $\mathcal{C}_{loc}$. Notice that $\lambda'(0) = 0$. Indeed, the fact that $h(q, p)$ is even and 2π-periodic in the q-variable suggests the symmetry $\mathfrak{g} : h(q, p) \mapsto h(q + \pi, p)$, with $\mathfrak{g}^2$ being the identity operator. If we set $h = H + w$ and parametrize the local bifurcation curve by $s \mapsto (\lambda(s), w^s)$ for $|s| < \varepsilon_0$, then $w^{-s} = \mathfrak{g}w^s$; cf. [72]. In the infinite-depth irrotational case we always have $\lambda''(0) \neq 0$ (cf. [38, Section 11.1]), but for irrotational water waves propagating over a flat bed, the sign of $\lambda''(0)$ changes as one passes from small average depth d to very large average depth (cf. [72]).

- The characterization of the boundary of the continuum of solutions: near to the boundary the supremum of the horizontal velocity in the fluid domain approaches the wave speed so that one expects the existence of waves with stagnation points. In the irrotational case the stagnation point has to occur at the wave crest—these are the Stokes waves of greatest height with a corner at their crest. For certain vorticity distributions the point of maximal horizontal fluid velocity is at the wave crest (see [73, 306]) so that one expects the existence of a wave similar to the Stokes wave of greatest height. However, the numerical simulations in [180, 181]) indicate that for some constant nonzero vorticities the point of maximal horizontal fluid velocity is located on the bed beneath the wave crest. Moreover, it appears that a variable vorticity might lead to the stagnation occurring in the interior of the fluid domain (beneath the wave crest).

- Throughout this chapter we fixed the wave speed $c > 0$, the wavelength L, and the relative mass flux p_0. While the average depth d does not remain constant along the continuum $\mathcal{S}^0$ or along the curve $\mathcal{K}$, the numerical simulations in [180, 181] indicate that it varies slightly. It is of interest to adapt the techniques developed here toward an approach that fixes c, L, and d, allowing p_0 to vary.
- It appears that certain features of the underlying current are reflected in the geometry of the surface wave. Is it possible to distinguish between rotational waves and irotational waves, or waves with constant positive/negative vorticity from the shape of the wave profile?
- Concerning symmetry the natural question is whether there exist steady waves with profiles that are not symmetric about the wave crest. Even in the irrotational case this is a challenging problem (see the discussion in [245]).
- Perhaps the most spectacular open problem is the possibility of steady waves with overhanging profiles: numerical simulations in [89, 277, 304] produced Ω-shaped profiles for waves with constant nonzero vorticity. Some field data (see [89]) seem to indicate that these phenomena could be observed so that the simulations are not mere numerical artifacts. Analytically this is a very difficult problem since these simulations indicate that starting from a flat profile and increasing the amplitude, one encounters first waves with stagnation points and a further increase seems to produce the overhanging wave profiles. The methods developed in this chapter cannot handle waves with stagnation points. Two local bifurcation approaches showing the existence of steady waves with constant nonzero vorticity that admit stagnation points in the flow were recently proposed in [309] and [76], but they both concern waves of small amplitude. The issue of global bifurcation and the possibility of overhanging profiles remain to be explored.

Other than these problems, there are interesting possible extensions of the presented theory outside the realm of gravity water waves propagating over a flat bed:

- Allowing discontinuous vorticity (the typical example being a sudden jump in the vorticity) is technically challenging since in this setting one has to investigate weak solutions to nonlinear elliptic partial differential equations with nonlinear boundary conditions; see [75].
- The theory of rotational deep water waves (infinite depth, with the velocity field and the vorticity decaying deep down to capture the realistic assumption that the wave motion dies out) is in its early stages (see [151, 54, 106]).
- A physically relevant problem is that of stratified flow. In this setting the density is not constant and for this reason these waves can never be irrotational (see the discussion in [310], where the existence of a global continuum is proved). It is of interest that in this case the extension of the symmetry approach presented in Section 3.3 is not that straightforward since one obtains elliptic operators with a zero-order term and to apply maximum principles (in accordance with the considerations in the Appendix) it seems that some restriction on the size of the free term cannot be avoided (see [311]). Since the free term is obtained by freezing the coefficients of a quasilinear elliptic equation, this places a restriction on the size of the solution. In other words, symmetry is proved only for waves of small amplitude although symmetric waves of large amplitude are known to exist.

- While pure capillary waves with vorticity are of limited physical interest, capillary-gravity waves present a wide range of problems worthy of investigation. We refer the reader to the discussion in [308] for the existence of waves of small amplitude. The first existence results for waves of this type that are not small perturbations of a flat free surface are very recent [312], and no symmetry results are available.

Chapter 4

Fluid kinematics for wave trains

"... the wave flees the place of its creation, while the water does not;
like the waves made in a field of grain by the wind,
where we see the waves running across the field
while the grain remains in its place."

Leonardo da Vinci

The subject of fluid kinematics concerns the study of how fluids flow and how to describe this motion. Proper to this subject is the Lagrangian viewpoint that we will use in describing the particle paths beneath a **Stokes wave**—a periodic two-dimensional traveling wave with no underlying current modelling swell generated by a distant storm and entering a region of still water. However, the Eulerian viewpoint is also convenient, as we shall see in investigating the pressure beneath an irrotational solitary water wave. Experimental measurements are generally more suited to the Eulerian description of a flow. The difference between the Lagrangian and Eulerian description is made clearer if we imagine a person standing on a bridge above a river and trying to measure properties of the water flow in the channel passing beneath the bridge: the Lagrangian viewpoint corresponds to throwing a probe that moves downstream with the water, while in the Eulerian approach one anchors the probe at a fixed location in the body of water.

A modern approach that proved very useful in fluid kinematics, both in physical experiments as well as in numerical solutions proper to computational fluid dynamics (CFD), is **fluid visualization**: the visual examination of flow field features. The fact that much can be learned from fluid visualization is beyond dispute as after all the human mind is designed to rapidly process visual information—as they say, a picture is worth a thousand words. However, without the support of advanced mathematics there are potentially serious pitfalls in a purely flow visualization approach, as we shall point out in the subsequent discussion. It is important to realize that whatever setting is chosen, it is only an approximation to the real-world phenomenon, and for this reason the hypotheses and limitations of the setting (as well as the various implications of these hypotheses) have to be well understood before making predictions.

4.1 Particle paths beneath a Stokes wave

While watching the sea or a lake it is often possible to trace a wave as it propagates on the water's surface. Contrary to a possible first impression, what one observes traveling across the sea is not the water but a pattern. An indication of what does happen can be seen by watching a small object, such as a cork, floating, since it will move more or less with the water. As the wave crest approaches, the object moves forward with the crest, while after the wave crest passes, the object moves backward nearly to its original position. During this time the object has risen to the top of the crest and fallen to the bottom of the trough, the combined motion being nearly circular in deep water and nearly elliptic in shallow water, where the "up-down" component of the motion appears to be "squeezed" by the nearby presence of the bottom. The aim of this section is to elucidate the pattern of the particle trajectories beneath a wave train moving at the surface of water in irrotational flow and with a flat bed, with no underlying current.

To present the mathematical formulation of the problem, let us write the governing equations in Cartesian coordinates (X,Y), with the Y-axis pointing vertically upward and the X-axis being the direction of wave propagation, while the origin lies on the mean water level $Y=0$. Let $(u(t,X,Y), v(t,X,Y))$ be the velocity field of the flow over the flat bed $Y=-d$ with $d>0$, and let $Y=\eta(t,X)$ be the water's free surface. Since $Y=0$ is the mean water level, at all times t we have

$$\int_0^L \eta(t,X)dX = 0, \tag{4.1}$$

where $L>0$ is the wave period. The functions η, u, and v are all L-periodic in the X-variable. The equation of mass conservation and Euler's equation read

$$u_X + v_Y = 0 \quad \text{in} \quad -d < Y < \eta(t,X), \tag{4.2}$$

$$\begin{cases} u_t + uu_X + vu_Y = -\dfrac{1}{\rho}P_X, \\ v_t + uv_X + vv_Y = -\dfrac{1}{\rho}P_Y - g, \end{cases} \quad \text{in} \quad -d < Y < \eta(t,X), \tag{4.3}$$

respectively, where $P(t,X,Y)$ is the pressure, ρ is the constant density of $1\,g/cm^3$, and $g = 9.8\,m/s^2$ is the gravitational constant of acceleration. Since the flow is assumed to be irrotational we have

$$u_Y = v_X \quad \text{in} \quad -d < Y < \eta(t,X). \tag{4.4}$$

The formulation of the governing equations is completed by the boundary conditions

$$v = \eta_t + u\eta_X \quad \text{on} \quad Y = \eta(t,X), \tag{4.5}$$

$$v = 0 \quad \text{on} \quad Y = -d, \tag{4.6}$$

$$P = P_{atm} \quad \text{on} \quad Y = \eta(t,X), \tag{4.7}$$

P_{atm} being the constant atmospheric pressure. We set $\rho = 1$.

By a **Stokes wave** we understand an irrotational smooth solution to the governing equations for which there exist a period $L>0$ and a wave speed c for which η, u, v, P have period L in the X-variable, η depends only on $(X-ct)$, while u, v, P depend only

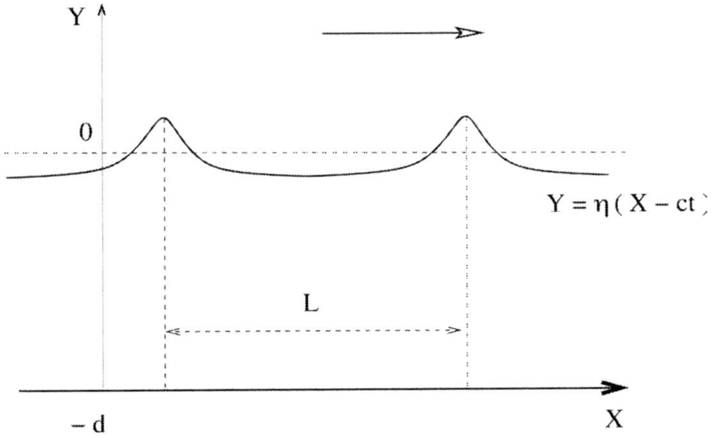

Figure 4.1. *A Stokes wave.*

on $(X - ct)$ and Y. Moreover, there is a single crest and trough per period, the profile is decreasing from crest to trough, and η, u, P are symmetric while v is antisymmetric about the crest. Moreover, we require that **there is no underlying current**—a hypothesis that will be elucidated below.

For convenience we take $c > 0$. In the study of Stokes waves that we pursued in Chapter 3, we eliminated time from the problem by passing to a moving frame

$$x = X - ct, \qquad y = Y. \tag{4.8}$$

In the moving frame Bernoulli's law holds, asserting that the expression

$$\frac{(u-c)^2 + v^2}{2} + g(y+d) + \frac{1}{\rho} P$$

is constant throughout the fluid domain

$$\Omega = \{(x,y) \in \mathbb{R}^2 : -d < y < \eta(x)\}.$$

Using (4.2) to define the **stream function** $\psi(x,y)$ up to a constant by

$$\psi_y = u - c, \quad \psi_x = -v, \tag{4.9}$$

we see that ψ is harmonic in Ω in view of (4.4), whereas (4.5) and (4.6) guarantee that ψ is constant on both boundaries of Ω; say, $\psi = 0$ on $y = \eta(x)$ while $\psi = m$ on $y = -d$. Thus

$$\psi(x,y) = m + \int_{-d}^{y} [u(x,s) - c]\, ds.$$

It follows that ψ has period L in the x-variable. Using Bernoulli's law, we can express the boundary condition (4.7) in an alternative form, thereby obtaining the following equivalent formulation for a Stokes wave. We seek a smooth solution $(\eta(x), \psi(x,y))$, periodic and even in the x-variable such that η has one maximum and one minimum per period, of the free boundary value problem

$$\begin{cases} \Delta\psi = 0 \quad \text{in} \quad -d < y < \eta(x), \\ \dfrac{|\nabla\psi|^2}{2} + g(y+d) = Q \quad \text{on} \quad y = \eta(x), \\ \psi = 0 \quad \text{on} \quad y = \eta(x), \\ \psi = m \quad \text{on} \quad y = -d. \end{cases} \tag{4.10}$$

Here m (**the relative mass flux**) and Q (related to the **hydraulic head**) are physical constants. Since $\eta(x)+d > 0$ we must have $Q > 0$. Moreover, it follows that $m \neq 0$ (unless $\psi \equiv 0$, in which case we have still water with a flat surface). We choose $m > 0$, so that the strong maximum principle applied to ψ and then to ψ_y yields

$$\psi_y = u - c < 0 \quad \text{in} \quad \overline{\Omega}, \tag{4.11}$$

unless $\psi \equiv 0$. This is consistent with experimental data that indicate in general that the horizontal motion of individual water particles is slower than the propagation speed of the wave. Moreover, according to the discussion in Chapter 3, for waves that are not near breaking, we have $u < c$ at the wave crest. The maximum of the harmonic function u in $\overline{\Omega}$ is attained at the wave crest. Indeed, unless u is constant,[81] the maximum of u in $\overline{\Omega}$ is attained on the boundary and the fact that $u_y = 0$ on $y = -d$, in view of (4.4) and (4.6), yields by Hopf's maximum principle that the maximum is attained on $y = \eta(x)$. On the other hand, from (4.3) we compute $\Delta P = -\psi_{xx}^2 - 2\psi_{xy}^2 - \psi_{yy}^2 \leq 0$, so that P is superharmonic in Ω. Since $P_y = -g < 0$ on $y = -d$, in view of (4.3) and (4.6), Hopf's maximum principle ensures that the minimum of P in $\overline{\Omega}$ is attained all along the free surface, where $P = P_{atm}$ by (4.7). From the monotonicity of the wave profile we deduce that $P_x(x,\eta(x)) \leq 0$ holds for $(x,\eta(x))$ on the free surface between crest and trough. But

$$\begin{aligned} P_x(x,\eta(x)) &= [c - u(x,\eta(x))]\,[u_x(x,\eta(x)) + \eta'(x)u_y(x,\eta(x))] \\ &= [c - u(x,\eta(x))]\,\partial_x u(x,\eta(x)) \end{aligned}$$

by the first equation in (4.3) and by (4.5). Consequently, as we move along the free surface $y = \eta(x)$ from crest to trough, the horizontal velocity component is nonincreasing as long as $u < c$. By assumption $u < c$ at the crest, so $u < c$ along the free surface between crest and trough; and, by symmetry, also between trough and crest.

Observe that in the moving frame the wave speed $c > 0$ itself does not appear in the system (4.10). We define the **mean current** κ on the bottom by

$$\kappa = \frac{1}{L}\int_0^L u(x,-d)\,dx < c, \tag{4.12}$$

where the inequality follows from (4.11). From the fact that u is harmonic it follows that κ also equals the average of u on any horizontal line below the trough. To express this in another way, notice that by Green's theorem

$$0 = \int_{-d}^{y_0}\int_0^L (u_y - v_x)\,dx\,dy = \int_0^L u(x,y_0)\,dx - \int_0^L u(x,-d)\,dx$$

[81] In which case v must also be constant, that is, zero by (4.6). But then (4.5) ensures that the free surface has to be flat since the possibility $u \equiv c$ throughout the flow was excluded.

at all depths y_0 below the wave trough level. Thus

$$\frac{1}{L}\int_0^L [u(x,y_0)-c]\,dx = \kappa - c < 0, \tag{4.13}$$

and κ is the average horizontal current in the water. Physically we can imagine that swell originating from a distant storm enters a region of water in uniform flow. Throughout this chapter we assume that there is no underlying current: $\kappa = 0$ or

$$\int_0^L u(x,-d)\,dx = 0. \tag{4.14}$$

This corresponds to swell entering a region of still water, the relation $\frac{1}{L}\int_0^L [c - u(x,-d)]\,dx = c$ being Stokes's original definition of the wave speed as the mean velocity in the moving frame of reference in which the wave is stationary.

With these specifications the trajectory of a particle located at the point (X_0, Y_0) in the fluid domain $\{(X,Y):\ -d \le Y \le \eta(t,X)\}$ is given by the unique solution $t \mapsto (X(t), Y(t))$ of the ordinary differential system[82]

$$\begin{cases} X'(t) = u(X(t),Y(t)), \\ Y'(t) = v(X(t),Y(t)), \end{cases} \tag{4.15}$$

with initial data $X(0) = X_0$, $Y(0) = Y_0$. To determine the form of the particle trajectory we cannot rely for guidance on some special exact solutions.[83] We will now present some theoretical and experimental investigations. Notice that in laboratory experiments with water waves, measurements close to the surface are generally expected to better reflect reality; special care is required deep down since in a water tank there are the effects of the undertow—the subsurface flow of water returning from the end of the channel and caused either by reflecting waves (if the channel has a vertical wall at the end) or by breaking waves (if the channel ends with a sloping beach-like bed profile).

Linear theory

In order to be able to treat the problem of particle paths, it is natural to first attempt to approximate the governing equations by ones which are more tractable. In many circumstances water-wave motions are surprisingly well described by linear theory [177], which is the core theory of ocean surface waves and in ocean and coastal engineering. It studies small deviations from a known particular exact solution (in our case, the solution with a flat free surface and no motion beneath, modelling still water) by selecting a dimensionless parameter (or parameters), say, ε, and assuming that the various functions entering into the problem may be expanded into power series in ε. The series are substituted into the problem and grouped according to powers of ε, the coefficient of ε giving the linearization.

[82] Since the right side of the system is smooth, the existence and uniqueness of a local solution are ensured by the Cauchy–Lipschitz theorem; cf. [79, 136, 265]. The fact that the solution is defined for all times t follows by analyzing the specific structure of this system, as we shall see.

[83] The only explicit gravity water-wave train is Gerstner's wave in water of infinite depth, presented in the Appendix to this chapter. While all particles beneath the Gerstner wave move in circles, the flow has nonzero vorticity.

Our aim now is to present the conclusions about particle paths that can be reached within the framework of linear theory. Since they do not capture the essence of the problem, we will subsequently discuss the nonlinear theory.

In deriving approximate equations by asymptotic methods it is necessary to first nondimensionalize the problem, so that the equations to be solved do not have any units and in this way, unambiguous small parameters can more readily be identified. The governing equations for Stokes waves are **nondimensionalized** using the wavelength λ and a typical amplitude of the wave a. That is, we define the set of nondimensional variables

$$X \mapsto \lambda x, \quad Y \mapsto d(y-1), \quad t \mapsto \frac{\lambda}{\sqrt{gd}}t, \quad u \mapsto u\sqrt{gd}, \quad v \mapsto v\frac{d\sqrt{gd}}{\lambda}, \quad \eta \mapsto a\eta,$$

it being understood that u is replaced by $u\sqrt{gd}$, so that afterward the symbol u stands for a nondimensional variable, etc., thus avoiding new notations. The pressure in the new nondimensional variables is

$$P = P_{atm} + \rho g d(1-y) + \rho g d\, p$$

with the nondimensional pressure variable p measuring the deviation from the hydrostatic pressure distribution $P_{atm} - \rho g Y$. We obtain the following boundary value problem in nondimensional variables:

$$\begin{cases} \begin{cases} u_t + uu_x + vu_y = -p_x, \\ \delta^2(v_t + uv_x + vv_y) = -p_y, \end{cases} & \text{in} \quad 0 < y < 1+\varepsilon\eta, \\ u_x + v_y = 0 \quad \text{and} \quad u_y = \delta^2 v_x & \text{in} \quad 0 < y < 1+\varepsilon\eta, \\ p = \varepsilon\eta \quad \text{on} \quad y = 1+\varepsilon\eta, \\ v = \varepsilon(\eta_t + u\eta_x) \quad \text{on} \quad y = 1+\varepsilon\eta, \\ v = 0 \quad \text{on} \quad z = 0, \end{cases} \tag{4.16}$$

where

$$\varepsilon = \frac{a}{d} \quad \text{and} \quad \delta = \frac{d}{\lambda}$$

are the amplitude parameter and the shallowness parameter, respectively. The absence of an underlying current, condition (4.14), is expressed in nondimensional variables by the requirement

$$\int_0^1 u(x,0)dx = 0,$$

while the fact that $Y = 0$ is the mean water level, condition (4.1), takes the form

$$\int_0^1 \eta(t,x)dx = 0.$$

From the fourth and fifth equations in (4.16) it is apparent that both v and p, if evaluated on $y = 1+\varepsilon\eta$, are essentially proportional to ε. This is consistent with the fact that as $\varepsilon \to 0$ we must have $v \to 0$ and $p \to 0$; i.e., with no disturbance the free surface becomes a horizontal line on which $v = p = 0$. This consideration leads us to the scaling of the nondimensional variables

$$p \mapsto \varepsilon p, \qquad (u,v) \mapsto \varepsilon(u,v),$$

avoiding again the introduction of new notation. The problem (4.16) becomes

$$\begin{cases} \left\{ \begin{array}{l} u_t + \varepsilon(uu_x + vu_y) = -p_x, \\ \delta^2 [v_t + \varepsilon(uv_x + vv_y)] = -p_y, \end{array} \right. \quad \text{in} \quad 0 < y < 1 + \varepsilon\eta, \\ u_x + v_y = 0 \quad \text{and} \quad u_y = \delta^2 v_x \quad \text{in} \quad 0 < y < 1 + \varepsilon\eta, \\ p = \eta \quad \text{and} \quad v = \eta_t + \varepsilon u \eta_x \quad \text{on} \quad y = 1 + \varepsilon\eta, \\ v = 0 \quad \text{on} \quad y = 0, \\ \displaystyle\int_0^1 \eta(t,x)\,dx = \int_0^1 u(x,0)\,dx = 0. \end{cases} \tag{4.17}$$

The linearized problem is now obtained by letting $\varepsilon \to 0$ in (4.17). We obtain

$$\begin{cases} u_t = -p_x \quad \text{and} \quad \delta^2 v_t = -p_y \quad \text{in} \quad 0 < y < 1, \\ u_x + v_y = 0 \quad \text{and} \quad u_y = \delta^2 v_x \quad \text{in} \quad 0 < y < 1, \\ v = \eta_t \quad \text{and} \quad p = \eta \quad \text{on} \quad y = 1, \\ v = 0 \quad \text{on} \quad y = 0, \\ \displaystyle\int_0^1 \eta(t,x)\,dx = \int_0^1 u(x,0)\,dx = 0. \end{cases} \tag{4.18}$$

Looking for solutions of (3.12) representing waves traveling at speed $c_0 > 0$, we impose that all functions u, v, p, and η have an (x,t)-dependence in the form of $x - c_0 t$. Furthermore, in seeking spatially periodic functions $\eta(x)$ of period one, that represent a wave train with a surface profile that is monotone between crests and troughs, and symmetric about the crest, we are led to the fundamental Fourier mode *Ansatz*

$$\eta(t,x) = \cos[2\pi(x - c_0 t)].$$

For this specific η the problem (4.18) has the solution

$$u(t,x,y) = 2\pi\, c_0\, \delta \,\frac{\cosh(2\pi\delta y)}{\sinh(2\pi\delta)} \cos[2\pi(x - c_0 t)],$$
$$v(t,x,y) = 2\pi\, c_0 \,\frac{\sinh(2\pi\delta y)}{\sinh(2\pi\delta)} \sin[2\pi(x - c_0 t)],$$
$$p(t,x,y) = \frac{\cosh(2\pi\delta y)}{\cosh(2\pi\delta)} \cos[2\pi(x - c_0 t)],$$

provided

$$c_0^2 = \frac{\tanh(2\pi\delta)}{2\pi\delta}.$$

Returning to the original physical variables, we perform the change of variables

$$x \mapsto \frac{X}{\lambda}, \quad y \mapsto 1 + \frac{Y}{d}, \quad t \mapsto t\,\frac{\sqrt{gd}}{\lambda}, \quad u \mapsto \frac{u}{\sqrt{gd}}, \quad v \mapsto \frac{\lambda}{d\sqrt{gd}}\,v, \quad \eta \mapsto \frac{\eta}{a}$$

(the variables on the right side being the physical variables) to obtain the linear wave solution

$$\begin{cases} \eta(t,X) = \varepsilon d \cos(kX - \omega t), \\ u(t,X,Y) = \varepsilon \omega d \, \dfrac{\cosh(k(Y+d))}{\sinh(kd)} \cos(kX - \omega t), \\ v(t,X,Y) = \varepsilon \omega d \, \dfrac{\sinh(k(Y+d))}{\sinh(kd)} \sin(kX - \omega t), \\ P(t,X,Y) = P_{atm} - \rho g Y + \varepsilon \rho g d \, \dfrac{\cosh(k(Y+d))}{\cosh(kd)} \cos(kX - \omega t) \end{cases} \tag{4.19}$$

of amplitude $\varepsilon d > 0$, wavelength $\lambda > 0$, over the flat bed $Y = -d$ and with mean water level $Y = 0$. Here $k = \frac{2\pi}{\lambda}$ and $\omega = \sqrt{gk \tanh(kd)}$, are the wavenumber, respectively, the frequency, and the dispersion relation

$$c = \frac{\omega}{k} = \sqrt{g \, \frac{\tanh(kd)}{k}}$$

determines the speed c of the linear wave.

If $(X(t), Y(t))$ is the path of the particle below the linear wave (4.19), then

$$\begin{cases} \dfrac{dX}{dt} = \varepsilon \dfrac{\omega d}{\sinh(kd)} \cosh(k(Y+d)) \cos(k(X-ct)), \\ \dfrac{dY}{dt} = \varepsilon \dfrac{\omega d}{\sinh(kd)} \sinh(k(Y+d)) \sin(k(X-ct)) \end{cases} \tag{4.20}$$

with initial data (X_0, Y_0). The right side of the differential system (4.20) is smooth so that the existence of a unique local smooth solution is ensured (see [136]). Also, since Y is bounded as $-d \leq Y \leq \varepsilon d$, the right side of (4.20) is bounded, and therefore this unique solution is defined globally, i.e., for all $t \geq 0$ (see [265]).

To describe the classical approach pursued in [192, 87, 177, 233, 286] we restrict our attention to the fixed time interval $[0,T]$, where $T = \lambda/c > 0$ is the wave period. Since Y belongs to a set bounded a priori as $-d \leq Y \leq \varepsilon d$, from (4.20) we readily obtain that $X(t) - X_0 = O(\varepsilon)$ and $Y(t) - Y_0 = O(\varepsilon)$ for $t \in [0,T]$. Using the mean-value theorem, we write (4.20) on $[0,T]$ as

$$\begin{cases} \dfrac{dX}{dt} = \varepsilon \dfrac{\omega d}{\sinh(kd)} \cosh(k(Y_0+d)) \cos(kX_0 - \omega t) + O(\varepsilon^2), \\ \dfrac{dY}{dt} = \varepsilon \dfrac{\omega d}{\sinh(kd)} \sinh(k(Y_0+d)) \sin(kX_0 - \omega t) + O(\varepsilon^2), \end{cases}$$

since $\omega = kc$. Neglecting terms of second order in ε, we find that

$$\begin{cases} \dfrac{dX}{dt} \approx \varepsilon \dfrac{\omega d}{\sinh(kd)} \cosh(k(Y_0+d)) \cos(kX_0 - \omega t), \\ \dfrac{dY}{dt} \approx \varepsilon \dfrac{\omega d}{\sinh(kd)} \sinh(k(Y_0+d)) \sin(kX_0 - \omega t). \end{cases}$$

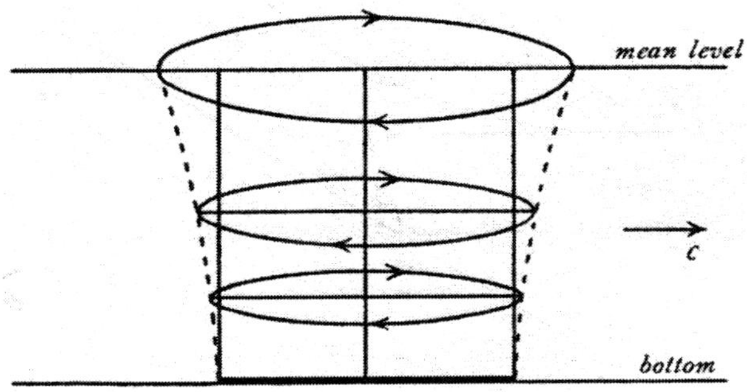

Figure 4.2. *This sketch [From "Theoretical Hydrodynamics" by L. M. Milne-Thomson, The Macmillan Co., New York,* 1960. *Reproduced with permission of Palgrave Macmillan* (2010)*] illustrates the conclusion obtained within the framework of linear theory. All particles above the flat bed appear to move on ellipses. The ellipses have the same distance between their foci, but the lengths of their axes decrease as we go downward into the water, and at the bottom the ellipse degenerates into a straight line. Each particle appears to describe its ellipse in a wave period, and they are all in the same phase. The motion of a line of particles after a half-period appears to be a translation and a bending of the line, illustrated by the dotted lines in the figure.*

By integration we obtain for $Y_0 > -d$ that

$$\begin{cases} X(t) \approx X_0 + \dfrac{\varepsilon d}{\sinh(kd)} \cosh(k(Y_0+d))\,[\sin(kX_0) - \sin(kX_0 - \omega t)], \\ Y(t) \approx Y_0 + \dfrac{\varepsilon d}{\sinh(kd)} \sinh(k(Y_0+d))\,[\cos(kX_0 - \omega t) - \cos(kX_0)]. \end{cases}$$

Thus

$$\frac{[X(t) - X_0^*]^2}{\cosh^2(k(Y_0+d))} + \frac{[Y(t) - Y_0^*]^2}{\sinh^2(k(Y_0+d))} \approx \frac{\varepsilon^2 d^2}{\sinh^2(kd)}$$

with

$$X_0^* = X_0 + \frac{\varepsilon d}{\sinh(kd)} \sin(kX_0), \quad Y_0^* = Y_0 - \frac{\varepsilon d}{\sinh(kd)} \cos(kX_0).$$

This is the equation of an ellipse: to a first-order approximation the water particles above the flat bed move clockwise in closed elliptic orbits,[84] the center of the ellipse being (X_0^*, Y_0^*). On the flat bed $Y = -d$ the particles appear to perform a back-and-forth motion as in this case

$$X(t) \approx X_0 + \frac{\varepsilon d}{\sinh(kd)} [\sin(kX_0) - \sin(kX_0 - \omega t)], \quad Y(t) = -d.$$

Analyzing the length of the major and minor axes of the ellipses, we see that the disturbance induced by the wave beneath the surface diminishes exponentially with depth. In this context

[84]Although the particles are presumed to move in closed ellipses, the particles themselves do not rotate: their motion is like that of the passenger cars on the Giant Wheel at the fairground. See the Java Applet for Coastal Engineering at http://www.coastal.udel.edu/faculty/rad/linearplot.html, which shows graphically the obtained elliptic orbital motions of the particles for given wave height, wave period, and average water depth.

Photograph 6. *Particle paths in a water wave.* [From A. Wallet & F. Ruellan, La Houille Blanche 5 (1950): 483–489. Reproduced with permission]

The predictions of linear theory seem verified in practice: the particles move up and down and forward and backward, and appear to return to their original position.

notice that (cf. [177]) for waves less than $3\,m$ high, wave action is of no practical importance at depth greater than $15\,m$; and submarines are hovering when they are submerged to $30\,m$ or more, even if monster waves $15\,m$ high occur at the surface of the sea.

Support for the particle trajectory pattern predicted by linear theory seems to be given by some experimental evidence, and there are even early films of small buoyant particles in laboratory wave tanks where closed elliptical paths appear to be recognizable [37]. An intriguing example is shown in Photograph 6: the water is kept in a narrow glass container with parallel walls, and neutrally buoyant white tracer particles are suspended in the water. Taking a time-exposure photograph for one complete wave period, one can clearly distinguish elliptical arcs induced beneath the surface wave train.

However, Longuet-Higgins carried out in 1986 the following simple experiment for the trajectory of a surface water particle for a steep wave. In a laboratory wave channel of width $60\,cm$ and total length $40\,m$ containing water of mean depth $35\,cm$, a small wooden bead, painted white, was floated on the water surface at a distance of about $20\,m$ from the wavemaker. This was viewed through a window in the side of the tank, and a time exposure of the bead's path was taken—see Photograph 7. This shows the beam at first nearly stationary. Then, as the waves arrive, the trajectory uncoils like a watch spring. As the steep wave passes, the fully developed trajectory is seen. After passage of the steep wave the bead becomes stationary once more. This experimental result indicates that a revision of the classical linear theory should be pursued.

Moreover, using (formally) quadratic quantities in the wave amplitude within the framework of linear theory, one can compute the average flow of energy and infer that the water particles in the fluid experience on average a net displacement in the direction in which the waves are propagating (see [206]). The corresponding mean rate of movement is known as the **Stokes drift**; Stokes [287] noticed this feature for infinitely deep water, and Ursell [303] examined it in water of finite depth. Consequently, a correction to the previously described linear theory is advisable.

The third incentive to look beyond linear theory comes from a purely mathematical point of view. Within the linear water-wave theory the explicit system (4.20) describing the motion of the particles is nevertheless nonlinear and a further linearization was performed. One can hardly expect an accurate description, especially since the outcome is closed orbits throughout the flow—a regular pattern that is easily destroyed by small perturbations! As a

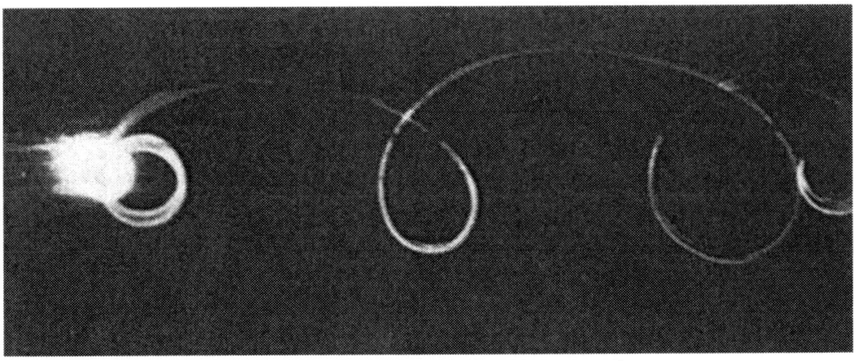

Photograph 7. *Photograph of the path of a surface particle.*
[From M. S. Longuet-Higgins, J. Fluid Mech. 173 (1986), pp. 683–707. Reproduced with permission from Cambridge University Press]
Surface particle trajectory for steep water waves.

matter of fact, if we perform just the linearization of the governing equations, without subsequently also linearizing the corresponding explicit system (4.20) for the particle paths, the same pattern as that presented in Photograph 7 emerges for particles above the flat bed. This was first proved by means of a detailed qualitative study of the system (4.20) in [77], and subsequently the paper [155] provided the explicit solution of the system (4.20). This approach is, however, clearly limited to waves of small amplitude (due to the linearization of the governing equations); what one gains by not linearizing the nonlinear explicit system (4.20) describing the motion of the particles is not primarily the ability to discuss large deviations from the rest state (still water), but rather the ability to discuss relatively small deviations over a long time scale.

Power series approach

We saw in Section 3.4 that for all Stokes waves except the wave of greatest height, at any instant t, the velocity field is real-analytic in a neighborhood of the closed fluid domain. Consequently, the Cauchy–Kowalewski theorem (cf. [186]) applies and we know that for any point in the fluid there exists a small neighborhood where (3.4) has a unique real-analytic solution. A possible approach which comes to mind is expansion of the solution in power series and identification of the series through substitution into the system. This was first proposed by Stokes in 1847. He assumed that the free surface elevation may be represented by an infinite Fourier series

$$\eta(x) = a_1 \cos(x) + a_2 \cos(2x) + \cdots$$

and that the velocity potential may be similarly represented by

$$\phi(x, y) = cx + b_1 \cosh(y) \sin(x) + b_2 \cosh(2y) \sin(2x) + \cdots,$$

where the coefficients c, a_n, and b_n are assumed to be power series expansions in the amplitude parameter ε such that the leading orders of a_n and b_n are ε^n. When these series are substituted into the boundary value problem, the coefficients in the power series can be found successively. However, this method (and subsequent alternative power series expansions) has a number of significant limitations. First, the actual computation of series coefficients by means of recursion formulas is tedious.[85] A more serious drawback is that the series may not converge in the full region where the solution is needed. The Cauchy–Kowalewski theorem ensures a nonzero radius of convergence, but to establish the size of the radius of convergence seems out of reach. Consequently, we can rely upon a power series approach only locally (in particular, for a short time) and for waves of small amplitude. Moreover, even for waves of small amplitude it turns out that if the water is shallow,[86] the contribution of the higher-order terms in the power series expansion will tend to dominate, and results obtained by truncation at a certain order will not be accurate (cf. [115]). Consequently, a power series approach is restricted to deep water waves of small amplitude.[87] Due to our interest in waves of moderate and large amplitude without restrictions on the average water depth, and in light of the previous discussion, we do not advocate a power series approach. Notice that the guiding principle that solutions of moderate and large amplitude hold few surprises, and that the insight provided by studies of waves of small amplitude reveals the main structure of the flow field, is sometimes erroneous.[88]

Nonlinear theory

We will now prove that the pattern depicted in Photograph 7 is valid for any particle trajectory above the flat bed in a Stokes wave. Consequently, the impression given by Photograph 6 is misleading, the explanation being that in that experiment the particle paths are almost closed, and the size of the used tracer particles together with the accuracy of the photography gives the impression of superposition of the particle location initially and after one wave period. This interpretation finds confirmation in the recent experiments carried out recently in a wave tank $25\,m$ long, $0.7\,m$ wide, and $1\,m$ deep, with tiny tracer particles having a grain size of $0.25\,mm$ and relative density with respect to water 1.01, using the modern reliable PIV (particle image velocimetry) experimental technique (see [302]). Even with this advanced modern technology some almost closed particle paths appear at first sight to be closed.

Let us first point out that several assumptions encompassed in the definition of a Stokes wave are not restrictive requirements. The free surface must always be a graph (cf. [305]). Moreover, the boundary must be a real-analytic curve with a parametrization $y = \eta(x)$ with the velocity components having harmonic extensions across it, and the symmetry of the free surface is actually guaranteed by the fact that the wave profile is monotone between crests and troughs (see Chapter 3). As a limiting case of the Stokes waves we have Stokes's famous wave of greatest height, for which the symmetric free surface is smooth except at the crest, where it is not continuously differentiable. Stokes's wave of greatest height fails to satisfy (4.11) precisely at the crest where $u = c$, and the profile has a corner with an angle of 120°

[85] Nevertheless, asymptotic solutions up to fifth order are available (see the discussion in [115]).

[86] A small *Ursell number* $\varepsilon\delta^{-2}$, where ε and δ are the amplitude and shallowness parameters defined in the above discussion of the linear theory, is required.

[87] Despite the issues of convergence, the power series approach provides insight; see, e.g., [42], where the pattern of Photograph 7 is obtained.

[88] For example, the supposition that important integral quantities such as speed, energy, and momentum would increase with wave height until the highest is reached has been disproved; the highest wave is not the fastest nor the most impulsive nor the most energetic [269].

(cf. [298]). The particle path pattern in Photograph 7 is actually valid for all irrotational wave trains which propagate at the surface of water with a flat bed and without underlying currents, having surface wave profiles that are monotone between crests and troughs, that is, also for Stokes's wave of greatest height (which is the only such wave train not covered by our definition of Stokes waves—see [298]). The approach presented below does not apply directly to the wave of greatest height but can be adapted to that setting, as we shall see.

Due to (4.11), the streamlines $\psi = k$ with $k \in [0,m]$ provide a foliation of the closure $\overline{\Omega}$ of the fluid domain Ω, where the free surface corresponds to $\psi = 0$ and the flat bed to $\psi = m$. Without loss of generality, we can normalize the wavelength $L = 2\pi$, and assume that in the moving frame the wave crest is at the point $(0, \eta(0))$ and the wave troughs at the points $(\pm\pi, \eta(\pm\pi))$. Denote

$$\begin{cases} \Omega_+ = \{(x,y) \in \mathbb{R}^2 : \ x \in (0,\pi),\ -d < y < \eta(x)\}, \\ \Omega_- = \{(x,y) \in \mathbb{R}^2 : \ x \in (-\pi,0),\ -d < y < \eta(x)\}, \end{cases}$$

and let

$$S_+ = \{\, x \in (0,\pi),\ y = \eta(x)\}, \quad S_- = \{\, x \in (-\pi,0),\ y = \eta(x)\}$$

be the two halves of the free surface S in one period of the fluid domain Ω, and let

$$B_+ = \{\, x \in (0,\pi),\ y = -d\}, \quad B_- = \{\, x \in (-\pi,0),\ y = -d\}$$

be the two halves of the flat bed B. The **crest line** is $\{x = 0, -d < y < \eta(0)\}$ while the **trough line** is $\{x = \pm\pi, -d < y < \eta(\pm\pi)\}$. By assumption the wave profile is symmetric about the crest and $\eta_x \le 0$ on S_+.

Let us first establish some additional basic properties of any Stokes wave. From (4.5) we know that $v = (u-c)\eta_x$ on $y = \eta(x)$ so that (4.11) and the assumption $\eta_x(x) \le 0$ for $x \in [0,\pi]$ ensure that $v(x,\eta(x)) \ge 0$ for $x \in [0,\pi]$, with equality only at the points where $\eta_x(x) = 0$. Notice that $v(x,-d) = 0$ for $x \in [0,\pi]$ by (4.6). On the other hand, $v = 0$ on the vertical fluid segments $x = 0$ and $x = \pi$ since v is odd and periodic in the x-variable throughout Ω. In view of (4.11) and (4.5), if the free surface is not flat, there is at least a point on S_+ where $v > 0$. Since $v \ge 0$ on the whole boundary of Ω_+, the maximum principle yields

$$v > 0 \quad \text{in} \quad \Omega_+, \quad \text{while} \quad v < 0 \quad \text{in} \quad \Omega_-, \tag{4.21}$$

recalling the fact that v is odd in the x-variable.

On the other hand, a direct calculation based on (4.3) shows that the function P is superharmonic in Ω as $\Delta P = -\psi_{xx}^2 - 2\psi_{xy}^2 - \psi_{yy}^2 \le 0$. Since $P_y = -g < 0$ on $y = -d$ by (4.3) and (4.6), Hopf's maximum principle yields that the minimum of P is attained only along the free surface $y = \eta(x)$, where $P = P_{atm}$ in view of (4.7). But then Hopf's principle ensures that $P_x(x,\eta(x)) \le 0$ along S_+, with $P_x(x,\eta(x)) < 0$ wherever $\eta_x(x) < 0$ for $x \in (0,\pi)$. But $\eta_x(x)$ has to be strictly negative for $x \in (0,\pi)$. Indeed, if $\eta_x(x_0) = 0$ for some $x_0 \in (0,\pi)$, then $v(x_0,\eta(x_0)) = 0$ and, if the minimum of v within the closure of Ω_+ were attained at $(x_0, \eta(x_0))$, Hopf's principle would yield $v_y(x_0, \eta(x_0)) < 0$. Thus $\eta_x(x_0) = 0$ implies $v(x_0, \eta(x_0)) = 0$ and $v_y(x_0, \eta(x_0)) < 0$. But then, using (4.3) and (4.2), we find that $P_x = -(c-u)v_y > 0$ at $(x_0, \eta(x_0))$ and we know that the opposite inequality is true. Thus we conclude that $\eta_x < 0$, $v > 0$, and $P_x < 0$ at all points on S_+. Since $(c-u)\,\partial_x[u(x,\eta(x))] = P_x(x,\eta(x))$ by (4.3) and (4.5), we can summarize our findings as

$$v(x,\eta(x)) > 0 \quad \text{and} \quad \partial_x[u(x,\eta(x))] < 0 \quad \text{for} \quad x \in (0,\pi). \tag{4.22}$$

Furthermore, the function u decreases strictly as we go from crest to trough along the lower three sides of Ω_+; that is, we have the strict inequalities

$$\begin{cases} u_y(0,y) > 0 & \text{for} \quad y \in (-d,\eta(0)), \\ u_x(x,-d) < 0 & \text{for} \quad x \in (0,\pi), \\ u_y(\pi,y) < 0 & \text{for} \quad y \in (-d,\eta(\pi)) \end{cases} \tag{4.23}$$

along these three sides. Indeed, $v = 0$ along the three sides so that (4.23) follows from Hopf's principle due to (4.21), (4.2), and (4.4).

Irrotationality allows us to define the velocity potential $\phi(x,y)$ by

$$\phi_x = u - c, \qquad \phi_y = v,$$

with $\phi = 0$ on the crest line. That is,

$$\phi(x,y) = \int_0^x [u(l,-d) - c]\,dl + \int_{-d}^y v(x,s)\,ds.$$

Then $\phi(x,y) + cx$ also has period 2π in x. Moreover, ϕ is odd in the x-variable and vanishes at $x = 0$, while

$$\phi(2\pi n, y) = \int_0^{2\pi n} [u(l,-d) - x]\,dl = -2c\pi n$$

for any y below the trough level, and for any integer n because of (4.14). By means of the stream function ψ and the velocity potential ϕ we can perform the conformal hodograph change of variables

$$\begin{cases} q = -\phi(x,y), \\ p = -\psi(x,y), \end{cases} \tag{4.24}$$

which transforms the free boundary problem (4.10) into a nonlinear boundary problem for the harmonic function

$$h(q,p) = y + d$$

in a fixed rectangular domain. The transformed boundary problem is

$$\begin{cases} \Delta_{q,p} h = 0 \quad \text{for} \quad -c\pi < q < c\pi, \quad -m < p < 0, \\ h = 0 \quad \text{on} \quad p = -m, \\ 2(Q - gh)(h_q^2 + h_p^2) = 1 \quad \text{on} \quad p = 0 \end{cases} \tag{4.25}$$

for h even and periodic of period $2\pi\, c$ in the q-variable, with $\int_{-\pi}^{\pi} h(q,0)\,dq = 0$.

It is convenient to denote the images of Ω_+, S_+, B_+ under the conformal hodograph transformation by

$$\hat{\Omega}_+ = (0,c\pi) \times (-m,0), \quad \hat{S}_+ = (0,c\pi) \times \{0\}, \quad \hat{B}_+ = (0,c\pi) \times \{-m\},$$

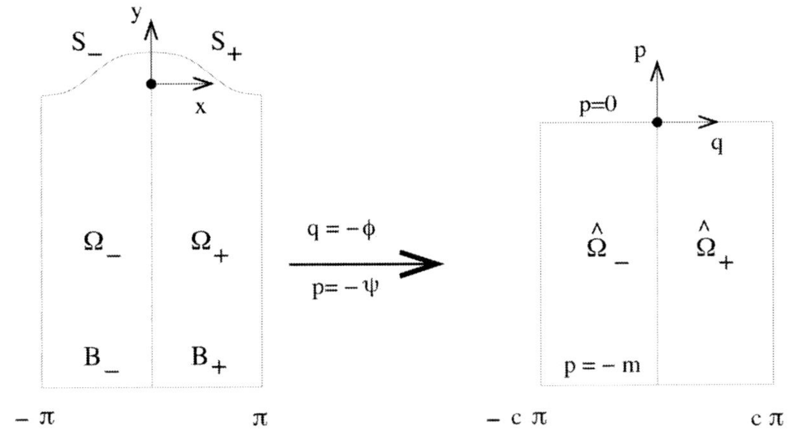

Figure 4.3. *The conformal change of variables.*

respectively. For further use notice that

$$\begin{cases} \partial_q = h_p\,\partial_x + h_q\,\partial_y, \\ \partial_p = -h_q\,\partial_x + h_p\,\partial_y, \end{cases} \qquad \begin{cases} \partial_x = (c-u)\,\partial_q + v\,\partial_p, \\ \partial_y = -v\,\partial_q + (c-u)\,\partial_p. \end{cases} \tag{4.26}$$

We also have

$$\begin{cases} h_q = -\dfrac{v}{(c-u)^2+v^2} = -\dfrac{\partial x}{\partial p} = \dfrac{\partial y}{\partial q}, \\ h_p = \dfrac{c-u}{(c-u)^2+v^2} = \dfrac{\partial x}{\partial q} = \dfrac{\partial y}{\partial p}. \end{cases} \tag{4.27}$$

If $(x, y(x))$ is the parametric equation of some streamline, then $\partial_x[\psi(x, y(x))] = 0$ by definition, so that $\frac{dy}{dx} = y' = -\frac{\psi_x}{\psi_y} = \frac{v}{u-c} = \frac{h_q}{h_p}$. By (4.26) it follows that

$$\partial_x[\,u(x, y(x))] = u_x + y'u_y = u_x + \frac{h_q}{h_p}\,u_y = \frac{1}{h_p}\,u_q. \tag{4.28}$$

Claim 1. *The horizontal fluid velocity u, even in x, is a strictly decreasing function of x along any streamline in $\overline{\Omega}_+$.*

Proof. Since the functions u and v are harmonic in the (x, y)-variables, they remain harmonic in the (q, p)-variables because (4.24) is a conformal change of variables. So u_q is also harmonic. Let us consider the restriction of u_q to the rectangle $\hat{\Omega}_+$. By (4.26)–(4.27) we have

$$u_q = \frac{1}{(c-u)^2+v^2}\,[(c-u)u_x - vu_y]. \tag{4.29}$$

From (4.29) we deduce that along the lateral sides of $\hat{\Omega}_+$, which are the images under (4.24) of the crest and trough lines, we have

$$u_q = \frac{1}{c-u}\,u_x = -\frac{v_y}{c-u} = 0 \quad \text{for} \quad q = 0, \tag{4.30}$$

and, similarly,

$$u_q = 0 \quad \text{for} \quad q = c\pi, \tag{4.31}$$

because $v = v_y = 0$ along these sides. On the lower boundary $\hat{B}_+ = \{p = -m\}$ of the rectangle $\hat{\Omega}_+$, the image of the flat bed, we also know by (4.6) that $v = 0$ so that

$$u_q = \frac{1}{c-u} u_x < 0 \quad \text{if} \quad p = -m \quad \text{and} \quad q \in (0, c\pi) \tag{4.32}$$

by (4.23). On the top boundary $\hat{S}_+$, the image of the free surface from crest to trough, we have $p = 0$ and

$$u_q = \frac{1}{(c-u)^2 + v^2}[(c-u)u_x - vu_y] = \frac{c-u}{(c-u)^2+v^2}\partial_x[u(x,\eta(x))] < 0 \tag{4.33}$$

for $p = 0$, $q \in (0, c\pi)$, in view of (4.5), (4.11), and (4.22).

From (4.30)–(4.33) and the strong maximum principle applied to the harmonic function u_q, we deduce that $u_q < 0$ in the open rectangle $\hat{\Omega}_+$. But then (4.28), together with $h_p > 0$, ensures that u is a strictly decreasing function of x along any streamline in $\overline{\Omega}_+$. Being even in the x-variable, the function u is then a strictly increasing function of x along any streamline in $\overline{\Omega}_-$. □

If (X_0, Y_0) is the initial location of a particle, its path $\{(X(t), Y(t))\}_{t\geq 0}$ is obtained as the solution of (4.15) with initial data $(X(0), Y(0)) = (X_0, Y_0)$. This path in the physical plane corresponds to the streamline in the moving frame given by $x(t) = X(t) - ct$, $y(t) = Y(t)$. Thus the system (4.15) is equivalent to the autonomous system

$$\begin{cases} x'(t) = u(x,y) - c, \\ y'(t) = v(x,y) \end{cases} \tag{4.34}$$

in the moving frame. Because $u - c < 0$ is bounded away from zero in view of (4.11), the horizontal component $x(t)$ goes from $+\infty$ to $-\infty$ as t goes from $-\infty$ to $+\infty$. Therefore, for any particle path there is surely a time, say, $t = 0$, when $x = \pi$, and another time, say, $t = \theta > 0$, when $x = -\pi$. We call θ the **elapsed time** per period of the streamline, that is, the time it takes $(x(t), y(t))$ to traverse one period in the moving plane.

Because $\partial_t \psi(x(t), y(t)) = 0$ by (4.9) and (4.34), the image of the streamline in the conformal frame is a line $p =$ constant: $p(t) = p(0)$ for all t. Moreover,

$$\frac{dq}{dt} = q' = -\phi_x x' - \phi_y y' = -(c-u)^2 - v^2 < 0,$$

and

$$q(0) = c\pi > 0, \quad q(\theta) = -c\pi < 0.$$

Claim 2. *If a streamline is given by $\psi(x,y) = -p$ or $y = y(x)$, the elapsed time is*

$$\theta(p) = \int_{-c\pi}^{c\pi} \left(h_q^2(q,p) + h_p^2(q,p)\right) dq = \int_{-\pi}^{\pi} \frac{dx}{c - u(x, y(x))} > \frac{2\pi}{c}. \tag{4.35}$$

Proof. Notice that along the particle path we have

$$y(\theta)-y(0)=\int_0^\theta v(x(t),\, y(t))\,dt=\int_{-c\pi}^{c\pi}\frac{v}{[c-u]^2+v^2}\,dq=0$$

since v is odd and u is even in the q-variable. On the other hand,

$$2\pi=x(0)-x(\theta)=\int_0^\theta[c-u(x(t),y(t))]\,dt=c\theta-\int_{-c\pi}^{c\pi}\frac{u}{[c-u]^2+v^2}\,dq$$

by (4.34). Converting to the conformal variables and taking (4.27) into account, we get

$$c\theta-2\pi=\int_{-c\pi}^{c\pi}\Big(c\,[h_q^2(q,\,p(0))+h_p^2(q,\,p(0))]-h_p(q,\,p(0))\Big)\,dq. \tag{4.36}$$

By the divergence theorem, integrating $0=\Delta h$ over the region $[-c\pi,\,c\pi]\times[-m,p]$ below a horizontal line, we get

$$\int_{-c\pi}^{c\pi}h_p(q,\,p)\,dq=\int_{-c\pi}^{c\pi}h_p(q,\,-m)\,dq$$

for every fixed $p\in[-m,0]$. But $h_p=\frac{1}{c-u}$ and $\frac{\partial q}{\partial x}=-\phi_x=c-u$ on the bottom $p=-m$, so that

$$\int_{-c\pi}^{c\pi}h_p(q,\,-m)\,dq=\int_{\pi}^{-\pi}\frac{u(x,-d)-c}{c-u(x,-d)}\,dx=2\pi.$$

Thus

$$\int_{-c\pi}^{c\pi}h_p(q,\,p)\,dq=2\pi,\qquad p\in[-m,0],$$

and (4.36) yields the first formula for the elapsed time

$$\theta(p)=\int_{-c\pi}^{c\pi}\Big(h_q^2(q,\,p)+h_p^2(q,\,p)\Big)\,dq,\qquad p\in[-m,0].$$

We can also write this formula in the moving frame. Indeed, (4.27) yields

$$h_q^2+h_p^2=\frac{1}{(c-u)^2+v^2},$$

and along the streamline

$$\frac{dq}{dx}=-\phi_x-\phi_y\frac{dy}{dx}=c-u+\frac{v^2}{c-u}=\frac{(c-u)^2+v^2}{c-u}.$$

This leads to the second formula for the elapsed time:

$$\theta(p)=\int_{-\pi}^{\pi}\frac{dx}{c-u(x,y(x))}.$$

Finally, to prove the inequality on $\theta(p)$, we have

$$4\pi^2 \le \left\{\int_{-\pi}^{\pi} \frac{dx}{c-u(x,y(x))}\right\}\left\{\int_{-\pi}^{\pi}[c-u(x,y(x))]dx\right\}$$

by the Cauchy–Schwarz inequality. Just as in the derivation of (4.13), but integrating over the region between $y=-d$ and a streamline $y=y(x)$,

$$2\pi c = \int_{-\pi}^{\pi}(c-u(x,y(x)))[1+(y'(x))^2]dx > \int_{-\pi}^{\pi}(c-u(x,y(x)))dx$$

unless we are on the flat bed $y=-d$, in which case equality holds. Combining the last two inequalities, we obtain

$$\theta(p) \ge \frac{4\pi^2}{\displaystyle\int_{-\pi}^{\pi}\Big(c-u(x,y(x))\Big)dx} > \frac{4\pi^2}{2\pi c} = \frac{2\pi}{c}.$$

Even on the flat bed the inequality is strict. Indeed, equality in Cauchy–Schwarz requires $c-u$ to be constant along the flat bed $y=-d$, but this is not the case in view of the second inequality in (4.23). □

By the **drift of a particle** we mean the net horizontal distance moved by the particle between its positions below two consecutive troughs (or crests), that is,

$$X(\theta)-X(0)=c\theta-2\pi = X(t+\theta)-X(t), \qquad t\in\mathbb{R}. \tag{4.37}$$

This corresponds to one period in the moving plane. The drift is positive if the particle always moves in the direction of wave propagation, and negative if it moves in the reverse direction. Zero drift characterizes a closed particle path that corresponds to a solution of (4.15) of period $\theta=\frac{2\pi}{c}$. Indeed, by (4.37) zero drift is equivalent to $\theta=\frac{2\pi}{c}$, in which case the relations $X(\theta)=X(0)$, $x(\theta)=x(0)-2\pi$, $Y(\theta)=Y(0)$, combined with the periodicity of both u and v ensure by uniqueness that $X(t+\theta)=X(t)$ and $Y(t+\theta)=Y(t)$ for all t.

Claim 3. *The drift of every particle strictly decreases with depth.*

Proof. In view of (4.37) it suffices to show that $\partial_p^2\theta>0$ for $p\in(-m,0)$ and that $\partial_p\theta(-m)=0$. From (4.35) we have

$$\partial_p\theta = 2\int_{-c\pi}^{c\pi}(h_q h_{qp}+h_p h_{pp})dq.$$

But

$$\int_{-c\pi}^{c\pi} h_q h_{qp}\,dq = -\int_{-c\pi}^{c\pi} h_p h_{qq}\,dq = \int_{-c\pi}^{c\pi} h_p h_{pp}\,dq$$

because h is harmonic and $2c\pi$-periodic in the q-variable. Thus

$$\partial_p\theta = 4\int_{-c\pi}^{c\pi} h_p h_{pp}\,dq = -4\int_{-c\pi}^{c\pi} h_p h_{qq} dq. \tag{4.38}$$

By (4.25) we have $h = h_q = h_{qq} = 0$ on the flat bed $p = -m$, so that $\partial_p \theta(-m) = 0$. But

$$\partial_p^2 \theta = 4 \int_{-c\pi}^{c\pi} (h_p h_{ppp} + h_{pp}^2) dq = 4 \int_{-c\pi}^{c\pi} (h_{qp}^2 + h_{pp}^2) dq \geq 0.$$

Therefore $\partial_p \theta \geq 0$. We claim that this inequality is strict. Indeed, if $\partial_p \theta(p^*) = 0$ for some $p^* > -m$, then $\partial_p^2 \theta(p^*) = 0$ for all $p \in [-m, p^*]$. So $h_{qp} = h_{pp} = 0$ and h_p is a constant in $R^* = [-c\pi, c\pi] \times [-m, p^*]$. But along the lower boundary of R^* we have $h_p = \frac{1}{c-u}$ by (4.27) and (4.6). The second inequality in (4.23) guarantees that h_p cannot be constant, which is a contradiction. □

Before discussing the particle paths in the physical variables, we provide some information about the corresponding streamlines in the moving frame.

Claim 4. *Along any fixed vertical line in the moving frame the absolute value of the slope of a streamline decreases with depth.*

Proof. The slope of a streamline $y = y(x)$ is given by $y' = \frac{v}{u-c}$. Its derivative is

$$\partial_y \frac{v}{u-c} = \frac{v_y(u-c) - u_y v}{(u-c)^2} = \frac{u_x(c-u) - u_y v}{(u-c)^2} = u_q \left(1 + \frac{v^2}{(c-u)^2}\right) < 0 \quad \text{in} \quad \Omega_+$$

in view of (4.2), (4.26), (4.28), and Claim 1. By (4.21) and (4.11), the slope is negative in Ω_+, positive in Ω_-, and zero on the crest and trough lines. □

Since v changes sign in accordance with (4.21), each particle lying above the flat bed has some vertical displacement in each half-period. From the previous result we infer the following useful fact.

Claim 5. *The vertical displacement of a particle decreases with depth.*

Proof. By (4.6), a particle on the bottom remains on the flat bed for all times. If the particle is above the flat bed, let $y = y(x)$ be the streamline to which the particle is confined in the moving frame. Then the amplitude of the vertical displacement of the particle is precisely $y(0) - y(\pi) = -\int_0^\pi y'(x)dx$, and we conclude applying Claim 4. □

We can now describe the particle trajectories above the flat bed as in Figure 4.4. First, the drift of any particle is positive because

$$c\theta - 2\pi > c\frac{2\pi}{c} - 2\pi = 0,$$

where the inequality is from Claim 2. The properties of u provided by (4.23) and Claim 1, in conjunction with (4.14), show that in Ω_+ the level set $\{u = 0\}$ consists of a continuous curve $\mathcal{C}_+$ in the moving frame that intersects each streamline $\psi = -p$ exactly once, where $p \in [-m, 0]$. In Ω_- the level set $u = 0$ is given by the reflection $\mathcal{C}_-$ of the curve $\mathcal{C}_+$ across the line $x = 0$. Between $\mathcal{C}_-$ and $\mathcal{C}_+$ we have $u > 0$ (including the top and lower boundaries),

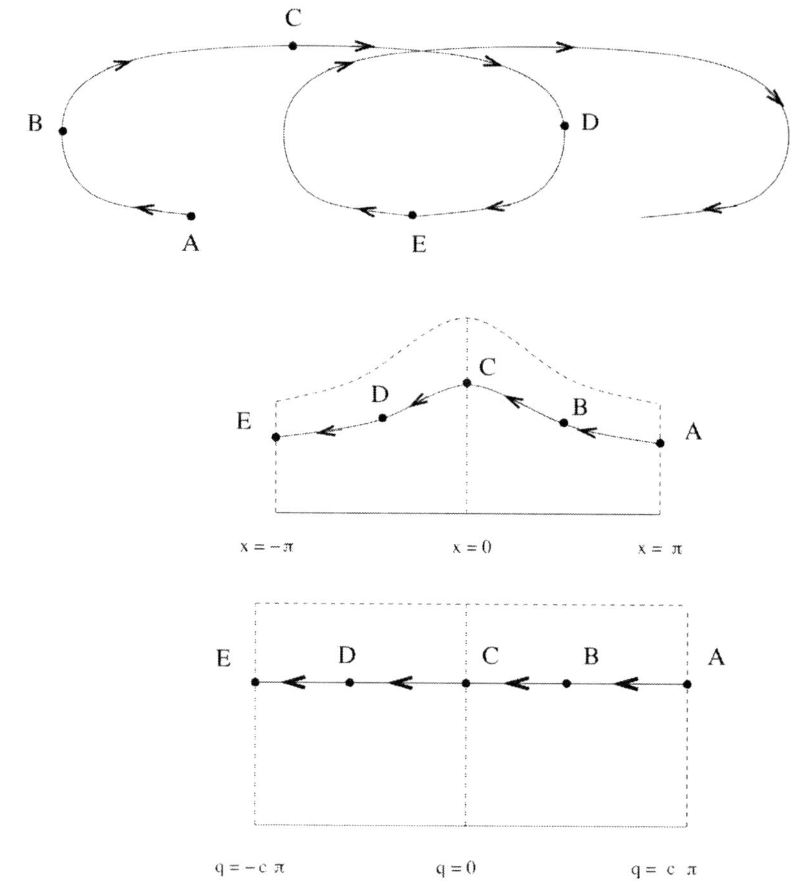

Figure 4.4. *Typical particle path above the flat bed for a Stokes wave. The top figure represents the physical frame: at A and E the wave trough is located right above the particle, while at C the wave crest is right above it. The middle and bottom figures depict the corresponding motion in the moving frame and in the conformal frame, respectively: in both cases the motion is to the left (the free surface and the flat bed are also drawn).*

while between $x = -\pi$ and $\mathcal{C}_-$, and between $\mathcal{C}_+$ and $x = \pi$, we have $u < 0$. In particular, $u < 0$ on the trough line and $u > 0$ on the crest line. In time $\theta = \theta(y_0)$ the particle in the moving frame reaches $x = -\pi$ with $y(\theta) = y_0$. In the time interval $(0, \theta)$ we know that the path $(x(t), y(t))$ intersects successively the curve $\mathcal{C}_+$ at a point B, the vertical segment $x = 0$ at a point C, and the curve $\mathcal{C}_-$ at D until it finally intersects $x = -\pi$ at $E = (-\pi, y_0)$ for $t = \theta$. In the time interval needed for $(x(t), y(t))$ to get in the moving frame from A to B and from D to E we know that $u < 0$ so that in the physical variables (X, Y) the particle $(X(t), Y(t))$ moves to the left. In the time interval needed for $(x(t), y(t))$ to get in the moving frame from B to D we have $u > 0$ so that $(X(t), Y(t))$ moves to the right. Also, between A and C we have $v > 0$ so that $(X(t), Y(t))$ moves up, while between C and E we know that $v < 0$ so that $(X(t), Y(t))$ moves down. The initial position is $(X(0), Y(0)) = (\pi, Y_0)$, while $(X(\theta), Y(\theta)) = (-\pi + c\theta, Y_0)$ as $x(t) = X(t) - ct$. From (4.37) and the inequality in Claim 2 we infer that $X(\theta) > \pi = X(0)$ so that E is strictly to the right of A and the looping motion depicted in Figure 4.4 (top figure) is confirmed. Thus all particles above the flat bed are looping to the right. The travel of the particle from A to E is slower than the wave speed c.

On the other hand, for a particle located on the flat bed at $(\pi, -d)$ the motion has a backward-forward pattern with a forward drift, mirroring the projection of the loop in the above figure to the flat bed.

On the particle trajectories in Stokes's wave of greatest height

In 1880, Stokes [288] pointed out that an irrotational periodic traveling wave solution with a stagnation point would mark a limit "beyond which no waves of the oscillatory irrotational kind can be propagated without change of form" and coined the phrase "wave of greatest height" to describe it.[89] He predicted that if $\eta(x)$ is the surface profile of an extreme wave with the wave crest located at $x = 0$, then $\lim_{\pm x \downarrow 0} \{\eta'(x)\} = \mp 1/\sqrt{3}$, and η is convex between its successive maxima. The existence of an extreme wave with a corner at its crest, each of the tangents to which makes an angle $\pi/6$ with the horizontal, was proved independently in [10] and [253], while the existence[90] of an extreme wave with a profile that is convex between successive crests was recently established [254]. The extreme waves are limiting cases of the regular Stokes waves considered before, and (4.11) breaks down as $u = c$ at the wave crest (elsewhere in the fluid we still have $u < c$). It is known that the free surface of an extreme wave is real-analytic except at the wave crest, where it is only Lipschitz continuous without being continuously differentiable (see the discussion in [298]). Moreover, the extreme wave is symmetric, and no irrotational wave trains exist beyond it (cf. [298]).

While the approach presented above for regular Stokes waves is not applicable, an important observation in the context of extreme waves is that **the wave crest is not a stagnation point of the flow**, despite the fact that $u = c$ and $v = 0$ there. Indeed, for an extreme wave we have $Q = g[\eta(0) + d]$ in view of the nonlinear boundary condition in (4.10). Since for an extreme wave the symmetric corner at the wave crest with angle $2\pi/3$ means

$$\lim_{x \to 0} [\eta'(x)]^2 = \lim_{x \to 0} \frac{v^2(x, \eta(x))}{[u(x, \eta(x)) - c]^2} = \frac{1}{3}, \tag{4.39}$$

and

$$[u(x, \eta(x)) - c]^2 + v^2(x, \eta(x)) = Q - 2g[\eta(x) + d] = 2g\,[\eta(0) - \eta(x)]$$

by the nonlinear boundary condition in (4.10), we deduce that

$$\lim_{x \to 0} \frac{[u(x, \eta(x)) - c]^2 + v^2(x, \eta(x))}{|x|} = \frac{2g}{\sqrt{3}}. \tag{4.40}$$

[89] Today one usually refers to these waves as "extreme waves"; there is no proof that these waves maximize the possible wave height of an irrotational wave train.

[90] The uniqueness question remains open; it is not known that every extreme wave has this shape.

From (4.39) and (4.40) we deduce that $\frac{[c-u(x,\eta(x))]^2}{|x|} \to \frac{g\sqrt{3}}{2}$ as $x \to 0$, and thus

$$\int_0^{\pi} \frac{ds}{c-u(s,\eta(s))} < \infty. \tag{4.41}$$

But then any free surface particle will reach the wave crest in finite time. Indeed, let the wave crest be $(0,\eta(0))$ at $t=0$, and the particle $(X(t),Y(t))$ occupies at time $t=0$ the position $(X_0,\eta(X_0))$ with $X_0 \in (0,\pi)$. Then, if for $t \in (0,\tau)$ the particle does not reach the wave crest, that is, $X(t) \neq ct$ for $t \in (0,\tau)$, we will have by (4.15) that

$$X'(t) - c = u(X(t),\eta(X(t))) - c < 0, \qquad t \in (0,\tau),$$

since $Y(t) = \eta(X(t))$ expresses the fact that the particle remains on the free surface. Integration yields

$$\tau = \int_{X(\tau)-c\tau}^{X_0} \frac{ds}{c-u(s,\eta(s))}.$$

Because of (4.41) we must have that the decreasing function $t \mapsto x(t) = X(t) - ct$ with $x(0) = X_0 > 0$ reaches zero in finite time; that is, the particle will reach the wave crest in finite time $\int_0^{X_0} \frac{ds}{c-u(s,\eta(s))}$. The particle cannot stay at the crest more than an instant since otherwise it would collide with other surface particles that occupy at $t=0$ a position close to $(X_0,\eta(X_0))$. Consequently, the wave crest of an extreme wave is only an apparent stagnation point of the flow; no particle can rest there.

4.2 Pressure beneath a Stokes wave

Understanding the behavior of the pressure beneath a surface water wave is of interest. After all, in scuba diving almost all techniques and knowledge center around pressure. The hydrostatic pressure

$$P_0 = P_{atm} - \rho g y \tag{4.42}$$

in still water with the flat free surface $y=0$, ρ being the constant density taken to be equal to $1\,g/cm^3$, can be computed by solving the governing equations with the Ansatz of the velocity field $(u,v) \equiv 0$, corresponding to still water. Notice that the hydrostatic pressure is the same at all points of the water body which are at the same height, which explains the so-called **hydrostatic paradox**: the pressure of still water on the bottom of a region does not depend upon the shape of the bottom or the amount of fluid; it depends only on the height of the water column. The air pressure at sea level is one atmosphere ($1\,atm$), defined as being equal to 101325 Pa, the Pascal unit being $1\,Pa = 1\,kg/(m\,s^2) = 10^{-5}\,bar$ so that $1\,Pa = 0.9869\,atm$, with the gravitational constant of acceleration $g = 9.869\,m/s^2$. Formula (4.42) shows that the hydrostatic pressure increases at the rate of $1\,atm$ every $10\,m$ of water depth. An observation is in order here. It is tempting to say that as we dive deep below the water surface an increase in pressure affects the density by crowding the water molecules together and thus reducing the volume occupied by a fixed mass of water. However, constant density is an accurate assumption so that the increase in pressure does not have any influence upon density: the hydrostatic pressure is a measure of the internal forces that keep the water under the constraint of homogeneity. With pressure understood as normal force per unit area, these considerations explain why deep diving is difficult: the

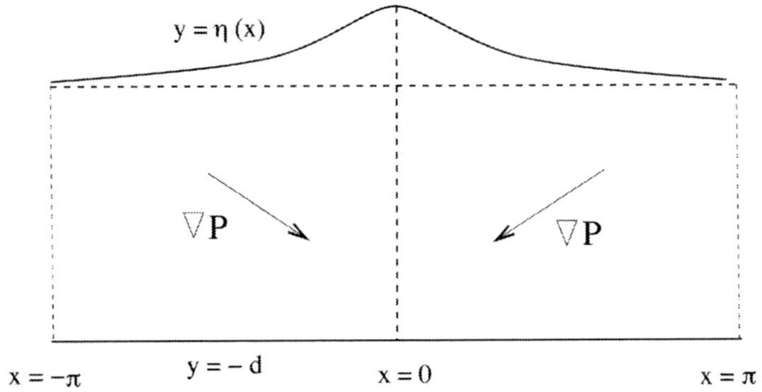

Figure 4.5. *Monotonicity properties of the pressure beneath the surface wave and between two consecutive wave troughs.*

human body is mostly water, and in recreational diving water pressure will be felt in the air spaces of the body (lungs, sinuses, and ear canals).

The presence of waves affects the pressure in the water, and it is of interest to understand to what effect. Throughout this section we use the notation $\Omega_\pm$, $B_\pm$, $S_\pm$ introduced in Section 4.1.

Claim. *For a smooth Stokes wave the sign of the first component P_x of the pressure gradient at a point in the fluid depends on the position of the point with respect to the crest line: $P_x = 0$ below the crest, $P_x < 0$ in $\Omega_+ \cup S_+ \cup B_+$, $P_x > 0$ in $\Omega_- \cup S_- \cup B_-$. The second component P_y is strictly negative throughout the fluid.*

Remark. An immediate consequence is that P_{atm} *is the minimum value of the pressure and is attained only on the free surface $y = \eta(x)$, while the maximum value of the pressure is attained on the flat bed directly below the crest.* Moreover, since a streamline decreases to the right of the crest line, being symmetric with respect to the crest line, we also have that *the maximum pressure on any internal streamline occurs directly underneath the wave crest.*

Proof. Throughout the proof we will take advantage of the fact (cf. the discusion in [298]) that the maximal slope of any Stokes wave is bounded:

$$\eta_x^2(x, \eta(x)) < 1, \qquad x \in \mathbb{R}. \tag{4.43}$$

Notice first that (4.3) and (4.29) yield

$$P_x = [(c-u)^2 + v^2]\, u_q. \tag{4.44}$$

In Claim 1 of Section 4.1 we found that $u_q < 0$ in $\hat{\Omega}_+ \cup \hat{S}_+ \cup \hat{B}_+$, so that (4.44) proves the assertion for P_x on $\Omega_+ \cup B_+ \cup S_+$. A similar argument shows that P_x is positive in $\Omega_- \cup B_- \cup S_-$.

We now claim that

$$\frac{d}{dx}\left((c-u)v \Big|_{y=\eta(x)} \right) < g, \quad x \in (-\pi, 0) \cup (0, \pi). \tag{4.45}$$

Indeed, using (4.2), (4.4), and (4.5), we find

$$\begin{aligned}\frac{d}{dx}\Big(\Big[c-u(x,\eta(x))\Big]\,v(x,\eta(x))\Big) &= -u_x v - u_y v\eta_x + (c-u)v_x + (c-u)v_y\eta_x \\ &= v_x(1+\eta_x^2)(c-u).\end{aligned} \tag{4.46}$$

On the other hand, differentiating the nonlinear boundary condition in (4.10) and using (4.5), we deduce that

$$\begin{aligned}0 &= (u-c)(u_x+u_y\eta_x)+v(v_x+v_y\eta_x)+g\eta_x \\ &= (u-c)u_x+u_y v+vv_x+(u-c)v_y\eta_x^2+g\eta_x.\end{aligned}$$

Invoking again (4.2) and (4.4), we get

$$v_y(c-u)(1-\eta_x^2)+2vv_x+g\eta_x=0 \quad \text{on} \quad y=\eta(x). \tag{4.47}$$

Let us first discuss (4.45) for $x\in(0,\pi)$, in which case $\eta'(x)<0$. Hence (4.5) and (4.47) imply that

$$(c-u)\Big(\frac{v_y}{\eta_x}(1-\eta_x^2)-2v_x\Big)+g=0 \quad \text{on} \quad S_+. \tag{4.48}$$

On the other hand, the second relation in (4.22) together with (4.2), (4.4), and (4.5) yield $\frac{v_y}{\eta_x}<v_x$ on S_+. Using (4.43) we infer that $\frac{v_y}{\eta_x}(1-\eta_x^2)<v_x(1-\eta_x^2)$ on S_+, which is equivalent to the inequality

$$\frac{v_y}{\eta_x}(1-\eta_x^2)-2v_x < -v_x(1+\eta_x^2) \quad \text{on} \quad S_+.$$

Combining this now with (4.48) and (4.46), we get (4.45) for $x\in(0,\pi)$. The symmetry of u and the antisymmetry of v with respect to the crest line $x=0$ give (4.45) for $x\in(-\pi,0)$.

We are now in a position to prove that $P_y<0$ in $\overline{\Omega}_+$. On the flat bed we have $v=0$ and thus $v_x=0$, so that the second component of (4.3) yields $P_y=-g<0$ on $y=-d$. To get the assertion on the free surface, we recall from Section 4.1 that the pressure P is superharmonic in the fluid domain, attainig its minimum everywhere on $y=\eta(x)$. Hence, Hopf's maximum principle implies that P_y is negative all along $y=\eta(x)$. It remains therefore to show that P_y is negative in Ω_+. To prove this we introduce the auxiliary function

$$f(x,y)=[c-u(x,y)]\,v(x,y)-gx, \qquad (x,y)\in\overline{\Omega}.$$

Using (4.2) and (4.4), we compute

$$\Delta_{(x,y)}f=(c-u)\Delta_{(x,y)}v-v\Delta_{(x,y)}u-2u_xv_x-2u_yv_y=0.$$

Since the change of variables (4.24) is conformal, f remains harmonic in the (q,p)-variables, and we conclude that f_q is harmonic in the strip $\hat{\Omega}_+$ as well. From (4.2) and (4.26) we get

$$f_q=v_x-\frac{g(c-u)}{(c-u)^2+v^2} \quad \text{in} \quad \mathbb{R}\times[-m,0]. \tag{4.49}$$

On the flat bed $y=-d$ we have $v=0$ by (4.6), and thus $v_x(x,-d)=0$ as well, and (4.49) implies that

$$f_q=-\frac{g}{c-u}<0 \quad \text{on} \quad p=-m. \tag{4.50}$$

On the free surface we have $v = (u-c)\eta_x$ by (4.5). Thus (4.49) yields

$$f_q = \frac{1}{(c-u)(1+\eta_x^2)}\Big(v_x(c-u)(1+\eta_x^2) - g\Big) \quad \text{on} \quad p = 0.$$

Invoking (4.46) and (4.45), we get

$$f_q(q,0) < 0 \quad \text{for all} \quad q \in (0, c\pi). \tag{4.51}$$

On $\{(c\pi, p) : -m < p < 0\}$ we have $v = 0$. As $v_x < 0$ by (4.4) and (4.23), we get from (4.51) that $f_q < 0$ there. Since f_q is even in the q-variable and harmonic in $(-c\pi, c\pi) \times (-m, 0)$, using (4.50) and (4.51), we obtain from the maximum principle that

$$f_q = v_x - \frac{g(c-u)}{(c-u)^2 + v^2} < 0 \quad \text{in} \quad [-c\pi, c\pi] \times [-m, 0]. \tag{4.52}$$

On the other hand, in proving Claim 1 we showed that

$$0 > u_q = \frac{(c-u)u_x - vu_y}{(c-u)^2 + v^2} \quad \text{in} \quad (0, c\pi) \times [-m, 0].$$

Using (4.2), (4.4), and (4.11), the previous inequality can be restated as $v_y > \frac{vv_x}{u-c}$ in $\Omega_+ \cup S_+ \cup B_+$. Combining this inequality with (4.21) and with the second component of (4.3), we get

$$P_y = -g + (c-u)v_x - vv_y < -g + \frac{v_x[(c-u)^2 + v^2]}{c-u}$$

in Ω_+. But in view of (4.52) this implies that $P_y < 0$ in Ω_+. Recall that on the crest line $x = 0$ and on the trough line $x = \pi$ we have $v = 0$. Thus the second components of (4.3) and (4.49) yield

$$P_y = (c-u)v_x - g = (c-u)f_q < 0 \quad \text{on} \quad x = 0 \text{ and on } x = \pi,$$

in view of (4.52) and (4.11). Therefore P_y is negative in $\overline{\Omega}_+$. The evenness of P in the x-variable completes the proof of the claim. □

4.3 Appendix for Chapter 4: Gerstner's wave

A remarkable early paper by Gerstner [125] gave the first exact nonlinear solution[91] for wave trains on deep water by requiring that each particle in the water describes a circle. Still today this is the only known explicit solution of the two-dimensional governing equations for gravity water waves with a nonflat free surface! The Gerstner flow is, however, not irrotational, and the fluid domain must be of infinite depth. Historically, Gerstner's wave found favor with naval architects, but today it is sometimes regarded more as a curiosity than as a result of practical importance, partly because the wave is not irrotational (cf. [85]) and partly because the direction of vorticity is opposite that of

[91] First published in 1802 in an obscure Bohemian journal, then reprinted a few years later [126], and rediscovered by Rankine in 1863 [258].

propagation of the wave[92] (cf. [192]). There is an open question as to whether or not the nonlinear interaction between a nonuniform current and Stokes's wave might result in something resembling Gerstner's wave (for some steps toward a confirmation that this might be the case we refer the reader to the simulation in [147]; see also the discussion in [235]). Be that as it may, Gerstner's wave remains a remarkable achievement.

The Lagrangian description of motion is intrinsically appropriate for describing Gerstner's wave. In the Lagrangian description of motion, the (now dependent) variables X and Y that denote the position (in the physical frame) of a particle at time t are functions of some labels (a,b). Notice that, in contrast to our discussion of the particle paths in a Stokes wave, using in this context the initial particle positions (X_0, Y_0) as labels is not convenient because the initial fluid domain, while explicitly known, has a complicated expression in the physical frame. Instead, we consider the labelling domain $\{(a,b):\ a \in \mathbb{R},\ b \le b_0\}$, where $b_0 \le 0$ is a fixed constant,[93] and we define the particle location at time t in the physical domain as the image of (a,b) under the map

$$(a,b) \mapsto \Big(X(t;a,b),\, Y(t;a,b) \Big). \tag{4.53}$$

The coordinate transformation from Eulerian to Lagrangian variables is

$$\begin{cases} u(t,X,Y) = \dfrac{\partial}{\partial t} X(t;a,b), \\ v(t,X,Y) = \dfrac{\partial}{\partial t} Y(t;a,b), \end{cases} \tag{4.54}$$

and the spatial derivatives in the two coordinate systems are related via

$$\begin{cases} \dfrac{\partial}{\partial a} = X_a \dfrac{\partial}{\partial X} + Y_a \dfrac{\partial}{\partial Y}, \\ \dfrac{\partial}{\partial b} = X_b \dfrac{\partial}{\partial X} + Y_b \dfrac{\partial}{\partial Y}, \end{cases} \qquad \begin{cases} \dfrac{\partial}{\partial X} = \dfrac{1}{J}\Big(Y_b \dfrac{\partial}{\partial a} - Y_a \dfrac{\partial}{\partial b}\Big), \\ \dfrac{\partial}{\partial Y} = \dfrac{1}{J}\Big(X_a \dfrac{\partial}{\partial b} - X_b \dfrac{\partial}{\partial a}\Big), \end{cases} \tag{4.55}$$

where J is the Jacobian of the transformation, given by

$$J = \left| \frac{\partial(X,Y)}{\partial(a,b)} \right| = X_a\, Y_b - Y_a\, X_b\,. \tag{4.56}$$

Using (4.55) in combination with (4.54), we express the continuity equation (4.2) in Lagrangian coordinates as

$$\begin{aligned} 0 = u_X + v_Y &= \frac{Y_b\, u_a - Y_a\, u_b + X_a\, v_b - X_b\, v_a}{J} \\ &= \frac{Y_b\, X_{at} - Y_a\, X_{bt} + X_a\, Y_{bt} - X_b\, Y_{at}}{J} = \frac{J_t}{J}, \end{aligned}$$

using in the last step (4.56) differentiated with respect to t. Consequently, the equation of mass conservation takes in Lagrangian coordinates the simple form

$$\frac{\partial}{\partial t}(X_a\, Y_b - Y_a\, X_b) = 0. \tag{4.57}$$

---

[92] Considering the wave motion as derived from a past wind would account for the fast decay of the induced vorticity in the water, but the sign of the vorticity should be positive. Indeed, when the wind blows in the fixed horizontal x-direction over water of unlimited horizontal extent that is otherwise undisturbed, neglecting Coriolis accelerations, the total momentum imparted to the water will be similarly aligned and leads to the development of an underlying unidirectional current [199]. With the wind velocity $(U(y),0)$ dependent only upon the vertical y-direction, the physically realistic assumption that U increases with the distance from the sea level means that for the underlying current $(u(y),0)$ in the water we have $u_y \ge 0$ so that the vorticity is positive.

[93] The reason for not allowing $b_0 > 0$ will become apparent later on.

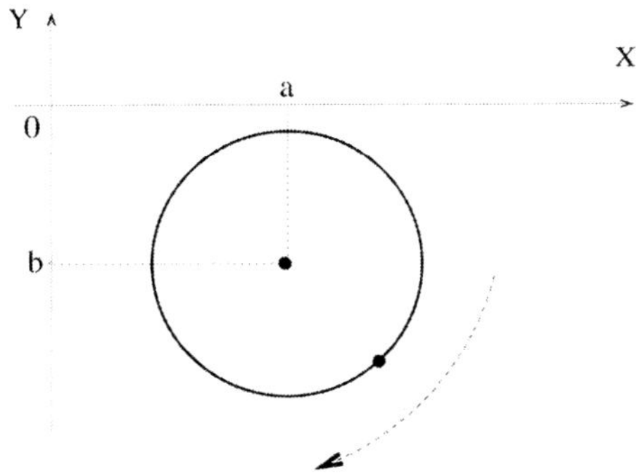

Figure 4.6. *Particle path in a Gerstner wave for which the surface wave train is propagating to the right.*

The Euler equations (4.3) take in Lagrangian coordinates the form

$$\begin{cases} X_{tt} = -\dfrac{1}{\rho}\dfrac{Y_b\,P_a - Y_a\,P_b}{J}, \\ Y_{tt} = -\dfrac{1}{\rho}\dfrac{X_a\,P_b - X_b\,P_a}{J} - g, \end{cases} \tag{4.58}$$

in view of (4.55) and (4.54). Using (4.56), we can express (4.58) equivalently as an equation for the gradient components of a scalar function in the label space,

$$\begin{cases} P_a = -\rho\,X_a\,X_{tt} - \rho\,Y_a\,(Y_{tt} + g), \\ P_b = -\rho\,X_b\,X_{tt} - \rho\,Y_b\,(Y_{tt} + g). \end{cases} \tag{4.59}$$

The label space being simply connected, the above equation is equivalent to the requirement $P_{ab} = P_{ba}$, or (ρ and g being constants)

$$X_a\,X_{btt} + Y_a\,Y_{btt} = X_b\,X_{att} + Y_b\,Y_{att}. \tag{4.60}$$

The complete formulation of the governing equations for a two-dimensional flow in Lagrangian coordinates consists of the continuity equation (4.57) and the Euler equation (4.60); the kinematic boundary condition (4.5) amounts to specifiying that the free surface corresponds to a fixed value b_0 of the label b, while the dynamic boundary condition (4.7) becomes $P(a,b_0) = P_{atm}$. If instead of a flat bed we have water of infinite depth, the kinematic boundary condition (4.6) is replaced by the assumption that at large depths the motion dies out. In addition, at any instant t the map (4.53) has to be a global diffeomorphism from the label domain to the fluid domain. In this context notice that $J \neq 0$ is only a sufficient condition for the particle labelling to be a local diffeomorphism.

Gerstner's wave is obtained by supposing that the position at time t of the particle with labels (a,b) is given by

$$\begin{cases} X(t;a,b) = a - \dfrac{e^{kb}}{k}\,\sin(ka - \sqrt{gk}\,t), \\ Y(t;a,b) = b + \dfrac{e^{kb}}{k}\,\cos(ka - \sqrt{gk}\,t), \end{cases} \tag{4.61}$$

where $k > 0$ is fixed: the path of this particle is a circle centered at (a,b) with radius $\frac{e^{kb}}{k}$, with the particle moving clockwise with constant angular speed $\sqrt{gk}$. A direct calculation yields $J = 1 - e^{2kb}$

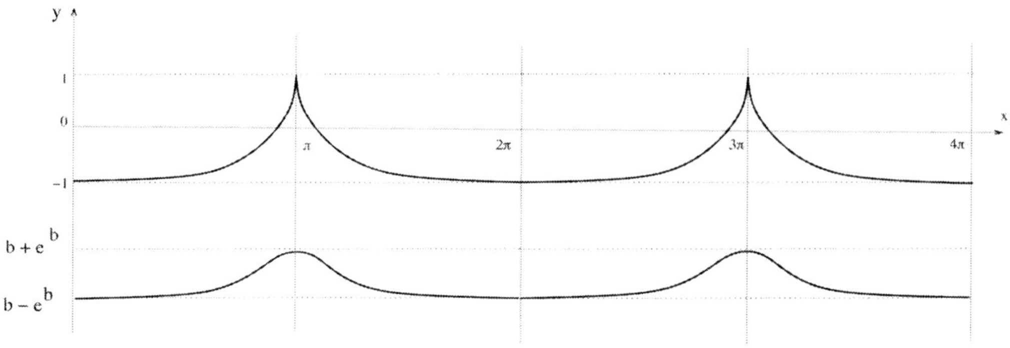

Figure 4.7. *The surface wave profile in Gerstner's wave with wavelength 2π: the upper curve is a cycloid and curves below are trochoids. Any curve could represent the profile of a surface wave train.*

so that (4.57) holds. The validity of (4.60) is easily checked, while (4.59) and the fact that $P = P_{atm}$ for $b = b_0$ yield

$$P(t;a,b) = P_{atm} - \rho g(b - b_0) + \frac{\rho g}{2k}(e^{2kb} - e^{2kb_0}). \tag{4.62}$$

We postpone for the moment the proof that the motion (4.61) is dynamically possible, that is, at any fixed instant t the label map is a diffeomorphism from the label set $\{(a,b): a \in \mathbb{R},\ b \le b_0\}$ to the fluid domain beneath the free surface given parametrically by

$$a \mapsto \Big(a - \frac{e^{kb_0}}{k}\sin(ka - \sqrt{gk}\,t),\ b_0 + \frac{e^{kb_0}}{k}\cos(ka - \sqrt{gk}\,t)\Big). \tag{4.63}$$

The surface wave profile (4.63) is a smooth reversed trochoid if $b_0 < 0$ and a cycloid with upward cusps at the wave crests if $b_0 = 0$. Indeed, given $k > 0$ and $b_0 \le 0$, the curve $Y = h_0(X)$, given parametrically by $s \mapsto (\frac{s}{k} - \frac{e^{kb_0}}{k}\sin(s), \frac{1}{k} - \frac{e^{kb_0}}{k}\cos(s))$, is called a trochoid if $b_0 < 0$ and a cycloid if $b_0 = 0$. It represents the path traced by a fixed point at a distance $\frac{1}{k}e^{kb_0}$ from the center of a circle of radius $\frac{1}{k}$ rolling without slipping along a horizontal line; these are the curves traced by a pedal when a bicycle is pedaled along a straight line.[94] The free surface (4.63) has the equation

$$Y = b_0 + \frac{1}{k} - h_0\Big(X - \sqrt{\frac{g}{k}}\,t\Big),$$

representing the profile of a wave train propagating to the right with wave speed $c = \sqrt{\frac{g}{k}}$. Notice that the fact that Gerstner's wave is a trochoid/cycloid turned upside down accounts for the relatively high and narrow crests and wide troughs of the wave profile. Since the function $s \mapsto h_0(s)$ is periodic with minimal period $\frac{2\pi}{k}$, Gerstner's wave has wavelength $\lambda = \frac{2\pi}{k}$.

To prove that the motion (4.61) is dynamically possible it suffices to analyze the map (4.61) at time $t = 0$, when it is of the form

$$\begin{cases} X = a - \dfrac{e^{kb}}{k}\sin(ka), \\[2ex] Y = b + \dfrac{e^{kb}}{k}\cos(ka). \end{cases} \tag{4.64}$$

[94]These curves also resemble the back arches of Stradivarius violins [252].

The general case follows by changing variables $(a,b) \to (a+t\sqrt{\frac{g}{k}}, b)$, performing (4.64), and subsequently shifting the horizontal variable by $t\sqrt{\frac{g}{k}}$. As a varies by the amount $2\pi/k$, the Y value reoccurs, while X is shifted linearly by $2\pi/k$. Therefore, it suffices to analyze the map (4.64) on the restricted half-strip $\{(a,b):\ a \in [0, 2\pi/k],\ b \le b_0\}$.

First, the image of the vertical half-lines $\{(0,b):\ b \le b_0\}$ and $\{(2\pi/k,b):\ b \le b_0\}$ under (4.64) are the vertical half-lines

$$\left\{(0,Y):\ Y \le b_0 + \frac{e^{kb_0}}{k}\right\} \quad \text{and} \quad \left\{(2\pi/k,Y):\ Y \le b_0 + \frac{e^{kb_0}}{k}\right\},$$

respectively, while the horizontal segment $\{(a,b_0):\ 0 \le a \le 2\pi/k\}$ is mapped into a part of the surface trochoid/cycloid. Since (4.64) yields $|(X,Y)-(a,b)| \mapsto 0$ as $b \to -\infty$, if we show that (4.64) is one-to-one, we deduce from the "Invariance of domain theorem"[95] that it is surjective onto the fluid domain. Notice that the differential of (4.64) at a fixed point (a,b) in the label set is precisely

$$\begin{pmatrix} 1-e^{kb}\cos(ka) & -e^{kb}\sin(ka) \\ -e^{kb}\sin(ka) & 1+e^{kb}\cos(ka) \end{pmatrix}$$

with determinant $1-e^{2kb} > 0$ for $b < 0$. Consequently, by the inverse function theorem (cf. [96]), (4.64) is a smooth local diffeomorphism onto its image if $b_0 < 0$. In the limiting case $b_0 = 0$ the surface cycloid has a cusp at the wave crest, so that the smoothness is valid only for interior points. The local diffemorphism is actually a global diffeomorphism from the closed/open label set to the closed/open fluid domain if $b_0 < 0$, respectively, if $b_0 = 0$ (in which case it extends to a global homeomorphism between the two closed sets). To prove these statements, it suffices to show that (4.64) is injective. Let $\xi = a+ib$ and notice that (4.64) corresponds to the map

$$\xi \mapsto \xi + \frac{i}{k}e^{ik\overline{\xi}}$$

defined for $\Im m(\xi) \le b_0 \le 0$. If the map fails to be injective in the closed label set, we can find $\xi_1 \ne \xi_2$ in the closed complex lower half-plane with

$$ik(\xi_1-\xi_2) = e^{ik\overline{\xi_1}} - e^{ik\overline{\xi_2}} = ik\,[\overline{\xi_1}-\overline{\xi_2}]\int_0^1 e^{ik[s\overline{\xi_1}+(1-s)\overline{\xi_2}]}ds, \tag{4.65}$$

so that

$$\frac{\xi_1-\xi_2}{\overline{\xi_1}-\overline{\xi_2}} = \int_0^1 e^{ik[s\overline{\xi_1}+(1-s)\overline{\xi_2}]}ds. \tag{4.66}$$

The fact that $[s\overline{\xi_1}+(1-s)\overline{\xi_2}]$ has nonnegative imaginary part means that the absolute value of the complex number $e^{ik[s\overline{\xi_1}+(1-s)\overline{\xi_2}]}$ is at most 1, the equality being possible only if the imaginary part of $[s\overline{\xi_1}+(1-s)\overline{\xi_2}]$ vanishes. Since the complex number on the left side of (4.66) has modulus 1, we see that the imaginary part of $[s\overline{\xi_1}+(1-s)\overline{\xi_2}]$ must vanish for every $s \in [0,1]$. But in this case both ξ_1 and ξ_2 must be real. Returning now to the first equality in (4.65), we deduce that

$$\begin{cases} \cos(k\xi_1) = \cos(k\xi_2), \\ \sin(k\xi_1) - \sin(k\xi_2) = k(\xi_1-\xi_2), \end{cases} \quad \text{with} \quad \xi_1,\xi_2 \in \mathbb{R}.$$

[95]Stating that if U is an open subset of $\mathbb{R}^n$ and $F:\overline{U} \to \mathbb{R}^n$ is an injective continuous map, then $F(U)$ is open and F is a bijection between $\overline{U}$ and $f(\overline{U})$ with $F(\partial U) = \partial F(U)$ (cf. [263]).

Writing the second relation in the equivalent form

$$\int_{k\xi_1}^{k\xi_2} \cos(\tau)\,d\tau = k\xi_2 - k\xi_1,$$

we infer that $\xi_1 = \xi_2$, which contradicts our initial assumption. We conclude that the motion (4.61) is dynamically possible for $b_0 \leq 0$. Notice that for $b > 0$ the curve given parametrically by (4.63) presents self-intersections and, as such, cannot be viewed as representing the profile of a surface wave train.

We conclude by pointing out some interesting features of Gerstner's wave.

- Using (4.55) we compute the vorticity of the flow as

$$\gamma = u_Y - v_X = \frac{X_a X_{bt} - X_b X_{at} - Y_b Y_{at} + Y_a Y_{bt}}{J} = -\frac{2\sqrt{gk}\,e^{2kb}}{1 - e^{2kb}}.$$

  Thus Gerstner's wave is rotational with a negative vorticity that decays rapidly as we descend into the liquid. Notice that in the case of the cusped waves, that is, for $b_0 = 0$, the vorticity becomes singular at the free surface.

- From (4.62) we infer that the pressure of a particle is always constant; it is of interest to point out that no wave with this property can exist in water of finite depth, as proved by Dubreil-Jacotin in 1934 [99].

- The explicit formula $c = \sqrt{\frac{g}{k}}$ for the speed of a Gerstner wave with wavelength $\lambda = \frac{2\pi}{k}$ yields $c = \sqrt{\frac{g}{2\pi}\lambda}$, which is precisely the dispersion relation for irrotational deep water-waves (see the discussion in Chapter 3).

- The flow velocity is given explicitly in Lagrangian variables by

$$\begin{cases} u(t;a,b) = \sqrt{\dfrac{g}{k}}\,e^{kb}\cos(ka - \sqrt{gk}\,t), \\ v(t;a,b) = \sqrt{\dfrac{g}{k}}\,e^{kb}\sin(ka - \sqrt{gk}\,t). \end{cases}$$

  This shows not only that the motion dies out exponentially with depth but also that the particles have maximum horizontal velocities at the trough (backward) and at the crest (forward). This replicates a feature that we elucidated in our discussion of the particle paths in a Stokes wave.

- To third order of the power series expansion in the small-amplitude parameter ε, the profile of the Stokes wave and that of Gerstner's wave are identical; only at fourth-order differences emerge (cf. [177, Section 5.2]). This illustrates the perils of relying too much on formal power series.

4.4 Notes for Chapter 4

The presented results on the particle trajectories and on the pressure beneath a Stokes wave are a combination of the approaches developed in [47] and [74], while our discussion of Gerstner's wave relies on [46] and [141].

The qualitative description of the flow beneath a smooth Stokes wave with no underlying current is almost complete. The only missing major aspect (other than understanding the wave of greatest height) is the behavior of the vertical velocity component v: we know how its sign depends on the location within the fluid domain, but how about its monotonicity? One would conjecture that

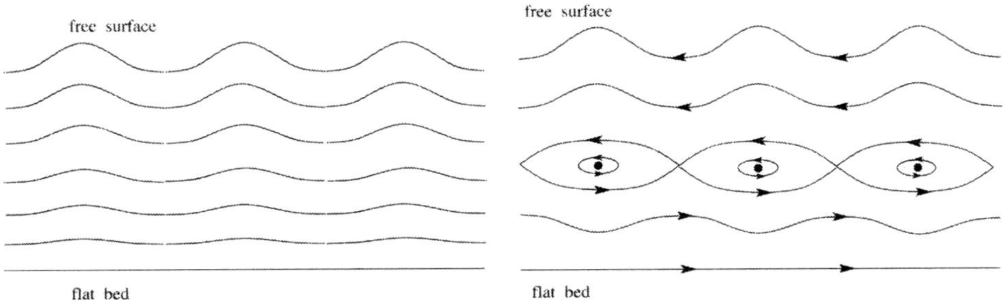

Figure 4.8. *Streamlines beneath a wave train propagating at the surface of a flow with constant nonzero vorticity: in the absence of stagnation points (left) and with stagnation points (right—Kelvin's "cat's eye" flow pattern).*

along each streamline, between crest and trough, v first increases with positive values away from the crest line and then decreases toward zero beneath the wave trough.

Results for the flow in wave trains with currents underneath the free surface present a considerable challenge. The case of irrotational flow (in which case the underlying current is uniform) was dealt with in [74]: weak currents do not affect the situation depicted in Figure 4.4, but a strong current will deform the loops therein, the outcome being dependent on whether the current is favorable or adverse. For rotational waves of small amplitude in a flow without stagnation points some recent results were obtained in [107, 108]. These results rely on the fact that a linearization will capture the main features if the waves are small perturbations of a flat free surface. Within the setting of waves of small amplitude, still relying on linearizing the governing equations, the analysis performed in the papers [309] and [76] permits an understanding of the motion beneath such waves. It turns out that the presence of stagnation points leads to fundamental changes in the dynamics. To illustrate this we refer the reader to Figure 4.8, where the streamlines are represented in the moving frame for a flow with constant nonzero vorticity. Near the stagnation points we observe the presence of closed streamlines, leading to Kelvin's "cat's eye" flow pattern [173].

For rotational waves of moderate and large amplitude, with or without stagnation points, the flow pattern beneath the wave (particle trajectories, behavior of the presssure and of the velocity field) is basically unknown. There is a lack of experimental, numerical, and theoretical results. Since wave trains at sea are the exception rather than the rule, why bother? The sheer beauty of wave trains is certainly compelling, but there are practical benefits as well. After all, most of our present understanding of water waves is based on results derived for small perturbations of a flat free surface flow. Understanding well the rotational wave trains opens up possibilities for studying perturbations of these special exact solutions of the governing equations, thus giving insight into wave motions that are far from flows with a flat free surface.

Chapter 5

Solitary water waves

"... that singular and beautiful phaenomenon."

J. Scott Russell

Some two-dimensional traveling water waves have a profile that is localized (a hump of water which drops smoothly back to a flat water level far ahead and far behind the wave crest). For these waves, called **solitary waves**, the initial profile $\eta(X,0) = \eta_0(X)$ moves at constant speed $c > 0$, without distortion, in the direction X of wave propagation: the profile $\eta(X,t)$ at a later time t is a translation of the initial profile by an amount ct in the positive X direction, $\eta(X,t) = \eta_0(X - ct)$. If a solitary wave profile of depression[96] is set up—a possibility within reach for laboratory experiments—one notices that it immediately collapses into small oscillatory waves.[97] Solitary waves were first observed by John Scott Russell in 1834 and described in a report to the British Association in 1844 [270]:

"I believe I shall best introduce the phaenomenon by describing the circumstances of my own first acquaintance with it. I was observing the motion of a boat which was rapidly drawn along a narrow channel by a pair of horses, when the boat suddenly stopped—not so the mass of water in the channel which it had put in motion; it accumulated round the prow of the vessel in a state of violent agitation, then suddenly leaving it behind, rolled forward with great velocity, assuming the form of a large solitary elevation, a rounded, smooth and well-defined heap of water, which continued its course along the channel apparently without change of form or diminution of speed. I followed it on horseback, and overtook it still rolling on at a rate of some eight or nine miles an hour, preserving its original figure some thirty feet long and a foot to a foot and a half in height. Such, in the month of August, 1834, was my first chance interview with that singular and beautiful phaenomenon."

The ability of this water wave to retain its shape for a long period of time is quite remarkable. Russell conducted numerous detailed experiments to investigate the nature

[96] The wave profile would be the reflection in a horizontal line of the profile of a wave of elevation.

[97] Solitary waves of depression are possible only if one allows for capillary effects, and consequently these are waves of very small amplitude. We consider gravity waves (of elevation) propagating at the free surface of a layer of water of average depth measuring from tens of *cm* to hundreds of *m*, with displacements of the free surface in excess of a few *mm*, as these are of interest from the point of view of waves at sea or in a canal.

Figure 5.1. *Sketch of Scott Russell's experiment.*

Photograph 8. *Recreation of the Scott Russell sighting.* [Reproduced with permission from the Department of Mathematics, Heriot-Watt University, Edinburgh, Scotland]

On July 12, 1995, an international gathering of scientist witnessed a recreation of Scott Russel's famous "first" sighting of a solitary wave on the Union Canal near Edinburgh.

of what he called "the great wave of translation," but which came to be known as the solitary wave.

Solitary waves can be easily obtained experimentally by dropping a weight at one end of long rectangular tank. Russell was the first to observe that an initial elevation might, depending on the relation between its height and length, evolve into any of a pure solitary wave of elevation, a single solitary wave of elevation with some residual oscillatory waves, or two or more solitary waves of elevation with or without residual oscillatory waves. He also attempted to produce solitary waves of depression but found that an initial depression is transformed into oscillatory waves of gradually increasing length and decreasing amplitude. From his laboratory instruments he concluded that the speed c of a solitary wave is given by

$$c = \sqrt{g(d+a)}, \tag{5.1}$$

where d is the average depth of the water above a flat bed, a is the wave amplitude, and g is the acceleration due to gravity. Notice that the above formula indicates that higher solitary waves travel faster.

In the early days, the existence of solitary waves excited some controversy since Russell's conclusions contradicted some predictions of Airy [6], considered at that moment to be the leading expert in hydrodynamics.[98] The conflict between Russell's observations and the accepted theory was resolved independently by Boussinesq in 1871 [28] and by Lord Rayleigh in 1876 [260]. They showed that suitable approximations—their insight was to incorporate weak nonlinear effects, in particular, an appropriate allowance for vertical acceleration which was neglected in the linear theory developed by Airy—to the governing equations for water waves lead, for wave amplitudes $a > 0$ that are sufficiently small, to the solitary wave solution

$$\eta(X,t) = \frac{a}{\cosh^2[\beta(X - ct)]} \tag{5.2}$$

with the wave speed c given by Russell's formula (5.1) and $\beta = \frac{\sqrt{3a}}{2d\sqrt{d+a}}$. They did not, however, derive a model equation (via an approximation procedure starting from the governing equations for water waves) which admits the above solitary wave solution. This final step was completed by Korteweg and de Vries in 1895 [184]. We will indicate in the Appendix to Chapter 7 how this equation arises as a model for the propagation of small-amplitude shallow-water waves at the surface of water with a flat bed; for present purposes we will consider not the equation written in physical variables but its normalized form

$$\eta_t + \eta\eta_X + \eta_{XXX} = 0. \tag{5.3}$$

We look for a solitary wave solution $\eta(X,t) = f(X - ct)$, where $c > 0$, and $f(s)$, $f'(s)$, and $f''(s)$ tend to zero as $|s| \to \infty$. Substituting this form, (5.3) can be integrated, but it is easier to check that for any constant $\alpha \in \mathbb{R}$ the wave profile

$$\eta_\alpha(X,t) = \frac{3c}{\cosh^2\left(\frac{\sqrt{c}}{2}(X - ct) + \alpha\right)} \tag{5.4}$$

represents a solitary wave solution of the Korteweg–de Vries (KdV) equation[99]: at any fixed time t, the graph of the solitary wave (5.4) is formed by a hump which drops smoothly and rapidly to zero away from its crest located at $X = ct - \frac{2\alpha}{\sqrt{c}}$. Russell observed in his experiments that solitary waves with greater amplitude, defined for a solitary wave of elevation as the maximal elevation of the wave profile above the asymptotic flat level,[100] travel faster and are narrower. This is borne out in the solution (5.4) as the amplitude a of the wave is three times the wave speed c and the width of the wave (defined as the distance between the points of height $a/2$) is inversely proportional to the square root of its amplitude. Note the similarity between (5.4) and (5.2).

The aim of this chapter is to discuss some aspects of solitary water waves. We will first study the motion beneath these waves in the setting of the governing equations. Subsequently we will illustrate the peculiar properties of the solitary waves in the shallow-water

[98] For more details on these fascinating historical aspects we refer the reader to [85]. Basically, the linear theory of waves of small amplitude fails to yield any approximation to solitary waves [286].

[99] It is not difficult to check that these are all the solitary wave solutions.

[100] Notice that this level is beneath the mean water level $Y = \int_{\mathbb{R}} \eta(x)\,dx$ since in the absence of the elevation the amount of water raised above it cannot disappear in view of mass conservation.

regime of waves of small amplitude, where KdV arises as a reasonable model equation for the propagation of unidirectional waves, elucidating thus the soliton phenomenon.

5.1 Particle paths beneath an irrotational solitary water wave

The rigorous theory of solitary water-wave solutions to the governing equations for irrotational flow started with an investigation by Friedrichs and Hyers [119] who proved the existence of small-amplitude solitary waves using a power series method. Subsequently, Beale [19] used the implicit function theorem of Nash–Moser to provide an alternative approach toward the existence of solitary water waves of small amplitude. The existence of large-amplitude solitary waves, including the existence of a solitary wave of greatest height, is due to Amick, Fraenkel, and Toland [10] and relies on global bifurcation theory.

We recall the governing equations for two-dimensional traveling waves $Y=\eta(X-ct)$ at the surface of water of constant density 1, in irrotational flow over the flat bed $Y=0$:

$$u_X+v_Y=0, \tag{5.5}$$

$$\begin{cases} (u-c)u_X+vu_Y=-P_X, \\ (u-c)v_X+vv_Y=-P_Y-g, \end{cases} \tag{5.6}$$

where $P(X-ct,Y)$ denotes the pressure and g is the gravitational constant of acceleration, with the condition

$$u_y=v_x, \tag{5.7}$$

expressing the absence of vorticity and with the boundary conditions

$$v=\eta_t+u\eta_X \quad \text{on} \quad Y=\eta(t,X), \tag{5.8}$$

$$v=0 \quad \text{on} \quad Y=0, \tag{5.9}$$

$$P=P_{atm} \quad \text{on} \quad Y=\eta(t,X), \tag{5.10}$$

P_{atm} being the constant atmospheric pressure. For a solitary wave we require the flow to be at rest for $X\to\pm\infty$, with the free surface approaching a height $h>0$ above the flat bed. The parameters $c>0$ and $h>0$ cannot be arbitrarily chosen: the inequality

$$c>\sqrt{gh} \tag{5.11}$$

must hold for nontrivial solutions (cf. [11]). Moreover, all solitary waves are always of positive elevation above their asymptotic limit h, symmetric about a single crest, and with a strictly monotone wave profile on either side of this crest, as shown by Craig and Sternberg [84].

In the moving frame

$$x=X-ct, \quad y=Y, \tag{5.12}$$

assume that the wave crest is located at $x=0$, and let

$$\Omega=\{(x,y)\in\mathbb{R}^2:\ 0<y<\eta(x)\}$$

be the fluid domain, with its two components

$$\Omega_- = \{(x,y) \in \mathbb{R}^2 : \ x < 0, \ 0 < y < \eta(x)\},$$
$$\Omega_+ = \{(x,y) \in \mathbb{R}^2 : \ x > 0, \ 0 < y < \eta(x)\},$$

the boundaries (top and bottom) of which we denote by

$$S_- = \{(x,y) \in \mathbb{R}^2 : \ x < 0, \ y = \eta(x)\}, \quad B_- = \{(x,y) \in \mathbb{R}^2 : \ x < 0, \ y = 0\},$$

respectively,

$$S_+ = \{(x,y) \in \mathbb{R}^2 : \ x > 0, \ y = \eta(x)\}, \quad B_+ = \{(x,y) \in \mathbb{R}^2 : \ x > 0, \ y = 0\}.$$

Defining the stream function $\psi(x,y)$ up to a constant by

$$\psi_y = u - c, \quad \psi_x = -v, \tag{5.13}$$

we see that ψ is harmonic in Ω in view of (5.7), whereas (5.8) and (5.9) ensure that ψ is constant on both boundaries of Ω, say, $\psi = 0$ on $y = \eta(x)$ and $\psi = m$ on $y = 0$. Throughout Ω we have $u < c$, with the inequality extending to the boundary[101] except for the wave of greatest height, in which case $u = 0$ at the crest with $u < c$ elsewhere; cf. [11]. That is,

$$\psi_y = u - c < 0 \quad \text{in} \quad \Omega \cup S_- \cup S_+ \cup B, \tag{5.14}$$

where $B = \{(x,0) : \ x \in \mathbb{R}\}$ is the flat bed, so that $m > 0$. Using the implicit function theorem we deduce that for all $\alpha \in (0,m]$ the level set $\psi = \alpha$ is a smooth curve $y = h_\alpha(x)$. Notice that $h_m \equiv 0$ while h_0 is the profile of the wave (with a corner at $x = 0$ in the case of the wave of greatest height).

Claim 1. *At any given time t the horizontal velocity component u is positive, while the sign of the vertical velocity component v at a point in the fluid depends on the position of the point with respect to the crest: $v = 0$ below the crest and on the flat bed, $v > 0$ if the crest is behind the particle located above the bed, and $v < 0$ if the point is above the bed and behind the crest.*

Proof. Since with respect to the frame of reference (5.12) the flow is steady and occupies a fixed region, it suffices to show that $u > 0$ throughout $\overline{\Omega}$, and $v(x,y) > 0$ if $(x,y) \in \Omega_+ \cup S_+$, while $v(x,y) < 0$ if $(x,y) \in \Omega_- \cup S_-$.

Notice that $v = 0$ on the flat bed $B = \{(x,0) : \ x \in \mathbb{R}\}$ in view of (5.9). The symmetry properties established in [84] ensure that u and η are symmetric with respect to the line $x = 0$, while v is antisymmetric. In particular, $v(0,y) = 0$ for all $y \in [0, \eta(0)]$. On the other hand, the convergence $u, v \to 0$ as $|x| \to \infty$ is exponentially fast [11]. Furthermore, since (5.8) and (5.14) ensure that $v < 0$ on S_- as the profile $x \mapsto \eta(x)$ of the free surface is strictly increasing for $x < 0$, while $v = 0$ at $x = -\infty$ and on B_- as well as for $x = 0$, we deduce by the maximum principle for the harmonic function v in Ω_- that $v < 0$ in Ω_-. Similarly one shows that $v(x,y) > 0$ for $(x,y) \in \Omega_+ \cup S_+$.

[101] The free surface is the graph of a real-analytic function and the fluid velocity components have harmonic extensions across it, except for the wave of greatest height where the curve is real-analytic everywhere but at the crest, where it is just continuous with a corner containing an angle of $2\pi/3$.

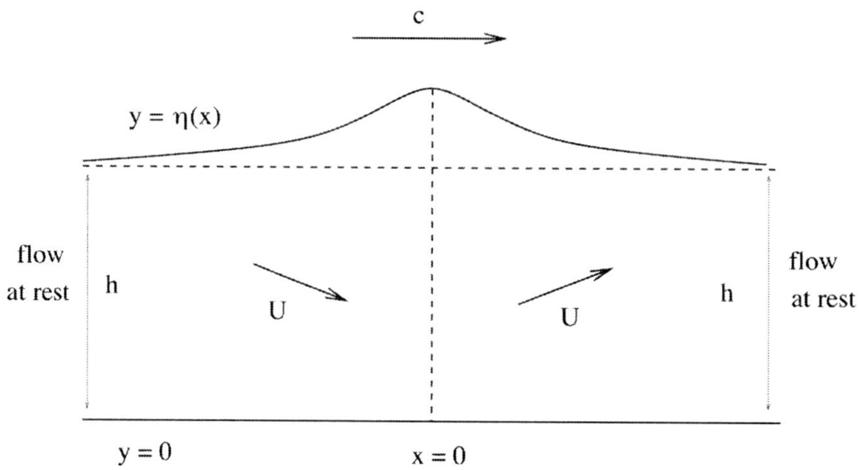

Figure 5.2. *Solitary wave (at time $t = 0$) with asymptotic height $h > 0$, moving at speed $c > 0$. The velocity at a point in the fluid is $U = (u, v)$.*

To prove that $u > 0$ in $\overline{\Omega}$ we proceed as follows. Notice that (5.6) yields Bernoulli's law: throughout the fluid domain Ω the expression $\frac{(u-c)^2+v^2}{2} + P + gy$ is constant. Evaluating the expression on the free surface as $|x| \to \infty$ we deduce that this constant value equals $\frac{c^2}{2} + P_{atm} + gh$. Consequently,

$$\frac{(u-c)^2+v^2}{2} + P + gy = \frac{c^2}{2} + P_{atm} + gh \quad \text{throughout} \quad \overline{\Omega}. \tag{5.15}$$

On the other hand, a direct calculation based on (5.6) and (5.7) yields that P is superharmonic in Ω:

$$\Delta P = -2\psi_{xy}^2 - 2\psi_{xx}^2 \le 0.$$

Therefore, in view of (5.10), the minimum of P in $\overline{\Omega}$ is attained on the flat bed or on the free surface since (5.15) yields

$$P(x,y) \to P_{atm} + g(h-y) \ge P_{atm} \quad \text{for} \quad |x| \to \infty.$$

But $P_y = -g$ on B by (5.6) and (5.9), so that Hopf's maximum principle ensures that the minimum of P is attained everywhere on the free surface (where $P = P_{atm}$) and $P > P_{atm}$ in Ω. Again by Hopf's maximum principle, we infer that $P_x(x,\eta(x)) > 0$ for $x < 0$, while $P_x(x,\eta(x)) < 0$ for $x > 0$ if we take into account the strict monotonicity of the graphs S_- and S_+. But $P_x = (c-u)[u_x + \eta_x u_y]$ on $y = \eta(x)$ in view of (5.6) and (5.8), so that (5.14) ensures $u_x(x,\eta(x)) + \eta_x(x) u_y(x,\eta(x)) > 0$ for $x < 0$, with the opposite sign for $x > 0$:

$$\frac{d}{dx} u(x,\eta(x)) > 0 \quad \text{for} \quad x < 0 \quad \text{and} \quad \frac{d}{dx} u(x,\eta(x)) < 0 \quad \text{for} \quad x > 0. \tag{5.16}$$

In other words, along the free surface u increases strictly from $x=-\infty$ to the crest $x=0$, and thereafter it is strictly decreasing. But the function u is harmonic in Ω and on the flat bed B we have $u_y = v_x = 0$ by (5.7) and (5.9), so that Hopf's maximum principle ensures that the minimum and maximum of u cannot occur on the bed. Since $u \to 0$ as $|x| \to \infty$, the monotonicity properties of u along the curve $y=\eta(x)$ encompassed in (5.16) ensure that $u>0$ throughout $\overline{\Omega}$, with u reaching its maximum value (less than or equal to the wave speed c) at the crest $(0,\eta(0))$. □

Claim 2. *To each fluid particle moving within the water there corresponds a unique time $t^* \in \mathbb{R}$ so that at $t=t^*$ the particle is exactly below the wave crest, while afterward it is located behind the wave crest, the wave crest being behind the particle for $t<t^*$.*

Proof. The path (past and future) $(X(t),Y(t))$ of a particle with location $(X(0),Y(0))$ at time $t=0$ is given by the solution of the differential system

$$\begin{cases} X' = u(X-ct,Y), \\ Y' = v(X-ct,Y). \end{cases} \tag{5.17}$$

Associate to (5.17) the Hamiltonian system

$$\begin{cases} x' = u(x,y)-c, \\ y' = v(x,y) \end{cases} \tag{5.18}$$

in the moving frame (5.12) in which the wave is stationary. The Hamiltonian function for (5.18) is $\psi(x,y)$ in view of (5.13). Notice that $x(t)$ describes precisely the horizontal position of the particle $(X(t),Y(t))$ with respect to the wave crest at time t, assuming that initially (at time $t=0$) the wave crest is located on $X=0$. All solutions of (5.18) are defined globally in time (in the past and in the future) since the boundedness of the right-hand side prevents blowup in finite time.

If we do not deal with a wave of greatest height, then for some $\varepsilon>0$ we have that $u \le c-\varepsilon$ throughout Ω. Therefore, $x(t) \le x(0)-\varepsilon t$ for $t>0$, while $x(t) \ge x(0)+\varepsilon t$ for $t<0$, and the statement of the lemma follows at once.

Let us now discuss the case of the wave of greatest height, in which case $u=c$ at the wave crest, with $u<c$ elsewhere in the fluid. Taking into account (5.14), we find that $x'(t) \le 0$ with an equality sign possible only if $x(t)=0$ and $y(t)=\eta(0)$. Notice that $(0,\eta(0))$ is the only critical point of the continuous autonomous system (5.18) with a right-hand side that is uniformly bounded in $\overline{\Omega}$ and smooth everywhere except at the critical point. Since $y(t)=h_\alpha(x(t))$ for some $\alpha \in [0,m]$ as $\psi(x(t),y(t))=\alpha$ for all $t \in \mathbb{R}$, we deduce that the only two possible scenarios for the statement not to hold are that $(x(0),y(0)) \in S_+ \cup S_-$ reaches $(0,\eta(0))$ in infinite time, or $(x(t),y(t))=(0,\eta(0))$ for all $t \in [0,T]$ with some $T>0$. The first possibility cannot occur since for, say, $(x(0),y(0)) \in S_+$ we have $y(0)=h_0(x(0))$, so that $y(t)=\eta(x(t))$ for all $t \ge 0$ and our claim follows as the inequality

$$\int_0^{x(0)} \frac{ds}{c-u(s,\eta(s))} < \infty \tag{5.19}$$

shows that the decreasing function $x(t)$ reaches 0 in finite (positive) time. The inequality (5.19) follows from the fact that $c-u(x,\eta(x)) = O(\sqrt{x})$ as $x \downarrow 0$. This last estimate is a direct consequence of the fact that $\eta'(x)$, existing for $x \neq 0$, is bounded away from zero as $x \downarrow 0$. Indeed, (5.8) and (5.15) yield

$$[c - u(x,\eta(x))]^2 + v^2(x,\eta(x)) = c^2 + 2g[h - \eta(x)] = O(x) \quad \text{as} \quad x \downarrow 0, \tag{5.20}$$

with the decay rate a consequence of the mean-value theorem as $\eta(0) = h + \frac{c^2}{2g}$ for the wave of greatest height in view of (5.15) evaluated at $(0,\eta(0))$. We know that if $\theta(x)$ is the angle between the wave profile $x \mapsto \eta(x)$ and the horizontal direction (1,0), then $\lim_{x\uparrow 0}\theta(x) = \pi/6$ while $\lim_{x\downarrow 0}\theta(x) = \pi - \pi/6$. But

$$\tan(\theta(x)) = \eta'(x) = \frac{v(x,\eta(x))}{u(x,\eta(x)) - c} \quad \text{for} \quad x \neq 0$$

if we take into account (5.8). In conjunction with (5.20), the above inequality yields

$$c - u(x,\eta(x)) = O(\sqrt{|x|}) \quad \text{and} \quad v(x,\eta(x)) = O(\sqrt{|x|}) \quad \text{for} \quad |x| \downarrow 0.$$

In particular, (5.19) holds true. This means (cf. [265]) that in the case of the wave of greatest height, uniqueness fails for the solution of (5.18) with initial data $(0,\eta(0))$. The possibility that a particle might stay at the wave crest for a positive period of time is ruled out by the following reasoning. The physically realistic solution of (5.18) with initial data $(0,\eta(0))$ is not the constant solution since this would mean that particles collide at the crest: the particle located at the crest would move at constant speed $c > 0$ to the right whereas, in view of (5.19), any particle located initially at some point $(x(0), y(0))$ with $x(0) > 0$ and $y(0) = \eta(x(0))$ would reach the crest in finite time. Therefore, in the moving frame a solution starting on $y = \eta(x)$ reaches the point $(0,\eta(0))$ in finite time and does not pause there but moves on with a decreasing x-coordinate as time goes by. This concludes the proof. □

Remark. The wave of greatest height is said to have a stagnation point at its crest since $u = c$ and $v = 0$ there. Claim 2 shows the somehow counterintuitive fact that this nevertheless does not mean that the particle at the crest moves along with the wave at speed c: the particle at the crest is left behind as the wave moves on.

5.2 Pressure beneath an irrotational solitary water wave

To discuss the behavior of the pressure beneath a solitary water wave we adapt to the present context the approach used in Chapter 4 to investigate the pressure beneath a Stokes wave. More precisely, we change variables performing a hodograph transform in terms of the velocity potential and the stream function. Recall (5.14): the solitary wave moves faster than the water particles; a wave is not a moving hump of water but a pattern representing a pulse of energy moving through water. For our considerations it is also important that (cf. [9]) the slope of the wave profile does not exceed 45 degrees,

$$\eta_x^2(x) < 1 \quad \text{for all} \quad x \in \mathbb{R}. \tag{5.21}$$

Let us introduce the velocity potential ϕ by

$$\phi_x = u - c, \qquad \phi_y = v, \qquad \phi(0,y) = 0, \tag{5.22}$$

that is, using (5.7),

$$\phi(x,y)=\int_0^x [u(l,0)-c]\,dl+\int_0^y v(x,s)\,ds, \quad (x,y)\in\overline{\Omega}.$$

Taking advantage of (5.5), we also define a stream function ψ by

$$\psi_x=-v, \qquad \psi_y=u-c, \qquad \psi(0,\eta(0))=0, \tag{5.23}$$

that is,

$$\psi(x,y)=m+\int_0^y [u(x,s)-c]\,ds, \quad (x,y)\in\overline{\Omega},$$

for a suitable constant $m\in\mathbb{R}$. It follows from the normalization in (5.23) and from (5.8)–(5.9) that ψ vanishes on the free surface $y=\eta(x)$ and that $\psi=m$ on the flat bed $y=-d$. In particular, we have

$$m=\psi(x,0)=-\int_0^{\eta(x)}\psi_y(x,s)\,ds=\int_0^{\eta(x)}[c-u(x,s)]\,ds, \qquad x\in\mathbb{R},$$

so that m is positive in view of (5.14). The constant m is called the **relative mass flux** of the flow. Summarizing, the stream function satisfies the boundary conditions

$$\psi(x,0)=m>0, \qquad \psi(x,\eta(x))=0, \quad x\in\mathbb{R}. \tag{5.24}$$

The Cauchy–Riemann equations and (5.22)–(5.23) imply that the complex function $(x+iy)\mapsto\phi(x,y)+i\,\psi(x,y)$ is holomorphic, so that the smooth functions ϕ and ψ are harmonic conjugated.

In the previous section we saw that the governing equations for irrotational two-dimensional solitary waves above a flat bed are equivalent to the study of the following elliptic free boundary problem:

$$\begin{cases} \Delta\psi=0, \quad x\in\mathbb{R}, \quad 0\le y\le\eta(x),\\ |\nabla\psi(x,\eta(x))|^2+2g\,\eta(x)=c^2+2gh, \quad x\in\mathbb{R},\\ \psi(x,\eta(x))=0, \quad x\in\mathbb{R},\\ \psi(x,0)=m, \quad x\in\mathbb{R}, \end{cases} \tag{5.25}$$

coupled with the asymptotic limits

$$\begin{cases} \lim_{|x|\to\infty}\eta(x)=h,\\ \lim_{|x|\to\infty}\nabla\psi(x,y)=(-c,0) \quad \text{uniformly for} \quad 0\le y\le\eta(x). \end{cases} \tag{5.26}$$

We introduce now the hodograph transform $\mathcal{H}$ induced by the pair (ϕ,ψ) as

$$\mathcal{H}:\overline{\Omega}\to\mathbb{R}^2, \qquad \mathcal{H}(x,y):=-\big(\phi(x,y),\psi(x,y)\big).$$

Notice that $\mathcal{H}$ is an analytic mapping on Ω, with Jacobian matrix

$$D\mathcal{H}(x,y)=\begin{bmatrix} c-u(x,y) & v(x,y)\\ -v(x,y) & c-u(x,y)\end{bmatrix}, \qquad (x,y)\in\overline{\Omega}.$$

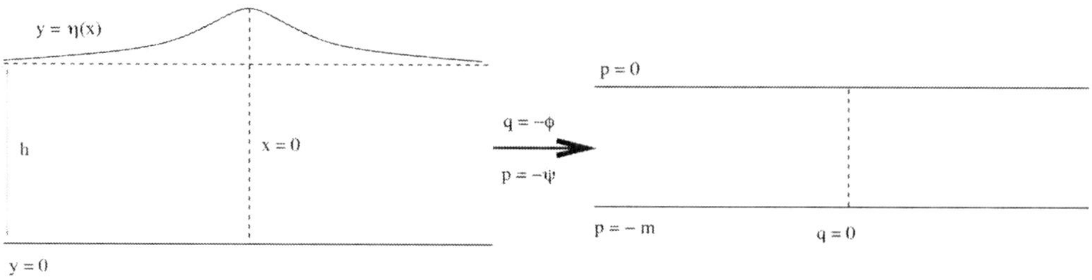

Figure 5.3. *The hodograph transform.*

We claim that $\mathcal{H}$ is a bijection from $\overline{\Omega}$ onto $\mathbb{R}\times[-m,0]$. Notice that in proving that $\mathcal{H}$ is one-to-one we cannot use the mean-value theorem globally in $\overline{\Omega}$, since the latter set in not convex (cf. [254]). We proceed as follows. Assume that (x_1,y_1), $(x_2,y_2)\in\overline{\Omega}$ are points with $\mathcal{H}(x_1,y_1)=\mathcal{H}(x_2,y_2)$. Then (x_1,y_1) and (x_2,y_2) lie on the same streamline $\psi=-p$ for some $p\in[-m,0]$. Due to (5.14), by the implicit function theorem, there is a smooth parametrization[102] $(x,y(x))$ of the streamline $\psi=-p$, with $y'(x)=\frac{v}{c-u}$. But then $\mathcal{H}(x_2,y_2)-\mathcal{H}(x_1,y_1)$ can be expressed as

$$\int_{x_1}^{x_2} D\mathcal{H}(x,y(x))\begin{pmatrix}1\\ y'(x)\end{pmatrix}dx=\int_{x_1}^{x_2}\begin{pmatrix}\dfrac{(c-u)^2+v^2}{c-u}\\ 0\end{pmatrix}dx$$

so $\mathcal{H}(x_1,y_1)=\mathcal{H}(x_2,y_2)$ yields $x_1=x_2$, which in turn yields $y_1=y(x_1)=y(x_2)=y_2$. Thus $\mathcal{H}$ is one-to-one in $\overline{\Omega}$. Using (5.22), (5.23) in combination with (5.14), it follows that $\mathcal{H}(S_\pm)=\mathbb{R}_\pm\times\{0\}$ and $\mathcal{H}(B_\pm)=\mathbb{R}_\pm\times\{-m\}$, and by Brouwer's domain invariance theorem (cf. [101]) we deduce that $\mathcal{H}$ is a smooth bijection from $\overline{\Omega}$ to $\mathbb{R}\times[-m,0]$. It is convenient to denote the coordinates in $\mathcal{H}(\overline{\Omega})=\mathbb{R}\times[-m,0]$ by (q,p), that is,

$$\begin{cases} q=-\phi(x,y),\\ p=-\psi(x,y).\end{cases}\tag{5.27}$$

Introducing the height function

$$h(q,p)=y\quad\text{for}\quad (q,p)\in\mathbb{R}\times[-m,0],\tag{5.28}$$

from the definition of $\mathcal{H}$ we see that (5.25) is equivalent to the following nonlinear boundary value problem:

$$\begin{cases}\Delta_{q,p}h=0,\quad q\in\mathbb{R},\quad -m<p<0,\\ h(q,-m)=0,\quad q\in\mathbb{R},\\ \left(h_q^2(q,0)+h_p^2(q,0)\right)\left(c^2-2gh(q,0)\right)=1,\quad q\in\mathbb{R},\end{cases}\tag{5.29}$$

[102]We do not deal with the extreme wave.

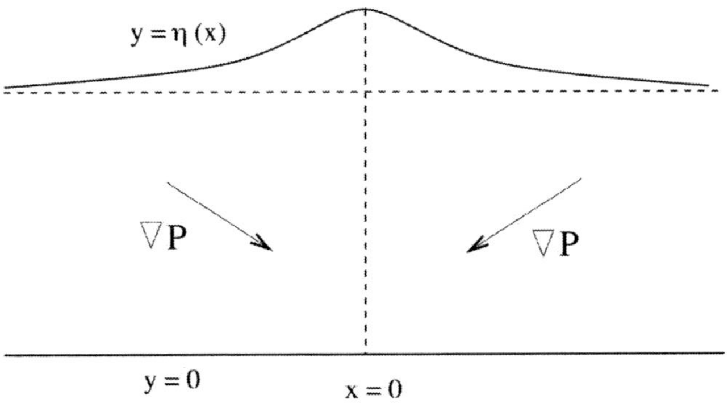

Figure 5.4. *Monotonicity properties of the pressure beneath a solitary wave.*

with the asymptotic limits

$$\begin{cases} \lim_{|q|\to\infty} h(q,0) = h_0, \\ \lim_{|q|\to\infty} \nabla_{q,p} h(q,p) = (0,\frac{1}{c}) \quad \text{uniformly for} \quad -m \le p \le 0. \end{cases} \tag{5.30}$$

Indeed, since $\mathcal{H}$ is conformal, it maps functions that are harmonic in the (x,y)-variables into functions that are harmonic in the (q,p)-variables. Now going through the same steps as we did in Chapter 4 for the case of the Stokes waves,[103] we deduce that the pressure strictly increases with depth and strictly decreases horizontally away from the vertical line beneath the crest. More precisely, *the sign of the first component P_x of the pressure gradient at a point in the fluid depends on the position of the point with respect to the crest line: $P_x = 0$ below the crest, $P_x < 0$ in $\Omega_+ \cup S_+ \cup B_+$, $P_x > 0$ in $\Omega_- \cup S_- \cup B_-$, while the second component P_y is strictly negative throughout the fluid.*

An immediate consequence is that *P_{atm} is the minimum value of the pressure and is attained only on the free surface $y = \eta(x)$, while the maximum value of the pressure is attained on the flat bed directly below the crest.* Moreover, since by (5.14) a streamline decreases to the right of the crest line, being symmetric with respect to the crest line, we also have that *the maximum pressure on any internal streamline occurs directly underneath the wave crest.*

The previous theoretical considerations about the behavior of the pressure beneath a solitary water wave in irrotational flow are supported by numerical simulations (cf. [113]) and by experiments that were carried out in April 2010 by Dr. Hung-Chu Hsu at the Department of Ocean Environmental Engineering of the National Sun Yat-Sen University, Taiwan. The water tank was $42\,m$ long, $1.5\,m$ deep, and $1\,m$ wide, with glass sidewalls to facilitate the camera recording of the evolution of the solitary wave. The solitary waves were generated at one end of the flume by a programmable high-resolution wavemaker. A plane beach with slope $1/20$ covered with a smooth layer of concrete starts $37.5\,m$ from the wave paddle. The wave recorded by 5 capacitance-type wave gauges located between $25\,m$ and $31\,m$ downstream of the wavemaker. To obtain the variations of the wave pressure at

[103]The only difference being that instead of $q \in [0,\pi]$ we now have $q \in [0,\infty)$, but otherwise the formulas and the reasoning carry over to the present setting.

different water depths associated with certain locations and at different horizontal positions along a constant water depth in the wave flume, a movable surveying sensor with a traversing carriage was equipped above the wave tank. The dynamic pressure was recorded by 7 tiny pressure transducers, each with a diameter of $3\,mm$, which were buried in the movable carriage. Pressure was measured at 7 locations along the vertical or horizontal line. In total, 5 wave gauges with different locations in the wave flume and 7 pressure transducers along the vertical or horizontal line were utilized in the 14 trials that were performed (with maximal wave elevations ranging from $0.04\,m$ to $0.11\,m$ in water with average depth $0.5\,m$, $0.6\,m$, and $0.7\,m$). The same wave conditions were conducted three times to make sure of experimental reappearance, and 15 to 30 minutes were needed to damp the remaining wave disturbances in the wave flume between two successive tests. The gathered experimental evidence confirms our conclusions with respect to the monotonicity of the pressure in the horizontal and vertical directions (see [57]).

5.3 Solitons

For the KdV equation larger amplitude solitary waves travel faster. The solitary waves for KdV have more peculiar properties: considering an initial profile like that in Figure 5.5, with the taller solitary wave preceding the smaller one, the taller one catches up to the other, and there is a period of complicated **nonlinear interaction**, but eventually both solitary waves emerge completely unscathed! This is what we would expect if the two solitary waves were to satisfy the superposition principle, but they do not (a careful examination shows actually that the waves are shifted relative to the positions they would have reached if the interaction were indeed linear): we have a very special type of nonlinear process at work here. Solitary waves which retain their identity upon collision are called **solitons**. The discovery of solitons occured about forty years ago. In 1965, Kruskal and Zabusky [189] reported the celebrated numerical computations of solutions to the KdV equation. They considered periodic traveling waves—a more complicated problem than that of solitary waves, but well suited to numerical computations—and observed the faster (taller) waves continuosly catching up and overtaking the slower (smaller) waves. Moreover, after a very long time a pattern very similar to the initial profile reappears. The persistence of these waves and their ability to apparently retain their identities in a collision, with each wave behaving like a "particle," led them to coin the name "soliton" (after photon). This observation stimulated theoretical research, and soon led to the discovery, by Gardner, Greene, Kruskal, and Miura [123], of exact "multisoliton" solutions describing the interaction of solitary waves as solitons. At the core of their discovery lies a relation between KdV and the spectrum of the second-order linear differential operator $\mathcal{L}(t) = (-\partial_x^2 - \frac{1}{6}\,\eta(x,t))$, in which time t is viewed as a parameter. Soon afterward Lax [195] proposed a more convenient and universal reformulation of their work, called now the **Lax formalism**. These first breaktroughs (described briefly in the Appendix) have led to an intense study over the last decades. Diverse branches of pure and applied mathematics (ranging from algebraic geometry to multiple-scale asymptotic theory) have had to be invoked to clarify aspects of this new subject, and presently solitons have a large impact on various branches of physics.

To provide a better understanding of solitons, we examine now in detail an illustrative example which will clarify the points mentioned above in relation to the solitary wave

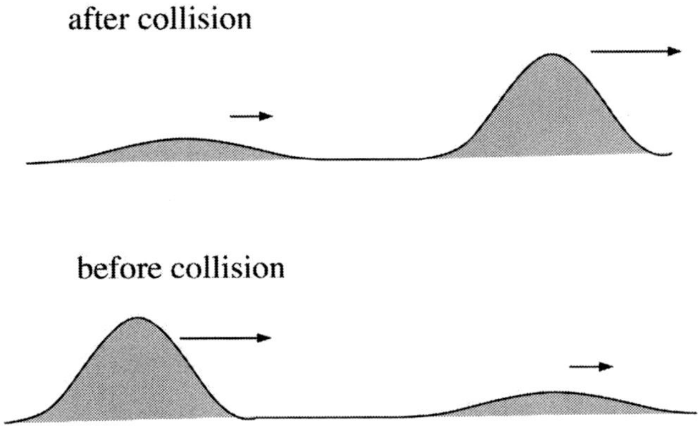

Figure 5.5. *The interaction of two solitons: the larger one overtakes the smaller, and both recover their initial shape and speed.*

solutions of KdV. A tedious computation confirms that

$$\eta(x,t) = 72\,\frac{3 + 4\cosh(2x - 8t) + \cosh(4x - 64t)}{[3\cosh(x - 28t) + \cosh(3x - 36t)]^2} \tag{5.31}$$

is a solution[104] of (5.4). Our aim is to show that (5.31) describes the nonlinear interaction of the two solitary wave solutions

$$\frac{12}{\cosh^2(x-4t)} \quad \text{and} \quad \frac{48}{\cosh^2(2x-32t)}$$

of (5.3), representing in view of (5.2) the solitary wave solutions (up to a shift) traveling with speed $c = 4$, respectively, $c = 16$. Since the solution (5.31) is defined for all $t \in \mathbb{R}$, we will examine the development of this wave both before and after the formation of the initial profile $x \mapsto \eta(x,0)$.

The character of the solution becomes transparent if we examine the asymptotic behavior of (5.31) as $t \to \pm\infty$. To gain some insight, we first examine the asymptotic behavior of this solution as $\xi = x - 4t$ is fixed, writing

$$\eta(x,t) = 144\,\frac{6 + 4e^{\xi} + 4e^{-\xi} + e^{4\xi-48t} + e^{48t-4\xi}}{\left(3e^{\xi-24t} + 3e^{24t-\xi} + e^{3\xi-24t} + e^{24t-3\xi}\right)^2},$$

so that

$$\eta(x,t) \approx 144\,\frac{e^{\mp 4\xi}}{\left(3e^{\mp\xi} + e^{\mp 3\xi}\right)^2} = \frac{12}{\cosh^2\left(\xi \pm \frac{\ln 3}{2}\right)} \quad \text{as} \quad t \to \pm\infty. \tag{5.32}$$

The interpretation of the above asymptotic limit (with $\xi = x - 4t$ fixed) is that we follow in (5.31) the development of the solitary wave of speed $c = 4$.

[104]This solution was not obtained by accident; see the discussion (in the Appendix) of the direct/inverse scattering problem for the two-soliton solution of KdV.

Figure 5.6. *At a fixed instant $t>0$ large enough, the deviation of the solution from the superposition (sum) of two solitary waves is negligible, according to the estimates.*

Similarly, with $\nu = x-16t$ fixed, we have

$$\eta(x,t) = 144\,\frac{6+4\,e^{2\nu+24t}+4\,e^{-2\nu-24t}+e^{4\nu}+e^{-4\nu}}{\left(3\,e^{\nu-12t}+3\,e^{12t-\nu}+e^{3\nu+12t}+e^{-3\nu-12t}\right)^2},$$

so that

$$\eta(x,t)\approx 144\,\frac{4\,e^{\pm 2\nu}}{\left(3\,e^{\mp\nu}+e^{\pm 3\nu}\right)^2} = \frac{48}{\cosh^2\left(2\nu\mp\frac{\ln 3}{2}\right)} \quad \text{as} \quad t\to\pm\infty. \tag{5.33}$$

In this second asymptotic limit performed to gain some insight, the interpretation of keeping $\nu = x-16t$ fixed is that we follow the development of a solitary wave moving with speed $c=16$.

We will now show that the above two asymptotic wave forms can be combined. Writing (5.31) as

$$\eta(x,t) = 144\,\frac{4\,[e^{x-4t}+e^{4t-x}]^2+[e^{32t-2x}-e^{2x-32t}]^2}{[3\,e^{x-28t}+3\,e^{28t-x}+e^{3x-36t}+e^{36t-3x}]^2},$$

the expression

$$\eta(x,t)-\frac{12}{\cosh^2\left(x-4t+\frac{\ln 3}{2}\right)}-\frac{48}{\cosh^2\left(2x-32t-\frac{\ln 3}{2}\right)}$$

takes the form

$$\begin{aligned}
&144\left\{\frac{[e^{32t-2x}-e^{2x-32t}]^2}{[3\,e^{x-28t}+3\,e^{28t-x}+e^{3x-36t}+e^{36t-3x}]^2}-\frac{1}{[3\,e^{x-4t}+e^{4t-x}]^2}\right\}\\
&+576\left\{\frac{[e^{x-4t}+e^{4t-x}]^2}{[3\,e^{x-28t}+3\,e^{28t-x}+e^{3x-36t}+e^{36t-3x}]^2}-\frac{1}{[e^{2x-32t}+3\,e^{32t-2x}]^2}\right\}\\
&=-144\,\frac{4\,[e^{3x-36t}+e^{x-28t}]\left(2e^{36t-3x}+6e^{28t-x}-2e^{3x-36t}+2e^{x-28t}\right)}{[3\,e^{x-28t}+3\,e^{28t-x}+e^{3x-36t}+e^{36t-3x}]^2\,[3\,e^{x-4t}+e^{4t-x}]^2}\\
&+576\,\frac{2\,[e^{36t-3x}-e^{x-28t}]\left(4\,e^{36t-3x}+6e^{28t-x}+2e^{3x-36t}+4e^{x-28t}\right)}{[3\,e^{x-28t}+3\,e^{28t-x}+e^{3x-36t}+e^{36t-3x}]^2\,[e^{2x-32t}+3\,e^{32t-2x}]^2}.
\end{aligned}$$

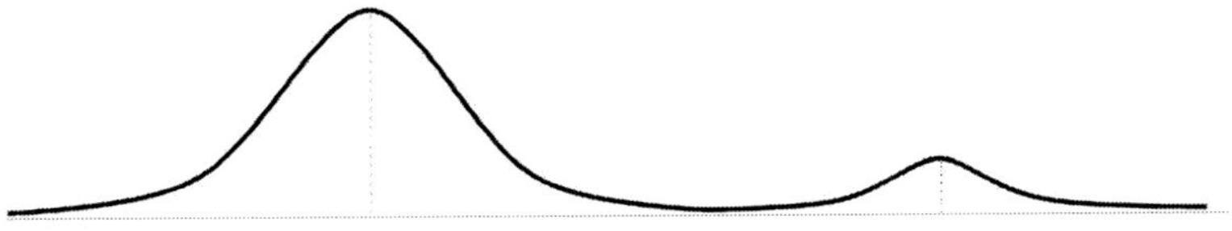

Figure 5.7. *At a fixed instant $t < 0$ with $|t|$ large enough, the deviation of the solution from the superposition (sum) of two solitary waves is negligible, according to the estimates.*

Since for any $x, t \in \mathbb{R}$,

$$\frac{|2e^{36t-3x}+6e^{28t-x}-2e^{3x-36t}+2e^{x-28t}|}{3\,e^{x-28t}+3\,e^{28t-x}+e^{3x-36t}+e^{36t-3x}} \le 2,$$

$$\frac{4\,e^{36t-3x}+6e^{28t-x}+2e^{3x-36t}+4e^{x-28t}}{3\,e^{x-28t}+3\,e^{28t-x}+e^{3x-36t}+e^{36t-3x}} \le 4,$$

$$\frac{e^{3x-36t}+e^{x-28t}}{[3\,e^{x-4t}+e^{4t-x}]^2} \le \frac{e^{3x-36t}}{9e^{2x-8t}}+\frac{e^{x-28t}}{6} \le e^{x-28t},$$

$$\frac{1}{3\,e^{x-28t}+3\,e^{28t-x}+e^{3x-36t}+e^{36t-3x}} \le \frac{1}{e^{28t-x}+e^{3x-36t}} \le \frac{1}{2\,e^{x-4t}},$$

$$\frac{|e^{36t-3x}-e^{x-28t}|}{[e^{2x-32t}+3\,e^{32t-2x}]^2} \le \frac{e^{36t-3x}}{9\,e^{64t-4x}}+\frac{e^{x-28t}}{6} \le e^{x-28t},$$

we conclude that

$$\left|\eta(x,t)-\frac{12}{\cosh^2\left(x-4t+\frac{\ln 3}{2}\right)}-\frac{48}{\cosh^2\left(2x-32t-\frac{\ln 3}{2}\right)}\right| \le 2880\,e^{-24t}$$

for all $x, t \in \mathbb{R}$. Therefore,

$$\eta(x,t) \approx \frac{12}{\cosh^2\left(x-4t+\frac{\ln 3}{2}\right)}+\frac{48}{\cosh^2\left(2x-32t-\frac{\ln 3}{2}\right)} \quad \text{as} \quad t \to \infty, \tag{5.34}$$

a result anticipated by (5.32). We proceed now along identical lines with the limiting behavior of (5.31) as $t \to -\infty$, obtaining the result suggested by our first asymptotic approximations (5.33),

$$\eta(x,t) \approx \frac{12}{\cosh^2\left(x-4t-\frac{\ln 3}{2}\right)}+\frac{48}{\cosh^2\left(2x-32t+\frac{\ln 3}{2}\right)} \quad \text{as} \quad t \to -\infty. \tag{5.35}$$

The asymptotic results (5.34) and (5.35) will now be used to describe some properties of the solution (5.31), explaining the soliton behavior. The interpretation of the results is that the initial profile obtained by setting $t = 0$ in (5.31) will evolve into two waves which move apart and progressively approach individual solitary waves as $t \to \infty$. Indeed, at a fixed time t, the crest of the first solitary wave in (5.34) is at $x = 4t - (\ln 3)/2$, whereas the second solitary wave has its crest at $x = 16t + (\ln 3)/4$. Therefore, since the function $s \mapsto \frac{1}{\cosh^2(s)}$ drops rapidly to zero away from its maximum located at $s = 0$, with increasing time t, we will recognize in (5.34) two separated humps with their crests a distance $[12t + (3\ln 3)/4]$ apart (cf. Figure 5.6). This is surprising but does not give us the whole picture. To capture another essential aspect, we consider the formation of this initial profile by looking at (5.35). In the limit $t \to -\infty$ we recognize two solitary waves, with the taller to the left of the shorter, the crests being a distance $[(3\ln 3)/4 - 12t]$ apart, since the taller has its crest at $x = 16t - (\ln 3)/4$ and the smaller at $x = 4t + (\ln 3)/2$; here $t < 0$ (see Figure 5.7). We expect the two solitary wave components to travel at different velocities, roughly proportional to their heights, as would be the case if each solitary wave would be undisturbed by the presence of the other. By performing an intricate analysis relying more on structural properties of the KdV equation (like its Hamiltonian structure and the associated integrals of motion) than on the explicit but quite complicated expression (5.31), the ongoing process can be described as follows (cf. [195]). There is an instant $t = t_0$ when the wave elevation (5.31) has one local maximum, while at any other instant $t \neq t_0$ there are precisely two local maxima located at $x = m_2(t)$, respectively, $x = m_1(t)$, with $m_2(t) > m_1(t)$, the corresponding wave amplitudes at these locations being $A_2(t)$, respectively, $A_1(t)$. Thus for $t \neq t_0$ we can call (5.31) a **two-wave solution** as there are two waves recognizable in it. As t goes from $-\infty$ to t_0, the distance $m_2(t) - m_1(t) > 0$ between the two local maxima decreases from ∞ to zero, and the amplitude $A_1(t)$ of the wave lagging behind decreases steadily from its supremum $\lim_{t\to-\infty} A_1(t) = 48$, while the amplitude $A_2(t)$ of the first wave increases from its infimum $\lim_{t\to-\infty} A_2(t) = 12$. As long as $t < t_0$ we have $A_2(t) < A_1(t)$. The two local points of maximal elevation $m_1(t)$ and $m_2(t)$ merge together at $t = t_0$, where $m_1(t_0) = m_2(t_0)$, and they separate again afterward. For $t > t_0$ we have $A_2(t) > A_1(t)$, with $A_2(t)$ growing toward its asymptotic limit $\lim_{t\to\infty} A_2(t) = 48$ and with $A_1(t)$ decaying toward its asymptotic limit $\lim_{t\to\infty} A_1(t) = 12$. Thus the taller wave catches up the smaller and absorbs it (at $t = t_0$ these two waves coalesce into a wave elevation with one local maximum) and then re-emits the smaller, and both recover their shape and speed, as depicted earlier.[105]

Note that two-soliton solutions exist for any two solitary waves traveling at speeds $c_2 > c_1$, and not only for $c_1 = 4$ and $c_2 = 16$. However, if $\frac{c_2}{c_1} < \frac{3+\sqrt{5}}{2}$, then the solitons exchange identities without ever coalescing: at each instant we have two local maxima, and the two solitary waves interchange their role without passing through each other (cf. [195]). The interaction of the two solitary waves might seem, at first sight, purely linear, and this would be the case if the two waves were to satisfy the superposition principle, but they do not. The appearance of the phase shifts (the taller wave moves forward by an amount $x = (\ln 3)/2$, and the shorter moves backward by $x = \ln 3$), relative to the positions they would have reached if the interaction were indeed linear, is a clear indication of the special type of nonlinear process at work here.

It is remarkable that this elastic collision is the prototype of the interaction of any n solitary waves (cf. [98]): as $t \to \infty$, the asymptotic solution corresponding to a profile formed by n solitary waves represents the same n separate solitary waves, ordered according to their speeds, with the tallest (and fastest) wave at the front followed by progressively

[105] It is instructive to consult the numerical simulations of two-soliton interactions available at http://www.ma.hw.ac.uk/solitons.

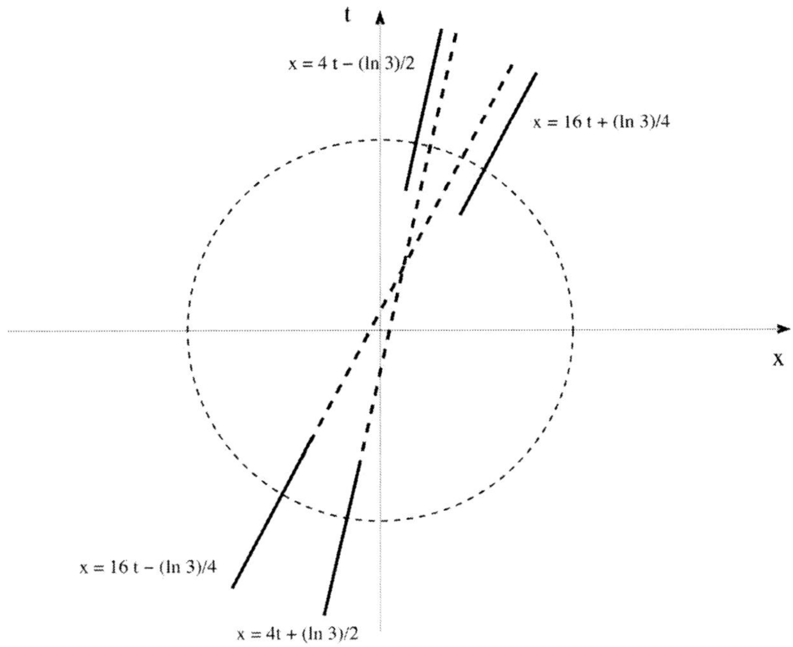

Figure 5.8. *Representation of the paths of the two wave crests of the solitary waves hidden in the solution. Inside the circle we find the region of dominant interaction, while the dotted lines represent the paths of the wave crests if linear superposition were to apply. The forward shift of the taller (faster) wave and the backward shift of the shorter (slower) are visible.*

shorter (and slower) solitons behind. Of course, "clean" interaction of solitary waves can also be obtained in linear theory, but the interesting feature is the complicated pattern obtained at $t = 0$, the instant when the solitary waves seem to have been lost in a messy waveform; however, they reappear! More than this is true: starting with arbitrary initial data that are smooth and sufficiently localized in space, the solution that evolves from these data is developing into a finite number of localized solitary waves (solitons), plus an oscillatory tail. Each solitary wave retains its localized identity, and taller waves travel faster than smaller ones, while the oscillatory tail disperses and spreads out in space; the proofs of these results rely on the fact that KdV is an integrable system (see the discussion in the Appendix to Chapter 5). Therefore, the solution evolves into an ordered set of solitons, with the tallest in front, followed by an oscillatory tail that gradually fades out. If the initial data that is picked up happens to be a multisoliton, an example of which is provided by evaluating (5.31) at time $t = 0$, then the solitary waves recognizable asymptotically present a "clean" interaction, without this oscillatory tail, and most of the time one observes just a number of humps, each signaling a solitary wave. In the region of dominant interaction of the solitary waves the wave pattern is however messier. It is of interest to point out that, unlike KdV, solitary waves for the governing equations for water waves do not collide elastically: after interactions there is a residual wave that trails the postcollision solitary waves, and the larger solitary wave exits from the interaction with slightly increased amplitude, while the smaller one loses amplitude. However, it is remarkable how small the departure from an elastic collision is (cf. the experimental and numerical evidence in [83]).

We conclude our discussion by pointing out that two major themes dominate the qualitative theory of dynamical systems: one represented by the remarkable degree of persistence of certain features under conditions that one might normally expect to destroy them (such as the soliton phenomenon described above), the other development being related to a behavior that appears to become erratic and unpredictable ("chaotic behavior," which occurs in turbulent fluid flows—an important aspect that we do not address).

5.4 Appendix for Chapter 5

5.4.1 Integrable systems

We attempt a brief introduction to the fascinating theory of integrable infinite-dimensional Hamiltonian systems by discussing the KdV equation. While integrable systems are rare, several such systems were discovered, and from many points of view the most relevant for hydrodynamics is KdV. Also, there is no general theory (that encompasses all known systems as special cases), but there are some general guiding principles, and to some extent each integrable equation offers something new.

The Hamiltonian formalism is the natural mathematical structure in which to develop the study of conservative mechanical systems (that is, systems which preserve energy—dissipative effects being neglected). KdV, being a partial differential equation, is an infinite-dimensional Hamiltonian system, in contrast to the classical (finite-dimensional) Hamiltonian systems that are described by ordinary differential equations. For this reason, before discussing the infinite-dimensional setting proper to KdV, we will describe some of the most relevant ideas from classical Hamiltonian systems. While the methods differ considerably, with new techniques required (most particularly in the extent to which functional analysis is used), the later considerations depend on the earlier conceptually.

Classical Hamiltonian systems

Let $n \geq 1$ be an integer representing the number of degrees of freedom of a system of differential equations describing the evolution in time t of the vectors $q = (q_1, \dots, q_n) \in \mathbb{R}^n$ and $p = (p_1, \dots, p_n) \in \mathbb{R}^n$. We say that the system is a **canonical conservative Hamiltonian system** if there is a smooth function $H : \mathbb{R}^{2n} \to \mathbb{R}$, called the **Hamiltonian function**, such that the system has the form

$$\begin{cases} q_i'(t) = \dfrac{\partial H}{\partial p_i}, \\ p_i'(t) = -\dfrac{\partial H}{\partial q_i} \end{cases} \tag{5.36}$$

for $i = 1, \dots, n$. The variables q and p are said to be **conjugate variables** and represent in many physical examples the position and momentum vectors. For a large class of systems the physical interpretation of the Hamiltonian function is the total energy of the system but this is not always the case (see [198]). An **integral of motion** for (5.36) is a smooth function $F : \mathbb{R}^{2n} \to \mathbb{R}$ which is constant along the solutions of (1.1). In many physical circumstances we expect to find several integrals of motion for (1.1), corresponding to energy, momentum, etc. The level sets $F^{-1}(c) = \{(q, p) \in \mathbb{R}^{2n} : F(q, p) = c\}$ of an integral of motion play an important role in understanding the dynamics since they are invariant sets: a solution starting in such a set remains in the set (by definition). In general, the level sets of an integral of motion are manifolds of dimension $(2n - 1)$, so that finding such an integral of motion reduces the degrees of freedom by constraining the motion to the manifold determined by the initial data. A possible approach is the following: if we were able to find $(2n-1)$ independent integrals of motion (in the sense that fixing the value of any $(2n-2)$ of them has no bearing whatsoever on the value of the remaining one[106]), then, holding all these integrals fixed would define a curve in $\mathbb{R}^{2n}$: the solution curve. The special structure of (5.36) ensures, as we shall see, that sometimes n such integrals of motion suffice to solve (5.36), in which case the system is said to be integrable in the classical sense; cf. [228].

The Poisson bracket operator plays a central role in regard to the integrals of motion for (5.36). For $F, G, H : \mathbb{R}^{2n} \to \mathbb{R}$ smooth maps we define the **canonical Poisson bracket** of $F(q,p)$ and $G(q,p)$ by

$$\{F,G\} = (\nabla F)^T J(\nabla G) = \sum_{i=1}^{n}\Big(\frac{\partial F}{\partial q_i}\frac{\partial G}{\partial p_i} - \frac{\partial F}{\partial p_i}\frac{\partial G}{\partial q_i}\Big), \tag{5.37}$$

where $(\nabla F)^T$ is the transpose of the $(2n)\times 1$ matrix ∇F and

$$J = \begin{pmatrix} 0 & I_n \\ -I_n & 0 \end{pmatrix}, \tag{5.38}$$

where I_n is the identity matrix. Clearly $\{F,G\}$ is a smooth map from $\mathbb{R}^{2n}$ to $\mathbb{R}$ as well, and one can easily verify that $\{\cdot,\cdot\}$ is **skew-symmetric**, i.e.,

$$\{F,G\} = -\{G,F\},$$

and **bilinear** as

$$\{F, aG + bH\} = a\{F,G\} + b\{F,H\}$$

for any constants $a, b \in \mathbb{R}$. A little calculation shows that **Jacobi's identity**

$$\{F,\{G,H\}\} + \{G,\{H,F\}\} + \{H,\{F,G\}\} = 0 \tag{5.39}$$

is also satisfied. Observe that we can write Hamilton's canonical form of the equations of motion (5.40) more concisely for $\mathbf{x} = (q,p) \in \mathbb{R}^{2n}$ as

$$\mathbf{x}' = J\nabla H. \tag{5.40}$$

The importance of the Poisson bracket is primarily due to the following results (cf. [228]):

- H is an integral of motion for (5.40), that is, H is constant along the solutions;
- F is an integral of motion for (5.40) if and only if $\{F,H\} = 0$;
- if F and G are both integrals of motion for (5.40), then so is $\{F,G\}$;
- $\{F,H\}$ is the time rate of change of F along the solutions of (5.40).

Often, rather than discussing problems in the entire space $\mathbb{R}^{2n}$, to avoid cumbersome conditions dealing with the possible behavior as $|\mathbf{x}| \to \infty$ that has to be kept under control, one restricts one's attention to problems in which the unknown function $t \mapsto \mathbf{x}(t) \in \mathbb{R}^{2n}$ is confined to compact subsets $M \subset \mathbb{R}^{2n}$. To investigate differential equations restricted to M, a simple topological structure (like compactness) is not sufficient and we have to endow M with a differentiable manifold

[106]We say that the C^1-maps $F_1,\dots,F_m : \mathbb{R}^{2n} \to \mathbb{R}$ are functionally independent if the $(2n)\times m$ matrix having the rows ∇F_i is of rank m, preventing the existence of some C^1-function $G : \mathbb{R}^m \to \mathbb{R}$ with $G(F_1(\mathbf{x}),\dots,F_m(\mathbf{x})) \equiv 0$, where $\mathbf{x} = (q,p) \in \mathbb{R}^{2n}$. If $F_1,\dots,F_m$ are functionally independent integrals of motion (here $m \le 2n-1$), we may assume (changing perhaps the ordering of the x_i's) that the determinant $\frac{D(F_1,\dots,F_m)}{D(x_1,\dots,x_m)}$ is not zero. Then the system $F_i(x) = c_i$, $i = 1,\dots,m$, of m equations can (at least locally) be solved implicitely with respect to $x_1,\dots,x_m$, and we obtain $x_i = \phi_i(x_{m+1},\dots,x_{2n},c_1,\dots,c_m)$ for $i = 1,\dots,m$.

structure, which enables us to deal with differential equations on M. Additional structures are often considered, e.g., on Riemannian manifolds we can define distances and angles, but to gain an understanding of the basic ideas of classical Hamiltonian systems we do not need to dwell into Riemannian geometry. The intuitive idea of a smooth k-dimensional manifold $M \subset \mathbb{R}^{2n}$, where $1 \le k \le 2n$, is that while the global structure of M may be quite complicated, locally the set M resembles the Euclidean space $\mathbb{R}^k$: each $\mathbf{x} \in M$ has a neighborhood $\mathcal{N}_\mathbf{x} \subset \mathbb{R}^{2n}$ such that $M \cap \mathcal{N}_\mathbf{x}$ is diffeomorphic to an open subset $\mathcal{O}$ of $\mathbb{R}^k$. Let $A : M \to \mathbb{R}^{2n}$ be a smooth vector field on M: at every point $\mathbf{x} \in M$ we are given a tangent vector $A(\mathbf{x}) \in T_\mathbf{x}M$, the tangent space to M at $\mathbf{x}$. With every such vector field we associate the **flow** $A^t : M \to M$ for which A is the velocity of the field, given by $\mathbf{x}_0 \mapsto A^t\mathbf{x}_0 = \mathbf{x}(t;\mathbf{x}_0)$ for $\mathbf{x}_0 \in M$ and $t \in \mathbb{R}$, where $\mathbf{x}(t;\mathbf{x}_0) \in \mathbb{R}^{2n}$ is the unique solution of the initial value problem given by the differential equation $\mathbf{x}'(t) = A(\mathbf{x}(t))$ with data $\mathbf{x}(0) = \mathbf{x}_0$. For any $\mathbf{x}_0 \in M$ we have that the solution curve $t \mapsto \mathbf{x}(t;\mathbf{x}_0)$ is defined for all $t \in \mathbb{R}$, since the boundedness of $\mathbf{x} \mapsto A(\mathbf{x})$ on the compact set M prevents blowup in finite time (cf. [265]), and represents a curve on the manifold M (cf. the discussion of flow-invariant sets in the Appendix to Chapter 2). $\{A^t\}_{t\in\mathbb{R}}$ is a one-parameter group of smooth diffeomorphisms of M, in the sense that the map $t \mapsto A^t$ from the additive group $\mathbb{R}$ to the group Diff(M) of smooth diffeomorphisms $f : M \to M$ of M (that is, bijective mappings with a smooth inverse) is a group homeomorphism: A^0 is the identity transformation and $A^t A^s = A^{t+s}$ for all $t, s \in \mathbb{R}$ (cf. [232]). Conversely, any group homeomorphism $\rho : \mathbb{R} \to \text{Diff}(M)$ arises this way: define $A(\mathbf{x}_0) = \frac{d}{dt}|_{t=0}\big(\rho(t)[\mathbf{x}_0]\big)$ for $\mathbf{x}_0 \in M$.

In trying to define the derivative of a vector field on a compact k-dimensional manifold $M \subset \mathbb{R}^m$ with $1 \le k < m$, we encounter an inherent difficulty. Inspired by the case of the Euclidean space it is natural to attempt to define a vector field's derivative as the limit of the difference between the vectors at nearby points divided by the distance between points. Leaving aside the issue of how to define the distance between two points on the manifold,[107] a far more serious problem is how to decide when two vectors at different points are "parallel" (and thus may be identified). In the Euclidean space this is a simple question to answer, but even within the framework of Riemannian geometry there is no intrinsic notion of parallelism between vectors defined at different points: the Levi-Civita connection (or covariant derivative) allows us to define a notion of parallel transport (i.e., moving a vector along a curve without changing its direction), but there is no global notion of parallelism. This is exemplified by considering in $\mathbb{R}^3$ the unit sphere centered at the origin, viewed as a compact two-dimensional manifold. At the north pole $N = (0,0,1)$ consider the unit tangent vector τ in the direction of the y-axis. Along the great half-circle joining N to the south pole $S = (0,0,-1)$ and passing through $(0,1,0)$, we transport τ by keeping it tangent to the circle and such that in terms of the two-dimensional tangent spaces the vector is not rotated. Transported along this curve τ winds up at S as a tangent unit vector τ_1 in the negative direction of the y-axis. If we transport τ along the great half-circle joining N to S and passing through $(1,0,0)$, the natural way to move it without twisting it is to keep it pointing in the direction of the y-axis, obtaining at S a tangent unit vector τ_2 pointing in the direction of the y-axis. From a three-dimensional perspective τ and τ_2 are parallel while τ and τ_1 are opposite, but from the intrinsic properties of the sphere, neither τ_1 nor τ_2 deserves to be called parallel to τ (cf. [268]). Ignoring the intrinsic properties of the manifold $M \subset \mathbb{R}^m$ and relying only on $\mathbb{R}^m$ without reference to M is not an option since in studying flows on M the shape/geometry of the manifold should be taken into account.[108] For this reason we refrain from trying to define derivatives of vector fields on compact k-dimensional manifolds $M \subset \mathbb{R}^m$ with $1 \le k < m$.

[107]The infimum over all curves joining the two points on the manifold comes to mind, meaning that one must add more structure (Riemannian) to be able to talk of geodesics.

[108]As an example, notice that the points $(0,\varepsilon)$ and $(0,-\varepsilon)$ are 2ε-close viewed as points in $\mathbb{R}^2$, but if we view them as points on the compact manifold $M = \{(x,y) \in \mathbb{R}^2 : \varepsilon^2 x^2 + y^2 = \varepsilon^2\}$, the distance between them on the manifold exceeds 2.

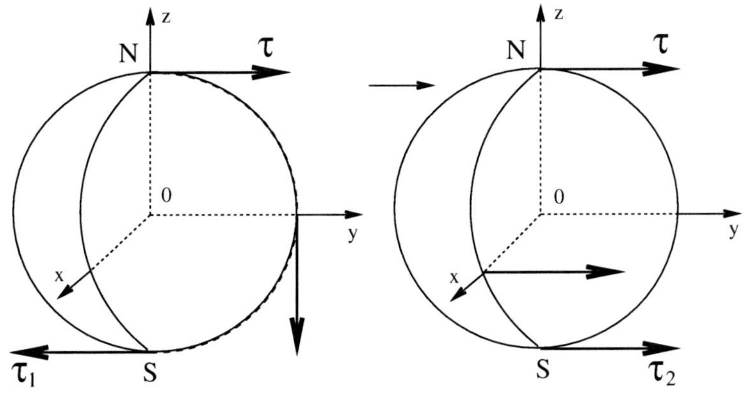

Figure 5.9. *Parallel transport of a tangent vector along different great half-circles from the north pole to the the south pole, with different results.*

While differentiation of a vector field $A : M \to \mathbb{R}^{2n}$ is inherently problematic, one can define the **Lie derivative** L_A, a first-order differential operator obtained by differentiating in the direction of the vector field A: for any smooth function $\phi : M \to \mathbb{R}$, let

$$(L_A\phi)(\mathbf{x}) = \frac{d}{dt}\bigg|_{t=0} \phi(A^t\mathbf{x}).$$

This way we associate to the smooth function $\phi : M \to \mathbb{R}$ another smooth function $L_A\phi : M \to \mathbb{R}$. Clearly the operator L_A is linear, that is, $L_A(a\phi_1 + b\phi_2) = a\,L_A\phi_1 + b\,L_A\phi_2$ for $a, b, \in \mathbb{R}$, and we have **Leibnitz's formula**

$$L_A(\phi_1\phi_2) = \phi_1\,L_A\phi_2 + \phi_2\,L_A\phi_1.$$

Let $(y_1, \ldots, y_k)$ be local coordinates on M: for $\mathbf{x_0} \in M$ we choose a neighborhood $\mathcal{N}_{\mathbf{x_0}}$ of $\mathbf{x_0}$ in $\mathbb{R}^{2n}$ such that there is a diffeomorphism φ from an open set $\mathcal{O} \subset \mathbb{R}^k$ to $\mathcal{N}_{\mathbf{x_0}} \cap M$; for convenience we may assume that $\varphi(0) = \mathbf{x_0}$. In this coordinate system the vector $A(\mathbf{x})$ is given by its components $(A_1(\mathbf{y}), \ldots, A_k(\mathbf{y}))$, and the flow A^t is given by the solution to the system of differential equations[109]

$$y_1' = A_1(\mathbf{y}), \ldots, y_k' = A_k(\mathbf{y}).$$

Therefore, $L_A\phi = A_1 \frac{\partial\phi}{\partial y_1} + \cdots + A_k \frac{\partial\phi}{\partial y_k}$ and we write $L_A = A_1 \frac{\partial}{\partial y_1} + \cdots + A_k \frac{\partial}{\partial y_k}$.

Given two smooth vector fields A and B on the manifold M, the corresponding flows A^t and B^s do not commute in general.[110] To measure the degree of noncommutativity of the two flows A^t and B^s we consider the points $A^tB^s\mathbf{x}$ and $B^sA^t\mathbf{x}$ for some fixed $\mathbf{x} \in M$. Without a Riemannian structure on M we cannot talk about the distance between these points on M, but we can compare the value at these two points of some smooth function $\phi : M \to \mathbb{R}$. The difference

$$\Delta_\phi(t,s) = \phi(A^tB^s\mathbf{x}) - \phi(B^sA^t\mathbf{x})$$

is clearly a differentiable function which is zero for $s = 0$ and $t = 0$. We can circumvent the need of endowing the manifold M with a Riemannian structure in order to talk about distances on M

[109] This simply means that $[(d\varphi)_\mathbf{y}]^{-1}A(\mathbf{x}) = (A_1(\mathbf{y}), \ldots, A_k(\mathbf{y})) \in \mathbb{R}^k$ for $\mathbf{x} = \varphi(\mathbf{y})$ with $\mathbf{y} \in \mathcal{O} \subset \mathbb{R}^k$, recalling that $[(d\varphi)_\mathbf{y}]^{-1}$ is by definition a linear isomorphism from $T_\mathbf{x}M$ to $\mathbb{R}^k$.

[110] That is, in general we expect $A^tB^s \neq B^sA^t$. For example, the vector fields $A(\mathbf{x}) = e_1$ and $B(\mathbf{x}) = x_1e_2$, defined for $\mathbf{x} = (x_1, x_2) = x_1e_1 + x_2e_2$ on the two-dimensional torus $\mathbb{T}^2 = \mathbb{R}^2/\mathbb{Z}^2$, are such that $A^tB^s\mathbf{x} = (t + x_1, sx_1 + x_2)$ and $B^sA^t\mathbf{x} = (t + x_1, ts + sx_1 + x_2)$, as $A^t\mathbf{x} = (t + x_1, x_2)$ and $B^s\mathbf{x} = (x_1, x_2 + sx_1)$.

since if for every smooth function $\phi : M \to \mathbb{R}$ the associated $\Delta_\phi(t,s) = 0$, then $A^t B^s \mathbf{x} = B^s A^t \mathbf{x}$. By direct calculation we see that for all smooth functions $\phi : M \to \mathbb{R}$ we have

$$\frac{\partial^2}{\partial t \partial s}\Big|_{t=s=0} \Delta_\phi(t,s) = (L_B L_A \phi - L_A L_B \phi)(\mathbf{x}), \quad \mathbf{x} \in M,$$

where the commutator operator of L_A and L_B, $L_B L_A - L_A L_B$, is a first-order[111] linear differential operator L_C. If $(A_1, \dots, A_k)$ and $(B_1, \dots, B_k)$ are the components of the vector fields A and B in the local coordinates $y_i, i = 1, \dots, k$, on M, then the vector field[112] $C = [A, B]$, called the **Poisson bracket** of A and B or the Lie derivative $L_A B$ of the vector field B along the vector field A, is given in the coordinates $y_1, \dots, y_k$, by

$$C_j = B_i \frac{\partial A_j}{\partial y_i} - A_i \frac{\partial B_j}{\partial y_i}, \quad j = 1, \dots, k.$$

A straightforward but messy calculation confirms that, given a smooth vector field A on M, the associated Lie derivative, whether acting on smooth functions $\phi : M \to \mathbb{R}$ or on smooth vector fields defined on M, is coordinate-independent. It therefore represents a meaningful notion of derivative in the context of smooth (compact) manifolds. Its importance is apparent from the fact that if A, B are smooth vector fields on a compact smooth k-dimensional manifold $M \subset \mathbb{R}^{2n}$, the two flows A^t and B^s commute if and only if the Poisson bracket $[A, B]$ is zero (cf. [16]). If the vector field A is of the form $A = J\nabla F$ for some smooth function $F : \mathbb{R}^{2n} \to \mathbb{R}$, with J given by (5.38), we say that A is **Hamiltonian** with **Hamiltonian function** F. A simple calculation shows that if $A = J\nabla F$ and $B = J\nabla G$, then $C = [A, B] = J\nabla\{F, G\}$.

Hamiltonian vector fields are very special. In particular, we expect only special changes of variables to preserve the class of Hamiltonian vector fields. One can prove (see [228]) that a smooth change of variables preserves the Hamiltonian form of all Hamiltonian equations[113] if and only if it is **symplectic**: the change of variables $\mathbf{x} \mapsto \xi = \Xi(\mathbf{x})$ is called symplectic if

$$\Big(\frac{\partial \Xi}{\partial \mathbf{x}}(\mathbf{x})\Big)^T J \Big(\frac{\partial \Xi}{\partial \mathbf{x}}(\mathbf{x})\Big) = J, \quad \mathbf{x} \in \mathbb{R}^{2n}.$$

From the above relation one can see at once that $\det(\frac{\partial \Xi}{\partial \mathbf{x}}(\mathbf{x})) = \pm 1$, the actual value being $+1$ (cf. [228]). Thus symplectic changes of variables are volume-preserving (see the Appendix to Chapter 2); in $\mathbb{R}^{2n}$ with $n > 1$ the class is much more restricted than that of volume- and orientation-preserving diffeomorphisms. In this context a useful observation is that every symplectic map can be locally represented in terms of a single scalar function, the so-called generating function (see [148]). We refrain from an in-depth pursuit of the interesting theory of symplectic transformations, limiting ourselves to pointing out that symplectic changes of coordinates preserve Poisson brackets. Also, if $\mathbf{x}(t;\zeta)$ is the unique solution of (5.40) with initial data $\mathbf{x}(0) = \zeta$, defined on the maximal interval of existence $[0, T)$ with $T > 0$, then for any fixed $t \in [0, T)$ the map $\zeta \mapsto \mathbf{x}(t;\zeta)$ is symplectic, and the converse is also true (cf. [228]).

If the Hamiltonian function $H : \mathbb{R}^{2n} \to \mathbb{R}$ for (5.36) is independent of one coordinate (called an **ignorable coordinate**), say, q_1, then $p_1'(t) = 0$, so that p_1 is an integral of motion and $p_1(t) = \alpha$ (a constant) for all times. Replacing in (5.36) the variable p_1 by the parameter α, the equations (5.36) become independent of q_1 and p_1, and the (remaining) equations for $q_2, \dots, q_n, p_2, \dots, p_n$

[111] At first glance this seems a second-order differential operator, being the difference of two second-order differential operators obtained as the composition of two first-order differential operators.

[112] By construction this is a vector field on M: if φ is the parametrization associated to these local coordinates, $(d\varphi)_\mathbf{y} : \mathbb{R}^k \to T_{\varphi(\mathbf{y})}M$ so that $(d\varphi)_\mathbf{y}(C(\mathbf{y})) \in T_\mathbf{x}M$ with $\mathbf{x} = \varphi(\mathbf{y})$.

[113] It might be possible that for a particular Hamiltonian system there is a nonsympectic transformation which transforms the system into another Hamiltonian system.

are Hamiltonian in $\mathbb{R}^{2n-2}$ with a parameter α. If the reduced system is solved explicitly in terms of t, the solutions can be substituted back into the equation

$$q_1'(t) = \frac{\partial H}{\partial \alpha}(q_2(t), \ldots, q_n(t), \alpha, p_2(t), \ldots, p_n(t))$$

of (5.36) and $q_1(t)$ can be found by a single integration or quadrature. Therefore, an ignorable coordinate reduces the problem of solving a Hamiltonian system to solving a Hamiltonian system in $\mathbb{R}^{2n-2}$ containing a parameter and a quadrature. This simple observation is indicative of a general procedure. It turns out that for a Hamiltonian system (5.36) an integral of motion gives rise to an ignorable coordinate in a suitable coordinate system, while k functionally independent integrals of motion in involution (with $2 \le k \le n$) produce k ignorable coordinates. More precisely, assume that the Hamiltonian system (5.36) admits an integral of motion $F : \mathbb{R}^{2n} \to \mathbb{R}$ in involution with H, that is, with zero Poisson bracket (5.37), $\{F, H\} = 0$. Then (cf. [228]), in a neighborhood of a regular point $\mathbf{x_0} \in \mathbb{R}^{2n}$ of F (a point where $\nabla F(\mathbf{x_0}) \neq \mathbf{0}$), there is a symplectic change of variables $\mathbf{x} \mapsto \mathbf{y}$ with $y_{n+1} = F(\mathbf{x})$: in the new coordinates the Hamiltonian system (5.36) becomes a Hamiltonian system in which y_1 is an ignorable coordinate. Actually (cf. [228]), if $\mathbf{x_0}$ is a regular point of the Hamiltonian function H, then near $\mathbf{x_0}$ the Hamiltonian system (5.36) has n independent integrals of motion in involution and there is a local symplectic change of variables so that near $\mathbf{x_0}$ the system (5.36) becomes

$$y_1' = 1, \qquad y_i' = 0 \quad \text{for} \quad 2 \le i \le 2n.$$

This remarkable result is, however, only a local result. Liouville's theorem (see [148]) is the global result: *if the Hamiltonian system* (5.40) *admits n functionally independent integrals of motion (with* $F_1 = H$*) in involution,* $\{F_i, F_j\} = 0$ *for* $i, j = 1, \ldots, n$*, and if we consider a level set* M_f *of the functions* F_i*,* $M_f = \{x : F_i(x) = f_i,\ i = 1, \ldots, n\}$ *with* $f = (f_1, \ldots, f_n) \in \mathbb{R}^n$*, such that the smooth manifold* M_f*, invariant under the flow of* (5.36)*, is compact and connected,*[114] *then there is an open neighborhood* $\mathcal{U}$ *of the compact set* M_f *in* $\mathbb{R}^{2n}$ *and a symplectic change of variables from* $\mathcal{U}$ *onto the open set* $\tilde{\mathcal{U}} \subset \mathbb{R}^{2n}$*, mapping* M_f *onto a torus and transforming* (5.40) *into a linear flow at constant speed*

$$\frac{dI}{dt} = 0, \qquad \frac{d\theta}{dt} = \omega.$$

The variables $(I, \theta) \in \mathbb{R}^n \times \mathcal{T}^n$ are called **action-angle variables**. Notice that the action-angle variables are not unique. For example, given a unimodular[115] matrix Z, a smooth function $w : \mathbb{R}^n \to \mathbb{R}$ and a constant vector $\theta_0 \in \mathbb{R}^n$, we find other action-angle variables (I_1, θ_1) by means of the symplectic diffeomorphism[116]

$$\begin{cases} I_1 = Z(I + \nabla_\theta w(\theta)), \\ \theta_1 = (Z^T)^{-1}\theta + \theta_0, \end{cases}$$

corresponding to a change of basis (mapping straight lines on $\mathcal{T}^n$ into straight lines) of the torus $\mathcal{T}^n$ by the matrix Z and a phase shift $\nabla_\theta w(\theta)$ of the action variables. Liouville's theorem shows that

[114]If M_f is compact but not connected, we can just restrict our attention to each connected component. If M_f is not compact, the flow is again given by a straight line motion at constant speed, but now the winding takes place on a cylinder $\mathcal{T}^k \times \mathbb{R}^{n-k}$ rather than a torus.

[115]This is an $n \times n$ matrix with integer entries and determinant ± 1. Equivalently, it is an integer square matrix that is invertible over the integers: there is an integer matrix which is its inverse. While a characterization of unimodular matrices is hard to get, for our purposes it suffices to notice that for any integer $n \in \mathbb{Z}$ the matrix $\begin{pmatrix} 2n+1 & 4n \\ n+1 & 2n+1 \end{pmatrix}$ is unimodular.

[116]Using the fact that the Hessian matrix with entries $(\frac{\partial^2 w}{\partial \theta_i \partial \theta_j})_{i,j=1,\ldots,n}$ is symmetric, it is easy to check that the Jacobian J_1 of this transformation satisfies $(J_1)^T J J_1 = J$.

in appropriate coordinates, the flow of the (nonlinear) dynamical system (5.40) is actually a linear flow at constant speed: the system is then said to be **integrable** since, up to the change of variables, the solutions can be determined by simple quadrature.

We conclude our discussion of classical Hamiltonian systems by pointing out the fact that there is no systematic way to determine whether a given system is integrable. In each individual case an analysis starting with a quest for integrals of motion is required. Finding integrals of motion in addition to the Hamiltonian H for a given Hamiltonian system (5.40) is a challenging problem, especially if the system (5.40) is nonlinear. A powerful tool relies on the idea that symmetries imply integrals of motion. To be more specific, given an open interval[117] $I \subset \mathbb{R}$ with $0 \in I$ and an open set $\Omega \subset \mathbb{R}^{2n}$, we say that a family of functions $\psi(t,\cdot) : \Omega \to \Omega$, which depends smoothly on the parameter $t \in I$, defines a Hamiltonian flow on $\mathbb{R}^{2n}$ if

- $\psi(0,\mathbf{x}) = \mathbf{x}$ for all $\mathbf{x} \in \Omega$;
- $\psi(t,\psi(s,\mathbf{x})) = \psi(t+s,\mathbf{x})$ for all $\mathbf{x} \in \Omega$ and all $t, s \in I$ with $t+s \in I$.
- for any fixed $t \in I$ the map $\mathbf{x} \to \psi(t,\mathbf{x})$ is symplectic in Ω.

The Hamiltonian flow $\{\psi(t,\cdot)\}_{t\in I}$ is a symplectic symmetry for the function $H : \Omega \to \mathbb{R}$ if

$$H(\psi(t,\mathbf{x})) = H(\mathbf{x}) \quad \text{for all} \quad \mathbf{x} \in \Omega.$$

If we define $F : \Omega \to \mathbb{R}$ by

$$F(\mathbf{x}) = \left.\frac{\partial \psi(t,\mathbf{x})}{\partial t}\right|_{t=0}, \qquad \mathbf{x} \in \Omega,$$

then one can show (cf. [228]) that $\psi(t,\mathbf{x})$ is the solution of the Hamiltonian system $\psi' = J\,\nabla F(\psi)$ with initial data $\psi(0) = \mathbf{x}$. This allows us to formulate **Noether's theorem**: *if the Hamiltonian flow $\{\psi(t,\cdot)\}_{t\in I}$ is a symplectic symmetry for H, then the corresponding function F is an integral of motion for the system (5.40), in involution with H.* In particular, Noether's theorem shows the following:

(i) translational invariance implies conservation of momentum; that is, if the Hamiltonian function $H : \mathbb{R}^6 \to \mathbb{R}$ is invariant under translations, $H(q+t\mathbf{v},p) = H(q,p)$ for all $t \in \mathbb{R}$ and $\mathbf{v} \in \mathbb{R}^3$, where $\mathbf{x} = (q,p) \in \mathbb{R}^3 \times \mathbb{R}^3$, then all three components of the momentum p are integrals of motion;

(ii) rotational invariance implies conservation of angular momentum; that is, if the Hamiltonian function $H : \mathbb{R}^6 \to \mathbb{R}$ is such that for any orthogonal 3×3 matrix A with determinant 1 (representing a rotation) we have $H(Aq,Ap) = H(q,p)$ for all $\mathbf{x} = (q,p) \in \mathbb{R}^3 \times \mathbb{R}^3$, then each component of the angular momentum $q \wedge p$ is an integral of motion.[118]

[117] It is customary to take $I = \mathbb{R}$ and $\Omega = \mathbb{R}^{2n}$ but we want to allow blowup in finite time, as is the case for the flow associated with the Hamiltonian system $\psi' = J\nabla F$ in $\mathbb{R}^2$ with $F(x_1,x_2) = \frac{x_1^4+x_2^2}{2}$, where $\psi(t,(1,1)) = (\frac{1}{1-t}, \frac{1}{(1-t)^2})$ blows up at time $t = 1$.

[118] The reason why we took $n = 3$ is related to the notion of angular momentum: this is the only dimension where it is possible to define a cross product of vectors in $\mathbb{R}^n$ so that it will have properties similar to the usual cross product of vectors in $\mathbb{R}^3$. More precisely, only for $n = 7 > 3$ can we assign to any two vectors $\mathbf{v}, \mathbf{w} \in \mathbb{R}^n$ a vector $\mathbf{v} \wedge \mathbf{w} \in \mathbb{R}^n$ such that $\mathbf{v} \wedge \mathbf{w}$ is a continuous bilinear function of $(\mathbf{v},\mathbf{w})$, the vector $\mathbf{v} \wedge \mathbf{w}$ is perpendicular to both $\mathbf{v}$ and $\mathbf{w}$, and the vector $\mathbf{v} \wedge \mathbf{w}$ is nonzero if $\mathbf{v}$ and $\mathbf{w}$ are linearly independent. But even $n = 7$ does not accomodate the requirement that for any rotation A (i.e., any orthogonal transformation of determinant 1) we have $A(\mathbf{v} \wedge \mathbf{w}) = A\mathbf{v} \wedge A\mathbf{w}$ (see the discussion in [219]). There are, however, abstract constructions of the angular momentum as elements of the duals of Lie algebras (see [17]).

KdV as an integrable infinite-dimensional Hamiltonian system

The persistence of the individuality of a soliton through the nonlinear interaction with other solitons suggests the existence of a large number of integrals of the motion, so that the individual solitary wave can, so to say, remember its shape and speed. Actually, there are infinitely many functionally independent integrals of motion for KdV, and they can be obtained from the Hamiltonian formulations of the equation.

Bi-Hamiltonian structure

The phase space is the space of square-integrable functions $L^2(\mathbb{R})$, the scalar product of which we denote by $(\cdot,\cdot)$. Thus functions H on the phase space are functionals on $L^2(\mathbb{R})$, mostly nonlinear and unbounded, reasons for which we introduce the subspace

$$X = \{u \in C^\infty(\mathbb{R}) : \sup_{x\in\mathbb{R}} \{|x|^n |\partial_x^k u(x)|\} < \infty \text{ for all integers } k, n \geq 0\} \subset L^2(\mathbb{R})$$

of rapidly decreasing smooth functions as suitable for defining these functionals. The *gradient* $G_F(u)$ of a functional F at $u \in X$ is a function that satisfies

$$\frac{d}{d\varepsilon} F(u+\varepsilon v)\Big|_{\varepsilon=0} = (v, G_F(u)), \qquad v \in X,$$

and the *second derivative* of F at $u \in X$ is the (possibly unbounded) linear operator S_F^u that satisfies

$$\frac{d}{d\delta} G_F(u+\delta w)\Big|_{\delta=0} = S_F^u[w], \qquad w \in X.$$

Since

$$\frac{d^2}{d\varepsilon d\delta} F(u+\varepsilon v+\delta w)\Big|_{\varepsilon=0,\delta=0} = (v, S_F^u[w]), \qquad v, w \in X,$$

we see that S_F^u is a symmetric operator. A Hamiltonian structure can be defined by means of the Poisson bracket[119]

$$\{F, H\}(u) = \Big(G_F(u), \partial_x G_H(u)\Big), \tag{5.41}$$

and the corresponding Hamiltonian flow is the partial differential equation

$$u_t = \partial_x G_H(u). \tag{5.42}$$

Notice that (5.41) ensures that the Poisson bracket is bilinear and antisymmetric. To check the Jacobi identity

$$\{\{F,H\},K\}+\{\{K,F\},H\}+\{\{H,K\},F\} = 0, \tag{5.43}$$

we first compute the gradient of $\{F,H\}$. Since

$$\begin{aligned}\frac{d}{d\varepsilon}\{F,H\}(u+\varepsilon v)\Big|_{\varepsilon=0} &= \frac{d}{d\varepsilon}\Big(G_F(u+\varepsilon v), \partial_x G_H(u+\varepsilon v)\Big)\Big|_{\varepsilon=0} \\ &= (S_F^u[v], \partial_x G_H(u)) + (G_F(u), \partial_x S_H^u[v]) = (v, S_F^u[\partial_x G_H(u)] - S_H^u[\partial_x G_F(u)]),\end{aligned}$$

[119] Here the unbounded linear operator ∂_x plays the role of the bounded operator (5.38) in the finite-dimensional classical setting.

we deduce that $G_{\{F,H\}}(u) = S_F^u\,[\partial_x\, G_H(u)]) - S_H^u\,[\partial_x\, G_F(u)]$. Therefore,

$$\begin{aligned}\{\{F,H\},K\}(u) &= (G_{\{F,H\}}(u), \partial_x\, G_K(u)) = (S_F^u\,[\partial_x\, G_H(u)] - S_H^u\,[\partial_x\, G_F(u)], \partial_x\, G_K(u)),\\ \{\{K,F\},H\}(u) &= (G_{\{K,F\}}(u), \partial_x\, G_H(u)) = (S_K^u\,[\partial_x\, G_F(u)] - S_F^u\,[\partial_x\, G_K(u)], \partial_x\, G_H(u)),\\ \{\{H,K\},F\}(u) &= (G_{\{H,K\}}(u), \partial_x\, G_F(u)) = (S_H^u\,[\partial_x\, G_K(u)] - S_K^u\,[\partial_x\, G_H(u)], \partial_x\, G_F(u)),\end{aligned}$$

and (5.43) follows from the symmetry of S_F^u, S_H^u, and S_K^u.

Choosing the Hamiltonian functional

$$H_1(\eta) = \frac{1}{2}\int_{\mathbb{R}} \eta_x^2\, dx - \frac{1}{6}\int_{\mathbb{R}} \eta^3\, dx, \tag{5.44}$$

we compute $G_{H_1}(\eta) = -\eta_{xx} - \frac{1}{2}\eta^2$, so that (5.42) becomes the KdV equation (5.3). KdV has a second Hamiltonian structure $\eta_t = \mathcal{U} G_{H_0}(\eta)$ with Hamiltonian functional

$$H_0(\eta) = -\frac{1}{2}\int_{\mathbb{R}} \eta^2\, dx \tag{5.45}$$

whose gradient is $G_{H_0}(u) = -u$, and Poisson bracket

$$\{F,H\}_0(u) = \Big(G_F(u), \mathcal{U}(u)\, G_H(u)\Big) \tag{5.46}$$

with Hamiltonian operator[120] $\mathcal{U}(u) = \partial_x^3 + \frac{2}{3}u\,\partial_x + \frac{1}{3}u_x$ whose coefficients depend on u. Indeed, the fact that (5.46) is bilinear and antisymmetric is clear, and the Jacobi identity is checked as above by relying on the explicit calculation of the gradient of $\{F,H\}_0$. For this, notice that $\frac{d}{d\varepsilon}\{F,H\}_0(u + \varepsilon v)\Big|_{\varepsilon=0}$ equals

$$\begin{aligned}&\frac{d}{d\varepsilon}\left(G_F(u+\varepsilon v), \left(\partial_x^3 + \frac{2}{3}(u+\varepsilon v)\,\partial_x + \frac{1}{3}(u_x + \varepsilon v_x)\right) G_H(u+\varepsilon v)\right)\Big|_{\varepsilon=0}\\ &= \left(S_F^u[v], \left(\partial_x^3 + \frac{2}{3}u\,\partial_x + \frac{1}{3}u_x\right) G_H(u)\right) + \left(G_F(u), (\partial_x^3 + \tfrac{2}{3}u\,\partial_x + \tfrac{1}{3}u_x)\, S_H^u[v]\right)\\ &\qquad + \left(G_F(u), \left(\frac{2}{3}v\,\partial_x + \frac{1}{3}v_x\right) G_H(u)\right)\\ &= \left(v, S_F^u\left[\left(\partial_x^3 + \frac{2}{3}u\,\partial_x + \frac{1}{3}u_x\right) G_H(u)\right] - S_H^u\left[\left(\partial_x^3 + \frac{2}{3}u\,\partial_x + \frac{1}{3}u_x\right) G_F(u)\right]\right)\\ &\qquad + \left(v, \frac{1}{3}\,G_F(u)\,\partial_x\, G_H(u) - \frac{1}{3}\,G_H(u)\,\partial_x\, G_F(u)\right),\end{aligned}$$

so that

$$\begin{aligned}G_{\{F,H\}_0}(u) = S_F^u&\left[\left(\partial_x^3 + \frac{2}{3}u\,\partial_x + \frac{1}{3}u_x\right) G_H(u)\right] - S_H^u\left[\left(\partial_x^3 + \frac{2}{3}u\,\partial_x + \frac{1}{3}u_x\right) G_F(u)\right]\\ &+ \frac{1}{3}\,G_F(u)\,\partial_x\, G_H(u) - \frac{1}{3}\,G_H(u)\,\partial_x\, G_F(u).\end{aligned}$$

Both H_0 and H_1 are integrals of motion for KdV since the respective Hamiltonian structures ensure that, just like in finite dimensions, $\{F,H\}$ is the rate of change of the functional F along the solutions of (5.42): for example, we have that

$$\frac{d}{dt}F(u(t)) = (u_t, G_F(u)) = (\partial_x\, G_H(u), G_F(u)) = \{F,H\}.$$

[120] Called this way since the corresponding Poisson bracket is bilinear and antisymmetric and satisfies the Jacobi identity.

The observation that $\partial_x G_{H_1}(\eta) = \mathcal{U}(\eta) G_{H_0}(\eta)$, as both expressions are equal to η_t, leads to the construction of an infinite sequence of Hamiltonian functionals of the form $H_k(\eta) = \int_{\mathbb{R}} p_k\, dx$ with p_k a polynomial in η, $\partial_x \eta, \ldots, \partial_x^k \eta$ defined recursively by

$$\partial_x G_{H_{k+1}}(u) = \mathcal{U}(u) G_{H_k}(u), \qquad k \geq 0,$$

(cf. [196]). These Hamiltonians are functionally independent and in involution with respect to either Poisson structure:

$$(G_{H_n}(u), \partial_x G_{H_m}(u)) = 0, \quad (G_{H_n}(u), \mathcal{U}(u) G_{H_m}(u)) = 0, \qquad m, n \geq 0,\ u \in X.$$

The fact that there are infinitely many integrals of motion in involution could permit us to declare that KdV is integrable, but the analogy to finite dimensions is not yet satisfactory: classically one needs n integrals in involution for a system in $\mathbb{R}^{2n}$ and half of infinity is not that precise. The way out of this conundrum is to try to prove the analogue of Liouville's theorem finding action-angle variables that transform the KdV flow into a linear flow at constant speed. The material presented below represents an attempt to present this theory in a self-contained and accessible way. While we follow to a large extent some classical approaches, we strive for simple arguments,[121] and for this reason, we are not concerned with the minimal requirements (for this direction we refer the reader to [94]).

The isospectral problem

It is now convenient to use a scaled version of the KdV equation (5.3): setting $\eta(x,t) = -6u(x,t)$, the new form of KdV is

$$u_t - 6uu_x + u_{xxx} = 0. \tag{5.47}$$

Assuming that the profile $x \mapsto u(x,0)$ at time $t = 0$ belongs to X, one can show (see the discussion of semigroup theory in Chapter 6) that there is a unique solution $u \in C^1(\mathbb{R}, X)$ to (5.47). This ensures the smoothness of u in the x-variable and its strong decay as $x \mapsto \pm\infty$, at any fixed instant $t \in \mathbb{R}$. It turns out that an infinite set of conserved quantities for (5.47) is associated with the Schrödinger operator

$$\mathcal{L}(t) = -\partial_x^2 + u, \tag{5.48}$$

in which t plays the role of a parameter. The problem (5.48) is referred to as the *isospectral problem* for the KdV equation (5.47) since if the potential $u(x,t)$ evolves according to (5.47), then the eigenvalues of $\mathcal{L}(t)$ are independent of t. Before proving this, let us first discuss the spectral problem (5.48) for a fixed potential $u \in X$. Let us consider the eigenvalue problem

$$\mathcal{L} f = k^2 f, \tag{5.49}$$

where $k \in \mathbb{C} \setminus \{0\}$ is such that $\Im m(k) \geq 0$ (thus k^2 covers $\mathbb{C} \setminus \{0\}$). Since $u(x) \to 0$ as $|x| \to \infty$, we expect that for $|x|$ large a bounded solution f of (5.49) is very nearly a superposition of e^{ikx} and e^{-ikx}. For this reason we introduce now the *Jost solutions* $f_\pm$ to (5.49), determined by

$$f_+(x,k) \approx e^{ikx} \quad \text{as} \quad x \to \infty, \quad f_-(x,k) \approx e^{-ikx} \quad \text{as} \quad x \to -\infty.$$

We will see that, with the possible exception of a finite number of values of k, the solutions $f_\pm$ are linearly independent. If we set for $x \in \mathbb{R}$ and $\Im m(k) \geq 0$, $k \neq 0$,

$$D_k(x) = \int_0^x e^{2iks}\, ds = \frac{e^{2ikx} - 1}{2ik}, \qquad g_\pm(x,k) = f_\pm(x,k) e^{\mp ikx},$$

[121] Some incorrect statements in [1, 97, 193] concerning the evolution in time of the eigenfunctions are also avoided (see the discussion in [7] for a detailed criticism).

then the *Faddeev function* g_+ solves the integral equation

$$g_+(x,k) = 1 + \int_x^\infty D_k(y-x)u(y)g_+(y,k)\,dy \tag{5.50}$$

and is given by

$$g_+(x,k) = 1 + \sum_{n=1}^\infty g_n(x,k),$$

where

$$g_n(x,k) = \int_{x\le x_1\le\cdots\le x_n} D_k(x-x_1)\cdots D_k(x_n-x_{n-1})u(x_1)\cdots u(x_n)\,dx_1\cdots dx_n, \quad n\ge 1.$$

Since $|D_k(x)| \le \frac{1}{|k|}$ for $\Im m(k)\ge 0$, $k\ne 0$, we have for $x\in\mathbb{R}$ and $\Im m(k)\ge 0$, $k\ne 0$, that

$$|g_n(x,k)| \le \frac{1}{|k|^n}\int_{x\le x_1\le\cdots\le x_n} |u(x_1)|\cdots|u(x_n)|\,dx_1\cdots dx_n = \frac{1}{|k|^n}\frac{1}{n!}\Big(\int_x^\infty |u(y)|\,dy\Big)^n.$$

On the other hand, setting $D_0(s)=s$, we define $f_+(x,0)=g_+(x,0)$ as the unique solution of the integral equation

$$g_+(x,0) = 1 + \int_x^\infty u(y)g_+(y,0)\,dy, \tag{5.51}$$

given by

$$g_+(x,0) = 1 + \sum_{n=1}^\infty \int_{x\le x_1\le\cdots\le x_n} (x-x_1)\cdots(x_n-x_{n-1})u(x_1)\cdots u(x_n)\,dx_1\cdots dx_n.$$

Since $|D_k(y)|\le y$ for $y\ge 0$ and $\Im m(k)\ge 0$, we get

$$\begin{aligned}|g_n(x,k)| &\le \int_{x\le x_1\le\cdots\le x_n} (x_1-x)\cdots(x_n-x_{n-1})\,|u(x_1)|\cdots|u(x_n)|\,dx_1\cdots dx_n\\ &= \frac{1}{n!}\Big(\int_x^\infty (y-x)\,|u(y)|\,dy\Big)^n, \qquad x\in\mathbb{R},\quad \Im m(k)\ge 0,\quad n\ge 1.\end{aligned}$$

These estimates show the locally uniform convergence of the series, so that $g_+(x,k)$ is smooth in $x\in\mathbb{R}$, analytic in $k\in\{\Im m(k)>0\}$, and continuous in $k\in\{\Im m(k)\ge 0\}$, with

$$|g_+(x,k)-1| \le \sum_{n=1}^\infty \frac{1}{n!}\sigma_1^n(x) \le e^{\sigma_1(x)}, \quad x\in\mathbb{R},\ \Im m(k)\ge 0, \tag{5.52}$$

$$|g_+(x,k)-1| \le \sum_{n=1}^\infty \frac{1}{n!}\frac{1}{|k|^n}\sigma^n(x) \le \frac{\sigma(x)}{|k|}e^{\sigma_1(x)},\ x\in\mathbb{R},\ \Im m(k)\ge 0, k\ne 0, \tag{5.53}$$

where

$$\sigma(x) = \int_x^\infty |u(y)|\,dy, \quad \sigma_1(x) = \int_x^\infty (y-x)\,|u(y)|\,dy = \int_x^\infty\int_y^\infty |u(s)|\,ds\,dy.$$

Thus, if we use (5.52) for $|k|\le 1$ and (5.53) for $|k|\ge 1$, we get

$$|g_+(x,k)-1| \le \frac{C}{1+|k|}, \qquad x\in\mathbb{R},\quad \Im m(k)\ge 0, \tag{5.54}$$

with

$$C = \left(1 + \int_{\mathbb{R}} (1+y^2)\,|u(y)|\,dy\right) \exp\left\{\int_{\mathbb{R}} \int_y^{\infty} (1+s^2)\,|u(s)|\,ds\,dy\right\} > 0.$$

Therefore, for each $x \in \mathbb{R}$, the function $x \mapsto [g_+(x,k)-1]$ belongs to the Hardy class H_+^2 of functions that are analytic in the upper half-plane $\{\Im m(k) > 0\}$ and whose square integrals along all lines $\Im m(k) = \text{constant} > 0$ are bounded by some constant.

We now show that $\dot{g}_+(x,k) = \frac{\partial g_+}{\partial k}(x,k)$ exists throughout $\mathbb{R} \times \{\Im m(k) \geq 0\}$. For this, notice first that the considerations leading us from (5.50) or (5.51) to (5.52) show that if $w \in X$ and $A(x,k)$ is continuous in $\mathbb{R} \times \{\Im m(k) \geq 0\}$, with $|A(x,k)|$ having polynomial growth as $|x| \to \infty$ for k in compact subsets of $\{\Im m(k) \geq 0\}$, then the integral equation

$$h(x,k) = A(x,k) + \int_x^{\infty} D_k(y-x)\,w(y)\,h(y,k)\,dy \tag{5.55}$$

has a solution given by

$$h(x,k) = A(x,k) + \sum_{n=1}^{\infty} h_n(x,k),$$

where for $x \in \mathbb{R}$, $\Im m(k) \geq 0$, $n \geq 1$, we set $h_n(x,k)$ equal to

$$\int_{x \leq x_1 \leq \cdots \leq x_n} D_k(x-x_1)\cdots D_k(x_n - x_{n-1})\,A(x_1,k)u(x_1)\cdots A(x_n,k)u(x_n)\,dx_1\cdots dx_n.$$

Moreover, for $(x,k) \in \mathbb{R} \times \{\Im m(k) \geq 0\}$ we have

$$|h(x,k)| \leq |A(x,k)| + \exp\left\{\int_x^{\infty} (y-x)\,|A(y,k)|\,|w(y)|\,dy\right\}. \tag{5.56}$$

A further useful fact is the following comparison result for integral equations of type (5.51). If $\varepsilon > 0$ is a constant and $h : \mathbb{R} \to [0,\infty)$ is a bounded continuous function satisfying

$$h(x) \leq \varepsilon + \int_x^{\infty} (y-x)\,w(y)\,h(y)\,dy, \qquad x \in \mathbb{R}, \tag{5.57}$$

for some $w : \mathbb{R} \to [0,\infty)$ in the function class X, then[122] $h(x) \leq 2\varepsilon\, h_0(x)$ for all $x \in \mathbb{R}$, where h_0 is the positive solution of the integral equation

$$h_0(x) = 1 + \int_x^{\infty} (y-x)\,w(y)\,h_0(y)\,dy, \qquad x \in \mathbb{R},$$

obtained by the iterative procedure analogous to that implemented above for (5.51).

For $(x,k) \in \mathbb{R} \times \{\Im m(k) \geq 0\}$ let us now consider the integral equation

$$\dot{g}_+(x,k) = \int_x^{\infty} D_k(y-x)\,u(y)\,\dot{g}_+(y,k)\,dy + \int_x^{\infty} \dot{D}_k(y-x)\,u(y)\,g_+(y,k)\,dy \tag{5.58}$$

for $\dot{g}_+(x,k)$, obtained by formally differentiating (5.54) with respect to k. Since

$$|\dot{D}_k(y-x)| \leq \left|\int_0^{y-x} 2is\,e^{2isk}\,ds\right| \leq (y-x)^2, \tag{5.59}$$

[122] From (5.57) we infer that $\limsup_{x\to\infty} h(x) \leq \varepsilon$, while $\liminf_{x\to\infty} h_0(x) \geq 1$. Therefore $h(x) \leq 2\varepsilon h_0(x)$ for x large enough. The existence of $x_0 = \sup\{x \in \mathbb{R} : \; h(x) = 2\varepsilon\, h_0(x)\}$ contradicts the inequality $h(x_0) - 2\varepsilon\, h_0(x_0) \leq -\varepsilon + \int_{x_0}^{\infty} (y-x_0)\,w(y)\,[h(y) - 2\varepsilon h_0(y)]\,dy \leq -\varepsilon$.

we deduce in view of (5.54) that

$$\left|\int_x^\infty \dot{D}_k(y-x)u(y)g_+(y,k)dy\right| \le \int_x^\infty (y-x)^2\,|u(y)|\,|g_+(y,k)|dy$$
$$\le 2(C+1)\int_x^\infty (x^2+y^2)\,|u(y)|\,dy \le 2(C+1)^2(1+x^2). \tag{5.60}$$

On the other hand, for $k \neq k_0$ with $\Im m(k) \ge 0$ and $\Im m(k_0) \ge 0$, from (5.50) and (5.51), we get

$$\frac{g_+(x,k)-g_+(x,k_0)}{k-k_0} = \int_x^\infty D_k(y-x)\,\frac{g_+(y,k)-g_+(y,k_0)}{k-k_0}\,u(y)\,dy$$
$$+\int_x^\infty \frac{D_k(y-x)-D_{k_0}(y-x)}{k-k_0}\,g_+(y,k_0)u(y)\,dy. \tag{5.61}$$

Setting $h(x) = \frac{|g_+(x,k)-g_+(x,k_0)|}{4(C+1)^2(1+x^2)\,|k-k_0|}$, since $|D_k(s)| \le s$ for $s \ge 0$, using (5.54), (5.59), and the mean-value theorem for complex-valued functions (see [96]), we deduce that

$$h(x) \le \frac{1}{2} + \int_x^\infty (y-x)(1+y^2)|u(y)|\,h(y)\,dy.$$

Since by (5.54) we have $\lim_{x\to\infty} h(x) = 0$, from the comparison theorem for (5.57) we infer that $h(x) \le C$ for all $x \in \mathbb{R}$, so that

$$|g_+(x,k)-g_+(x,k_0)| \le 4C(C+1)^2\,|k-k_0|\,(1+x^2), \qquad \Im m(k) \ge 0,\ \Im m(k_0) \ge 0. \tag{5.62}$$

These considerations indicate that we should seek solutions of (5.58) with

$$|\dot{g}_+(x,k)| \le 4C(C+1)^2\,(1+x^2), \qquad (x,k) \in \mathbb{R}\times\{\Im m(k) \ge 0\}. \tag{5.63}$$

The case $\Im m(k) > 0$ being settled as the existence of $\dot{g}_+(x,k)$ is ensured by the analyticity of $k \mapsto g_+(x,k)$ for $\Im m(k) > 0$, we have only to prove the existence and continuity of $\dot{g}_+(x,k)$ for $k \in \mathbb{R}$. For $\Im m(k_0) \ge 0$, setting $h(x,k_0) = \frac{\dot{g}_+(x,k_0)}{4(C+1)^2(1+x^2)}$, since $|D_{k_0}(s)| \le s$ for $s \ge 0$, we deduce from (5.58) and (5.60) that

$$h(x,k_0) = A(x,k_0) + \int_x^\infty D_{k_0}(y-x)u(y)h(y,k_0)\,dy$$

with $A(x,k_0)$ continuous and satisfying $|A(x,k_0)| \le 1$ for all $(x,k_0) \in \mathbb{R}\times\{\Im m(k_0) \ge 0\}$. The above considerations on (5.55) ensure the existence of a solution $h(x,k_0)$ that is continuous in $\mathbb{R}\times\{\Im m(k_0) \ge 0\}$. Moreover, by (5.56), $|h(x,k_0)| \le C$ for all $(x,k_0) \in \mathbb{R}\times\{\Im m(k_0) \ge 0\}$. Therefore, the expression $4(C+1)^2(1+x^2)h(x,k_0)$ provides us with a candidate for $\dot{g}_+(x,k_0)$, with the proper growth conditions (5.63), for all k_0 with $\Im k_0 \ge 0$. To show that this is indeed $\dot{g}_+(x,k)$, fix k_0 with $\Im k_0 \ge 0$. For $\Im m(k) \ge 0$ with $k \neq k_0$, we combine (5.63) and (5.58) to deduce that the bounded continuous function

$$h(x) = \frac{\left|\dfrac{g_+(x,k)-g_+(x,k_0)}{k-k_0} - \dot{g}_+(x,k_0)\right|}{4C(C+1)^2(1+x^2)}, \qquad x \in \mathbb{R},$$

satisfies

$$h(x) \le \int_x^\infty (y-x)(1+y^2)\,|u(y)|\,h(y)\,dy$$
$$+\frac{1}{4C(C+1)}\int_x^\infty \left|\frac{D_k(y-x)-D_{k_0}(y-x)}{k-k_0} - \dot{D}_{k_0}(y-x)\right|\,|u(y)|\,dy,$$

in view of (5.54). The identity

$$\frac{D_k(s)-D_{k_0}(s)}{k-k_0}-\dot{D}_{k_0}(s)=\frac{e^{2i(k-k_0)s}-1-2is\,(k-k_0)}{k-k_0}\,e^{2ik_0s},$$

in combination with the mean-value theorem (cf. [96]), first for the complex-valued function $s\mapsto e^{2i(k-k_0)s}-1-2is\,(k-k_0)$ and subsequently for $s\mapsto e^{2i(k-k_0)s}-1$, yields

$$\left|\frac{D_k(y-x)-D_{k_0}(y-x)}{k-k_0}-\dot{D}_{k_0}(y-x)\right|\le 2\sup_{0\le s\le y-x}\{|e^{2i(k-k_0)s}-1|\}\le 4\,|k-k_0|$$

for $|k-k_0|<\varepsilon$; we infer that

$$h(x)\le\varepsilon+\int_x^{\infty}(y-x)(1+y^2)\,|u(y)|\,h(y)\,dy,\qquad x\in\mathbb{R}.$$

Since we already know that h is bounded on $\mathbb{R}$, the comparison theorem for (5.57) yields $h(x)\le 2\varepsilon C$ for all $x\in\mathbb{R}$. Recalling the way we defined h, this proves the existence of $\dot{g}_+(x,k)$ for all $(x,k)\in\mathbb{R}\times\{\Im m(k)\ge 0\}$.

We estimate now $g_+'(x,k)$. From (5.50) and (5.51) we get

$$g_+'(x,k)=-e^{-2ikx}\int_x^{\infty}e^{2iky}u(y)\,g_+(y,k)\,dy,\qquad x\in\mathbb{R},\quad \Im m(k)\ge 0,\tag{5.64}$$

so that $g_+'(x,k)\to 0$ as $x\to\infty$: if $\Im m(k)=0$ the limit is immediate since $u\in L^1(\mathbb{R})$, and for $\Im m(k)>0$ we use l'Hôpital's rule. Notice that (5.54), (5.50), and (5.51) yield

$$|g_+(x,k)-1|\le C\int_x^{\infty}y\,|u(y)|\,dy,\qquad x\ge 0,\quad \Im m(k)\ge 0.\tag{5.65}$$

Since $f_+'(x,k)=e^{ikx}\,g_+'(x,k)+ik\,e^{ikx}\,g_+(x,k)$, we deduce that

$$f_+(x,k)\approx e^{ikx}\quad\text{and}\quad f_+'(x,k)\approx ik\,e^{ikx}\quad\text{as}\quad x\to\infty\quad\text{for}\quad \Im m(k)\ge 0.\tag{5.66}$$

Similar considerations for $g_-(x,k)$ lead to

$$f_-(x,k)\approx e^{-ikx}\quad\text{and}\quad f_-'(x,k)\approx -ik\,e^{-ikx}\quad\text{as}\quad x\to-\infty\quad\text{for}\quad \Im m(k)\ge 0.\tag{5.67}$$

Clearly,

$$\overline{f_+(x,k)}=f_+(x,-k),\quad \overline{f_-(x,k)}=f_-(x,-k),\qquad k\in\mathbb{R}\setminus\{0\}.\tag{5.68}$$

Any three solutions to (5.49) being linearly dependent, there must be relations of the form

$$f_-(x,k)=a_+(k)\,\overline{f_+(x,k)}+b_+(k)\,f_+(x,k),\qquad k\in\mathbb{R}\setminus\{0\},\tag{5.69}$$

and

$$f_+(x,k)=a_-(k)\,\overline{f_-(x,k)}+b_-(k)\,f_-(x,k),\qquad k\in\mathbb{R}\setminus\{0\}.\tag{5.70}$$

The Wronskian bilinear form, defined by

$$W(g,h)=\begin{vmatrix} g & h\\ g' & h'\end{vmatrix}=gh'-g'h,$$

is clearly antisymmetric, and for a pair of solutions to (5.49) it is a constant for $x \in \mathbb{R}$. In particular, we have

$$W(f_+(x,k), \overline{f_+(x,k)}) = -2ik, \quad W(f_-(x,k), \overline{f_-(x,k)}) = 2ik, \qquad k \in \mathbb{R}\setminus\{0\}, \tag{5.71}$$

by evaluating the expressions at $\pm\infty$, taking into account the asymptotic behavior of $f_\pm$. Computing the Wronskian of (5.69) with $f_+(x,k)$ and with $\overline{f_+(x,k)}$, we get by (5.71) that

$$W(f_-(x,k), f_+(x,k)) = 2ik\,a_+(k), \quad W(f_-(x,k), \overline{f_+(x,k)}) = -2ik\,b_+(k) \tag{5.72}$$

for $k \in \mathbb{R}\setminus\{0\}$, while from (5.70) combined with (5.71) we get

$$W(f_-(x,k), f_+(x,k)) = 2ik\,a_-(k), \quad W(\overline{f_-(x,k)}, f_+(x,k)) = -2ik\,b_-(k) \tag{5.73}$$

for $k \in \mathbb{R}\setminus\{0\}$. As consequences of (5.72)–(5.73) we get

$$b_+(k) + \overline{b_-(k)} = 0, \qquad k \in \mathbb{R}\setminus\{0\}, \tag{5.74}$$

and $a_-(k) = a_+(k)$ for $k \in \mathbb{R}\setminus\{0\}$, henceforth denoted by $a(k)$. Since $f_\pm(x,k)$ and $f'_\pm(x,k)$ are analytic in the upper half-plane, setting

$$2ik\,a(k) = W(f_-(x,k), f_+(x,k)), \qquad \Im m(k) \ge 0,\ k \ne 0, \tag{5.75}$$

we see that $a(k)$ will be analytic in $\{\Im m(k) > 0\}$ and continuous on $\Im m(k) \ge 0$, $k \ne 0$. Notice that, in general, $b_\pm$ do not extend off $\mathbb{R}\setminus\{0\}$ (see the discussion in [94]).

Suppose now that $a(k) = 0$ for some k with $\Im m(k) > 0$. Then the solutions $f_\pm$ are linearly dependent as $W(f_-(x,k), f_+(x,k)) = 0$, so that for some $d \in \mathbb{R}$ we have $d\,f_+(x,k) = f_-(x,k))$ for all $x \in \mathbb{R}$. Consequently, $f_+(x,k)$ decays exponentially for $|x| \to \infty$, so that $f_+(\cdot,k) \in L^2(\mathbb{R})$. From (5.49) we now infer that $f_+(\cdot,k) \in H^2(\mathbb{R})$ so that $f_+(\cdot,k)$ is an eigenfunction of $\mathcal{L}$ with eigenvalue k^2. Since $\mathcal{L}$ is self-adjoint, its eigenvalues are real (cf. [140]) so that all zeros of $a(k)$ in the upper half-plane lie on the imaginary axis. We now show that all eigenvalues of $\mathcal{L}$ arise in this way. Clearly if $\psi \in H^2(\mathbb{R})$ is an eigenfunction with eigenvalue k^2, where $k \in \{\Im m(k) > 0\}$, then ψ is asymptotically a linear combination of $e^{\pm ikx}$, so that, looking in the limit $x \to \infty$, ψ must be proportional to $f_+(x,k)$, while looking at $x \to -\infty$ yields that ψ must be proportional to $f_-(x,k)$. Consequently, $f_+(x,k)$ and $f_-(x,k)$ are linearly dependent, so their Wronskian is zero and k is a root of a. It turns out that $k^2 = 0$ is not an eigenvalue of $\mathcal{L}$. Assuming that it were, for the corresponding eigenfunction $\psi \in H^2(\mathbb{R})$, we would have from (5.49) that $\psi_{xx} = u\psi$, and thus

$$\psi(x) = \int_{-\infty}^{x}\int_{-\infty}^{y} u(s)\,\psi(s)\,ds\,dy = \int_{-\infty}^{x} (x-y)\,u(y)\,\psi(y)\,dy, \qquad x \in \mathbb{R}.$$

If $x_0 < 0$ is such that $\int_{-\infty}^{x_0} |y|\,u(y)|\,dy < 1$, then the previous inequality yields

$$\sup_{x \le x_0}\{|\psi(x)|\} \le \sup_{x \le x_0}\{|\psi(x)|\} \int_{-\infty}^{x_0} |y|\,|u(y)|\,dy,$$

so that $\psi(x) = 0$ for $x \le x_0$. Using this in the above integral equation for ψ, we get

$$|\psi(x)| \le \int_{x_0}^{x} |y|\,|u(y)|\,|\psi(y)|\,dy, \qquad x_0 \le x \le 0,$$

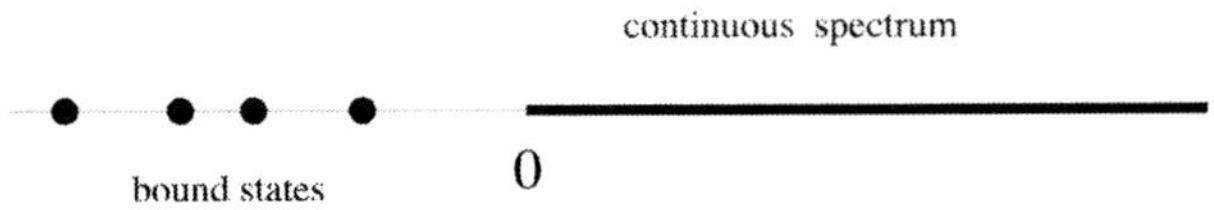

Figure 5.10. *The spectrum of $\mathcal{L}$.*

and Gronwall's inequality [79] yields $\psi(x) = 0$ for $x \leq 0$. A similar argument shows that $\psi(x) = 0$ for $x \geq 0$. Therefore, zero is not an eigenvalue of $\mathcal{L}$. If $\Im m(k) = 0$, then all solutions of (5.49) are asymptotically sinusoidal or constant as $x \to \pm\infty$ and are therefore bounded on $\mathbb{R}$ but do not belong to $L^2(\mathbb{R})$. The states $\lambda = k^2$ with $k \in \mathbb{R}$ are called *unbound states* and form the continuous spectrum of $\mathcal{L}$. Consequently, all eigenvalues of $\mathcal{L}$, called *bound states*, correspond, by means of $k \mapsto k^2$, to zeros of $a(k)$ in $\{\Im m(k) > 0\}$.

We now show that there are at most finitely many bound states. Notice that if $\lambda < 0$ is a bound state, then[123]

$$\lambda > \min_{x \in \mathbb{R}} \{u(x)\}. \tag{5.76}$$

In particular, if $u(x) \geq 0$ on $\mathbb{R}$, then there are no bound states. To prove that we cannot have infinitely many bound states we rely on the fact that if $\lambda_2 < \lambda_1$ are bound states, with corresponding eigenfunctions $\psi_i \in H^2(\mathbb{R})$, then[124] ψ_2 has strictly more zeros than ψ_1: if $x_1 \in \mathbb{R} \cup \{-\infty\}$ and $x_2 \in \mathbb{R} \cup \{\infty\}$ are two consecutive zeros of ψ_1, then ψ_2 has at least one zero in (x_1, x_2). For this, let us assume that $\psi_1(x) > 0$ for $x \in (x_1, x_2)$ and, say, $\psi_2(x) > 0$ for $x \in (x_1, x_2)$. Multiplying the equation for ψ_1 by ψ_2 and subtracting from this the equation for ψ_2 multiplied by ψ_1, we get after integration by parts that

$$\psi_1'(x)\psi_2(x) - \psi_1(x)\psi_2'(x) = \psi_1'(x_1)\psi_2(x_2) + (\lambda_1 - \lambda_2)\int_{x_1}^{x} \psi_1(y)\psi_2(y)\,dy > 0, \quad x \in (x_1, x_2),$$

as $\psi_1'(x_1) \geq 0$ and $\psi_2(x_1) \geq 0$. Consequently, $(\frac{\psi_2}{\psi_1})' > 0$ on (x_1, x_2), which is impossible unless $\psi_2(x_1) = \psi_2(x_2) = 0$. This, however, contradicts the previously displayed relation evaluated for $x \uparrow x_2$. We conclude that ψ_2 has at least one zero in (x_1, x_2). We now infer, from (5.65) and the corresponding inequality for g_-, the existence of a constant $M > 0$, independent of k with $\Im m(k) \geq 0$, such that

$$g_+(x,k) \geq \frac{1}{2} \quad \text{for} \quad x \geq M \quad \text{and} \quad g_-(x,k) \geq \frac{1}{2} \quad \text{for} \quad x \leq -M.$$

Since we must have $f_-(x,k) = e^{-ikx} g_-(x,k) = d\, f_+(x,k) = d\, e^{ikx}\, g_+(x,k)$ for some $d \neq 0$ if k^2 is an eigenvalue of $\mathcal{L}$, no eigenfunction f can have a finite zero[125] outside the interval $[-M, M]$. We now claim that, independent of the eigenvalue λ, the distance between two consecutive finite zeros $x_1 < x_2$ of f is at least[126] $\alpha = \frac{\pi}{\sqrt{\max_{x\in\mathbb{R}}\{|u(x)|\} - \min_{x\in\mathbb{R}}\{u(x)\}}}$. Indeed, multiplying (5.49) by f and integrating on $[x_1, x_2]$, we obtain

$$\int_{x_1}^{x_2} f_x^2(x)\,dx = \int_{x_1}^{x_2} [\lambda - u(x)]\, f^2(x)\,dx \leq \left(|\lambda| + \max_{x \in \mathbb{R}}\{|u(x)|\}\right) \int_{x_1}^{x_2} f^2(x)\,dx.$$

[123] If $\psi \in H^2(\mathbb{R})$ is the eigenfunction, then multiplying (5.49) by ψ and integrating on $\mathbb{R}$ yields $\int_{\mathbb{R}} \psi_x^2\,dx = \int_{\mathbb{R}} (\lambda - u)\psi^2\,dx$, so $\lambda \leq \min_{x\in\mathbb{R}}\{u(x)\}$ would force $\psi \equiv 0$.

[124] Interpreting (5.49) with $k^2 = -\lambda$ as the equation $f'' - (\lambda + u)f = 0$ describing the oscillatory motion of f, this type of *Sturm oscillation theorem* (see [144]) expresses mathematically the fact that the larger force $-\lambda_2 - u$ produces more oscillations than the smaller force $-\lambda_1 - u$.

[125] Since $f \in H^2(\mathbb{R})$, we have $\lim_{|x|\to\infty} f(x) = \lim_{|x|\to\infty} f'(x) = 0$.

[126] Notice that by (5.76), $\min_{x\in\mathbb{R}}\{u(x)\} < 0$ is a necessary condition for the existence of bound states.

Using Wirtinger's inequality (cf. [103]), we infer that

$$\int_{x_1}^{x_2} f_x^2(x)\,dx \le \Big(|\lambda| + \max_{x\in\mathbb{R}}\{|u(x)|\}\Big)\frac{(x_2-x_1)^2}{\pi^2}\int_{x_1}^{x_2} f_x^2(x)\,dx.$$

Using (5.76), we can now easily infer that $x_2 - x_1 \ge \alpha$. Consequently, on $[-M, M]$ the function f has at most a finite number of zeros N, independent of λ. Together with the previous oscillation result on the eigenfunctions, this completes the proof that there are at most finitely many bound states.

For $k \in \mathbb{R}\setminus\{0\}$ we deduce by (5.68), (5.69), and (5.70) that

$$\overline{a(k)} = a(-k), \quad \overline{b_-(k)} = b_-(-k), \quad \overline{b_+(k)} = b_+(-k). \tag{5.77}$$

Expressing f_+ by means of (5.70) in (5.71) and using (5.77), we find that

$$1 = |a(k)|^2 - |b_-(k)|^2, \qquad k\in\mathbb{R}\setminus\{0\}, \tag{5.78}$$

$$1 = |a(k)|^2 - |b_+(k)|^2, \qquad k\in\mathbb{R}\setminus\{0\}. \tag{5.79}$$

We now define the *reflection coefficient* and the *transmission coefficient* by

$$R_\pm(k) = \frac{b_\pm(k)}{a(k)}, \quad T(k) = \frac{1}{a(k)}, \qquad k\in\mathbb{R}\setminus\{0\}. \tag{5.80}$$

By (5.78)–(5.79) we see that $a(k)\neq 0$ for $k\in\mathbb{R}\setminus\{0\}$, so that $R_\pm(k)$ and $T(k)$ are well defined. The formulas (5.77) show that $R_\pm$ and T are skew-symmetric,

$$R_\pm(-k) = \overline{R_\pm(k)}, \quad T(-k) = \overline{T(k)}, \qquad k\in\mathbb{R}\setminus\{0\}, \tag{5.81}$$

while (5.74) yields

$$R_+(k)T(-k) + R_-(-k)T(k) = 0, \qquad k\in\mathbb{R}\setminus\{0\}. \tag{5.82}$$

Dividing (5.69) by $a(k)$, we get

$$T(k)\,f_-(x,k) = \overline{f_+(x,k)} + R_+(k)\,f_+(x,k), \qquad x\in\mathbb{R},\ k\in\mathbb{R}\setminus\{0\}, \tag{5.83}$$

while from (5.70) we obtain

$$T(k)\,f_+(x,k) = \overline{f_-(x,k)} + R_-(k)\,f_-(x,k), \qquad x\in\mathbb{R},\ k\in\mathbb{R}\setminus\{0\}. \tag{5.84}$$

Therefore,

$$f_-(x,-k) = \overline{f_-(x,k)} = T(k)\,f_+(x,k) - R_-(k)\,f_-(x,k), \qquad x\in\mathbb{R},\ k\in\mathbb{R}\setminus\{0\}. \tag{5.85}$$

Relation (5.85) is interpreted as follows: a solution which behaves like a plane wave of frequency k traveling to the right is, for $x\to\infty$, the sum of a "transmitted" part traveling in the same direction, and a "reflected" part traveling in the opposite direction, with the amplitudes related to the coefficients $R_-(k)$ and $T(k)$. The idea is that the interaction due to u has no effect asymptotically (far in the future/past): one can think of the nonlinear term $u(x)$ as being "turned off" for $x\to\pm\infty$. The scattering data $R_\pm(k)$ and $T(k)$ connect these distinguished states. Notice that by (5.78), (5.79), and (5.80) we get

$$|T(k)|^2 + |R_\pm(k)|^2 = 1, \qquad k\in\mathbb{R}\setminus\{0\}. \tag{5.86}$$

This is a form of conservation of energy for the transmitted and reflected waves.

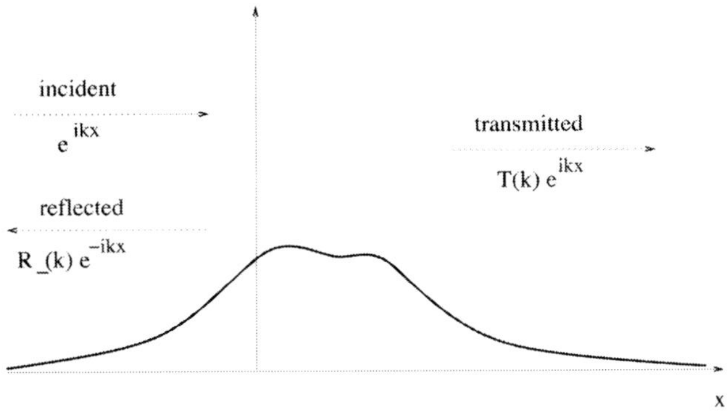

Figure 5.11. *Symbolic sketch of scattering.*

We are now in a position to show that the bound states of $\mathcal{L}(t)$ are invariant under the KdV flow (5.47): if the change of u in time is governed by (5.47), then at each instant t, the spectrum of $\mathcal{L}(t)$ is the same as initially (at time $t=0$).

We first show that the bound states and the corresponding eigenfunctions (suitably renormalized) present a C^1-dependence on the parameter t if u solves (5.47). For this, fix some t_0. We first claim that if the bounded open set $O \subset \mathbb{C}$ is such that $\overline{O}$ is a subset of the resolvent set $\rho(\mathcal{L}(t_0))$ of the operator $\mathcal{L}(t_0)$ on $L^2(\mathbb{R})$, then for $t \in \mathbb{R}$ with $|t-t_0|$ sufficiently small, $O \subset \rho(\mathcal{L}(t))$ and for any $z \in O$, the map $t \mapsto [z-\mathcal{L}(t)]^{-1}$ is C^1 from $\mathbb{R}$ to the space $\mathbb{B}$ of bounded linear operators on $L^2(\mathbb{R})$. Indeed, for $z \in O$ we have

$$z-\mathcal{L}(t) = z-\mathcal{L}(t_0)-[\mathcal{L}(t)-\mathcal{L}(t_0)] = \left(1-[\mathcal{L}(t)-\mathcal{L}(t_0)]\cdot[z-\mathcal{L}(t_0)]^{-1}\right)[z-\mathcal{L}(t_0)].$$

Choosing $\varepsilon>0$ so that

$$\|u(\cdot,t)-u(\cdot,t_0)\|_{L^\infty(\mathbb{R})} < \frac{1}{2\,\|[z-\mathcal{L}(t_0)]^{-1}\|_{\mathbb{B}}} \quad \text{for} \quad |t-t_0|<\varepsilon,$$

since $[\mathcal{L}(t)-\mathcal{L}(t_0)]$ is the multiplication operator by $[u(\cdot,t)-u(\cdot,t_0)]$, we infer that

$$\|[\mathcal{L}(t)-\mathcal{L}(t_0)]\cdot[z-\mathcal{L}(t_0)]^{-1}\|_{\mathbb{B}} < \frac{1}{2}.$$

This ensures the existence of

$$[z-\mathcal{L}(t)]^{-1} = [z-\mathcal{L}(t_0)]^{-1}\sum_{j=0}^{\infty}\left([\mathcal{L}(t)-\mathcal{L}(t_0)]\cdot[z-\mathcal{L}(t_0)]^{-1}\right)^j$$

with the right-hand side absolutely convergent in $\mathbb{B}$, proving the claim.

Let now $\lambda_0<0$ be a bound state of $\mathcal{L}(t_0)$. The previous considerations enable us (by choosing an open annulus) to find $\varepsilon>0$ and $r>0$ such that

$$C_r = \{z\in\mathbb{C}:\ |z-\lambda_0|=r\} \subset \rho(\mathcal{L}(t)) \quad \text{for all} \quad |t-t_0|<\varepsilon,$$

and such that λ_0 is the only bound state of $\mathcal{L}(t_0)$ enclosed by the contour C_r. We now define the Riesz projections

$$P(t) = \frac{1}{2\pi}\oint_{C_r}[z-\mathcal{L}(t)]^{-1}\,dz, \qquad |t-t_0|<\varepsilon.$$

It is known (see [140]) that $P(t)$ is the orthogonal projection onto the direct sum of the orthogonal spaces $ker(\mathcal{L}(t)-\xi_j)$, where ξ_j are the finitely many eigenvalues of $\mathcal{L}(t)$ in the open disk D_r enclosed by C_r. Since the difference of two orthogonal projections with finite-dimensional ranges clearly has norm at least one if the ranges do not have the same dimension, the fact that $\|P(t)-P(t_0)\|_{\mathbb{B}} \to 0$ for $t \to t_0$ ensures that for some $\varepsilon_0 \in (0,\varepsilon)$, all ranges of $P(t)$, with $|t-t_0|<\varepsilon_0$, are one-dimensional. This means that for $|t-t_0|<\varepsilon_0$ the operator $\mathcal{L}(t)$ has exactly one eigenvalue $\lambda(t)$ in D_r and $P(t)$ is the orthogonal projection onto $ker[\mathcal{L}(t)-\lambda(t)]$. For the eigenvalue $\lambda_0<0$ of $\mathcal{L}(t_0)$ we choose now an eigenfunction $w \in H^2(\mathbb{R}))$ with $\|w\|_{L^2(\mathbb{R})}=1$. $P(t)$ being the orthogonal projection onto the one-dimensional space $ker[\mathcal{L}(t)-\lambda(t)]$, we have

$$\mathcal{L}(t)\,P(t)\,w=\lambda(t)\,P(t)\,w \quad \text{for} \quad |t-t_0|<\varepsilon_0.$$

Since any $z \neq \lambda(t)$ in D_r belongs to $\rho(\mathcal{L}(t))$, we deduce that

$$[z-\mathcal{L}(t)]^{-1}\,P(t)\,w=\frac{1}{z-\lambda(t)}\,P(t)w, \qquad z \in D_r, \quad z \neq \lambda(t).$$

Taking the $L^2(\mathbb{R})$-inner product of the previous equation with w, we get

$$\int_{\mathbb{R}} w\,[z-\mathcal{L}(t)]^{-1}\,P(s)\,w\,ds=\frac{1}{z-\lambda(t)}\int_{\mathbb{R}} w\,[P(s)w]\,ds.$$

The fact that $t \mapsto [z-\mathcal{L}(t)]^{-1}$ is C^1, uniformly for $z \in C_r$, yields the C^1-dependence of $P(t)$ on t with $|t-t_0|<\varepsilon_0$. Thus both integral terms in the above equality present a C^1-dependence on t with $|t-t_0|<\varepsilon_0$. Since $P(t)w \to P(t_0)w=\lambda_0 w$ in $L^2(\mathbb{R})$ for $t \to t_0$, we have $\int_{\mathbb{R}} w\,[P(s)w]\,ds \to \lambda_0$ as $t \to t_0$, so that $\frac{1}{z-\lambda(t)}$ presents a C^1-dependence on t for $|t-t_0|$ small enough. This proves the C^1-dependence of the bound states on t, if the map $t \mapsto u(\cdot,t)$ is C^1 from $\mathbb{R}$ to $H^1(\mathbb{R})$. Notice that the previous considerations show that if $\lambda=k^2$ with $\Im m(k)>0$ is a bound state of $\mathcal{L}$, then we can normalize the corresponding eigenfunction ψ by imposing the asymptotic behavior $[\psi(x)-e^{ikx}] \to 0$ as $x \to \infty$; that is, $\psi=f_+(\cdot,k)$. Since $f_+(\cdot,k)$ depends analytically on k with $\Im m(k)>0$, and we elucidated the dependence of the bound states on t, we deduce that if u solves (5.47), then the bound states λ and the corresponding eigenfunctions ψ both present a C^1-dependence on t. Since $u(\cdot,t) \in X$ for all t if $u(\cdot,0) \in X$, and $\mathcal{L}\psi=\lambda\psi$, with $\lambda<0$ and $\psi \in H^2(\mathbb{R})$, we deduce first that $\psi \in H^3(\mathbb{R})$ and subsequently that $\psi \in H^n(\mathbb{R})$ for any integer $n \geq 2$. Moreover, using the series representation for $g_\pm$ and the explicit relations relating $f_\pm$ to $g_\pm$, one can easily check that $\psi_t \in H^2(\mathbb{R})$.

For the subsequent considerations, rather than setting the eigenfunction $\psi=f_+$, it is more convenient to normalize it by setting it equal to αf_+, with $\alpha>0$ chosen such that $\|\psi\|_{L^2(\mathbb{R})}=1$. This does not alter the fact that the map $t \mapsto \psi(\cdot,t)$ is C^1 from $\mathbb{R}$ to $H^5(\mathbb{R})$, with $\psi_t \in H^2(\mathbb{R})$. Notice that if we associate to the unbounded self-adjoint operator $\mathcal{L}(t)$, acting on $L^2(\mathbb{R})$ with domain $H^2(\mathbb{R})$, the skew-adjoint unbounded operator

$$\mathcal{B}(t)=-4\partial_x^3+6u\partial_x+3u_x,$$

with domain $H^3(\mathbb{R})$, then the commutator equation

$$\mathcal{L}_t=\mathcal{B}\mathcal{L}-\mathcal{L}\mathcal{B} \tag{5.87}$$

is the renormalized KdV equation (5.47). Differentiating the equation $\mathcal{L}\psi=\lambda\psi$ with respect to t, we get

$$\mathcal{L}_t\psi+\mathcal{L}\psi_t=\lambda_t\psi+\lambda\psi_t.$$

Using (5.87), we obtain

$$\lambda_t \psi = (\mathcal{L} - \lambda)(\psi_t - \mathcal{B}\psi).$$

Taking the $L^2(\mathbb{R})$-inner product of this relation with ψ and using the fact that $(\mathcal{L} - \lambda)$ is self-adjoint with $(\mathcal{L} - \lambda)\psi = 0$, we see that $\lambda_t = 0$. Thus

$$\mathcal{L}(\psi_t - \mathcal{B}\psi) = \lambda(\psi_t - \mathcal{B}\psi),$$

with $(\psi_t - \mathcal{B}\psi) \in H^2(\mathbb{R})$. Since the eigenspace of the eigenvalue λ is one-dimensional, we must have that $(\psi_t - \mathcal{B}\psi) = c\psi$ for some $c(t) \in \mathbb{R}$. Taking the $L^2(\mathbb{R})$-inner product with ψ, we get $c(t) = 0$ since $\mathcal{B}$ is skew-adjoint and $\|\psi\|_{L^2(\mathbb{R})} = 1$. In conclusion, we can associate to the KdV equation (5.47) the *Lax pair* $(\mathcal{L}, \mathcal{B})$, with the eigenvalues of $\mathcal{L}$ independent of time (so that $\mathcal{L}$ provides us with the associated *isospectral problem*) and the normalized eigenfunctions ψ evolving according to $\psi_t = \mathcal{B}\psi$.

The scattering data

The *scattering data* associated to the KdV equation (5.47) at time t are the reflection and transmission coefficients $T(k,t)$ and $R(k,t)$ defined by (5.80) for $k > 0$, together with the *norming constants* $c_j(t) > 0$ associated to the bound state λ_j (with $1 \le j \le n$ for some integer $n \ge 0$, unless there are no bound states), by means of the asymptotic behavior

$$\psi_j(x) \approx c_j e^{-\kappa_j x} \quad \text{for} \quad x \to \infty, \tag{5.88}$$

for the eigenfunction ψ_j corresponding to the eigenvalue $\lambda_j = -\kappa_j^2$ with $\kappa_j > 0$, normalized by setting $\|\psi_j\|_{L^2(\mathbb{R})} = 1$. Note that since $f_+(x, i\kappa_j)$ is real-valued with $f(x, i\kappa_j) \approx e^{-\kappa_j x}$ as $x \to \infty$, we have $\psi_j = c_j f_+(\cdot, i\kappa_j)$.

Before we show that the time evolution of the scattering data can be computed explicitly, we establish that at a fixed time t, all the zeros of $a(k)$ in the upper half-plane $\{\Im m(k) > 0\}$ are simple.

Let now $k = i\eta$ with $\eta > 0$ be a zero of $a(k)$. Differentiating (5.75) with respect to k, if we denote $\partial_k f = \dot{f}$, we get

$$\begin{aligned} \dot{a}(i\eta) &= -\frac{1}{2\eta}\Big[W\big(\dot{f}_-(x,k), f_+(x,k)\big) + W\big(f_-(x,k), \dot{f}_+(x,k)\big)\Big] \\ &= -\frac{1}{2\eta}\Big[\frac{1}{d} W\big(\dot{f}_-(x,k), f_-(x,k)\big) + d\, W\big(f_+(x,k), \dot{f}_+(x,k)\big)\Big], \end{aligned}$$

where $d \neq 0$ is such that $d\, f_+(x,k) = f_-(x,k)$ for all $x \in \mathbb{R}$. Differentiating now (5.49) with $f = f_+(x,k)$, we get

$$-\partial_x^2 \dot{f} + u \dot{f} = k^2 \dot{f} + 2kf, \qquad x \in \mathbb{R}.$$

Multiplying both sides by f and subtracting from this (5.49) multiplied by $\dot{f}$, we obtain

$$\partial_x[\dot{f}\,(\partial_x f) - f\,(\partial_x \dot{f})] = 2k\, f^2, \qquad x \in \mathbb{R}.$$

On the left side we have $\partial_x W(\dot{f}, f)$. Recalling that $k = i\eta$ with $\eta > 0$, differentiating (5.64) with respect to k, and using (5.54), (5.63), and the fact that $f_+(x,k) = e^{ikx} g_+(x,k)$, we see that $\lim_{x\to\infty} \dot{f}(x,k) = \lim_{x\to\infty} \dot{f}'(x,k) = 0$. Since $f(\cdot,k) \in H^2(\mathbb{R})$, we infer that $\lim_{x\to\infty} W\big(\dot{f}_+(x,k), f_+(x,k)\big) = 0$, so that integrating the previously displayed relation on $[x,\infty)$ leads to

$$W\big(\dot{f}_+(x,k), f_+(x,k)\big) = -2k \int_x^\infty f_+^2(y,k)\,dy.$$

Similarly,

$$W\big(\dot{f}_-(x,k),\, f_-(x,k)\big) = 2k \int_{-\infty}^{x} f_-^2(y,k)\,dy.$$

Substituting these values in the above formula for $\dot{a}(i\eta)$, we end up with

$$\dot{a}(i\eta) = -id \int_{\mathbb{R}} f_+^2(x,i\eta)\,dx \neq 0. \tag{5.89}$$

This shows that all zeros of $a(k)$ in the upper half-plane $\{\Im m(k) > 0\}$ are simple.

The evolution of the scattering data in time

Let ψ_j be the eigenfunction corresponding to the bound state $\lambda_j = -\kappa_j^2$ with $\kappa_j > 0$, normalized by (5.88). By evaluating $\partial_t \psi_j = \mathcal{B}\psi_j$ at leading order for $x \to \infty$, we deduce that

$$c_j'(t) = 4\kappa_j^3\, c_j(t) \tag{5.90}$$

since $u(x,t) = o(1)$, $u_x(x,t) = o(1)$, and $\psi_j(x,t) = c_j(t)e^{-\kappa_j x} + o(1)$ for $x \to \infty$, and suitable estimates allow us to differentiate these asymptotic relations with respect to t as well as with respect to x (the latter even several times). The evolution of the reflection and trasmission coefficients is only slightly more complicated. Let us fix an unbound state $\lambda > 0$. Differentiating (5.49), with $k > 0$ such that $k^2 = \lambda$ and $f = f_+$, with respect to t, we get $\mathcal{L}_t f + \mathcal{L} f_t = \lambda f_t$ since $\lambda_t = 0$. Using (5.87), we deduce that $(\mathcal{L} - \lambda)(f_t - \mathcal{B} f) = 0$. Consequently, $(f_t - \mathcal{B})$ is a solution to (5.49) and hence can be expressed as a linear combination of the two linearly independent solutions $f_+(x,k,t)$ and $f_-(x,k,t)$, with coefficients $q(k,t)$ and $p(k,t)$, respectively:

$$\begin{aligned} &\partial_t f_+(x,k,t) - \big(-4\partial_x^3 + 6u(x,t)\partial_x + 3u_x(x,t)\partial_x\big) f_+(x,k,t) \\ &= q(k,t) f_+(x,k,t) + p(k,t) f_-(x,k,t), \qquad x \in \mathbb{R},\ t \geq 0. \end{aligned}$$

For each fixed t the coefficients $q(k,t)$ and $p(k,t)$ can be evaluated by letting $x \to \infty$ and using (5.83). At leading order for $x \to \infty$, we obtain

$$\partial_t e^{ikx} + 4\partial_x^3 e^{ikx} = q(k,t)e^{ikx} + p(k,t)\left(\frac{1}{T(k,t)}e^{-ikx} + \frac{R_+(k,t)}{T(k,t)}e^{ikx}\right).$$

Comparing the coefficients of e^{ikx} and e^{-ikx}, we get $p(k,t) = 0$ and $q(k,t) = -4ik^3$. Thus, the time evolution of $f_+(x,k,t)$ is determined by

$$\partial_t f_+ - \big(-4\partial_x^3 + 6u\,\partial_x + 3u_x\,\partial_x\big) f_+ = -4ik^3 f_+, \qquad x \in \mathbb{R},\ t \geq 0.$$

For a fixed t we evaluate in the limit $x \to -\infty$, using (5.84), to obtain at leading order

$$\partial_t\left(\frac{1}{T(k,t)}e^{ikx} + \frac{R_-(k,t)}{T(k,t)}e^{-ikx}\right) = (-4\partial_x^3 - 4ik^3)\left(\frac{1}{T(k,t)}e^{ikx} + \frac{R_-(k,t)}{T(k,t)}e^{-ikx}\right).$$

Comparing the coefficients of e^{ikx} and e^{-ikx} on both sides, we obtain

$$\partial_t T(k,t) = 0, \quad \partial_t R_-(k,t) = -8ik^3 R_-(k,t). \tag{5.91}$$

Proceeding in a similar manner with the solution f_- instead of f_+, we obtain that $\partial_t R_+(k,t) = 8ik^3 R_+(k,t)$. This, together with (5.90) and (5.91), allows us to deduce that the transmission coefficient remains unchanged, while the reflection coefficients and the norming constants undergo a simple phase change as t progresses:

$$\begin{cases} T(k,t) = T(k,0) \quad \text{and} \quad R_\pm(k,t) = R(k,0)e^{\pm 8ik^3 t} \quad \text{for} \quad k > 0, \\ c_j(t) = c_j(0)e^{4k_j^3 t} \quad \text{for} \quad j = 1,\dots,n. \end{cases} \tag{5.92}$$

Direct scattering

The *scattering matrix* associated to (5.49) is given by

$$S(k)=\begin{pmatrix} T(k) & R_+(k) \\ R_-(k) & T(k) \end{pmatrix}, \qquad k\in\mathbb{R}\setminus\{0\}.$$

The *forward problem* of scattering theory is to determine the properties of $S(k)$ from those of the potential u. For example, in view of (5.81) and (5.82), $S(k)$ is unitary. Moreover, for potentials $u\in X$, the transmission and reflection coefficients $T(k)$, $R_\pm(k)$ admit a continuous extension to the entire $\mathbb{R}$. Indeed, from (5.50) we can express $[g_+(x,k)-1]$ in the form

$$\frac{e^{-2ikx}}{2ik}\int_{\mathbb{R}} e^{2iky}u(y)g_+(y,k)\,dy-\frac{1}{2ik}\int_{\mathbb{R}} u(y)g_+(y,k)\,dy-\int_{-\infty}^{x} D_k(y-x)u(y)g_+(y,k)\,dy,$$

so that by (5.65) and the fact that $|D_k(s)|\le s$ for $s\ge 0$, we get

$$\lim_{x\to-\infty}\Big[g_+(x,k)-1-\frac{e^{-2ikx}}{2ik}\int_{\mathbb{R}} e^{2iky}u(y)g_+(y,k)\,dy+\frac{1}{2ik}\int_{\mathbb{R}} u(y)g_+(y,k)\,dy\Big]=0.$$

On the other hand, by (5.84) and (5.67), we have

$$\Big[g_+(x,k)-\frac{1}{T(k)}-\frac{R_-(k)}{T(k)}e^{-2ikx}\Big]\to 0 \quad\text{as}\quad x\to-\infty.$$

The previous two relations yield the integral representations

$$\frac{1}{T(k)}=1-\frac{1}{2ik}\int_{\mathbb{R}} u(y)g_+(y,k)\,dy,\qquad \frac{R_-(k)}{T(k)}=\frac{1}{2ik}\int_{\mathbb{R}} e^{2iky}u(y)g_+(y,k)\,dy \tag{5.93}$$

for the scattering coefficients, and similarly

$$\frac{1}{T(k)}=1-\frac{1}{2ik}\int_{\mathbb{R}} u(y)g_-(y,k)\,dy,\qquad \frac{R_+(k)}{T(k)}=\frac{1}{2ik}\int_{\mathbb{R}} e^{-2iky}u(y)g_-(y,k)\,dy. \tag{5.94}$$

There are two possibilities:[127] either $\int_{\mathbb{R}} u(y)g_+(y,0)\,dy=0$ or $\neq 0$. In the former case we infer from (5.93) that

$$\frac{1}{T(k)}=1-\frac{1}{2ik}\int_{\mathbb{R}} u(y)[g_+(y,k)-g_+(y,0)]\,dy,$$

and (5.62) and (5.63) yield by the dominated convergence theorem that $\lim_{k\to 0}\frac{1}{T(k)}$ exists and is finite. Since $\frac{1}{|T(k)|}\ge 1$ along $\mathbb{R}\setminus\{0\}$, in accordance with (5.86), we deduce that $\lim_{k\to 0}\frac{1}{T(k)}\ge 1$. Consequently, $T(k)$ is continuous on $\mathbb{R}$ and $\inf_{k\in\mathbb{R}}|T(k)|\ge 1$, while $\inf_{\Im m(k)\ge 0}|T(k)|>0$ since by (5.93) and (5.54) we have

$$T(k)=1+O(1/k)\quad\text{as}\quad |k|\to\infty,\quad \Im m(k)\ge 0. \tag{5.95}$$

The continuity of $R_\pm(k)$ at $k=0$ follows now similarly from (5.93) and (5.94). On the other hand, if $\int_{\mathbb{R}} u(y)g_+(y,0)\,dy=\nu\neq 0$, then (5.93) yields

$$\begin{aligned}\frac{1}{T(k)}&=1-\frac{\nu}{2ik}-\int_{\mathbb{R}} u(y)\frac{g_+(y,k)-g_+(y,0)}{2ik}\,dy,\\ &=1-\frac{\nu}{2ik}-\frac{1}{2i}\int_{\mathbb{R}} u(y)\dot g_+(y,0)\,dy+O(k)\quad\text{as}\quad k\to 0,\ \Im m(k)\ge 0,\end{aligned}$$

[127] Both can occur; cf. the discussion in [94].

since in the considerations shortly after (5.63) we proved that

$$\left|\frac{g_+(y,k)-g_+(y,0)}{2ik}-\dot{g}_+(y,0)\right|\leq 8\,|k|\,C^2(C+1)^2(1+y^2),\qquad (y,k)\in\mathbb{R}\times\{\Im m(k)\geq 0\}.$$

Consequently, $\frac{k}{T(k)}=\frac{i\nu}{2}+O(k)$ and

$$T(k)=-\frac{2i}{\nu}\,k+O(k^2)\quad\text{as}\quad k\to 0,\quad \Im m(k)\geq 0. \tag{5.96}$$

For $k\in\mathbb{R}\setminus\{0\}$ we deduce from (5.93) that

$$\frac{1+R_-(k)}{T(k)}=1+\int_{\mathbb{R}}\frac{e^{2iky}-1}{2ik}\,u(y)\,g_+(y,k)\,dy=1+\int_{\mathbb{R}}y\,u(y)\,g_+(y,0)\,dy+O(k)$$

in view of (5.62) and the fact that by the mean-value theorem for the complex-valued function $s\mapsto e^{2iks}$ (cf. [96]),

$$\left|\frac{e^{2iky}-1}{2ik}-y\right|=\left|\int_0^y(e^{2iks}-1)\,ds\right|\leq 2|k|\int_0^{|y|}s\,ds=|k|\,y^2.$$

Taking into account (5.96), we deduce that $1+R_-(k)=O(k)$ as $k\to 0$ with $\Im m(k)\geq 0$. The case of $[1+R_+(k)]$ is similar, so that $S(k)$ has a continuous extension for $k\in\mathbb{R}$.

A more subtle property of $S(k)$ is that, apart from the bound states, one of the reflection coefficients $R_\pm(k)$ with $k>0$ contains all the information about $S(k)$: $S(k)$ can be constructed in terms of one of the reflection coefficients $R_\pm(k)$ with $k\in(0,\infty)$, and of the bound states $\lambda_j=(ik_j)^2$ with $k_j>0$ $(1\leq j\leq n)$. To see this, suppose the bound states $\lambda_j=(ik_j)^2$ with $k_j>0$ $(1\leq j\leq n)$ and $R_(k)$ for $k>0$ are given. Since $R_-(k)=-R_+(-k)\,\frac{T(k)}{T(-k)}$ by (5.82), in view of (5.81), it suffices to reconstruct $T(k)$. Since ik_j are the only zeros of $a(k)$ in $\{\Im m(k)\geq 0\}$, and we proved that they are simple, we deduce that the function

$$h(k)=T(k)\prod_{j=1}^{n}\frac{k-ik_j}{k+ik_j}$$

is analytic in $\{\Im m(k)>0\}$ and continuous in $\{\Im m(k)\geq 0\}$. The only possible zero of h in $\{\Im m(k)\geq 0\}$ is at $k=0$, in which case (5.96) yields

$$|h(k)|\geq A\,|k|\quad\text{for}\quad |k|\leq\delta,\ \Im m(k)\geq 0, \tag{5.97}$$

for some $A>0$ and $\delta>0$. Moreover, due to (5.95) and (5.86), by the maximum modulus principle (cf. [39]) we have that $|h(k)|\leq 1$ in $\{\Im m(k)\geq 0\}$. Consequently, $h\in H_+^\infty$, the Hardy space of functions that are analytic and bounded in $\{\Im m(k)>0\}$ (cf. [183], Chapter VI). We now show that h is actually an *outer function* (cf. [102, 183]) since

$$\log|h(x+iy)|=\frac{y}{\pi}\int_{\mathbb{R}}\frac{\log|h(s)|}{(s-x)^2+y^2}\,ds,\qquad x\in\mathbb{R},\quad y>0. \tag{5.98}$$

Let $\varepsilon>0$ and define $h_\varepsilon(k)=h(k+i\varepsilon)$ for $\Im m(k)\geq 0$, so that for $\varepsilon>0$ sufficiently small h_ε is analytic in $\{\Im m(k)>-\varepsilon\}$, continuous in $\{\Im m(k)\geq-\varepsilon\}$, and such that $\inf_{\Im m(k)\geq 0}\{|h_\varepsilon(k)|\}>0$. Since by (5.95) we have $h(k)=1+O(1/k)$ for $|k|\to\infty$ with $\Im m(k)\geq 0$, the Cauchy integral theorem yields

$$\log h_\varepsilon(k)=\frac{1}{2\pi i}\int_{\mathbb{R}}\frac{\log h_\varepsilon(s)}{s-k}\,ds,\quad 0=\frac{1}{2\pi i}\int_{\mathbb{R}}\frac{\log h_\varepsilon(s)}{s-\bar{k}}\,ds,\qquad \Im m(k)>0.$$

Therefore,

$$\log h_\varepsilon(k) = \frac{1}{2\pi i}\int_{\mathbb{R}} \frac{\log h_\varepsilon(s)}{s-k}\,ds + \frac{1}{2\pi i}\int_{\mathbb{R}} \frac{\overline{\log h_\varepsilon(s)}}{s-k}\,ds = \frac{1}{\pi i}\int_{\mathbb{R}} \frac{\log|h_\varepsilon(s)|}{s-k}\,ds.$$

Taking the real part we deduce the validity of (5.98) for h_ε:

$$\log|h_\varepsilon(k)| = \frac{y}{\pi}\int_{\mathbb{R}} \frac{\log|h_\varepsilon(s)|}{(s-x)^2+y^2}\,ds, \qquad k = x+iy \quad \text{with} \quad y>0.$$

Since h is continuous in $\{\Im m(k)\ge 0\}$, we have $|h_\varepsilon(k)| \to |h(k)|$ as $\varepsilon \downarrow 0$. On the other hand, since $1 \ge \sup_{\Im m(k)\ge 0}\{|h_\varepsilon(k)|\} \ge \inf_{\Im m(k)\ge 0,\,|k|\ge\delta/2}\{|h_\varepsilon(k)|\} = \mu > 0$, we have

$$\lim_{\varepsilon\downarrow 0}\int_{|s|\ge\delta/2} \frac{\log|h_\varepsilon(s)|}{(s-x)^2+y^2}\,ds = \int_{|s|\ge\delta/2} \frac{\log|h(s)|}{(s-x)^2+y^2}\,ds$$

whenever $y>0$. Since (5.97) and $|h(k)|\le 1$ in $\{\Im m(k)\ge 0\}$ yield $|\log|h_\varepsilon(s)|| \le -\log A - \log|s|$ for $\varepsilon \le \delta/2$ and for $s\in\mathbb{R}$ with $|s|\le\delta/2$, by the dominated convergence theorem we obtain

$$\lim_{\varepsilon\downarrow 0}\int_{|s|\le\delta/2} \frac{\log|h_\varepsilon(s)|}{(s-x)^2+y^2}\,ds = \int_{|s|\le\delta/2} \frac{\log|h(s)|}{(s-x)^2+y^2}\,ds$$

whenever $y>0$. The previous three displayed relations yield (5.98). If we now define

$$g(k) = \exp\left\{\frac{1}{\pi i}\int_{\mathbb{R}} \frac{\log|T(s)|}{s-k}\,ds\right\}, \qquad \Im m(k) > 0,$$

then $g(k)$ is analytic in $\{\Im m(k)>0\}$. Indeed, $\log|T(s)| = O(1/|s|)$ for $s\to\pm\infty$ in view of (5.95), while $s\mapsto T(s)$ is continuous on $\mathbb{R}$, having only $s=0$ as a possible zero, in which case (5.96) ensures that the function $s\mapsto \log|T(s)|$ is locally integrable near $s=0$. Since

$$\log|g(x+iy)| = \frac{y}{\pi}\int_{\mathbb{R}} \frac{\log|T(s)|}{(s-x)^2+y^2}\,ds, \qquad x\in\mathbb{R}, \quad y>0,$$

from (5.98) we deduce that $|g(k)| = |h(k)|$ for $\Im m(k)>0$. Since $k\mapsto \frac{g(k)}{h(k)}$ is an analytic function of constant modulus in $\{\Im m(k)>0\}$, there exists a constant $z_0\in\mathbb{C}$ such that $h(k) = z_0\,g(k)$ for all k with $\Im m(k)>0$. Consequently,

$$T(k) = z_0\exp\left\{\frac{1}{2\pi i}\int_{\mathbb{R}} \frac{\log[1-|R_+(s)|^2]}{s-k}\,ds\right\}\prod_{j=1}^{n}\frac{k+ik_j}{k-ik_j}, \qquad \Im m(k)>0,$$

if we recall (5.86) and the definition of h. In the limit $y\to\infty$ with $k=iy$ we have by (5.95) that $T(k)\to 1$, which shows that $z_0=1$. Thus

$$T(k) = \exp\left\{\frac{1}{2\pi i}\int_{\mathbb{R}} \frac{\log[1-|R_+(s)|^2]}{s-k}\,ds\right\}\prod_{j=1}^{n}\frac{k+ik_j}{k-ik_j}, \qquad \Im m(k)>0.$$

We know that except for the simple poles at $k=ik_j$, $T(k)$ is continuous in $\{\Im m(k)\ge 0\}$. The value $T(s)$ of the transmission coefficient (with $s\in\mathbb{R}$) is therefore obtained from the previously displayed relation as $\lim_{\varepsilon\downarrow 0} T(s+i\varepsilon)$. As pointed out in the beginning of this discussion, knowing $R_+(k)$ and $T(k)$ for $k\in\mathbb{R}$, we reconstruct the entire scattering matrix $S(k)$. Similar considerations show that $S(k)$ can be constructed in terms of the bound states and the reflection coefficient $R_-(k)$ with $k>0$.

Inverse scattering

Of particular interest is how to determine the potential u in (5.49) from the scattering data. In the absence of bound states, either one of the reflection coefficients $R_\pm$ uniquely determines the potential. When there are bound states, in addition to them, one must also specify the corresponding norming constants c_j.

The iterative procedure for the construction of the solution to (5.50) yields

$$g_+(x,k) - 1 + \frac{1}{2ik}\int_x^\infty u(y)\,dy = O\Big(\frac{1}{|k|^2}\Big), \qquad \Im m(k) > 0. \tag{5.99}$$

Consequently, for each $x \in \mathbb{R}$, $k \mapsto [g_+(x,k) - 1]$ and $k \mapsto -ik[g_+(x,k)-1] + \frac{1}{2}\int_x^\infty u(y)\,dy$ belong to the Hardy class H_+^2. According to the Paley–Wiener theorem (see [183, Chapter VI]), the Fourier transform of the restrictions[128] of these functions to $\mathbb{R}$ are supported on the positive axis. More precisely, if we define

$$G_N(x,y) = \frac{1}{2\pi}\int_{-N}^{N} [g_+(x,k)-1]e^{-iky}\,dk,$$

then, as $N \to \infty$, the functions $G_N(x,\cdot)$ converge in $L^2(\mathbb{R})$ to the Fourier transform $G(x,\cdot) \in L^2(\mathbb{R})$, and $G(x,y) = 0$ a.e. for $y \le 0$. Since by (5.68) the function $s \mapsto [f_+(x,s) - e^{isx}]$ is skew-symmetric, its Fourier transform is real and supported on $[x,\infty)$: $G(x,y-x) \in \mathbb{R}$ and

$$f_+(x,k) = e^{ikx} + \int_x^\infty G(x,y-x)e^{iky}\,dy, \qquad k \in \mathbb{R}, \tag{5.100}$$

(see [118]). Interpreting $G_y(x,\cdot)$ in the sense of distributions, and denoting by δ the Dirac unit measure concentrated at $\{0\} \in \mathbb{R}$ and by H the Heaviside step function (that is, the characteristic function of $[0,\infty) \subset \mathbb{R}$), the Fourier transform of the continuous function $s \mapsto -is[g_+(x,s)-1] - \frac{1}{2}\int_x^\infty u(\tau)\,d\tau$, regarded as a tempered distribution on $\mathbb{R}$, is

$$G_y(x,\cdot) - \Big(\frac{1}{2}\int_x^\infty u(\tau)\,d\tau\Big)\delta = G_y(x,\cdot) - \Big(\frac{1}{2}\int_x^\infty u(y)\,dy\Big)H_y,$$

in the class of tempered distributions (cf. [118, 290]). Since the above expression must be an $L^2(\mathbb{R})$-function supported on $[0,\infty)$, we deduce the existence of a continuous function $h(x,\cdot) : [0,\infty) \to \mathbb{R}$, such that

$$G(x,y) = \begin{cases} \dfrac{1}{2}\displaystyle\int_x^\infty u(\tau)\,d\tau + \int_0^y h(x,s)\,ds, & y \ge 0, \\ 0, \quad y < 0. \end{cases}$$

Consequently, $y \mapsto G(x,y)$ is not continuous on $\mathbb{R}$, but it presents a jump discontinuity at $y = 0$. On the other hand, the smooth dependence of $g_+(x,k)$ on x ensures a smooth dependence of $G(x,y)$ on x for each fixed $y \in \mathbb{R}$. Differentiating the above formula evaluated at $y = 0$, we get

$$u(x) = -2\,\partial_x\big(\lim_{y\downarrow 0}\{G(x,y)\}\big), \qquad x \in \mathbb{R}. \tag{5.101}$$

Let us now show how to reconstruct the potential u from the reflection coefficient $R_+(k)$ in the absence of bound states. First, since (5.95) yields $1 - |T(s)|^2 = O\big(\frac{1}{s^2}\big)$ for $|s| \to \infty$, from (5.86) we deduce that

$$|R_+(s)|^2 = O\Big(\frac{1}{s^2}\Big) \quad \text{for} \quad |s| \to \infty. \tag{5.102}$$

[128]Both functions are continuous in $\{\Im m(k) \ge 0\}$, so that instead of the usual nontangential limit in the setting of Hardy spaces (cf. [183]), we can simply consider the classical boundary value.

Consequently $R_+ \in L^2(\mathbb{R})$. In view of (5.68), its Fourier transform $r \in L^2(\mathbb{R})$, given by

$$r(y) = \lim_{N\to\infty} \frac{1}{2\pi} \int_{-N}^{N} R_+(k) e^{-iky}\, dk, \qquad y \in \mathbb{R},$$

is real. We claim that $r \in L^1(\mathbb{R}) \cap L^\infty(\mathbb{R})$, with

$$\int_{-\infty}^{0} \int_{-\infty}^{0} |r(y+s)|^2\, dy\, dt < \infty. \tag{5.103}$$

Indeed, since $T(k) = \frac{1}{a(k)}$, from (5.75) and (5.94) we express $R_+(k)$ as

$$\frac{T(k)}{2ik} \int_{\mathbb{R}} e^{-2iky} u(y) g_-(y,k)\, dy = \frac{1}{W(f_-(x,k),\, f_+(x,k))} \int_{\mathbb{R}} e^{-2iky} u(y) g_-(y,k)\, dy.$$

Using (5.54) in combination with the inequality for $\dot{g}_-(x,k)$ that corresponds to (5.63), we obtain $R_+(k) \in L^1(\mathbb{R})$ and $\partial_k R_+(k) \in L^1(\mathbb{R}) \cap L^2(\mathbb{R})$. Consequently, $y\, r(y) \in L^2(\mathbb{R}) \cap L^\infty(\mathbb{R})$ (cf. [284, Chapter I]), so that $r \in L^1(\mathbb{R}) \cap L^\infty(\mathbb{R})$. Moreover, for $y < 0$ we have

$$\begin{aligned} |r(y)|^2 &= \left(\int_{\mathbb{R}} R_+(k) e^{iky}\, dk \right) \left(\int_{\mathbb{R}} \overline{R_+(s)} e^{-isy}\, ds \right) \\ &= \left(\int_{\mathbb{R}} R_+(k+s) e^{i(k+s)y}\, dk \right) \left(\int_{\mathbb{R}} R_+(-s) e^{-isy}\, ds \right) \\ &= \int_{\mathbb{R}} \int_{\mathbb{R}} R_+(k+s) R_+(-s) e^{iky}\, ds dk = \int_{R} \left(\int_{\mathbb{R}} R_+(k+s) R_+(-s)\, ds \right) e^{iky}\, dk. \end{aligned}$$

Since $\partial_k R_+(k) \in L^1(\mathbb{R}) \cap L^2(\mathbb{R})$, we have

$$\partial_k^2 \int_{\mathbb{R}} R_+(k+s) R_+(-s)\, ds = \int_{\mathbb{R}} R_+'(k+s) R_+'(-s)\, ds \in L^2(\mathbb{R})$$

by Young's inequality (see [284, Chapter I]) and the properties of the convolution. Invoking the properties of the Fourier transform (cf. [300, Chapter X]), we deduce that

$$|r(y)|^2 = \frac{1}{y^2} \int_{\mathbb{R}} \alpha(k) e^{iky}\, dk, \qquad y < 0,$$

for some $\alpha \in L^2(\mathbb{R})$. By the Cauchy–Schwarz inequality and the fact that the Fourier transform is bounded on $L^2(\mathbb{R})$, we get

$$\int_{-\infty}^{0} |r(y+s)|^2\, ds \leq M \left(\int_{-\infty}^{0} \frac{1}{|y+s|^4}\, ds \right)^{1/2} = \frac{M}{\sqrt{3}\, |y|^{3/2}}, \qquad y < 0,$$

for some constant $M > 0$ independent of y. Since $r \in L^\infty(\mathbb{R})$, this proves (5.103). From (5.100) and

$$R_+(k) = \int_{\mathbb{R}} r(y) e^{iky}\, dy, \qquad y \in \mathbb{R},$$

we see that (5.83) has the form

$$\begin{aligned} \left(e^{-ikx} + \int_x^\infty G(x, y-x) e^{-iky}\, dy \right) + \left(\int_{\mathbb{R}} r(y) e^{iky}\, dy \right) \left(e^{ikx} + \int_x^\infty G(x, s-x) e^{iks}\, ds \right) \\ = T(k) e^{-ikx} g_-(x,k). \end{aligned}$$

We rewrite this equation as

$$\int_{-\infty}^{-x} G(x,-x-y)e^{iky}\,dy + \int_{\mathbb{R}} r(y-x)e^{iky}\,dy + \int_{\mathbb{R}} \left(\int_{\mathbb{R}} r(y-s)G(x,s-x)\,ds \right) e^{iky}\,dy$$

$$= [T(k)-1]e^{-ikx} + T(k)[g_-(x,k)-1]e^{-ikx}. \tag{5.104}$$

We claim that the Fourier transform of the right side is supported on $[-x,\infty)$. Indeed, we saw earlier that $[T(k)-1] \in H_+^2$ so that by the Paley–Wiener theorem (cf. [183, Chapter VI]), the Fourier transform of its restriction to $\mathbb{R}$ is a square integrable function supported on $[0,\infty)$. On the other hand, we also know that $[g_-(x,k)-1] \in H_+^2$. Since $|T^2(k)| \le 1$ for $k \in \mathbb{R}$, in view of (5.86), we deduce that $T(k)[g_-(x,k)-1] \in H_+^2$, so the Fourier transform of its restriction to $\mathbb{R}$ is supported on $[0,\infty)$. Multiplication by e^{-ikx} shifts the support of the Fourier transform to $[-x,\infty)$. Consequently, the Fourier transform of the left side of (5.104) is a function in $L^2(\mathbb{R})$ that is zero on $(-\infty,-x)$. Therefore,[129]

$$G(x,-x-y) + r(y-x) + \int_x^{\infty} r(y-s)G(x,s-x)\,ds = 0, \qquad y < -x.$$

If we replace y by $-y$, and denote $P(s) = r(-s)$, we obtain

$$G(x,y-x) + P(x+y) + \int_x^{\infty} P(y+s)G(x,s-x)\,ds = 0, \qquad x < y. \tag{5.105}$$

This equation is the celebrated *Gelfand–Levitan–Marchenko* equation: an *inhomogeneous Fredholm integral equation* for $G(x,\cdot) \in L^2(\mathbb{R})$, with parameter x. It is remarkable that this reconstruction process of u by means of (5.101) and (5.105) is linear.

Let us now discuss the case when $n \ge 1$ bound states $\lambda_j = (i\eta_j)^2 < 0$ with $\eta_j > 0$, for $j = 1,\dots,n$, are present. In this case $T(k)$ has simple poles at $k = i\eta_j$ with $j = 1,\dots,n$. As before, we obtain (5.104). To compute the Fourier transform

$$\frac{1}{2\pi} \int_{\mathbb{R}} [T(k)\,f_-(x,k) - e^{-ikx}]e^{-iky}\,dy$$

of the right side of (5.104), we shift the path of integration from $k \in \mathbb{R}$ to the horizontal line $\Im m(k) = K \to \infty$. The residue of the integrand at $k = i\eta$, where $\eta \in \{\eta_1,\dots,\eta_n\}$, is $\frac{1}{\dot a(i\eta)} f_-(x,i\eta)e^{\eta y}$, with $\dot a(i\eta)$ provided by (5.89) and with $d\,f_+(x,i\eta) = f_-(x,i\eta)$. Consequently, the above Fourier transform equals

$$B(x,y+x) - \sum_{j=1}^{n} \frac{f_+(x,i\eta_j)}{\int_{\mathbb{R}} f_+^2(s,i\eta_j)\,ds} e^{\eta_j y},$$

where for each fixed $x \in \mathbb{R}$ the function $B(x,\cdot) \in L^2(\mathbb{R})$ is zero on $(-\infty,0)$. Since

$$c_j = \left(\int_{\mathbb{R}} f_+^2(s,i\eta_j)\,ds \right)^{-1/2}, \qquad j = 1,\dots,n, \tag{5.106}$$

by (5.88), substituting the Fourier representation (5.100), we obtain instead of (5.105) an equation of the same form, except that P is now redefined as

$$P(s) = r(-s) + \sum_{j=1}^{n} c_j^2 e^{-\eta_j s}. \tag{5.107}$$

This is the complete form of the *Gelfand–Levitan–Marchenko* equation.

[129] By Young's inequality (cf. [284, Chapter I]), for each $x \in \mathbb{R}$, the convolution $r * G(x,\cdot) \in L^2(\mathbb{R})$ since $r \in L^1(\mathbb{R})$ and $G(x,\cdot) \in L^2(\mathbb{R})$.

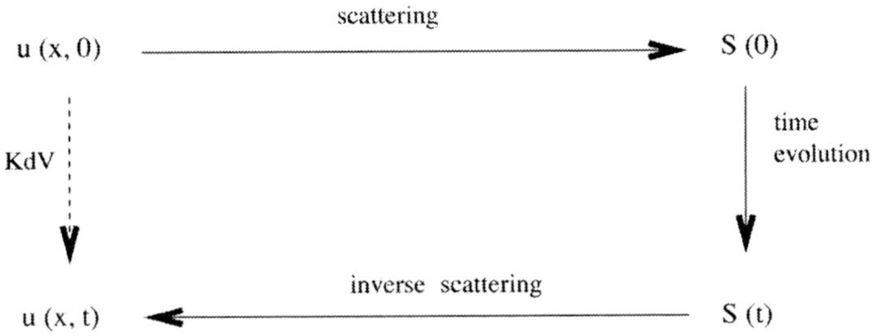

Figure 5.12. *Representation of the inverse scattering transform for KdV. Note the parallel to the scheme using Fourier transforms to solve the linearized KdV equation* $u_t + u_{xxx} = 0$ *with initial data* $u(x,0) = u_0(x) \in X$*, in which case the solution is given explicitly by* $u(x,t) = \int_{\mathbb{R}} \hat{u}_0(k) e^{ik(x+k^2 t)}\, dk$*, with* $\hat{u}_0$ *being the Fourier transform of* u_0*.*

Setting $w(s) = G(x,s)$ and $p(s) = P(2x+s)$, the solvability of (5.105), with P defined by (5.107), amounts to solving for $w \in L^2(0,\infty)$ the integral equation

$$w(s) + p(s) + \int_0^\infty p(s+\tau)\, w(\tau)\, d\tau = 0, \quad s > 0.$$

We refrain from discussing the solvability issue in its full generality.[130] Instead, we investigate two important particular cases:

(1) the case when bound states are absent;

(2) the case of *reflectionless potentials*, that is, potentials u for which $R_+ \equiv 0$ on $\mathbb{R}$. It follows from (5.86) that for such potentials we also have $R_- \equiv 0$ on $\mathbb{R}$, while (5.92) shows that this class of potentials is preserved by the KdV flow.

In the absence of bound states, since $r \in L^2(\mathbb{R})$ and (5.103) holds, we have

$$\int_0^\infty \int_0^\infty |p(s+\tau)|^2\, ds d\tau < \infty$$

so that (cf. [262, Section VI.6]) the operator

$$(Aw)(s) = \int_0^\infty p(s+\tau) w(\tau)\, d\tau, \quad s > 0,$$

acting on the real Hilbert space $L^2(0,\infty)$, is a self-adjoint Hilbert–Schmidt operator. Now

$$(w,\, Aw)_{L^2(0,\infty)} = 2\pi \int_{\mathbb{R}} [\hat{w}(-k)]^2 R_+(k) e^{-2ikx}\, dk, \qquad w \in L^2(0,\infty),$$

where we denoted by $\hat{w} \in L^2(\mathbb{R})$ the Fourier transform of the extension of w to $\mathbb{R}$, obtained by setting $w(s) = 0$ for $s \le 0$. Therefore,

$$|(w,\, Aw)_{L^2(0,\infty)}| \le \|w\|^2_{L^2(0,\infty)},$$

[130] For this we refer the reader to [94].

so that $\|A\|_{L^2(0,\infty)} \le 1$. Notice that the existence of $w \in L^2(0,\infty)$ with $Aw = \pm w$ yields

$$(w, Aw)_{L^2(0,\infty)} = \pm \int_{\mathbb{R}} w^2(s)\,ds = \pm 2\pi \int_{\mathbb{R}} |\hat{w}(k)|^2\,dk.$$

But since (5.86) and the fact that $|T(k)| > 0$ for $k \in \mathbb{R}\setminus\{0\}$ ensure that $|R_+(k)| < 1$ for $k \in \mathbb{R}\setminus\{0\}$, the previous two displayed relations are possible only if $\hat{w}(k) = 0$ a.e. on $\mathbb{R}$, in which case $w \equiv 0$. Consequently, ± 1 are not eigenvalues of the compact self-adjoint operator A, and for this reason $\|A\|_{L^2(0,\infty)} < 1$. This ensures that the Neumann series $(1+A)^{-1} = \sum_{n=0}^{\infty}(-A)^n$ converges in the space of bounded linear operators on $L^2(0,\infty)$. The solution we seek is $w = -(1+A)^{-1}p \in L^2(0,\infty)$.

The most interesting setting is that of reflectionless potentials, when in (5.107) we have only a finite linear combination of exponentials in y, and therefore (5.105) shows that so is $G(x, y-x)$. To maintain symmetry, we set

$$G(x,y) = \sum_{j=1}^{n} c_j\, g_j(x)\, e^{-\eta_j(y+x)}, \quad y > 0. \tag{5.108}$$

Substituting this into (5.105) and carrying out the integration, we obtain

$$\sum_{j=1}^{n} c_j\, g_j(x) e^{-\eta_j y} + \sum_{j=1}^{n} c_j^2 e^{-\eta_j(y+x)} + \sum_{j=1}^{n}\sum_{l=1}^{n} \frac{c_j{}^2 c_l}{\eta_j+\eta_l}\, g_l(x) e^{-\eta_j y-(\eta_j+\eta_l)x} = 0, \quad 0 < y.$$

Equating to zero the exponential terms $e^{-\eta_j y}$ that occur shows, after division by $c_j > 0$, that (5.105) can be reduced in this case to the finite system of linear equations ($1 \le j \le n$)

$$g_j(x) + c_j e^{-\eta_j x} + \sum_{l=1}^{n} \frac{c_j c_l}{\eta_j+\eta_l}\, g_l(x) e^{-(\eta_j+\eta_l)x} = 0. \tag{5.109}$$

The solution can be expressed by Cramer's rule in the form

$$g_j(x) = -\sum_{l=1}^{n} \frac{D_{jl}(x)}{D(x)}\, c_l\, e^{-\eta_l x}, \qquad j = 1,\dots,n,$$

where $D(x)$ and $D_{jl}(x)$ are the determinant, respectively, the signed minors, of order $(n-1)$ of the $n \times n$ matrix with entries $(\delta_{jl} + \frac{c_j c_l}{\eta_j+\eta_l}\, e^{-(\eta_j+\eta_l)x})$, with $\delta_{jl} = 1$ if $j = l$ and $\delta_{jl} = 0$ if $j \ne l$. Setting this into (5.108), we get

$$\lim_{y \downarrow 0} \{G(x,y)\} = -\frac{1}{D(x)} \sum_{j=1}^{n}\sum_{l=1}^{n} D_{jl}(x)\, c_j c_l\, e^{-(\eta_j+\eta_l)x}.$$

By differentiating $D(x)$ and expanding by columns, a lengthy but straightforward calculation shows that the expression on the right side is precisely $\frac{d}{dx}\log D(x)$. In view of (5.101), we obtain the following remarkable formula for the potential:[131]

$$u(x) = -2\,\frac{d^2}{dx^2}\log D(x). \tag{5.110}$$

For KdV a reflectionless potential describes the n-soliton interaction (see [98]). We exemplify this by means of the solitary wave solution and the two-soliton solution.

[131] It turns out that (5.110) also holds for potentials that are not reflectionless, but in this case $D(x)$ has to be interpreted (see [196, 223]) as the Fredholm determinant of the Fredholm operator on the left of (5.105).

Example 5.1 (solitary wave solution). For the initial profile $u_0(x) = -\frac{2}{\cosh^2(x)}$, the isospectral problem (5.48) at time $t = 0$ becomes

$$f_{xx} + \left(\lambda + \frac{2}{\cosh^2(x)}\right) f = 0, \qquad x \in \mathbb{R}.$$

The substitution $s = \tanh(x)$ transforms the above equation into the Legendre equation

$$\frac{d}{ds}\left((1-s^2)\frac{df}{ds}\right) + \left(2 + \frac{\lambda}{1-s^2}\right) f = 0, \qquad s \in (-1,1). \tag{5.111}$$

The general solution of (5.111) can be expressed in terms of the hypergeometric function (see [98]). This confirms that for $\lambda < 0$, the only bounded solution occurs when $\lambda = -1$, in which case the eigenfunction for (5.111) is a Legendre polynomial, corresponding to the eigenfunction

$$f_+(x,i) = \frac{1}{2\cosh(x)}.$$

On the other hand, for $\lambda = k^2$ with $k > 0$, we get (cf. [98, Chapter 3])

$$f_-(x,k) = [2\cosh(x)]^{ik} F\left(2-ik, -1-ik; 1-ik, \frac{1}{2} + \frac{1}{2}\tanh(x)\right), \qquad x \in \mathbb{R},$$

where F is the hypergeometric function. This shows that for $k > 0$ we have

$$f_-(x,k) \approx \frac{1+ik}{ik-1} e^{-ikx} \quad \text{as} \quad x \to \infty.$$

From (5.83) we get $R_+(k) = 0$ for $k > 0$, so that the potential is reflectionless. Since $\int_{\mathbb{R}} \frac{dy}{\cosh^2(y)} = \frac{1}{2}$, (5.88) provides us with the value $c_1 = \sqrt{2}$ of the norming constant at time $t = 0$. By (5.92), we have

$$c_1(t) = \sqrt{2}e^{4t}, \qquad t \geq 0.$$

Equation (5.109) becomes, at time t,

$$g_1(x,t) + \sqrt{2}e^{4t} e^{-x} + e^{8t} g_1(x) e^{-2x} = 0,$$

so that

$$g_1(x,t) = -\frac{\sqrt{2}e^{4t-x}}{1+e^{8t-2x}}.$$

From (5.108) we get

$$\lim_{y \downarrow 0}\{G(x,y,t)\} = -\frac{2e^{8t-4x}}{1+e^{8t-2x}},$$

so that (5.101) yields

$$u(x,t) = -\frac{2}{\cosh^2(x-4t)}, \qquad x \in \mathbb{R},\ t \geq 0.$$

By means of $\eta(x,t) = -6u(x,t)$, we recover the solitary wave solution (5.4) to (5.3), of amplitude 12 and speed of propagation 4, considered in Section 5.1. ∎

Example 5.2 (two-soliton solution). For the initial profile $u_0(x) = -\frac{6}{\cosh^2(x)}$, the problem (5.48), at time $t=0$, becomes

$$f_{xx} + \left(\lambda + \frac{6}{\cosh^2(x)}\right) f = 0, \qquad x \in \mathbb{R}.$$

By means of the substitution $s = \tanh(x)$ we again transform this into a Legendre equation on $(-1,1)$, whose solutions are explicit in terms of the hypergeometric function (cf. [98, Chapter 3]). We find two eigenvalues, $\lambda_1 = -1$ and $\lambda_2 = -4$, with the eigenfunctions

$$f_+(x,i) = \frac{\sinh(x)}{2\cosh^2(x)}, \qquad f_+(x,2i) = \frac{1}{4\cosh^2(x)},$$

respectively. Since

$$\int_{\mathbb{R}} f_+^2(x,i)\,dx = \frac{1}{6}, \qquad \int_{\mathbb{R}} f_+^2(x,2i)\,dx = \frac{1}{12},$$

we find the values $c_1(0) = \sqrt{6}$ and $c_2(0) = 2\sqrt{3}$ of the norming constants at time $t=0$, and (5.92) yields

$$c_1(t) = \sqrt{6}\,e^{4t}, \qquad c_2(t) = 2\sqrt{3}\,e^{32t}, \qquad t \ge 0.$$

On the other hand, for $\lambda = k^2$ with $k>0$, we get (cf. [98, Chapter 3])

$$f_-(x,k) = [2\cosh(x)]^{ik}\, F\left(3-ik, -2-ik; 1-ik, \frac{1}{2} + \frac{1}{2}\tanh(x)\right), \qquad x \in \mathbb{R},$$

where F is the hypergeometric function. This shows that for $k>0$ we have

$$f_-(x,k) \approx \frac{(ik-2)(ik-1)}{(-ik-2)(-ik-1)}\, e^{-ikx} \quad \text{as} \quad x \to \infty.$$

From (5.83) we get $R_+(k) = 0$ for $k>0$, so that the potential is reflectionless. The system (5.109) becomes, at time t,

$$\begin{cases} g_1(x,t) + \sqrt{6}e^{4t}\, e^{-x} + 3e^{8t} g_1(x,t)e^{-2x} + 2\sqrt{2}e^{36t} g_2(x,t)e^{-3x} = 0, \\ g_2(x,t) + 2\sqrt{3}e^{32t}\, e^{-2x} + 2\sqrt{2}e^{36t} g_1(x,t)e^{-3x} + 3e^{64t} g_2(x,t)e^{-4x} = 0. \end{cases}$$

We compute $D(x,t) = 1 + 3e^{64t-4x} + 3e^{8t-2x} + e^{72t-6x}$, and find the solutions

$$g_1(x,t) = -\sqrt{6}\,\frac{e^{4t-x} - e^{68t-5x}}{D(x,t)}, \qquad g_2(x,t) = -2\sqrt{3}\,\frac{e^{32t-2x} + e^{40t-4x}}{D(x,t)},$$

so that by (5.108) we get

$$\lim_{y \downarrow 0} \{G(x,y,t)\} = -\frac{6e^{8t-4x} + 12e^{64t-4x} + 6e^{72t-6x}}{1 + 3e^{64t-4x} + 3e^{8t-2x} + e^{72t-6x}}.$$

Now (5.101) yields

$$u(x,t) = -12\,\frac{3 + 4\cosh(2x-8t) + \cosh(4x-64t)}{[3\cosh(x-28t) + \cosh(3x-36t)]^2}, \qquad x \in \mathbb{R}.$$

As $\eta(x,t) = -6u(x,t)$, we recover the two-soliton solution discussed in Section 5.3. ∎

Similarly to these two examples, for an integer $n \geq 3$, an n-soliton solution will develop from the initial profile $u_0(x) = -\frac{n(n+1)}{\cosh^2(x)}$ for $x \in \mathbb{R}$ (cf. [98, Chapter 4]).

Numerical methods

Initial profiles of the form $u_0(x) = -\frac{U}{\cosh^2(x)}$, for a constant $U \in \mathbb{R}$ that is not of the form $n(n+1)$ for an integer $n \geq 0$, will generally give rise to nonzero reflection coefficients. Unfortunately, solving (5.105) explicitly (that is, in a closed form that is accessible) when $R_+ \not\equiv 0$ appears not to be within reach. A possible way out is to use numerical simulations. For example, in a finite-difference scheme derivatives are approximated by some differences to give a difference equation instead of a differential equation to be solved. More precisely, the solution $\eta(x,t)$ is specified numerically as $f(q,p)$ at the discrete points $x = qh$ and $t = pl$ for $q = 1,\ldots,n$ and $p = 0,\ldots,m$, where $h = 2\pi/n$, say, is the space step and $l > 0$ the time step (for computation, we take finite boundaries, albeit distant ones), and then we use, for example, the leap-frog scheme

$$f(q,p+1) = f(q,p) + 2l\,N_h(q,p)$$

to compute the solution at the $(p+1)$th time step explicitly in terms of the solution at time p. Here we expressed (5.3) in the form $\eta_t = N(\eta)$ with $N(\eta) = -\eta\eta_x - \eta_{xxx}$, and we let $N_h(q,p)$ be an approximation to $N(\eta)$ at $x = qh$, $t = pl$. For example, Kruskal and Zabusky [189] chose the scheme

$$\begin{aligned}
\eta_t(ah,pl) &\approx \frac{f(q,p+1) - f(q,p-1)}{2l},\\
\eta_x(qh,pl) &\approx \frac{f(q+1,p) - f(q-1,p)}{2h},\\
\eta(qh,pl) &\approx \frac{f(q+1,p) + f(q,p) + f(q-1,p)}{3},\\
\eta_{xxx}(qh,pl) &\approx \frac{f(q+2,p) - 2f(q+1,p) + 2f(q-1,p) - f(q-2,p)}{6h^3}.
\end{aligned}$$

This scheme preserves in some sense certain features of (5.3). For example, the conservation of mass, expressed by the fact that $\int_{\mathbb{R}} \eta(x,t)\,dx$ is a constant of motion (i.e., time independent along solutions) is replicated by the fact that $\sum_{q=0}^{n-1} f(q,p)$ is independent of p. Also, the choice of the average

$$\frac{f(q+1,p) + f(q,p) + f(q-1,p)}{3}$$

to approximate η in the nonlinear term $\eta\eta_x$ was made to conserve the energy to second order: instead of the integral of motion $\frac{1}{2}\int_{\mathbb{R}} [\eta(x,t)]^2\,dx$ of (5.3), we have

$$\frac{1}{2}\sum_{q=0}^{n-1} f^2(q,p+1) - \frac{1}{2}\sum_{q=0}^{n-1} f^2(q,p-1) = O(l^3) \quad \text{as} \quad l \to 0.$$

Another standard class of numerical methods is that of spectral methods whereby the solution is approximated by some finite linear combination of a suitable set of functions (for these matters we refer the reader to the discussion in [98, Chapter 7]). The numerical simulations indicate that from an initial profile that is not reflectionless emerge some solitary waves that interact like solitons (the number being equal to the number of eigenvalues at time $t = 0$) and propagate to the right, followed by an oscillatory tail which propagates to the left with a decaying amplitude (as depicted in Figure 5.13); we refer the reader to [98] for examples.

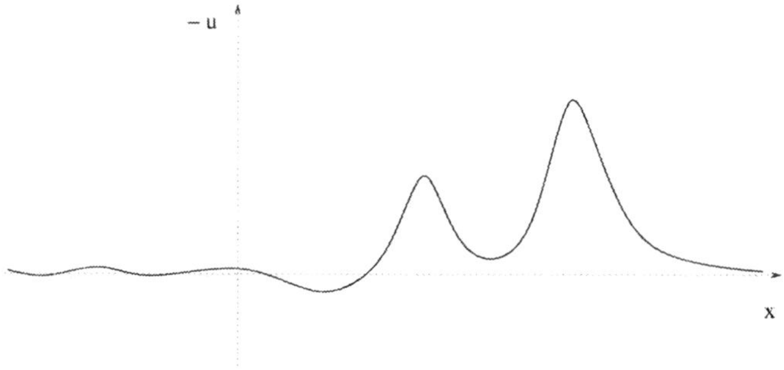

Figure 5.13. *The initial data* $u(x,0) = -4\cosh^{-2}(x)$ *presents two bound states and a nonzero reflection coefficient (cf. [98]). Consequently, the corresponding solution to KdV will eventually split into two solitons traveling to the right, with the larger and faster one first, and behind them a dispersive tail traveling to the left. Note that* $-u$ *is plotted against* x*, in accordance with the physical form of KdV.*

Riemann–Hilbert problems

The first mathematical results showing that the numerical evidence that an arbitrary initial profile $u_0 \in X$ will eventually split into a number of solitons traveling to the right and a decaying part traveling to the left with amplitude diminishing toward zero as $t \to \infty$ were based on an in-depth study of (5.105). Since these considerations are beyond the scope of this text, we refer the reader to the discussion in [267]. We conclude the discussion of the relevance of solitary waves to the asymptotic behavior of solutions to (5.3) that develop from an initial localized smooth profile by pointing out an alternative powerful approach. Recall that $T(k)$ is meromorphic in the upper half-plane $\{\Im m(k) > 0\}$, with simple poles at $i\kappa_1, \ldots, i\kappa_n$ with $k_j > 0$ (corresponding to the eigenvalues $\lambda_j = -\kappa_j^2 < 0$), and is continuous up to the real line $\mathbb{R}$. From (5.85), (5.68), (5.86), (5.81), (5.82), and (5.92) we deduce that for any $(x,t) \in \mathbb{R} \times [0,\infty)$, we have

$$T(k)\,g_-(x,k,t) = [1 - |R_+(k,0)|^2]\,g_+(x,-k,t) + R_+(k,0)\,e^{2ikx+8ik^3t}\,g_-(x,-k,t), \quad k \in \mathbb{R},$$

while considering (5.84) instead of (5.85) leads to

$$g_+(x,k,t) = -\,\overline{R_+(k,0)}\,e^{-2ikx-8ik^3t}\,g_+(x,-k,t) + T(-k)\,g_-(x,-k,t), \quad k \in \mathbb{R}.$$

Therefore, if for $(x,t) \in \mathbb{R} \times [0,\infty)$ fixed we introduce the vector function

$$m(x,k,t) = \begin{cases} \big(T(k)\,g_-(x,k,t) \quad g_+(x,k,t)\big) & \text{for} \quad \Im m(k) > 0, \\ \big(g_+(x,-k,t) \quad T(-k)\,g_-(x,-k,t)\big) & \text{for} \quad \Im m(k) < 0, \end{cases} \tag{5.112}$$

and for $k \in \mathbb{R}$ the 2×2 matrix

$$\mathfrak{G}(x,k,t) = \begin{pmatrix} 1 - |R_+(k,0)|^2] & -\,\overline{R_+(k,0)}\,e^{-2ikx-8ik^3t} \\ R_+(k,0)\,e^{2ikx+8ik^3t} & 1 \end{pmatrix}, \tag{5.113}$$

then for every fixed $(x,t) \in \mathbb{R} \times [0,\infty)$, the map $k \mapsto m(x,k,t)$ is meromorphic with at most a finite number of simple poles in the upper and in the lower half-plane; in both half-planes it has continuous extensions $m_\pm$ to the real line $\mathbb{R}$, the line across which the jump condition[132] takes the form

$$m_+(x,k,t) = m_-(x,k,t)\mathfrak{G}(x,k,t) \quad \text{for} \quad k \in \mathbb{R}. \tag{5.114}$$

This is the Riemann–Hilbert problem reformulation of the inverse scattering problem for KdV: find the vector-valued sectional analytic function m that satisfies the explicit jump condition (5.114) and approaches the constant vector $(1\ 1)$ as $|k| \to \infty$. The main benefit of this reformulation arises in asymptotic analysis. Note that at each fixed $(x,t) \in \mathbb{R} \times [0,\infty)$ we have a Riemann–Hilbert problem, and the way the parameters (x,t) enter the problem through $\mathfrak{G}(x,k,t)$ is in the form of the oscillatory multipliers $e^{\pm(2ikx+8ik^3t)}$, so one may expect that the evaluation of the solution $m(x,k,t)$ of the Riemann–Hilbert problem (5.114) for $x,\, t \to \infty$ is similar to the asymptotic evaluation of oscillatory integrals[133] by the classical method of stationary phase. Using (5.99), we can evaluate asymptotically $u(x,t)$ if we know the behavior of $m(x,k,t)$ as $x,\, t \to \infty$. Indeed, in complete analogy to the classical method, one can deform the jump contour, such that all jumps will be moved into regions away from the stationary points of the phase $\phi(k) = 8ik^3 + 2ik\frac{x}{t}$ (with t representing the large parameter), where the oscillatory terms will decay exponentially. This way one can prove that in the region $\frac{x}{t} > C$ for some $C > 0$, the solution $u(x,t)$ is asymptotically given by the sum of one-soliton solutions (cf. the discussion in [134])

$$u(x,t) \approx -2\sum_{j=1}^{n} \frac{\kappa_j^2}{\cosh^2(\kappa_j x - 4\kappa_j^3 t - p_j)},$$

where the phase shifts are given by

$$p_j = \frac{1}{2}\log\left(\frac{c_j^2(0)}{2\kappa_j}\prod_{l=j+1}^{n}\left(\frac{\kappa_l - \kappa_j}{\kappa_l + \kappa_j}\right)^2\right), \qquad j = 1,\dots,n.$$

While the general analytic framework for Riemann–Hilbert problems was satisfactorily developed[134] in the period 1973–1992 after efforts by several authors (see the survey [93]), its implementation is not at all algorithmic and, depending on the context, might use quite sophisticated ideas peculiar to that setting (cf. the discussion in [158]). For a comprehensive discussion of the applicability of Riemann–Hilbert problems to the long-time asymptotics of solutions to KdV we refer the reader to [134]; see [29, 58, 60] for its applicability to other integrable nonlinear models for water waves.

[132]If the function m were continuous across $\mathbb{R}$ for $k \in \mathbb{C}$, from the fact that it has a finite number of simple poles, and is asymptotic to $(1\ 1)$ as $|k| \to \infty$ in view of (5.95), (5.54), and the analogous inequality for g_-, by Liouville's theorem each of its two components would have to be equal to a rational function throughout $\mathbb{C}$. But this is not generally the case (e.g., the one-soliton solution shows that we have to expect a more complicated functional dependence on k).

[133]We exemplify the method of stationary phase by considering the integral $I(\lambda) = \int_{\mathbb{R}} e^{i\lambda\theta(x)}A(x)\,dx$, interpreted as an average for oscillations of amplitude A and phase θ. We assume that both A and θ are smooth, with A having compact support in $[\alpha,\beta]$. If $\theta'(x) \neq 0$ for $x \in [\alpha,\beta]$, successive integrations by parts show that $I(\lambda) = O(\lambda^{-j})$ as $\lambda \to \infty$, for any integer $j \geq 0$. On the other hand, if we assume that x_0 is the only stationary point of ϕ in $[\alpha,\beta]$, that is, $\theta'(x_0) = 0$ but $\theta'(x) \neq 0$ for $x \in [\alpha,\beta]$, $x \neq x_0$, then if λ is very large, we expect that the contribution to $I(\lambda)$ from any interval not containing x_0 will be very small, since the rapid oscillations of the exponential will tend to average to zero. On the other hand, the contribution from a neighborhood of x_0 need not be small since there the oscillations slow down as the phase becomes stationary. Indeed (cf. [283, Chapter VIII]), if the stationary point x_0 is nondegenerate, that is, $\theta''(x_0) \neq 0$, then $I(\lambda) = \lambda^{-1/2}\left(\frac{2\pi}{-i\,\theta''(x_0)}\right)^{1/2} A(x_0) + O(\lambda^{-1})$ as $\lambda \to \infty$. More generally, the character of the asymptotic expansion depends on the smallest integer $j \geq 2$ for which $(\partial_x^j\theta)(x_0) \neq 0$.

[134]The precise sense of the limits at ∞ and from the two sides of $\mathbb{R}$ and the allowable functional classes for the map $\mathfrak{G}$ are highly nontrivial questions (see [93, 158]).

KdV and algebraic geometry

The integrability of KdV for spatially periodic solutions is of interest since wave trains are easier to observe than solitary waves. However, the periodic setting is more intricate due to a lack of simple asymptotic limit of the isospectral operator,[135] but, using methods of algebraic geometry, one can actually prove the analogue of Liouville's theorem: the periodic KdV flow is a linear flow at constant speed on a (mostly infinite-dimensional) torus. To discuss the integrability of KdV for (spatially) periodic waves with wavelength L, it is convenient to transform (5.3) by means of the scaling $q(x,t) = -\frac{L^2}{6}\,\eta(Lx, \frac{L^3}{2}t)$ into the equivalent form

$$q_t - 3qq_x + \frac{1}{2}q_{xxx} = 0, \tag{5.115}$$

to be solved for smooth functions of (spatial) period 1.

Cnoidal waves

To bring algebraic geometry into the picture, notice that the simplest nontrivial solution of (5.115) is obtained by seeking wave trains $q(x-ct)$: we obtain the **cnoidal wave**[136]

$$q(x,t) = 2\wp(x-ct+\omega_2)+q_0,$$

where $q_0 \in \mathbb{R}$ is a constant and

$$\wp(z) = \frac{1}{z^2} + \sum_{\omega\in\Lambda\setminus\{0\}} \left(\frac{1}{(z-\omega)^2} - \frac{1}{\omega^2}\right) \tag{5.116}$$

is the Weierstrass function with primitive periods $2\omega_1 = 1$ and $2\omega_2 \in i\mathbb{R}\setminus\{0\}$, $\Lambda = \{m\omega_1+n\omega_2 : m,n\in\mathbb{Z}\}$ being the lattice generated by ω_1 and ω_2; the translation by ω_2 ensures that q is real and smooth.

Floquet theory

Let us first recall Floquet's theory (see [215]) for the operator $\mathcal{L} = -\partial_x^2 + q(x)$ with $q \in C^3(\mathbb{R})$ of period 1. We denote by $y_1(x,\lambda)$ and $y_2(x,\lambda)$ the solutions of $\mathcal{L}y = \lambda y$ with initial data $y_1(0) = 1$, $y_1'(0) = 0$, and $y_2(0) = 0$, $y_2'(0) = 1$, respectively, and we define the *discriminant*

$$\Delta(\lambda) = \frac{y_1(1,\lambda)+y_2'(1,\lambda)}{2}.$$

The roots of $\Delta(\lambda) = \pm 1$ are the eigenvalues of $\mathcal{L}$ with periodic/antiperiodic boundary conditions $y(x+1) = \pm y(x)$. They comprise a simple periodic ground state λ_0 followed by alternating antiperiodic and periodic pairs $\lambda_0 < \lambda_1 \le \lambda_2 < \lambda_3 \le \lambda_4 < \cdots \uparrow \infty$ of simple or double eigenvalues, the corresponding eigenfunctions f_n with $n \ge 0$ being such that f_0 never vanishes while f_{2n-1} and f_{2n} have n roots apiece in $[0,1)$ and are periodic/antiperiodic when $n \ge 1$ is even/odd. We have

$$\lambda_{2n-1}, \lambda_{2n} = n^2\pi^2 + \int_0^1 q(x)\,dx + O(n^{-2}), \qquad n \ge 1. \tag{5.117}$$

[135] In the nonperiodic case, the operator was asymptotic to a constant coefficient operator, and this fact was essential in defining the scattering matrix.

[136] The name stems from the fact that the solution can also be expressed in terms of Jacobi's basic elliptic function cn: $q(x,t) = q_1 + q_2\,\mathrm{cn}^2(\frac{x}{4K(k)};k)$ for suitable constants $q_1, q_2 \in \mathbb{R}$ and $k \in (0,1)$, where $4K(k) > 2\pi$ is the real period of $s \mapsto \mathrm{cn}(s;k)$; see [14].

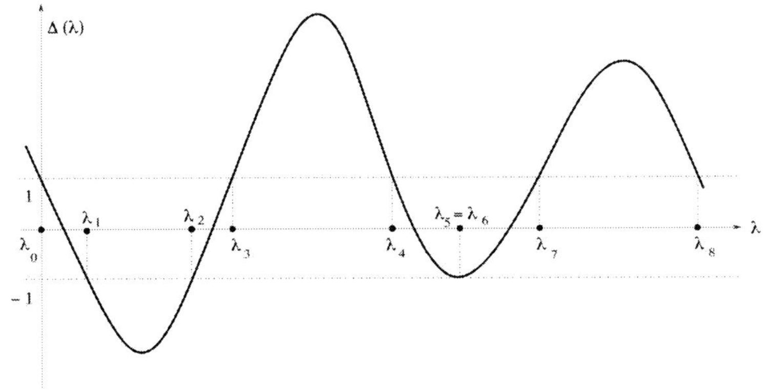

Figure 5.14. *The discriminant.*

The intervals $(-\infty, \lambda_0)$ and $(\lambda_{2n-1}, \lambda_{2n})$ with $n \geq 1$ are called forbidden bands since only the zero solution of $\mathcal{L}y = \lambda y$ is bounded on $\mathbb{R}$ if λ belongs to one of these intervals. Except for the first forbidden band, all others may collapse; we say that q is a *finite-gap potential* if all but a finite number of bands are collapsed. The roots $\{\mu_n\}_{n\geq 1}$ of $y_2(1,\mu) = 0$ comprise the Dirichlet spectrum of $\mathcal{L}$, with the boundary conditions $y(0) = y(1) = 0$, and interlace the periodic/antiperiodic spectrum: $\mu_n \in [\lambda_{2n-1}, \lambda_{2n}]$ for $n \geq 1$. The eigenfunction $y_2(x, \mu_n)$ has exactly $n-1$ (simple) roots in $(0,1)$. Applying Picard's method of iteration (see [215, 146]) one obtains the estimates[137]

$$\begin{cases} y_1(x,\lambda) = \cos(\sqrt{\lambda}\,x)[1+o(1)], \\[2ex] y_2(x,\lambda) = \dfrac{\sin(\sqrt{\lambda}\,x)}{\sqrt{\lambda}}[1+o(1)] \end{cases} \tag{5.118}$$

for $\lambda \downarrow -\infty$ (real), while

$$\begin{cases} y_1(x,\lambda) = \cos(\sqrt{\lambda}\,x) + \dfrac{\sin(\sqrt{\lambda}\,x)}{2\sqrt{\lambda}} \displaystyle\int_0^x q(s)\,ds \\[2ex] \qquad + \dfrac{\cos(\sqrt{\lambda}\,x)}{4\lambda}\left(q(x) - q(0) - \dfrac{1}{2}\left(\displaystyle\int_0^x q(s)\,ds\right)^2\right) + O(\lambda^{-3/2}), \\[2ex] y_2(x,\lambda) = \dfrac{\sin(\sqrt{\lambda}\,x)}{\sqrt{\lambda}} - \dfrac{\cos(\sqrt{\lambda}\,x)}{2\lambda} \displaystyle\int_0^x q(s)\,ds \\[2ex] \qquad + \dfrac{\sin(\sqrt{\lambda}\,x)}{4\lambda\sqrt{\lambda}}\left(q(x) + q(0) - \dfrac{1}{2}\left(\displaystyle\int_0^x q(s)\,ds\right)^2\right) + O(\lambda^{-2}) \end{cases} \tag{5.119}$$

for $\lambda \uparrow \infty$ (real), and $y_1(1,\lambda)$, $y_2(1,\lambda)$, and $\Delta(\lambda)$ are entire functions of order[138] 1/2. Therefore, since $\sum_{n\geq 0:\ \lambda_n \neq 0} \frac{1}{|\lambda_n|} < \infty$ and $\sum_{n\geq 1:\ \mu_n \neq 0} \frac{1}{|\mu_n|} < \infty$ in view of (5.117), Hadamard's factorization

[137] These can be differentiated with respect to x and/or λ, uniformly for $x \in [0,1]$.

[138] A function $f : \mathbb{C} \to \mathbb{C}$ is entire of order 1/2 if f is analytic in $\mathbb{C}$ and its order of growth for $|\lambda| \to \infty$ (where, unless f is a polynomial, f has an essential singularity) is exactly 1/2; that is, there exists a constant $M > 0$ such that $|f(\lambda)|\, e^{-M\sqrt{|\lambda|}}$ is bounded on $\mathbb{C}$, while for any $\alpha < 1/2$ we have $\limsup_{|\lambda|\to\infty} \{|f(\lambda)|\, e^{-m|\lambda|^{\alpha}}\} = \infty$ for any $m > 0$.

theorem (see [23]) yields[139]

$$y_2(1,\lambda)=\prod_{n=1}^{\infty}\frac{\mu_n-\lambda}{n^2\pi^2},\qquad \Delta^2(\lambda)-1=(\lambda_0-\lambda)\prod_{n=1}^{\infty}\frac{(\lambda_{2n-1}-\lambda)(\lambda_{2n}-\lambda)}{n^4\pi^4}. \tag{5.120}$$

Trace formula for finite-gap potentials

Consider now a finite-gap potential q for which $\lambda_{2n}=\lambda_{2n-1}$ for all $n>N$ with $N\geq 1$ fixed. The relations (5.120) can be used to prove the following *trace formula*:

$$q(0)=\lambda_0+\sum_{n=1}^{N}(\lambda_{2n-1}+\lambda_{2n}-2\mu_n). \tag{5.121}$$

Indeed, since $\lambda_{2n}=\lambda_{2n-1}=\mu_n$ for $n>N$, from (5.120) we get

$$f(\lambda)=\frac{-\lambda\, y_2^2(1,\lambda)}{\Delta^2(\lambda)-1}=\frac{\lambda}{\lambda-\lambda_0}\prod_{n=1}^{N}\frac{(\lambda-\mu_n)^2}{(\lambda-\lambda_{2n})(\lambda-\lambda_{2n-1})}.$$

Consider this meromorphic function inside the circle C_R centered at the origin and with radius $R>\sup_{0\leq k\leq 2N}\{|\lambda_k|\}$, and denote $\dot f=\partial f/\partial\lambda$. Since there is no pole and zero of f on the circle C_R, by the residue theorem we have

$$\frac{1}{2\pi i}\int_{C_R}\frac{\dot f(\lambda)}{f(\lambda)}\lambda\, d\lambda=-\lambda_0+\sum_{n=1}^{N}(2\mu_n-\lambda_{2n-1}-\lambda_{2n}).$$

Notice that for $\lambda\to\infty$ (complex) we have $f(\lambda)=1+z_0\lambda^{-1}+O(|\lambda|^{-2})$ for some constant $z_0\in\mathbb{C}$, an estimate which implies that for R sufficiently large the function $\log(f)$ is single-valued and analytic in a neighborhood of C_R, with $\log(f(\lambda))=z_0\lambda^{-1}+O(|\lambda|^{-2})$. Integration by parts then yields

$$\int_{C_R}\frac{\dot f(\lambda)}{f(\lambda)}\lambda\, d\lambda=\int_{C_R}\lambda\,\partial_\lambda\log(f(\lambda))\,d\lambda=-\int_{C_R}\log(f(\lambda))\,d\lambda=-2\pi i\, z_0+O(R^{-1}).$$

Letting $R\to\infty$ we get that the right side of (5.121) equals $z_0=\lim_{\lambda\to\infty}\lambda[f(\lambda)-1]$. The precise asymptotic estimates (5.119) for $\lambda\uparrow\infty$ (real) enable us to evaluate this limit (along the real line) as being precisely $q(0)$. This proves (5.121).

Spectral effect of translation

Notice that if we translate $q(x)$ by $0\leq t<1$, the periodic or antiperiodic eigenvalues do not change (it suffices to translate by the same amount the corresponding eigenfunctions), while $\mu_n(t)$ will generally move inside the interval $[\lambda_{2n-1},\lambda_{2n}]$. This simple observation shows by means of the trace formula (5.121) that finite-gap potentials q with just one genuine gap, namely, $(-\infty,\lambda_0)$,

[139] Actually, the factorization theorem yields the expressions in (5.120) up to a multiplicative constant that is determined using (5.118). For example, start with $y_2(1,\lambda)=A\prod_{n=1}^{\infty}\frac{\mu_n-\lambda}{n^2\pi^2}$. Since $\sin(z)=z\prod_{n=1}^{\infty}\frac{n^2\pi^2-z^2}{n^2\pi^2}$ (cf. [23]), from (5.118) we get $\lim_{\lambda\to-\infty}\prod_{n=1}^{\infty}\frac{\mu_n-\lambda}{n^2\pi^2-\lambda}=\frac{1}{A}$. Choosing $M>0$ such that $-M<\mu_n$ for all $n\geq 1$, we infer from (5.117) that the series $\sum_{n=1}^{\infty}\left|\frac{\mu_n-\lambda}{n^2\pi^2-\lambda}-1\right|$ converges uniformly on $(-\infty,-M]$. Therefore (see [264, Chapter 15]) the product $\prod_{n=1}^{\infty}\frac{\mu_n-\lambda}{n^2\pi^2-\lambda}$ will converge uniformly on $(-\infty,-M]$. Thus we can commute product and limit for $\lambda\to-\infty$, obtaining $A=1$.

correspond to constant potentials $q(x) \equiv \lambda_0$. A nonconstructive result by Borg [27] ensures that for a finite-gap potential q with $g \geq 1$ finite gaps, the simple spectrum $\{\lambda_k : 0 \leq k \leq 2g\}$, the Dirichlet eigenvalues[140] $\{\mu_n\}_{1\leq n\leq g}$, and the signs of the numbers $\left(1 - [y_2'(1,\mu_n)]^2\right)$, $1 \leq n \leq g$, specify the potential q. Before proving this, let us first show that *the simple spectrum of a finite-gap potential determines the double spectrum.* Since $\Delta(\lambda) - 1$ is an entire function of order $1/2$, and we know all its zeros, we have

$$\Delta(\lambda) - 1 = C_0(\lambda_0 - \lambda) \prod_{k=0}^{\infty} \frac{(\lambda_{4k+3} - \lambda)(\lambda_{4k+4} - \lambda)}{(2k+2)^4\pi^4}$$

for some constant $C_0 \in \mathbb{R}$. Consequently,

$$\frac{\dot{\Delta}(\lambda)}{\Delta(\lambda) - 1} = \sum_{k=0}^{\infty} \frac{1}{\lambda - \lambda_{4k+3}} + \sum_{k=0}^{\infty} \frac{1}{\lambda - \lambda_{4k}}. \tag{5.122}$$

If $\lambda = a + ib$ is a zero of $\dot{\Delta}$, then considering the imaginary part of the expression on the left side of the equation

$$\sum_{k=0}^{\infty} \frac{a - \lambda_{4k+3} - ib}{(a - \lambda_{4k+3})^2 + b^2} + \sum_{k=0}^{\infty} \frac{a - \lambda_{4k} - ib}{(a - \lambda_{4k})^2 + b^2} = 0,$$

we infer that $b = 0$, so that all zeros of $\dot{\Delta}$ are real. Now, differentiating the relation (5.122) we find, for real λ,

$$\partial_\lambda \left(\frac{\dot{\Delta}(\lambda)}{\Delta(\lambda) - 1} \right) = -\sum_{k=0}^{\infty} \frac{1}{(\lambda - \lambda_{4k+3})^2} - \sum_{k=0}^{\infty} \frac{1}{(\lambda - \lambda_{4k})^2}.$$

This shows that $\lambda \mapsto \frac{\dot{\Delta}(\lambda)}{\Delta(\lambda)-1}$ is a decreasing function and therefore the zeros of $\dot{\Delta}$ are separated by those of $\Delta - 1$. Similarly, we see that the zeros of $\dot{\Delta}$ are separated by those of $\Delta + 1$. Therefore, $\dot{\Delta}$ has precisely one (simple) zero in each lacunae $[\lambda_{2k-1}, \lambda_{2k}]$ with $k \geq 1$, at the point $\lambda = \dot{\lambda}_k$ where Δ attains its local maximum or minimum. The absence of other zeros shows that Δ is strictly monotone on each of the forbidden bands. Since in each of the forbidden bands $(\lambda_0, \lambda_1), \ldots, (\lambda_{2g-2}, \lambda_{2g-1})$ is mapped by Δ bijectively onto $(-1, 1)$, we may set $\Delta(\lambda) = \cos\theta(\lambda)$ on the closure of each of these intervals, with θ of class C^1 and such that $\theta(\lambda_{4k}) = 0$ and $\theta(\lambda_{4k+1}) = \pi$ for all integers $0 \leq k \leq (g-1)/2$, while $\theta(\lambda_{4k+2}) = \pi$ and $\theta(\lambda_{4k+3}) = 2\pi$ for all integers $0 \leq k \leq g/2 - 1$. Differentiating $\Delta = \cos\theta$ yields $\dot{\Delta} = -\dot{\theta}\sin\theta$, so that

$$\int_{\lambda_{4k}}^{\lambda_{4k+1}} \frac{\dot{\Delta}(\lambda)}{\sqrt{1 - \Delta^2(\lambda)}} d\lambda = -\int_{\lambda_{4k}}^{\lambda_{4k+1}} \dot{\theta}(\lambda)\, d\lambda = -\pi, \tag{5.123}$$

while

$$\int_{\lambda_{4k+2}}^{\lambda_{4k+3}} \frac{\dot{\Delta}(\lambda)}{\sqrt{1 - \Delta^2(\lambda)}} d\lambda = \int_{\lambda_{4k}}^{\lambda_{4k+1}} \dot{\theta}(\lambda)\, d\lambda = \pi. \tag{5.124}$$

On the other hand, from (5.120) we infer that

$$1 - \Delta^2(\lambda) = (\lambda - \lambda_0) \prod_{n=1}^{g} \frac{(\lambda_{2n-1} - \lambda)(\lambda_{2n} - \lambda)}{\hat{n}^4\pi^4} \prod_{\substack{double \\ spectrum}} \frac{(\lambda_j - \lambda)^2}{\hat{j}^4\pi^4}$$

---

[140] Earlier we allowed double periodic or antiperiodic eigenvalues among those indexed by $1 \leq n \leq N$, but from now on the index n keeps track of the open gaps (which may interlace with the collapsed gaps), while $\hat{n} \geq n$ is the actual index (counting also double eigenvalues).

for $\lambda \in [\lambda_0,\lambda_1] \cup \cdots \cup [\lambda_{2g-2},\lambda_{2g-1}]$, whereas for these values of λ, the knowledge of the zeros of the entire function $\dot{\Delta}$ of order $1/2$ ensures that

$$\dot{\Delta}(\lambda) = C_1 \prod_{n=1}^{g} \frac{\lambda - \dot{\lambda}_n}{\hat{n}^2\pi^2} \prod_{\substack{double \\ spectrum}} \frac{\lambda_j - \lambda}{\hat{j}^2\pi^2}$$

for some $C_1 \in \mathbb{R}$. Since (5.119) yields $\Delta(\lambda) \approx \cos\sqrt{\lambda}$ as $\lambda \to \infty$ (real), we infer that

$$\dot{\Delta}(\lambda) \approx -\frac{\sin\sqrt{\lambda}}{2\sqrt{\lambda}} = \prod_{j=1}^{\infty} \frac{\lambda - j^2\pi^2}{2j^2\pi^2}$$

as $\lambda \to \infty$ (real). Therefore, $C_1 = 1/2$ so that, by cancellation of the double spectrum in $\dot{\Delta}$ and $\sqrt{1-\Delta^2}$,

$$\frac{\dot{\Delta}(\lambda)}{\sqrt{1-\Delta^2(\lambda)}} = \pm \frac{\prod_{n=1}^{g}(\lambda - \dot{\lambda}_n)}{2\sqrt{(\lambda-\lambda_0)\prod_{n=1}^{g}(\lambda-\lambda_{2n-1})(\lambda-\lambda_{2n})}} \tag{5.125}$$

for $\lambda \in [\lambda_0,\lambda_1] \cup \cdots \cup [\lambda_{2g-2},\lambda_{2g-1}]$. From (5.123) and (5.124) we get a system of g linear equations for the elementary symmetric functions[141] of $\dot{\lambda}_n$ $(n = 1,\dots,g)$, with determinant

$$\det\Big[\int_{\lambda_{2k}}^{\lambda_{2k+1}} \frac{\lambda^{j-1}}{\sqrt{\prod_{n=0}^{2g}(\lambda-\lambda_n)}}\,d\lambda \quad \text{with} \quad 1 \le j \le g,\ 0 \le k \le g-1\Big]$$

$$= \int_{\lambda_0}^{\lambda_1} \cdots \int_{\lambda_{2g-2}}^{\lambda_{2g-1}} \begin{vmatrix} \xi_1 & \cdots & \xi_1^{g-1} \\ & \vdots & \\ \xi_g & \cdots & \xi_g^{g-1} \end{vmatrix} \frac{d\xi_1}{\sqrt{\prod_{n=0}^{2g}(\xi_1-\lambda_n)}} \cdots \frac{d\xi_g}{\sqrt{\prod_{n=0}^{2g}(\xi_g-\lambda_n)}} \neq 0,$$

as in the last expression the integrand is a Vandermonde-type determinant that is always nonzero (see [197]). Therefore, the elementary symmetric functions of $\dot{\lambda}_n$ $(n = 1,\dots,g)$ are uniquely determined. Using Viète's relations, the roots $\dot{\lambda}_n$ of $\dot{\Delta}$ $(n = 1,\dots,g)$ are found as the g distinct real roots of a polynomial of degree g. Now, the reasoning leading to (5.123) shows that for $\lambda \in (\lambda_0,\lambda_1)$,

$$\theta(\lambda) = \int_{\lambda_0}^{\lambda} \frac{\dot{\Delta}(\xi)}{\sqrt{1-\Delta^2(\xi)}}\,d\xi = (-1)^g \int_{\lambda_0}^{\lambda} \frac{\prod_{n=1}^{g}(\xi-\dot{\lambda}_n)}{2\sqrt{(\xi-\lambda_0)\prod_{n=1}^{g}(\xi-\lambda_{2n-1})(\xi-\lambda_{2n})}}\,d\xi$$

in view of (5.125) since for these values of λ we have $\dot{\Delta}(\lambda) > 0$ and $\prod_{n=1}^{g}(\lambda-\dot{\lambda}_n)$ has the sign of $(-1)^g$ as $\dot{\lambda}_n \in (\lambda_{2n-1},\lambda_{2n})$ for $n = 1,\dots,g$. Consequently, the simple spectrum $\{\lambda_k\}_{0\le k\le 2g}$ determines $\{\dot{\lambda}_n\}_{1\le n\le g}$ and $\Delta = \cos\theta$ on (λ_0,λ_1). Since $\lambda \mapsto \Delta(\lambda)$ is an analytic function, the entire function is thus determined, particularly the roots of $\Delta(\lambda) = \pm 1$ which comprise the double spectrum.

Let us now prove that

$$\frac{d\mu_n}{dt} = \pm 2\frac{\sqrt{(\lambda_0-\mu_n)\prod_{k=1}^{g}(\lambda_{2k-1}-\mu_n)(\lambda_{2k}-\mu_n)}}{\prod_{k=1,k\neq n}^{g}(\mu_n-\mu_k)}, \qquad n = 1,\dots,g, \tag{5.126}$$

[141] These are $\sum_{j=1}^{g}\dot{\lambda}_j,\ \sum_{1\le i<j\le n}\dot{\lambda}_i\dot{\lambda}_j,\dots,\prod_{j=1}^{g}\dot{\lambda}_j$.

where the sign is to be specified. Before we check the validity of (5.126), notice that the structure of the operator $\mathcal{L} = -\partial_x^2 + q(x)$ ensures that

$$y_2'(x,\lambda)y_1(x,\lambda) - y_2(x,\lambda)y_1'(x,\lambda) = 1, \qquad x \in \mathbb{R}, \tag{5.127}$$

the above Wronskian expression being constant on $[0,1]$, and, by definition, equal to 1 at $x = 0$. In particular,

$$y_2'(1,\mu_n)y_1(1,\mu_n) = 1, \qquad n \geq 1. \tag{5.128}$$

Since $\{\mu_n\}_{n\geq 1}$ are the simple roots of the entire function $\lambda \mapsto y_2(1,\lambda)$, a function that can be expressed by means of the Picard iterative procedure as a power series in λ with coefficients consisting of integrals involving q (cf. [215]), from the Hurwitz convergence theorem (cf. [5]) we infer that the map $t \mapsto \mu_n(t)$ is continuous for every $n \geq 1$. If we denote by a superscript y_2^{+h} the function y_2 computed for q translated in the amount h, the continuous dependence of $y_2^{+h}(x,\mu_n(h))$ on h, uniformly for $x \in [0,1]$, is now ensured as $[q(x) - q(x+h)]$ is uniformly small if $|h|$ is small. Multiplying the differential equation for $y_2^{+h}(x,\mu_n(h))$ by $y_2(x,\mu_n)$ and substracting from the result the differential equation for $y_2(x,\mu_n)$ multiplied by $y_2^{+h}(x,\mu_n(h))$, we get by integrating the outcome on $[0,1]$ and performing some integrations by parts that $\mu_n(t)$ has a derivative given by

$$\frac{d\mu_n}{dt} = \frac{\displaystyle\int_0^1 q'(x)\, y_2^2(x,\mu_n)\,dx}{\displaystyle\int_0^1 y_2^2(x,\mu_n)\,dx} = -\frac{\displaystyle 2\int_0^1 q(x)\, y_2(x,\mu_n)\, y_2'(x,\mu_n)\,dx}{\displaystyle\int_0^1 y_2^2(x,\mu_n)\,dx}. \tag{5.129}$$

Multiplying the differential equation for $y_2(x,\mu_n)$ by $y_2'(x,\mu_n)$, an integration yields

$$2\int_0^1 q(x)\, y_2(x,\mu_n)\, y_2'(x,\mu_n)\,dx = [y_2'(1,\mu_n)]^2 - 1. \tag{5.130}$$

On the other hand, if we differentiate $\mathcal{L}y_2 = \lambda y_2$ with respect to λ, multiply by y_2, and substract from the result the equation $\mathcal{L}y_2 = \lambda y_2$ multiplied by $\dot{y}_2$, integrating over $[0,1]$ the obtained outcome evaluated at $\lambda = \mu_n$, we get

$$\int_0^1 y_2^2(x,\mu_n)\,dx = y_2'(1,\mu_n)\,\dot{y}_2(1,\mu_n) \tag{5.131}$$

since $\dot{y}_2(0,\mu_n) = 0$. Combining (5.129)–(5.131) leads us to

$$\frac{d\mu_n}{dt} = -\frac{[y_2'(1,\mu_n)]^2 - 1}{y_2'(1,\mu_n)\,\dot{y}_2(1,\mu_n)} = \pm\frac{2\sqrt{\Delta^2(\mu_n) - 1}}{\dot{y}_2(1,\mu_n)}, \tag{5.132}$$

since $4[\Delta^2(\mu_n) - 1] = [y_2'(1,\mu_n) - \frac{1}{y_2'(1,\mu_n)}]^2$ by (5.128) and the definition of the discriminant. Now, since (5.120) yields $\dot{y}_2(1,\lambda) = -\sum_{m=1}^{\infty} \frac{1}{m^2\pi^2} \prod_{k=1,k\neq m}^{\infty} \frac{\mu_k - \lambda}{k^2\pi^2}$, we have

$$\frac{2\sqrt{\Delta^2(\mu_n) - 1}}{\dot{y}_2(1,\mu_n)} = \pm\frac{2\sqrt{(\lambda_0 - \mu_n)\prod_{k=1}^{g}(\lambda_{2k-1} - \mu_n)(\lambda_{2k} - \mu_n)}}{\prod_{k=1,k\neq n}^{g}(\mu_n - \mu_k)},$$

with the sign on the right side of the last two displayed relations to be determined. By (5.129), (5.130), and (5.131), we see that $\frac{d\mu_n(t)}{dt} > 0$ if and only if $[y_2'(1,\mu_n)]^2 < 1$.

The previous considerations suggest Figure 5.15, in which the gap $[\lambda_{2n-1},\lambda_{2n}]$ is opened up into a circle and $\mu_n(t)$ is placed at the point $p_n(t)$ on the upper/lower semicircle according to the

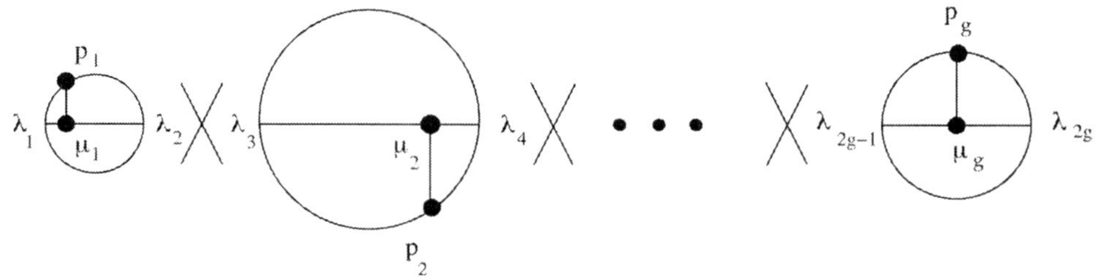

Figure 5.15. *Opening up the forbidden bands.*

$\pm$ sign of $\frac{d\mu_n(t)}{dt} > 0$ in (5.126). We claim that $p_n(t)$ *runs clockwise around the circle changing banks when it hits* λ_{2n-1} *or* λ_{2n}, *and making* $\hat{n}$ *complete revolutions in unit time*. Moreover, at each root $x_k \in (0,1)$ of $y_2(x,\mu_n(0))$ we have $\mu_n(x_k) = \mu_n(0)$ for $k = 1,\dots,\hat{n}-1$.

Indeed, from (5.126) it is plain that $\mu_n(t)$ cannot pause at some point in $(\lambda_{2n-1},\lambda_{2n})$, moving clockwise. We now show that it cannot pause at λ_{2n-1} or λ_{2n}. For this, notice first that

$$y_2^{+t}(x,\lambda) = y_1(t,\lambda)\,y_2(x+t,\lambda) - y_1(x+t,\lambda)\,y_2(t,\lambda), \qquad x,t\in[0,1], \quad \lambda\in\mathbb{R},$$

since for fixed t and λ, both functions of x solve $-z''(x)+q(x+t)z(x) = \lambda z(x)$ with initial data $z(0) = 0$ and $z'(0) = 1$, in view of (5.127). In particular,

$$y_2^{+t}(1,\lambda) = y_1(t,\lambda)\,y_2(1+t,\lambda) - y_1(1+t,\lambda)\,y_2(t,\lambda), \qquad t\in[0,1], \quad \lambda\in\mathbb{R}. \tag{5.133}$$

Assume now that $\mu_n(t) = \lambda_{2n}$ for $t\in[a,b]$, where $0\le a<b\le 1$. Then $y_2^{+t}(1,\lambda_{2n}) = 0$ on $[a,b]$, and since $y_2(x,\lambda_{2n})$ has only simple zeros, it will have finitely many on $[0,2]$, so that we find $c<d$ with $[c,d]\subset[a,b]$ such that $y_2(t,\lambda_{2n})\,y_2(1+t,\lambda_{2n}) \neq 0$ whenever $t\in[c,d]$. Using (5.133) we get

$$\frac{y_1(1+t,\lambda_{2n})}{y_2(1+t,\lambda_{2n})} = \frac{y_1(t,\lambda_{2n})}{y_2(t,\lambda_{2n})}, \qquad t\in[c,d],$$

and differentiation with respect to t yields

$$y_2^2(t,\lambda_{2n}) = y_2^2(1+t,\lambda_{2n}), \qquad t\in[c,d],$$

on account of (5.127). By construction $y_2(t,\lambda_{2n})$ and $y_2(1+t,\lambda_{2n})$ have constant sign on $[c,d]$, so that the previous two relations imply

$$\big(y_1(t,\lambda_{2n}),\,y_2(t,\lambda_{2n})\big) = \pm\big(y_1(1+t,\lambda_{2n}),\,y_2(1+t,\lambda_{2n})\big), \qquad t\in[c,d].$$

Consequently,

$$\begin{pmatrix} y_1(c,\lambda_{2n}) & y_2(c,\lambda_{2n}) \\ y_1'(c,\lambda_{2n}) & y_2'(c,\lambda_{2n}) \end{pmatrix} = \pm\begin{pmatrix} y_1(1+c,\lambda_{2n}) & y_2(1+c,\lambda_{2n}) \\ y_1'(1+c,\lambda_{2n}) & y_2'(1+c,\lambda_{2n}) \end{pmatrix}.$$

Since $y_1(x,\lambda_{2n})$ and $y_2(x,\lambda_{2n})$ are linearly independent, this would mean that the equation

$$-z''(x)+q(x+c)z(x) = \lambda_{2n}z(x)$$

has two linearly independent periodic (or antiperiodic) solutions: $y_1(x+c,\lambda_{2n})$ and $y_2(x+c,\lambda_{2n})$. But this is impossible since λ_{2n} is a simple eigenvalue. The procedure can be repeated for λ_{2n-1} so that $\mu_n(t)$ never rests. Since at $t=1$ it must come back to its initial position on the circle (by periodicity), we see that before returning it hits both λ_{2n-1} and λ_{2n} at least once.

Let now $x_k \in (0,1)$ be a root of $y_2(x,\mu_n(0))$ and observe that the function $z(x) = y_2(x+x_k,\mu_n(0))$ satisfies

$$-z''(x)+q(x+x_k)z(x) = \mu_n(0)z(x)$$

with $z(0)=0$. If we set $w(x) = y_2(1+x,\mu_n(0))$ for $x \in \mathbb{R}$, then $w(x) = c\, y_2(x,\mu_n(0))$ on $\mathbb{R}$, where $c = y_2'(1,\mu_n(0))$, as both functions solve

$$-f''(x)+q(x)f(x) = \mu_n(0)f(x)$$

with the same initial conditions at $x=0$. We deduce that

$$z(1) = w(x_k) = c\, y_2(x_k,\mu_n(0)) = 0.$$

Since $z(0) = z(1) = 0$ with $z(x) \not\equiv 0$ on $\mathbb{R}$, the differential equation satisfied by $z(x)$ yields $\mu_n(x_k) = \mu_n(0)$, as $\mu_n(0) \in [\lambda_{2n-1},\lambda_{2n}]$ must be a Dirichlet eigenvalue for the potential $x \mapsto q(x+x_k)$ and in this interval lies the single eigenvalue $\mu_n(x_k)$. For $\mu_n(0) = \lambda_{2n-1}$ or $\mu_n(0) = \lambda_{2n}$ this fact ensures that $\mu_n(t)$ makes $\hat{n}$ complete revolutions in unit time. Even if $\mu_n(0)$ is not in one of these positions the result remains true. Indeed, we established before that it will hit one of the endpoints; say, it hits first λ_{2n-1} in finite time $\tau \in (0,1)$. For $t \in [\tau,\tau+1]$ we know that $p_n(t)$ will make precisely $\hat{n}$ revolutions, and by the periodicity of q we have $p_n(t) = p_n(t+1)$ for $t \in (0,\tau)$, without hitting λ_{2n-1} or λ_{2n}. Consequently, $\mu_n(t)$ always makes $\hat{n}$ complete revolutions. This also means that the sign of the right side of (5.126) at $t=0$ determines the sign at any $t \in (0,1)$, with alternations between ± 1 due to changes every time $\mu_n(t)$ hits λ_{2n-1} or λ_{2n}.

We are now in a position to show that the system (5.126) has a unique periodic solution which does not pause at simple eigenvalues. Place in the complex plane open disjoint disks D_k over the g gaps, the centers of which coincide with the midpoints of the gaps, such that the closed gap lies in the interior of D_k, and rewrite (5.126) as

$$\frac{d\mu_n}{dt} = \pm h_n(\mu_1,\dots,\mu_n)\sqrt{(\lambda_{2n}-\mu_n)(\mu_n-\lambda_{2n-1})}, \qquad n=1,\dots,g,$$

where h_n is analytic in $\mu_k \in D_k$, $k=1,\dots,g$, and positive for $\mu_k \in [\lambda_{2n-1},\lambda_{2n}]$. The sign of the radical is determined by (5.126), but this is taken care of by letting

$$\mu_n(t) = \lambda_{2n-1} + (\lambda_{2n}-\lambda_{2n-1})\sin^2(x_n(t)), \qquad n=1,\dots,g,$$

since with

$$H(x_1,\dots,x_n) = (\lambda_{2n}-\lambda_{2n-1})h_n(\mu_1,\dots,\mu_n),$$

the previous two relations yield

$$\frac{dx_n}{dt} = H_n(x_1,\dots,x_n), \qquad n=1,\dots,g. \tag{5.134}$$

Indeed, the clockwise motion without rest of $p_n(t)$, making $\hat{n}$ complete trips around the circle in unit time, translates into $x_n(1) - x_n(0) = 2\hat{n}\pi$, with $x_n(t)$ strictly increasing on $[0,1]$. The uniqueness of solutions to (5.134) once we specify $x_n(0) \in [0,2\pi)$ for $n=1,\dots,g$, is plain (cf. [265]) since the right-hand side is smooth. We saw before that the role of the sign on the right side of (5.126) is to specify whether $p_n(0)$ lies on the upper or lower semicircle; this amounts now to specifying whether $x_n(0) \in [0,\pi]$ or $x_n(0) \in [\pi,2\pi]$.

The previous considerations show that the motion of $\mu_n(t)$, $n = 1, \ldots, g$, for $t \in [0,1]$ is fully determined by specifying $\{p_n(0)\}_{1\le n\le g}$. Moreover, from (5.121) we get

$$q(x) = \lambda_0 + \sum_{n=1}^{g} \big(\lambda_{2n-1} + \lambda_{2n} - 2\mu_n(x)\big), \qquad x \in [0,1]. \tag{5.135}$$

The ambiguity of the sign on the right side of (5.126) is removed by interpreting (5.126) on the Riemann surface associated to the hyperelliptic curve

$$s(\lambda) = \sqrt{-\prod_{k=0}^{2g}(\lambda - \lambda_k)},$$

with a change of sheet each time μ_m hits λ_{2m-1} or λ_{2m}, so as to keep $\lambda_{2m-1} \le \mu_m(t) \le \lambda_{2m}$ for $1 \le m \le g$ (see below).

Riemann surfaces and hyperelliptic curves

In turning our attention to integrating the system (5.126) we encounter abelian integrals of the form $\int_{x(0)}^{x(t)} r(x,y)\,dx$, where $r(x,y)$ is a rational function (e.g., $r(x,y) = \frac{1}{y}$) and y is the (generally) multivalued function of x defined by

$$y^2 = (\lambda_0 - x)\prod_{k=1}^{g}(\lambda_{2k-1} - x)(\lambda_{2k} - x).$$

We saw above that x by itself does not suffice as a parameter and that it is advantageous to incorporate the determination of the sign of the square root defining y. The choice of the proper branch of the square root is clarified by constructing the *associated Riemann surface* $\boldsymbol{M}$, on which y "becomes" single-valued, and to the integral $\int_{x(0)}^{x(t)} r(x,y)\,dx$ corresponds a differential form that we integrate along paths on $\boldsymbol{M}$. Although we start from real expressions, it is convenient to work with complex variables. Given $g \ge 1$ and complex numbers $\{\lambda_k\}_{0\le k\le 2g}$ such that $\lambda_j \ne \lambda_k$ if $j \ne k$, the solution set[142] $(x,y) \in \mathbb{C}^2$ of the equation

$$y^2 - \prod_{k=0}^{2g}(\lambda_k - x) = 0 \tag{5.136}$$

represents a hyperelliptic[143] curve. To take care of "points at infinity" one introduces homogeneous coordinates so that the point (x,y) is represented by the set of triples $\{(z, zx, zy) :\ z \ne 0\}$ and (5.136) becomes the homogeneous equation

$$y^2 z^{2g-1} - \prod_{k=0}^{2g}(\lambda_k z - x) = 0 \tag{5.137}$$

that is dehomogenized by setting $z = 1$. In this process the affine curve (5.136) corresponds to the projective algebraic curve (5.137). The projective space $\mathbb{P}^n$ is the set of complex lines through the origin $\mathbf{0} \in \mathbb{C}^{n+1}$, a typical element $[z_0 : z_1 : \cdots : z_n]$ (with not all complex numbers z_k zero) standing

[142]We consider this equation in $\mathbb{C}^2$ since we do not expect all solutions to be real numbers even if all λ_k are real.

[143]For $g = 1$ the curve is called elliptic because it can be parametrized by elliptic functions [132].

for the equivalence class $\{(\xi z_0, \xi z_1, \ldots, \xi z_n) : \xi \neq 0\}$. As a quotient of a compact, connected space (the unit sphere in $\mathbb{C}^{n+1}$), the projective space is compact and connected. The isomorphism

$$[z_0 : z_1 : z_2] \mapsto \begin{cases} \left(\frac{z_1}{z_0}, \frac{z_2}{z_0}\right) \in \mathbb{C}^2, & z_0 \neq 0, \\ [z_1 : z_2] \in \mathbb{P}^1, & z_0 = 0, \end{cases}$$

shows that $\mathbb{P}^2$ compactifies $\mathbb{C}^2$ by adding the line $\mathbb{P}^1$ to $\mathbb{C}^2$. Similarly, $\mathbb{P}^1$ compactifies $\mathbb{C}$ by adding a point at infinity[144] and can be identified with the extended complex plane $\mathbb{C} \cup \{\infty\}$ (the model being the unit sphere in $\mathbb{R}^3$, from which it is obtained by stereographic projection). A singular point (singularity) of the affine algebraic curve $f(x,y) = 0$, or the projective algebraic curve $F(z,x,y) = 0$, is a point on the curve where f_x and f_y, respectively F_x, F_y, and F_z, are all zero. Thus (5.136) has no singularities while (5.137) has the singularity $[0 : 0 : 1]$ "at infinity" if $g \geq 2$. The importance of the singular points arises from the fact that by the implicit function theorem for analytic functions,[145] near a nonsingular point the solution set is a smooth curve. To understand what happens at the singularity $[0 : 0 : 1]$ of (5.137), we set $y = 1$ in the left side of (5.137) and write the outcome $z^{2g-1} - \prod_{k=0}^{2g}(\lambda_k z - x)$ as the sum of homogeneous polynomials in ascending degrees, selecting the polynomial of lowest degree z^{2g-1}. The solution to $z^{2g-1} = 0$ is $z = 0$ with multiplicity $2g - 1 \geq 3$ if $g \geq 2$, so that at $[0 : 0 : 1]$ we have a cusp singularity (cf. [132, Chapter II]). For $g = 1$ the curve (5.137) has the line at infinity $[0 : z_1 : z_2]$ as a tangent line at the regular point $[0 : 0 : 1]$. Therefore, unless $g = 1$, by considering the closed set (5.137) in the compact space $\mathbb{P}^2$, we compactified the curve (5.136), but we also introduced a cusp-type singularity. Topologically the elliptic curve (5.137) with $g = 1$ looks like a torus. Indeed, considering for a fixed complex number $\omega \in \mathbb{C} \setminus \{0\}$ the lattice $\Lambda_\omega = \{n + m\omega : n, m \in \mathbb{Z}\}$, the associated Weierstrass function $\wp_\omega$ given by (5.116) is analytic on the complement $\mathbb{C} \setminus \Lambda_\omega$ of the lattice Λ_ω and satisfies $\wp_\omega(z) = \wp_\omega(z + z_0)$ for $z \in \mathbb{C} \setminus \Lambda_\omega$ and $z_0 \in \Lambda_\omega$. Moreover, there exist complex numbers a_ω and b_ω such that $[\wp'_\omega(z)]^2 = 4[\wp_\omega(z)]^3 + a_\omega \wp_\omega(z) + b_\omega$, with the cubic $4x^3 + a_\omega x + b_\omega$ having different roots in $\mathbb{C}$. These facts translate into the geometric statement that the map $z \mapsto (\wp_\omega(z), \wp'_\omega(z))$ from $(\mathbb{C} \setminus \Lambda_\omega)/\Lambda_\omega$ to the elliptic curve $\{(x,y) \in \mathbb{C}^2 : y^2 = 4x^3 + a_\omega x + b_\omega\}$ is an isomorphism of complex manifolds. The set $(\mathbb{C} \setminus \Lambda_\omega)/\Lambda_\omega$ is "most" of the torus $\mathbb{C}/\Lambda_\omega$: since $\wp_\omega$ is not defined at the lattice points (where it has a pole), one point is missing, and to compensate for this we append the elliptic curve $y^2 = 4x^3 + a_\omega x + b_\omega$ in $\mathbb{P}^2$ by the point $[0 : 0 : 1]$, obtaining an analytic isomorphism between the torus $\mathbb{C}/\Lambda_\omega$ and the smooth elliptic curve $y^2 = 4x^3 + a_\omega x + b_\omega$ in $\mathbb{P}^2$. The converse result is also true: all elliptic curves (5.137) with $g = 1$ arise in this way from tori.[146] We have touched upon the intimate relationship between the study of compact Riemann surfaces[147] and that of algebraic curves: the smooth elliptic curves (5.137) with $g = 1$ are the realization of tori as subsets of $\mathbb{P}^2$. Actually, by the normalization theorem (cf. [132]), for any curve $\mathcal{C} \subset \mathbb{P}^2$ given by (5.137), there exist a compact Riemann surface M and a holomorphic map $\sigma : M \to \mathcal{C}$ called a *desingularization* of the curve $\mathcal{C}$ (cf. [275]), such that $\sigma(M) = \mathcal{C}$ and σ is injective on the inverse image of the set of regular points of $\mathcal{C}$. The surface M is a two-sheeted cover of $\mathbb{P}^1$ except over the points λ_k and

[144]Being a complex manifold, it admits a metric, so compactness is equivalent to every sequence having a convergent subsequence. Clearly $z_n = n$ has no limit in $\mathbb{C}$, but $z_n = [1 : n] = [\frac{1}{n} : 1]$ converges to the "point at infinity" $[0 : 1]$ in $\mathbb{P}^1$.

[145]Applicable also in $\mathbb{P}^2$ as this is a complex manifold, so locally it looks like an open set in $\mathbb{C}^2$.

[146]By the linear change of variables $x \mapsto \frac{x+\lambda_0+\lambda_1+\lambda_2}{2}$, $y \mapsto \frac{y}{2}$, the elliptic curve (5.136) with $g = 1$ is indistinguishable from an elliptic curve of the form $y^2 = 4x^3 + ax + b$.

[147]An abstract Riemann surface M is (cf. [282]) a one-dimensional connected complex manifold (that is, locally it is just an open set in $\mathbb{C}$ and the changes of charts are analytic). Viewed over $\mathbb{R}$, M will be an orientable, connected, smooth, two-dimensional manifold. From the topological viewpoint the compact Riemann surfaces are characterized by their *genus* $\mathfrak{g} \geq 0$: M is homeomorphic to a sphere with $\mathfrak{g}$ handles.

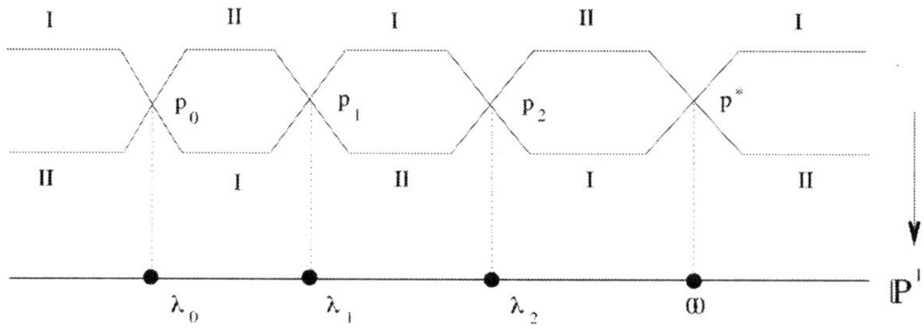

Figure 5.16. *Schematic representation of the elliptic curve as a two-sheeted covering of $\mathbb{P}^1$: at each branch point λ_0, λ_1, λ_2, $\infty = [0:0:1]$, we change sheets, compactness being captured by identifying on each of the three levels points at the extreme ends.*

$\infty = [0:0:1]$, where a branching of the radical takes place. There the cover is ramified over the base. Both cover and base are complex manifolds: near λ_k the local parameter is $\xi \in \mathbb{C}$ downstairs and $\sqrt{\xi - \lambda_k}$ upstairs, while at ∞ we use $\frac{1}{\xi}$ downstairs and $\frac{1}{\sqrt{\xi}}$ upstairs.

Unfortunately, the two-sheeted surface M associated to the hyperelliptic curve (5.137) cannot be realized in our three-dimensional Euclidean space, and it is this fact that lends an air of mystery to this surface. However, M is topologically[148] a sphere with g handles.

We illustrate this in Figure 5.17 for $g = 3$. To each $x \in \mathbb{C}$ correspond two values of y defined by $y^2 = (\lambda_0 - x)(\lambda_1 - x)\cdots(\lambda_6 - x)$, one the negative of the other (unless $x = \lambda_k$). We go from one to the other by continuing $y(x)$ over any closed path winding around just one of the roots λ_k. If we cut the z-plane from λ_1 to λ_2, we cannot wind around either λ_1 or λ_2 alone without crossing the cut. If we choose a path which winds around both λ_1 and λ_2 without intersecting the cut, the circuit produces two changes of sign and so no change at all. Notice that the radical $y(x)$ will have different signs on the two banks of the cut. Thus, if we separate the branch points into pairs (λ_1, λ_2), (λ_3, λ_4), (λ_5, λ_6), (λ_0, ∞) and make cuts joining λ_1 to λ_2, etc., we obtain a single-valued branch of $y(x)$ in $\mathbb{P}^1$ with the cuts, since we can never wind around an odd number of branch points. Take now two copies of the cut $\mathbb{P}^1$ space (identified with the Riemann sphere with three circular holes), place them next to each other, rotate the spheres until the holes face each other, and pull the edges of the cuts outward to make little tubes, with the $+$ edges of the tubes on one sphere opposite the $-$ edges of the tubes on the other sphere. Thus we can join together the ends of the tubes to form the next to last surface in Figure 5.17. The two spheres with the one tube joining the cuts between λ_0 and ∞ represent topologically a sphere, and each of the remaining three tubes looks like a handle on the sphere: thus we get the topological model of the Riemann surface associated to (5.137) as a sphere with g handles, g being the genus of the surface.

To clarify the topology of a sphere M with g handles, recall (cf. [282, Chapter 6]) that differential forms are an approach to calculus on manifolds that is independent of coordinates, providing a suitable setting for integration theory on manifolds.[149] A complex-valued function $f : M \to \mathbb{C}$ is called differentiable/analytic if the transition functions relative to local coordinates on an open cover are differentiable/analytic. This way one can define the tangent spaces T_pM at $p \in M$ and the *tangent bundle* $TM = \bigcup_{p \in M} T_pM$. One has a manifold structure on M and a projection

[148]That is, it can be identified by means of a continuous map that is bijective with a continuous inverse.

[149]One has to address two issues: integrals should not depend upon the selection of local coordinates, and, for regions which do not lie within one single coordinate disk in $\mathbb{C}$, the pieces have to be patched together. For this reason, one does not integrate functions on manifolds, but differential forms. These problems did not arise in the definition of integrals in $\mathbb{C}$ (or in any Euclidean space $\mathbb{R}^n$) because we had a single (macroscopic) coordinate system over the whole space.

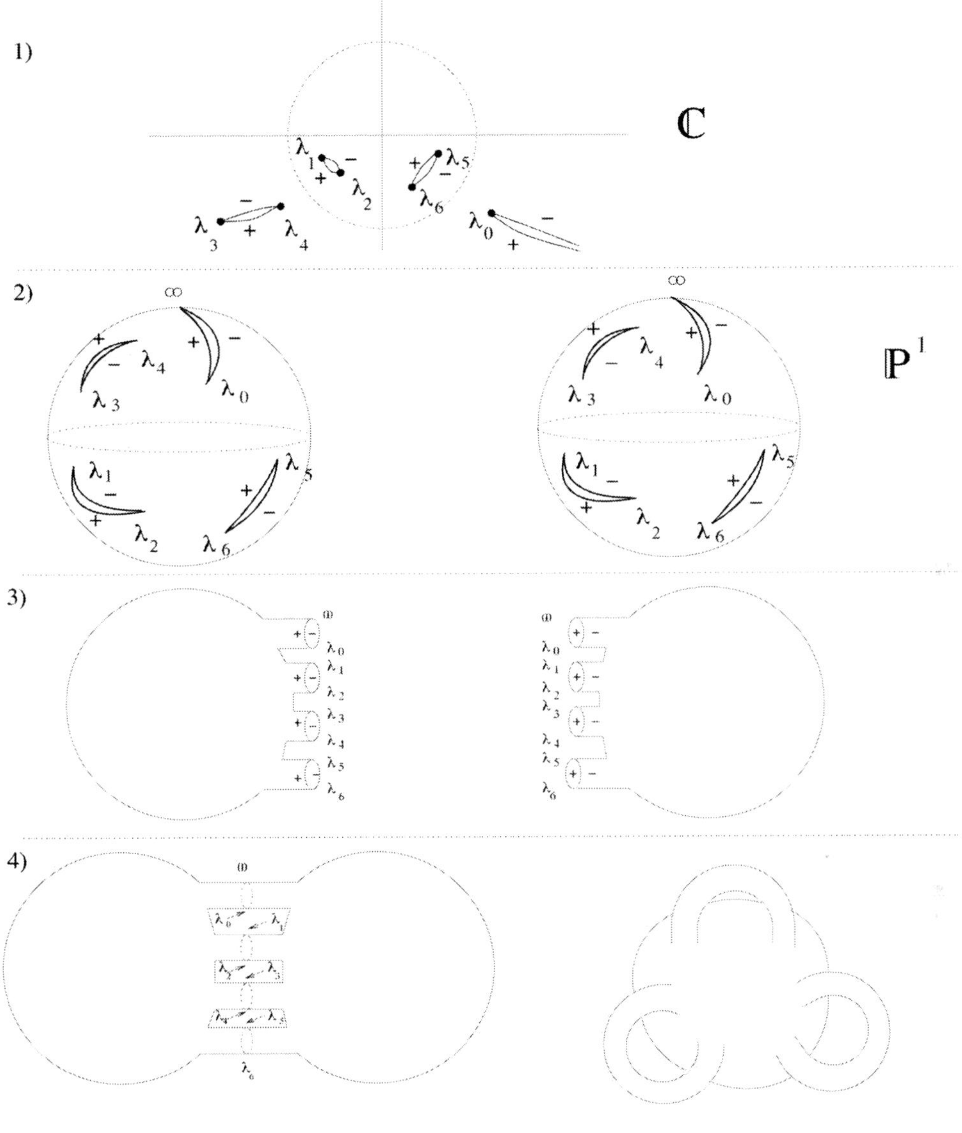

Figure 5.17. *The various stages in the topological construction of the Riemann surface for $g = 3$ (a sphere with three handles).*

map $\Pi : TM \to M$ with $\Pi^{-1}p = T_pM$ (cf. [268]). A *global section* of TM (that is, a smooth map $V : M \to TM$ with $\Pi(V(p)) = p$ on M) is called a *vector field*, with the understanding that $V(p) \in T_pM$. Since integration is dual to differentiation, differentials are duals to tangent vectors: a complex-valued function on TM which is linear on each space T_pM is called a *differential form*, or simply a *differential*. If $z = x + iy$ is a local coordinate at a point $p \in M$, then the partial derivatives $\frac{\partial f}{\partial x}(p)$ and $\frac{\partial f}{\partial y}(p)$ of differentiable functions $f : M \to \mathbb{C}$, written in the local coordinates x and y, determine differentiations $\frac{\partial}{\partial_x}|_p$ and $\frac{\partial}{\partial_y}|_p$ at p which form a basis of the tangent space T_pM, and a dual basis (for the dual cotangent vector space T_p^*M) is $dx|_p, dy|_p$ with

$$dx\left(\frac{\partial}{\partial_x}\right)=dy\left(\frac{\partial}{\partial_y}\right)=1,\ dx\left(\frac{\partial}{\partial_y}\right)=dy\left(\frac{\partial}{\partial_x}\right)=0.$$

Global sections of the *cotangent bundle* $T^*M=\bigcup_{p\in M}T_p^*M$ are the *differential* 1*-forms*, or simply the *differentials*, on M. In local coordinates a differential can be written as $f\,dx+g\,dy$, where f and g are complex-valued functions on an open set $U\subset M$, and $z=x+iy$ is a local coordinate on U, with the differential called *differentiable* if for every local representation, the functions f and g are differentiable. Defining

$$\frac{\partial}{\partial z}=\frac{1}{2}\left(\frac{\partial}{\partial x}-i\frac{\partial}{\partial y}\right),\quad \frac{\partial}{\partial \bar z}=\frac{1}{2}\left(\frac{\partial}{\partial x}+i\frac{\partial}{\partial y}\right),$$

we obtain another basis of T_pM, with the corresponding dual basis dz, $d\bar z$ of T_p^*M. We say that a differential is *holomorphic* or a *differential of the first kind* if locally it is of the form $h\,dz$, where h is a holomorphic function. If $\omega=\{h_\alpha(z_\alpha)dz_\alpha\}$ is a holomorphic form, where α is an index for the local parametrizations covering M, and $\gamma=\bigcup_\alpha\gamma_\alpha\subset M$ is a piecewise C^1-curve[150] on M, then we define

$$\int_\gamma\omega:=\sum_\alpha\int_{\gamma_\alpha}h_\alpha(z_\alpha)dz_\alpha,$$

the right-hand side being independent of choices of local coordinates and the partition of γ into local pieces. Of interest are also analytic differentials which have the local representation

$$\omega=(a_nz^n+a_{n+1}z^{n+1}+\cdots)dz,\qquad a_n\neq 0,$$

near $p_0\in M$, in which case we say that ω has a zero of order n at p_0 if $n>0$, and a pole of order $-n$ if $n<0$. If $n<0$, we say that the analytic differential is meromorphic; it will be analytic near p_0 except at a finite number of poles determined by its singular (or principal) part $a_nz^n+\cdots+a_{-1}z^{-1}$. If γ is the image of a parametric disk in which the principal part has no singularity other than $z=0$, then $\int_\gamma\omega=2\pi i a_{-1}$ is invariant under changes of the local parameter, so that a_{-1} is an invariant, called the *residue* of ω at p_0. A meromorphic differential all of whose singularities are poles of order no less than two will be called a *differential of the second kind*, while the *differentials of the third kind* are taken to be all differentials (including those of the first and second kinds). One can prove analogues of results from classical complex analysis (cf. [282, Chapter 6]):

- a bounded analytic function $f:M\to\mathbb{C}$ has to be constant;
- the sum of the residues of a meromorphic differential on M is zero;
- for a meromorphic function $f:M\to\mathbb{C}$ (that is, analytic except at a finite number of points where it has poles) the sum of the orders of the zeros is equal to the sum of the orders of the poles;[151]
- a meromorphic function $f:M\to\mathbb{C}$ assumes every value[152] the same number of times (counting multiplicity).

[150]That is, a continuous curve on M that is continuously differentiable except at finitely many points.

[151]Since df/f is an analytic differential with the residue at a zero or pole of f precisely the order of the zero or the negative order of the pole.

[152]By applying the previous result to the meromorphic function $f-z_0$, this value can be seen to be the same as the sum of the order of poles of f.

A basis of the g-dimensional linear space of differentials of the first kind is given by

$$\omega_k = \frac{z^{k-1}\,dz}{y(z)}, \qquad k = 1,\ldots,g,$$

where $y(z) = \sqrt{\prod_{j=0}^{2g}(\lambda_j - z)}$. Notice that these are everywhere regular differentials on M. Indeed, at any point $p \in M$, other than those corresponding to the g branch points in the representation of M as a two-sheeted cover of $\mathbb{P}^1$, we can select z as the local parameter. Near a point on M corresponding to λ_m we can take $\xi = \sqrt{z - \lambda_m}$ as the local parameter; then $\xi^2 = z - \lambda_m$ or $2\xi\,d\xi = dz$. In the neighborhood of $\xi = 0$ we have

$$\frac{z^{k-1}\,dz}{y(z)} = \frac{2\,(\xi^2 + \lambda_m)^{k-1}\,d\xi}{\sqrt{-\prod_{j=0,\,j\neq m}^{2g}[\lambda_j - \xi^2 - \lambda_m]}},$$

which is regular at $\xi = 0$. Finally, $\xi = \frac{1}{\sqrt{z}}$ is the local parameter near ∞ and $z = \frac{1}{\xi^2}$ or $dz = -\frac{2}{\xi^3}\,d\xi$, so that

$$y^2 = \left(\lambda_0 - \frac{1}{\xi^2}\right)\cdots\left(\lambda_{2g} - \frac{1}{\xi^2}\right) = \frac{1}{\xi^{2(2g+1)}}(\lambda_0\xi^2 - 1)\cdots(\lambda_{2g}\xi^2 - 1)$$

and

$$\frac{z^{k-1}\,dz}{y(z)} = \frac{-2\,\xi^{2g+1}\,d\xi}{\xi^{2k+1}\sqrt{\prod_{j=0}^{2g}(\lambda_j\xi^2 - 1)}}$$

is regular about $\xi = 0$ as $2g+1 \geq 2k+1$ for $1 \leq k \leq g$. If $\mathfrak{a}_j$ and $\mathfrak{b}_j$ $(1 \leq j \leq g)$ are closed cycles of M, with $\mathfrak{a}_j$ running once about the jth hole and $\mathfrak{b}_j$ running once about the jth handle, then for any smooth closed path $\gamma \subset M$ there exists a closed set $\Gamma \subset M$ with boundary $\partial\Gamma$ such that $\gamma = \sum_{j=1}^{g} m_j\mathfrak{a}_j + \sum_{j=1}^{g} n_j\mathfrak{b}_j + \partial\Gamma$, with the integers m_j and n_j uniquely determined by γ. We say that the $2g$ cycles form a basis of the *homology group* since the integral of a differential of the first kind over a closed path γ in M depends only on the homology class of γ: $\int_\gamma \omega = \sum_{j=1}^{g} m_j \int_{\mathfrak{a}_j}\omega_k + \sum_{j=1}^{g} n_j \int_{\mathfrak{b}_j}\omega_k$ for any differential ω_k of the first kind, with the values of the integrals of ω_k over cycles being called the *periods* of ω_k. For fixed points $\mathfrak{o}_1,\ldots,\mathfrak{o}_g \in M$ the map of the *divisor* $\mathfrak{P} = (\mathfrak{p}_1,\ldots,\mathfrak{p}_g)$ of points in M to the sum

$$\sum_{j=1}^{g}\int_{\mathfrak{o}_j}^{\mathfrak{p}_j}(\omega_1,\ldots,\omega_g) = \mathfrak{X}(\mathfrak{P})$$

covers $\mathbb{C}^g$; in fact, the map is onto but ambiguous, the paths of integration $\mathfrak{o}_j\mathfrak{p}_j$ being unspecified. The periods form a lattice $\mathbb{L}_\mathbb{C} \subset \mathbb{C}^g$, and the map of divisors is rendered unambiguous by viewing the image of $\mathfrak{X}(\mathfrak{P})$ as a point of the quotient $\mathbb{C}^g/\mathbb{L}_\mathbb{C}$. The quotient space is the *Jacobi variety* $J_\mathbb{C}$ of $\mathfrak{X}$: a g-dimensional complex torus. If we let $\mathfrak{p}_j$ be in *real position*, meaning that the projection lies in the lacuna $[\lambda_{2j-1},\lambda_{2j}]$, then the map is onto $\mathbb{R}^g$ provided the path $\mathfrak{o}_j\mathfrak{p}_j$ is restricted to the cycle covering $[\lambda_{2j-1},\lambda_{2j}]$; also, the period lattice $\mathbb{L}_\mathbb{R}$ obtained from closed paths is now real, and the quotient space $J = \mathbb{R}^g/\mathbb{L}_\mathbb{R}$ is a g-dimensional real torus, called the *real part* of the Jacobi variety, with the map from $\mathfrak{P}$ in real position to $\mathfrak{X}(\mathfrak{P}) \in J$ injective (see [114, 276]).

The periodicity condition

From (5.126) we find for $1 \leq k \leq g$ that

$$\frac{1}{2}\sum_{j=1}^{g}\frac{\mu_j^{k-1}}{\sqrt{(\lambda_0-\mu_j)(\lambda_1-\mu_j)\cdots(\lambda_{2g}-\mu_j)}}\,\frac{d\mu_j(t)}{dt} = \sum_{j=1}^{g}\frac{\mu_j^{k-1}}{\prod_{n=1,n\neq j}^{g}(\mu_j-\mu_n)} \tag{5.138}$$

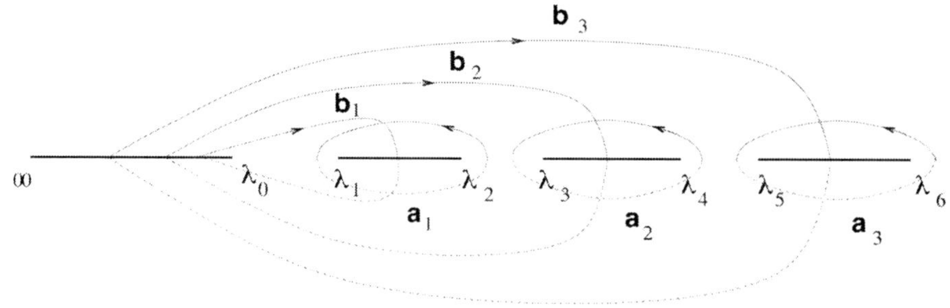

Figure 5.18. *An example of a canonical basis of cycles for $g = 3$. The parts of the cycles on the "lower" sheet of the covering are marked by dotted lines.*

with the sign $\pm$ absorbed in the definition of the radical $s(\mu_j(t))$ on the left side, so that it is more convenient to think of the left side as a differential of the first kind. Since for the rational functions

$$f_k(\lambda) = \frac{\lambda^{k-1}}{\prod_{n=1}^{g}(\lambda - \mu_n)}, \qquad 1 \le k \le g,$$

we have

$$\operatorname{Res}_{f_k}(\mu_j) = \frac{\mu_j^{k-1}}{\prod_{n=1, n\neq j}^{g}(\mu_j - \mu_n)}, \qquad 1 \le j, k \le g,$$

and

$$\lim_{R\to\infty} \int_{|\lambda|=R} f_k(\lambda)\,d\lambda = \begin{cases} 0, & 1 \le k \le g-1, \\ 2\pi i, & k = g, \end{cases}$$

by the residue theorem we can evaluate the right side of (5.138) as

$$\lim_{R\to\infty} \left(\frac{1}{2\pi i} \int_{|\lambda|=R} f_k(\lambda)\,d\lambda\right) = \begin{cases} 0, & 1 \le k \le g-1, \\ 1, & k = g. \end{cases}$$

Integration of (5.138) over $0 \le t \le 1$ yields the periodicity condition[153]

$$\sum_{m=1}^{g} \hat{m} \int_{\lambda_{2m-1}}^{\lambda_{2m}} \frac{\xi^{k-1}}{s(\xi)}\,d\xi = \begin{cases} 0, & 1 \le k \le g-1, \\ 1, & k = g, \end{cases} \tag{5.139}$$

where $s(\xi) = \sqrt{(\lambda_0 - \xi)(\lambda_1 - \xi)\cdots(\lambda_{2g} - \xi)}$. *The periodicity condition* (5.139) *is not only necessary but also sufficient for $\lambda_0 < \lambda_1 < \cdots < \lambda_{2g}$ to be the simple spectrum of the operator $\mathcal{L} = -\partial_x^2 + q(x)$ with $q \in C^\infty(\mathbb{R})$ of period* 1. Indeed, let us solve the differential equations (5.126) with the understanding that μ_j does not rest at λ_{2j-1} or λ_{2j} (the signature of the radical flips and μ_j turns back) for $1 \le j \le g$. Let the motion on the Riemann surface M associated to $s(\xi)$ begin at $p_j^0 = [\lambda_{2j-1}, 0] \in M$ at time $t = 0$ and end at $p_j = [\mu_j(1), s(\mu_j(1))] \in M$ at time $t = 1$, for $1 \le j \le g$. From (5.126), as above, we infer that

$$\frac{1}{2} \sum_{j=1}^{g} \frac{\mu_j^{k-1}(t)}{s(\mu_j(t))} \mu_j'(t) = \begin{cases} 0, & 1 \le k \le g-1, \\ 1, & k = g. \end{cases}$$

[153]Recall that $\mu_m(t)$ hits both λ_{2m-1} and λ_{2m} exactly $\hat{m}$ times in a full period $0 \le t < 1$, $\hat{m}$ being the index (counting also the collapsed gaps) of the gap $[\lambda_{2m-1}, \lambda_{2m}]$. As $\mu_m(t)$ moves in $[\lambda_{2m-1}, \lambda_{2m}]$ to the right/left, the sign of $s(\mu_m(t))$ is $\pm$.

Thus, for $1 \le k \le g$, we have

$$\frac{1}{2}\sum_{j=1}^{g}\int_{p_j^0}^{p_j}\frac{\xi^{k-1}}{s(\xi)}d\xi = \frac{1}{2}\sum_{j=1}^{g}\int_0^1 \frac{\mu_j^{k-1}(t)}{s(\mu_j(t))}\mu_j'(t)dt = \begin{cases} 0, & 1 \le k \le g-1, \\ 1, & k = g. \end{cases}$$

Using the periodicity condition (5.139), we get

$$\sum_{j=1}^{g}\int_{p_j^0}^{p_j}\frac{\xi^{k-1}}{s(\xi)}d\xi = 2\sum_{j=1}^{g}\hat{j}\int_{\lambda_{2j-1}}^{\lambda_{2j}}\frac{\xi^{k-1}}{s(\xi)}d\xi, \quad 1 \le k \le g,$$

or, what is better for the purpose,

$$\sum_{j=1}^{g}\int_{p_j^0}^{p_j}\frac{\xi^{k-1}}{s(\xi)}d\xi = 0, \qquad 1 \le k \le g,$$

by choice of real paths on M. If, say, the paths $p_j^0 p_j$ are trivial for $m < j \le g$ and nontrivial for $1 \le j \le m$, then

$$\det\left[\int_{p_j^0}^{p_j}\frac{\xi^{k-1}}{s(\xi)}d\xi \ : \quad 1 \le j \le j, k \le m\right]$$

$$= \int_{p_1^0}^{p_1}\cdots\int_{p_m^0}^{p_m}\begin{vmatrix} 1 & \xi_1 & \cdots & \xi_1^{m-1} \\ & \vdots & & \\ 1 & \xi_m & \cdots & \xi_m^{m-1}\end{vmatrix}\frac{d\xi_1}{s(\xi_1)}\cdots\frac{d\xi_m}{s(\xi_m)} \ne 0,$$

since the Vandermonde determinant is never zero (cf. [197]) as $\lambda_{2j-1} \le \xi_j \le \lambda_{2j}$ for $1 \le j \le m$, all paths on M being real. This contradiction proves that $p_j^0 = p_j$ for $1 \le j \le g$, so all $\mu_j(t)$ are of period 1 for $1 \le j \le g$. The function

$$q(x) = \sum_{j=0}^{2g}\lambda_j - 2\sum_{j=1}^{g}\mu_j(x), \tag{5.140}$$

suggested by the trace formula (5.121), is now seen to be smooth and of period 1; it is to be proved that λ_j $(0 \le j \le 2g)$ is the simple spectrum associated with this potential. The key step is to prove for

$$F(x,\lambda) = \prod_{j=1}^{g}(\lambda - \mu_j(x))$$

the identity[154]

$$s^2(\lambda) = \frac{1}{4}(F')^2 - \frac{1}{2}FF'' + F^2(q-\lambda). \tag{5.141}$$

This is equivalent to proving that for $\lambda \ne \mu_j$ $(1 \le j \le g)$, we have

$$\frac{1}{2}\left(\frac{F'}{F}\right)' + \frac{1}{4}\left(\frac{F'}{F}\right)^2 = q - \lambda - \frac{s^2}{F^2}. \tag{5.142}$$

Note that (5.126) takes the form

$$\mu_m' = \frac{2s(\mu_m)}{\dot{F}(\mu_m)}, \qquad 1 \le m \le g. \tag{5.143}$$

[154]Here $' = \partial/\partial_x$ and $\dot{} = \partial/\partial_\lambda$.

Since for $1 \le m \le g$,

$$\dot{F}(\mu_m) = \prod_{l \neq m} (\mu_m - \mu_l),$$

$$\partial_x \left(\frac{1}{\dot{F}(\mu_m)} \right) = - \frac{\sum_{j \neq m} \prod_{k \neq m,j} (\mu_m - \mu_k)(\mu_m' - \mu_j')}{\prod_{l \neq m} (\mu_m - \mu_l)^2}$$

$$= - \sum_{j \neq m} \prod_{k \neq m,j} \frac{1}{\mu_m - \mu_k} \left[\frac{2s(\mu_m)}{\dot{F}(\mu_m)} - \frac{2s(\mu_j)}{\dot{F}(\mu_j)} \right] \frac{1}{(\mu_m - \mu_j)^2},$$

$$\partial_x\, s(\mu_m) = \frac{2s(\mu_m)\dot{s}(\mu_m)}{\dot{F}(\mu_m)},$$

we deduce that

$$\frac{1}{2}\frac{F'}{F} = -\frac{1}{2} \sum_{m=1}^{g} \frac{\mu_m'}{\lambda - \mu_m} = - \sum_{m=1}^{g} \frac{1}{\lambda - \mu_m} \frac{s(\mu_m)}{\dot{F}(\mu_m)} \tag{5.144}$$

and

$$\frac{1}{2}\left(\frac{F'}{F}\right)' = - \sum_{m=1}^{g} \frac{1}{(\lambda - \mu_m)^2} \frac{2s^2(\mu_m)}{\dot{F}^2(\mu_m)} - \sum_{m=1}^{g} \frac{1}{\lambda - \mu_m} \frac{2s(\mu_m)\dot{s}(\mu_m)}{\dot{F}^2(\mu_m)}$$

$$+ \sum_{m=1}^{g} \frac{s(\mu_m)}{\lambda - \mu_m} \sum_{j \neq m} \prod_{k \neq m,j} \frac{1}{\mu_m - \mu_k} \left[\frac{2s(\mu_m)}{\dot{F}(\mu_m)} - \frac{2s(\mu_j)}{\dot{F}(\mu_j)} \right] \frac{1}{(\mu_m - \mu_j)^2}$$

$$= -2 \sum_{m=1}^{g} \frac{1}{(\lambda - \mu_m)^2} \frac{s^2(\mu_m)}{\dot{F}^2(\mu_m)} - 2 \sum_{m=1}^{g} \frac{1}{\lambda - \mu_m} \frac{s(\mu_m)\dot{s}(\mu_m)}{\dot{F}^2(\mu_m)}$$

$$+2 \sum_{m=1}^{g} \frac{1}{\lambda - \mu_m} \sum_{j \neq m} \left[\frac{s^2(\mu_m)}{\dot{F}^2(\mu_m)} - \frac{s(\mu_m)}{\dot{F}(\mu_m)} \frac{s(\mu_j)}{\dot{F}(\mu_j)} \right] \frac{1}{\mu_m - \mu_j}.$$

Taking advantage of the algebraic identity

$$\frac{1}{\lambda - \mu_m} \frac{1}{\mu_m - \mu_j} + \frac{1}{\lambda - \mu_j} \frac{1}{\mu_j - \mu_m} = \frac{1}{\lambda - \mu_m} \frac{1}{\lambda - \mu_j}, \qquad j \neq m,$$

and of the fact that

$$\ddot{F}(\mu_m) = 2 \sum_{j \neq m} \prod_{k \neq m,j} (\mu_m - \mu_k)$$

yields

$$\frac{\ddot{F}(\mu_m)}{\dot{F}(\mu_m)} = \sum_{j \neq m} \frac{2}{\mu_m - \mu_j}, \tag{5.145}$$

we obtain that

$$\begin{aligned}\frac{1}{2}\left(\frac{F'}{F}\right)' = &-2\sum_{m=1}^{g}\frac{1}{(\lambda-\mu_m)^2}\frac{s^2(\mu_m)}{\dot F^2(\mu_m)} - 2\sum_{m=1}^{g}\frac{1}{\lambda-\mu_m}\frac{s(\mu_m)\dot s(\mu_m)}{\dot F^2(\mu_m)}\\ &+\sum_{m=1}^{g}\frac{1}{\lambda-\mu_m}s^2(\mu_m)\frac{\ddot F(\mu_m)}{\dot F^3(\mu_m)} - 2\sum_{m=1}^{g}\sum_{j\neq m}\frac{1}{\lambda-\mu_m}\frac{1}{\lambda-\mu_j}\frac{s(\mu_m)}{\dot F(\mu_m)}\frac{s(\mu_j)}{\dot F(\mu_j)}.\end{aligned}$$

The previous relation and (5.144) enable us to express the left side of (5.142) as

$$-\sum_{m=1}^{g}\frac{1}{(\lambda-\mu_m)^2}\frac{s^2(\mu_m)}{\dot F^2(\mu_m)} + \sum_{m=1}^{g}\frac{1}{\lambda-\mu_m}\left[\frac{s^2(\mu_m)\ddot F(\mu_m)}{\dot F^3(\mu_m)} - \frac{2s(\mu_m)\dot s(\mu_m)}{\dot F^2(\mu_m)}\right],$$

and this is to be compared to $q-\lambda-s^2/F^2$. Both are rational functions of λ vanishing at ∞, in view of (5.140). We now claim that they have the same principal parts at their common poles $\lambda=\mu_m$ $(m=1,\dots,g)$, and therefore they are the same. Indeed, since

$$\frac{\dot F}{F} = \sum_{m=1}^{g}\frac{1}{\lambda-\mu_m},$$

we find that

$$\frac{s^2}{F^2} = \frac{s^2}{\dot F^2}\left\{\frac{1}{(\lambda-\mu_m)^2} + \frac{2}{\lambda-\mu_m}\sum_{k\neq m}\frac{1}{\lambda-\mu_k} + \cdots\right\}$$

so that the coefficient of $\frac{1}{(\lambda-\mu_m)^2}$ in the Laurent series expansion of the rational function $\lambda\mapsto\frac{s^2}{F^2}$, near $\lambda=\mu_m$, is $\frac{s^2(\mu_m)}{\dot F^2(\mu_m)}$, while the coefficient of $\frac{1}{\lambda-\mu_m}$ is

$$\begin{aligned}\frac{s^2(\mu_m)}{\dot F^2(\mu_m)}\sum_{k\neq m}\frac{2}{\mu_m-\mu_k} + \partial_\lambda\left(\frac{s^2}{\dot F^2}\right)\Big|_{\lambda=\mu_m} &= \frac{s^2(\mu_m)}{\dot F^2(\mu_m)}\frac{\ddot F(\mu_m)}{\dot F(\mu_m)} + \partial_\lambda\left(\frac{s^2}{\dot F^2}\right)\Big|_{\lambda=\mu_m}\\ &= -\frac{s^2(\mu_m)\ddot F(\mu_m)}{\dot F^3(\mu_m)} + \frac{2s(\mu_m)\dot s(\mu_m)}{\dot F^2(\mu_m)},\end{aligned}$$

in view of (5.145). The proof of the identity (5.141) is finished. By direct calculation we infer from (5.141) that

- $f_k(x)=\sqrt{F(x,\lambda_k)}$ satisfies $(-\partial_x^2+q)f_k=\lambda_k f_k$ for $k=0,\dots,2g$;
- for $\lambda\ge\lambda_0$ outside[155] the intervals $[\lambda_{2k-1},\lambda_{2k}]$ with $1\le k\le g$, we have

$$\begin{aligned}y_1(x,\lambda) = &\sqrt{\frac{F(x,\lambda)}{F(0,\lambda)}}\cos\left(i\,s(\lambda)\int_0^x F^{-1}(y,\lambda)\,dy\right)\\ &-\frac{1}{2}F'(0,\lambda)\sqrt{\frac{F(x,\lambda)}{F(0,\lambda)}}\,\frac{\sin\left(i\,s(\lambda)\int_0^x F^{-1}(y,\lambda)\,dy\right)}{i\,s(\lambda)},\end{aligned}$$

$$y_2(x,\lambda) = \pm\sqrt{F(0,\lambda)\,F(x,\lambda)}\,\frac{\sin\left(i\,s(\lambda)\int_0^x F^{-1}(y,\lambda)\,dy\right)}{i\,s(\lambda)},$$

where $i=\sqrt{-1}\in\mathbb{C}$, and the sign of $F(0,\lambda)$ determines the sign in the second formula.

[155] So that $F(x,\lambda)\neq 0$ for all $x\in[0,1]$, as $\mu_k(x)\in[\lambda_{2k-1},\lambda_{2k}]$.

Therefore λ_k, $k = 0,\ldots,2g$, are eigenvalues of $\mathcal{L} = -\partial_x^2 + q$ with q given by (5.140). They must be simple eigenvalues. Indeed, if some λ_k were a double eigenvalue for some $k \in \{0,\ldots,2g\}$, then at this point we would have a collapsing spectral lacunae with $\mu_k = \lambda_k$ not changing as we perform translations of the potential. But we know that $\mu_k(x)$ moves without rest for $x \in [0,1)$. It remains to show that the rest of the spectrum is double. For $\lambda \geq \lambda_0$ outside the intervals $[\lambda_{2k-1}, \lambda_{2k}]$ with $1 \leq k \leq g$, from the above formulas for $y_1(x,\lambda)$ and $y_2(x,\lambda)$ we compute

$$\Delta(\lambda) = \frac{y_1(1,\lambda) + y_2'(1,\lambda)}{2} = 2\cos\Big(i\,s(\lambda)\int_0^1 F^{-1}(y,\lambda)\,dy\Big).$$

For these values of λ we have $i\,s(\lambda) \in \mathbb{R}$, so that $|\Delta(\lambda)| \leq 1$ and therefore we can have only double eigenvalues in this range. Moreover, λ is a double eigenvalue $\lambda_{2\hat{n}-1} = \lambda_{2\hat{n}}$ if and only if $y_2(x,\lambda)$ has precisely $\hat{n}$ roots in a period $0 \leq x < 1$, that is, if and only if

$$i\,s(\lambda)\int_0^1 \frac{dx}{(\lambda-\mu_1(x))\cdots(\lambda-\mu_g(x))} = \pm\hat{n}\pi.$$

Since $s(\lambda_0) = 0$, the limit of the left side as $\lambda \downarrow \lambda_0$ is zero, and from the oscillatory behavior of the eigenfunctions we infer that λ_0 is the first eigenvalue. By the periodicity condition, $\mu_m(x)$ hits λ_{2m-1} and λ_{2m} exactly $\hat{m}$ times as $x \in [0,1)$, so that $[\lambda_{2m-1}, \lambda_{2m}]$ is the $\hat{m}$th lacunae (counting also the collapsed gaps). The upshot is an *identification of the isospectral class determined by the simple spectrum $\lambda_0 < \lambda_1 < \cdots < \lambda_{2g}$ with the real part J of the Jacobi variety of the hyperelliptic curve* $s(\lambda) = \sqrt{-\prod_{k=0}^{2g}(\lambda-\lambda_k)}$, the map $q \mapsto (p_1,\ldots,p_g)$, with $p_j = [\mu_j, s(\mu_j)]$ for $1 \leq j \leq g$, being $1:1$ and onto the g-dimensional torus J.

These considerations show also that *the finite-gap potential q is of period $1/k$ if and only if k divides $\hat{m}_j$ for all $j = 1,\ldots,g$*, where $\hat{m}_j$ is the actual index of the jth gap (counting also the collapsed gaps corresponding to double eigenvalues).

Straightening out the periodic KdV flow

We now show that if $\mathfrak{X}(t)$ is the point of the Jacobi variety assigned to the potential $q(\cdot,t)$, and if $t \mapsto q(\cdot,t)$ moves under the KdV flow, then $\frac{d\mathfrak{X}}{dt} = \mathfrak{s}_0$ is independent of t, so that $\mathfrak{X}(t)$ *moves in a straight line at constant speed,*[156] $\mathfrak{X}(t) = \mathfrak{X}(0) + t\,\mathfrak{s}_0$. To prove this, let $f(x) = c_n\, y_2(x,\mu_n)$ be the normalized eigenfunction[157] of the Dirichlet eigenvalue $\mu_n(t) \in [\lambda_{2n-1}, \lambda_{2n}]$ of the operator $(-\partial_x^2 + q(\cdot,t))$, for some index $n = 1,\ldots,g$, that fixes the noncollapsed lacunae, and some fixed

[156] This is not all transparent in the auxiliary coordinates p_j $(1 \leq j \leq g)$, the situation being analogous to the addition theorem of hyperelliptic integrals. Even in the simplest nontrivial case of genus $g = 1$, the fact that elliptic curves have purely algebraic properties is quite remarkable: one can define an operation on the points of an elliptic curve that turns the whole curve into an abelian group. Namely, if $g = 1$ and $s(x) = \sqrt{-\prod_{k=1}^{3}(x-\lambda_k)}$, the addition formula says that $\int_0^{\mathfrak{p}} \frac{dx}{s(x)} + \int_0^{\mathfrak{q}} \frac{dx}{s(x)} = \int_0^{\mathfrak{p}+\mathfrak{q}} \frac{dx}{s(x)}$, where $\mathfrak{p}+\mathfrak{q}$ refers to the group structure on the elliptic curve. The visualization of the addition formula in the (x,y)-plane is more intricate. To add two points P and Q on the elliptic curve, we first draw the straight line through P and Q (if $P = Q$ we draw through P the tangent line to the curve), and the third point of intersection of this line with the elliptic curve is reflected across the x-axis, by changing the sign of its y-coordinate, to provide us with the point $P+Q$. If the line through P and Q does not intersect the elliptic curve, we define the intersection point as the point at infinity, its mirror reflection being the same point at infinity. The additive inverse of a point is its mirror reflection with respect to the x-axis, and the point at infinity is the neutral element for the group addition (cf. [14]).

[157] With $c_n^2 = (\int_0^1 y_2^2(x,\mu_n)\,dx)^{-2}$ we have $\int_0^1 f^2(x)\,dx = 1$. Notice that $f(0) = f(1) = 0$ and f has $\hat{n}-1$ additional (simple) zeros in $(0,1)$.

time t. If f_ε is the corresponding eigenfunction for the potential $q(\cdot,t+\varepsilon)$, we have

$$\begin{cases} -f\varepsilon''(x)+q(x,t+\varepsilon)\,f_\varepsilon(x)=\mu_n(t+\varepsilon)\,f_\varepsilon(x), \\ \quad -f''(x)+q(x,t)\,f(x)=\mu_n(t)\,f(x). \end{cases} \tag{5.146}$$

Multiplying the first equation by f and subtracting from the outcome the second equation multiplied by f_ε, an integration on $[0,1]$ yields

$$\int_0^1 [q(x,t+\varepsilon)-q(x,t)]\,f_\varepsilon(x)\,f(x)\,dx=[\mu_n(t+\varepsilon)-\mu_n(t)]\int_0^1 f^2(x)\,dx.$$

First dividing by ε, and subsequently letting $\varepsilon\to 0$, we get

$$\frac{d\mu_n(t)}{dt}=\int_0^1 q_t(x,t)\,f^2(x)\,dx.$$

If q solves (5.115), the above relation becomes

$$\begin{aligned}\frac{d\mu_n(t)}{dt}&=\int_0^1\left(3qq'-\frac{1}{2}q'''\right)f^2\,dx=-3\int_0^1 q^2ff'\,dx+\int_0^1 q''ff'\,dx\\ &=-3\int_0^1 q^2ff'\,dx-\int_0^1 q'(f')^2\,dx-\int_0^1 q'ff''\,dx\\ &=-3\int_0^1 q^2ff'\,dx-\int_0^1 q'(f')^2\,dx-\int_0^1 qq'f^2\,dx+\mu_n\int_0^1 q'f^2\,dx\\ &=-2\int_0^1 q^2ff'\,dx-\int_0^1 q'(f')^2\,dx-2\mu_n\int_0^1 qff'\,dx.\end{aligned}$$

In the second equality of the first line we integrated by parts in each integral, the second line is obtained after integrating by parts in the last integral of the first line, and the third line is obtained by expressing f'' from the second relation in (5.146). Now the third integral in the expression on line three combines, after integration by parts, with the first integral, so that performing an integration by parts in the last integral produces the last line. Another integration by parts yields

$$\int_0^1 q'(f')^2\,dx=q\,(f')^2\Big|_0^1-2\int_0^1 q\,f'f''\,dx,$$

which, by expressing again f'' from the second relation in (5.146), leads us to

$$\int_0^1 q'(f')^2\,dx=q\,(f')^2\Big|_0^1+2\mu_n\int_0^1 q\,ff'\,dx-2\int_0^1 q^2\,ff'\,dx.$$

Therefore,

$$\frac{d\mu_n(t)}{dt}=-q\,(f')^2\Big|_0^1-4\mu_n\int_0^1 q\,ff'\,dx.$$

Expressing in the above integral qf from the second relation in (5.146), we obtain

$$\frac{d\mu_n(t)}{dt}=-q\,(f')^2\Big|_0^1-2\,\mu_n\,(f')^2\Big|_0^1=-[q(0,t)+2\mu_n]\Big([f'(1)]^2-[f'(0)]\Big).$$

Recalling now the motion of μ_n if we perform translations in the spatial variable of the potential, (5.126), (5.131), and the first equality in (5.132), we deduce that

$$\frac{d\mu_n(t)}{dt} = [q(0,t) + 2\mu_n]\frac{2\sqrt{-\prod_{m=0}^{2g}(\mu_n - \lambda_m)}}{\prod_{k=1,k\neq n}^{g}(\mu_n - \mu_k)}. \tag{5.147}$$

Let us now consider the point of the Jacobi variety J assigned to $q(\cdot,t)$, given by

$$\mathfrak{X}(t) = \sum_{j=1}^{g}\int_{\mathfrak{o}_j}^{\mathfrak{p}_j(t)}(\omega_1,\dots,\omega_g),$$

where $\mathfrak{o}_j = [\lambda_{2j-1}, 0]$, $\mathfrak{p}_j = [\mu_j(t), s(\mu_j(t))]$, and $\omega_j = \frac{\lambda^{j-1}\,d\lambda}{\sqrt{-\prod_{m=0}^{2g}(\lambda-\lambda_m)}}$ are the differentials of the first kind, with the path $\mathfrak{o}_j\mathfrak{p}_j$ restricted to the cycle covering the lacuna $[\lambda_{2j-1}, \lambda_{2j}]$ for $j = 1,\dots,g$. Using (5.147) we find that the motion of the kth component of $\mathfrak{X}(t)$ under the KdV flow (5.115) is governed by

$$\frac{d\mathfrak{X}_k(t)}{dt} = \sum_{j=1}^{g}\big[q(0,t) + 2\mu_j(t)\big]\frac{2\mu_j^{k-1}(t)}{\prod_{m=1,m\neq j}^{g}(\mu_j - \mu_m)}, \qquad 1 \le k \le g. \tag{5.148}$$

Consider the rational functions

$$f_k(z) = \big[q(0,t) + 2z\big]\frac{2\,z^{k-1}}{\prod_{m=1}^{g}(z-\mu_m)}, \qquad 1 \le k \le g,$$

for which

$$\operatorname{Res}_{f_k}(\mu_j) = \big[q(0,t) + 2\mu_j\big]\frac{2\,\mu_j^{k-1}}{\prod_{m=1,m\neq j}^{g}(\mu_j - \mu_m)}, \qquad 1 \le j,k \le g,$$

and

$$\operatorname{Res}_{f_k}(\infty) = -\operatorname{Res}_{z\mapsto \frac{1}{z^2} f_k\left(\frac{1}{z}\right)}(0) = \begin{cases} 0, & 1 \le k \le g-2, \\ -4, & k = g-1, \\ -2q(0,t) - 4\sum_{m=1}^{g}\mu_m, & k = g, \end{cases}$$

since for $|z|$ small we have

$$\begin{aligned}\frac{1}{z^2} f_k\Big(\frac{1}{z}\Big) &= 2\,[q(0,t)z + 2]\frac{z^{g-2-k}}{\prod_{m=1}^{g}(1 - z\mu_m)} = 2\,[q(0,t)z + 2]\frac{z^{g-2-k}}{1 - z\sum_{m=1}^{g}\mu_m - O(z^2)} \\ &= 2\,[q(0,t)z + 2]z^{g-2-k}\Big(1 + z\sum_{m=1}^{g}\mu_m + O(z^2)\Big).\end{aligned}$$

The sum of the residues of a rational function being zero (cf. [39]), we can evaluate the right side of (5.148), obtaining

$$\frac{d\mathfrak{X}_k(t)}{dt} = \begin{cases} 0, & 1 \le k \le g-2, \\ 4, & k = g-1, \\ 2q(0,t) + 4\sum_{m=1}^{g}\mu_m(t), & k = g. \end{cases} \tag{5.149}$$

Taking into account the trace formula (5.121), we find that $\mathfrak{s}_0 = (0,\dots,0,4,2\sum_{m=0}^{2g}\lambda_m)$ is independent of $\mathfrak{X}(t)$. While $q(\cdot,t)$ moves in a complicated nonlinear way under KdV, the corresponding point $\mathfrak{X}(t)$ moves in a straight line at constant speed; in the proper variables we can linearize and integrate KdV!

5.4.2 Alternative model equations within the same regime

The existence of solitary wave solutions for KdV is possible because the equation describes a balance between nonlinear effects (that normally lead to wave breaking) and dispersion effects (that cause waves to spread out due to the dependence of the propagation speed of a wave train on the wavelength).

Indeed, the inviscid Burgers equation

$$\eta_t + \eta\eta_x = 0 \tag{5.150}$$

can be solved by the method of characteristics to show that any initial profile $\eta_0 \in X$, $\eta \not\equiv 0$, develops into a breaking wave.[158] More precisely, introducing the characteristic curves $(x,t) \mapsto \xi(x,t)$ as the solutions to the ordinary differential equation $\xi_t = \eta(\xi,t)$ with data $\xi(x,0) = x$, we see that $\partial_t\,\eta(\xi(x,t),t) = 0$ along a classical C^1-solution of (5.150), so that $\eta(\xi(x,t),t) = \eta_0(x)$, as $\eta(x,0) = \eta_0(x)$. This enables us to deduce that the characteristic curves are actually straight lines: $\eta(x,t) = x + t\,\eta_0(x)$. Consequently, we obtain the implicit formula

$$\eta(x + t\,\eta_0(x),t) = \eta_0(x) \tag{5.151}$$

for the solution to (5.150). Note that the necessary and sufficient condition for two characteristic lines to intersect for $t > 0$ is the existence of $x_1 < x_2$ with $\eta_0(x_1) > \eta_0(x_2)$. Since the only nondecreasing initial profiles $\eta_0 \in X$ are $\eta_0 \equiv 0$, we deduce that for $\eta_0 \not\equiv 0$ we have characteristic lines that intersect in finite time

$$\tau = \inf_{\{x_1<x_2:\,\eta_0(x_1)>\eta_0(x_2)\}} \left\{ -\frac{x_2 - x_1}{\eta_0(x_2) - \eta_0(x_1)} \right\}.$$

By the mean-value theorem (cf. [96]) we see that $\tau = \inf_{\{x\in\mathbb{R}:\,\eta_0'(x)<0\}}\{\frac{-1}{\eta_0'(x)}\}$ is the breaking time of the wave: differentiating (5.151) with respect to x, we get

$$\eta_x(x + t\,\eta_0(x),t) = \frac{\eta_0'(x)}{1 + t\eta_0'(x)},$$

so that the slope of the solution becomes unbounded at time τ. Notice that η remains bounded in view of (5.151).

On the other hand, let us briefly elucidate the dispersive character of the *linearized KdV equation*

$$\eta_t + \eta_{xxx} = 0. \tag{5.152}$$

Let us find for a given $k > 0$ the condition on $\omega \in \mathbb{R}$ ensuring that (5.152) has a solution $\eta(x,t) = f(kx - \omega t)$ for some C^3-function $f : \mathbb{R} \to \mathbb{R}$ that is nonconstant and periodic with (minimal) period 2π. Let $s = kx - \omega t$. For $\omega = 0$ the general solution is a second-order polynomial in s, while for $\omega > 0$ it is, up to an additive constant, a linear combination of the functions $s \mapsto e^{\sqrt{\omega/k^3}\,s}$ and

[158] By a breaking wave we understand a wave profile that remains bounded but develops a vertical slope in finite time (see Chapter 6). However, (5.150) should not be regarded as an accurate model for water waves, the purpose of this discussion being actually a better understanding of the structure of KdV.

$s \mapsto e^{-\sqrt{\omega/k^3}\,s}$. These solutions are not periodic (unless they are constants). On the other hand, for $\omega < 0$ we have the general solution

$$f(s) = \alpha + \beta \sin\left(\sqrt{\frac{-\omega}{k^3}}\, s\right) + \gamma \cos\left(\sqrt{\frac{-\omega}{k^3}}\, s\right), \qquad s \in \mathbb{R},$$

with $\alpha, \beta, \gamma \in \mathbb{R}$ constants. If $\beta \neq 0$, the above function can be written as

$$f(s) = \alpha + \frac{\beta}{\cos(\arctan(\frac{\gamma}{\beta}))} \cos\left(\sqrt{\frac{-\omega}{k^3}}\, s - \arctan\left(\frac{\gamma}{\beta}\right)\right), \qquad s \in \mathbb{R}.$$

We see that the minimal period of f is 2π if and only if $\omega = -k^3$. The dependence of $c = \omega/k = -k^2$ upon k means that the equation is dispersive: wave trains of smaller wavelength $L = 2\pi/k$ will move faster than those of larger wavelength. Dispersion is a property of the equation (and not of some particular solution), and in general it is easier to visualize its effect on a localized initial profile rather than on a (periodic) wave train. To exemplify this, consider the Airy function, introduced by Airy back in 1838 in his study of light intensity, and defined by the improper integral

$$\mathrm{Ai}(x) = \frac{1}{\pi} \lim_{S \to \infty} \int_0^S \cos\left(\frac{s^3}{3} + sx\right) ds, \quad x \in \mathbb{R}.$$

As a tempered distribution, Ai is the Fourier transform of $k \mapsto e^{-ik^3/3}$, and we take for granted the following facts:[159] the Airy function $\mathrm{Ai}(x)$ is a bounded smooth function that is strictly positive and convex for $x > 0$, while for $x < 0$ it oscillates around zero with increasing frequency and decreasing amplitude as $x \to \infty$, with

$$\mathrm{Ai}(x) \approx \frac{e^{-\frac{2}{3}x^{3/2}}}{2\sqrt{\pi}\, x^{1/4}} \quad \text{as} \quad x \to \infty, \qquad \mathrm{Ai}(x) \approx \frac{\sin\left(\frac{2}{3}|x|^{3/2} + \frac{\pi}{4}\right)}{\sqrt{\pi}\, |x|^{1/4}} \quad \text{as} \quad x \to -\infty.$$

The profile of the Airy function is depicted in Figure 5.19. Consider now a nonnegative smooth function $x \mapsto \eta_0(x)$ of compact support, $\eta_0 \not\equiv 0$, but with $\eta_0(x) = 0$ for $|x| \geq a$ for some $a > 0$. Solving (5.152) with initial data $\eta(x,0) = \eta_0(x)$ using Fourier transforms, we obtain the solution

$$\eta(x,t) = \int_{\mathbb{R}} \hat{\eta}_0(k) e^{ik(x+k^2 t)}\, dk, \tag{5.153}$$

where $\hat{\eta}_0$ is the Fourier transform of η_0, defined by

$$\hat{\eta}_0(k) = \frac{1}{2\pi} \int_{\mathbb{R}} \eta_0(x) e^{-ikx}\, dx, \qquad x \in \mathbb{R}.$$

We can write (5.153) as

$$\eta(x,t) = (3t)^{-1/3} \int_{\mathbb{R}} \mathrm{Ai}\Big(\frac{x-y}{(3t)^{1/3}}\Big)\, \eta_0(y)\, dy.$$

Since η_0 has its support in the interval $[-a,a]$, we have

$$\eta(x,t) = (3t)^{-1/3} \int_{-a}^{a} \mathrm{Ai}\Big(\frac{x-y}{(3t)^{1/3}}\Big)\, \eta_0(y)\, dy.$$

[159]They can be looked up in any comprehensive treatise on special functions; see also [315].

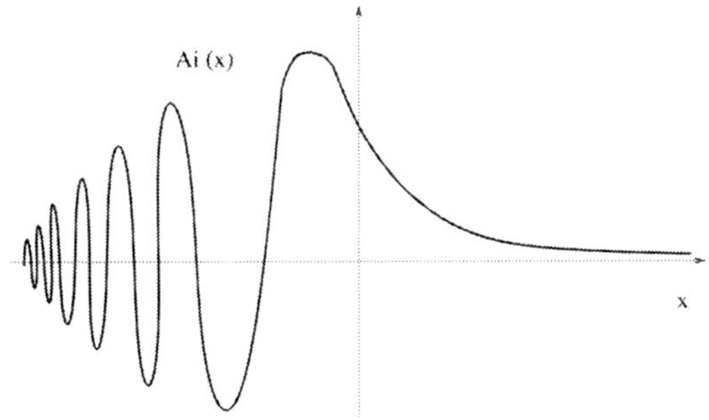

Figure 5.19. *The profile of the Airy function.*

Therefore, for every fixed $t > 0$ we have $\eta(x,t) > 0$ as soon as $x > a$, since then $\frac{x-y}{(3t)^{1/3}} > 0$ for all $y \in [-a,a]$. Also,

$$\max_{x\in\mathbb{R}} |\eta(x,t)| = O(t^{-1/3})$$

by the boundedness of the Airy function. The information we gathered about $\eta(x,t)$ shows that the dispersive property leads to a drastic change in the shape of the initial profile, as predicted. Indeed, at time $t = 0$ the wave is supported on $[-a,a]$, but at any later time $t > 0$ the disturbance is spread out (it can be noticed at any $x > a$), and its maximal amplitude decreases to zero as $t \to \infty$.

While there are many equations admitting solitary wave solutions, only solitary waves that emerge completely unscathed from the interaction with other solitary waves are solitons. An example of a nonlinear equation relevant to the study of shallow-water waves and having *solitary-wave solutions that are not solitons* is the BBM equation (sometimes also called the regularized long-wave equation or the Peregrine equation)

$$\eta_t - \eta_{txx} + \eta_x + 2\eta\eta_x = 0 \tag{5.154}$$

for the displacement $\eta(x,t)$ of the water's surface, introduced independently by Peregrine in 1966 [249] and by Benjamin, Bona, and Mahoney in 1972 [20] as an alternative model to KdV. When first encountered, the situation we refer to might seem perplexing: in talking about the propagation of shallow-water waves, in addition to the KdV equation there exists another nonlinear equation, the BBM equation, which provides an approximation of the governing equations for water wave equations in the shallow-water regime of the same accuracy as the KdV equation, whose solitary waves have similar expressions, but the solitary waves of KdV are solitons while those of the BBM equation are not since a solitary waves does not emerge completely unscathed from the interaction with another solitary wave (cf. [218]). Let us try to explain how this apparent paradox is resolved.

Considerable progress in understanding solitary waves based on the governing equations for water waves has been made. However, an in-depth study of solitary wave interactions using the governing equations is not yet within reach. To make progress one has therefore to make approximations that lead to simplified model equations. Since the linear theory of waves of small amplitude fails to yield any approximation to solitary waves [286], weakly nonlinear approximations to the governing equations for water waves have to be made, and this is how the KdV and BBM equations arise. In assessing the relative importance of these two model equations, the benchmark is provided by the degree to which they provide a description/explanation of specific important water-wave phenomena encountered in nature. Soliton interactions of water waves occur in nature and can be reproduced in the laboratories, with the predictions made by the KdV equation in close agreement with experimental measurements (cf. the discussion in [83]). It is thus no accident that

KdV plays a more important role in water-wave theory than the BBM equation (which remains a valid model equation but of more limited interest).

5.5 Notes for Chapter 5

We principally used [55] and [57] for the results on particle trajectories and pressure beneath an irrotational solitary gravity water wave. While the irrotational theory is well understood, there is a general lack of results for solitary waves on flows with vorticity. The only available existence results are for rotational solitary water waves of small amplitude [153, 133], and the absence of stagnation points throughout the flow is known to ensure that a rotational solitary gravity wave must have a symmetric and monotone profile (cf. [154]). No results (experimental, numerical, or analytical) are available concerning the particle paths or the pressure.

The principal sources for our survey of the direct/inverse scattering approach for KdV are [94, 98, 196]. As for the integrability of KdV in the class of spatially periodic functions, we followed [223, 224, 301] to a large extent. In addition to the presented results, as a further example of the rich structure tied into KdV, we point out that a spatially periodic finite-gap solution of KdV must be quasi-periodic[160] in time (cf. [203]). We only discuss the finite-gap case partly for historical reasons[161] and partly because the general case is technically more involved. The fact that finite-gap potentials form a dense family in the space of continuous periodic potentials (cf. [217]) shows the relevance of the presented results. Note that in the case of a general smooth periodic initial data (with infinitely many genuine lacunae), one can generalize notions of classical algebraic geometry, such as projective curves, holomorphic differentials, Jacobi varieties, etc., to the infinite-dimensional case [226, 301]. Other than the realization that these tools of algebraic geometry are robust, having a meaning in circumstances very remote from the classical setting, a substantial amount of qualitative information is gained using this approach. For example, the potential q is of spatial period $1/k$ if and only if all intervals $[\lambda_{2n-1}, \lambda_{2n}]$ collapse whenever n is not a multiple of k, and, on the scale of C^k-spaces, the smoothness in the x-variable of spatially periodic solutions of KdV is encoded in the asymptotics of the widths $\lambda_{2n} - \lambda_{2n-1}$. In particular, on this scale there is no gain of smoothness in time, and q is real-analytic if and only if the widths are exponentially decreasing.

The subject of infinite-dimensional integrable systems is relevant to physics and applied mathematics while also exhibiting a beautiful structure which appeals to pure mathematicians. In the context of infinite-dimensional systems, KdV provided an ideal concrete problem for which the approach was pioneered: not so wide that the hidden structure can be unveiled, not so narrow that no extension of any consequence can be accommodated. Integrability, in the sense that one can change coordinates to convert the original flow into straight-line motion at constant speed, is elusive: there are effectively no general theorems—only several beautiful examples. However, these examples are numerous among physically important systems, and an in-depth qualitative understanding of integrable systems permits the study in great detail of systems close to integrable ones. While the integrability property is structurally unstable, some of the structure of an integrable problem

[160] A continuous function $f : \mathbb{R} \to \mathbb{R}$ is called quasi-periodic if there exist an integer $n \geq 1$, some positive constants c_j $(j = 1, \dots, n)$, and a continuous function $F : \mathbb{R}^n \to \mathbb{R}$, periodic (with period 1) with respect to each variable, so that $f(t) = F(c_1 t, c_2 t, \dots, c_n t)$ for all $t \in \mathbb{R}$. If the real numbers c_j $(j = 1, \dots, n)$ are incommensurate, i.e., linearly independent over the rational numbers, then the function f is not periodic; the simplest example of a quasi-periodic function being the sum of two periodic functions with incommensurate periods. Such functions occur naturally in Hamiltonian mechanics to describe multiperiodic motions of integrable systems.

[161] The simple cnoidal wave solution, found by Korteweg and de Vries, was the only known spatially periodic solution of the KdV equation for almost a century [221]. Since it corresponds to a single open lacuna, it was natural to expect that the finite-gap potentials might represent the kind of initial data for which one could expect to develop an approach. The breakthrough was due to several researchers (cf. [100, 159, 224, 244]).

persists under perturbations, so that integrable models provide insight about the structure of nearby nonintegrable models. If one accepts that science is a search for structure, then the extraordinary structure of integrable systems compels one to their study. In physical contexts when a nonlinear partial differential equation that is an integrable system arises as a valid model (e.g., KdV as an equation for shallow-water waves of small amplitude), the appealing feature that the initial value problem can be solved exactly, with the detailed long-time behavior predicted from knowledge of the initial data, enhances the importance of these equations. For example, the study of periodic solutions to KdV did bring cnoidal waves to the attention of coastal engineers, who now use them regularly for engineering calculations.

Chapter 6

Breaking waves

"Truth in science can be defined as the working hypothesis best suited to open the way to the next better one."
K. Lorenz

Of all the magnificent scenes presented by water waves, breaking is surely among the most impressive. Not all waves break in the same way; it is possible to distinguish, roughly, two types of breaking phenomena.

Spilling breakers are a familiar sight on beaches with a gentle slope, but they can also appear in water waves propagating over a flat bed. They arise from long waves of low steepness: the crests of the wave increase in height, become peaky, and start to develop whitecaps—a turbulent mass of air and water running down the front slope of the wave, a process evident over considerable distances as the wave travels forward. Spilling breakers are not, as a rule, highly destructive.

The second type of breaking is usually restricted to shallow water, typically arising as steep waves move over moderately sloped beaches: the leading slope along the crest line of the wave becomes vertical, the top leaps forward and a veritable cascade tumbles down in front of the wave—we have a **plunging breaker**. Smaller plunging breakers are sometimes referred to as **collapsing breakers** if the plunge starts below the highest point of the wave. When the top of the wave crest plunges into the trough in front of it, air is trapped beneath it, becoming subject to compression so its pressure rises. The effect of this high pressure air can be seen because it causes spray to be projected upward: after breaking, the trapped air expands, forcing the aerated water upward. The weight of the plunging water combined with the force produced by the compressed air can be quite destructive.

When the beach slope exceeds wave steepness (but the shore is not close to vertical, so that reflection is ruled out), **surging breakers** are favored. The wave does not actually break but rapidly runs up the shore, producing foam and large surges of water.

In the first snapshot of Photograph 9 we see the verticality developing in the slope of the plunging breaker. In the second snapshot we clearly see the overturning, plunging breaker with an overhanging profile, while in the third snapshot the wave has collapsed, in the sense that the plunging section of the wave has fallen down onto the surface that exists between this wave and the preceding one. In all three snapshots, the spilling breaker is essentially unchanged. For this latter wave, the amplitudes and slopes are much less than

Photograph 9. *Spilling breaker in front of a plunging breaker [Hawaii].*

for the next wave, thus inhibiting the production of a plunging breaker. The wind, from the beach toward the ocean, causes the spray to be blown "backwards."

The experiments reported in Photograph 10 were performed in a glass-walled channel $25\,m$ long, $0.7\,m$ wide, with a beach of slope 1 : 10 beginning at $19.5\,m$ from the paddle, and filled to a depth of $0.6m$ with water. In the spilling case breaking occurs just ahead of the wave crest and evolves over a distance of almost a wavelength, while in the plunging case the breaking initiates on the steep front face. An important conclusion from the experimental data provided in [259] was that while an increase in steepness triggers wave breaking, a "snapshot" of the local wave geometry just before breaking is insufficient to predict spilling or plunging; the growth rate of the steepness is apparently needed to predict the type of breaking.

Although the mathematical description of the processes of breaking could hardly be regarded as complete, some theoretical investigations offer insight into this fundamental aspect of water waves. We consider waves at the surface of water with a flat bed so that the effects of the topography of the bed on the wave profile are not relevant; we refer the reader to [162] and [313] for some aspects of waves propagating over a sloping bed. For waves at sea (or in a channel) the assumption of a flat bed is realistic over large distances, and, although in this context spilling breakers are usually encountered, in special circumstances large-amplitude waves become plunging breakers. We assume that the two-dimensional waves under investigation have travelled some distance and have developed into a fairly regular pattern. Without necessarily restricting our attention to traveling waves, we assume that the wave pattern under investigation represents a weakly irregular perturbation of a traveling wave in the sense that averages over suitable times/distances approximate a traveling wave.

The present mathematical understanding of wave breaking is very limited. Following Whitham [313], we make the following specification.

Definition 6.1. *For a two-dimensional surface wave $\eta(x,t)$,* **wave breaking** *occurs at time $T > 0$ if the wave remains bounded but its slope becomes unbounded, that is,*

$$\sup_{(x,t)\in\mathbb{R}\times[0,T)} \{|\eta(x,t)|\} < \infty \quad \textit{and} \quad \limsup_{t\uparrow T} \left\{ \sup_{x\in\mathbb{R}} |\eta_x(x,t)| \right\} = \infty.$$

Whether after time $T > 0$ the wave breaks down as a plunging or as a spilling breaker is not within reach of our approach. As pointed out above, the experimental data gathered in [259] indicates that a local analysis is not sufficient to decide whether one is dealing with a plunging or a spilling breaker. However, notice that, in the case of a plunging breaker, verticality clearly has to precede the development of an overhanging profile; correspondingly, spilling breakers become peaky so that, at the instant before developing whitecaps, the wave slope becomes unbounded at the peak (even if this appears to occur very locally).

6.1 Long waves of small amplitude

The study of breaking waves, using the governing equations for water waves, has proved intractable. To make progress, one has therefore to make approximations that lead to simplified model equations. Within the confines of the linear theory, one cannot cope with the wave-breaking phenomenon so that nonlinear approximations to the governing equations for water waves have to be made. To this end, we need to use the length scales, time scales,

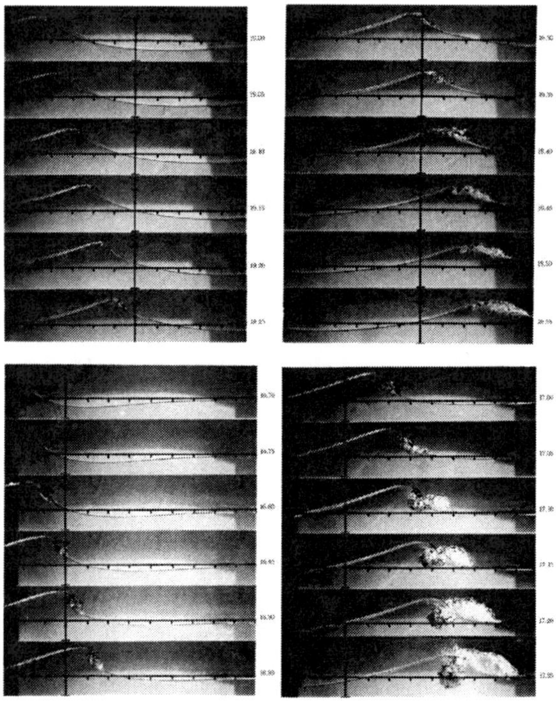

Photograph 10. *Photographs of a single spilling/plunging wave in the laboratory.* [From R. J. Rapp and W. K. Melville, Phil. Trans. Roy. Soc. London A, 331 (1990), 735–800. Reproduced with permission from The Royal Society]

The snapshots on the top (left and right) indicate the evolution of a spilling breaker, with a plunging breaker on the bottom (left and right). The thick marks are $10cm$ intervals and times (in seconds) at right are referenced from paddle start.

etc., that appear naturally by defining a set of nondimensional variables. The significance of this process lies in the fact that some parameters naturally arise in a particular problem; these enable us to define approximate linear or weakly nonlinear versions of the governing equations. In order to derive approximations to the governing equations, it is useful, first, to write them in nondimensional form so that the terms can be compared and so one can give a meaning to "small with respect to." We now describe the underlying principles of the process of nondimensionalizing and scaling that will lead to a suitable form of the governing equations for water waves. In particular, we introduce the two fundamental parameters (associated with amplitude and with wavelength) and show how they are used to represent different approximations (with corresponding different interpretations and applications).

The governing equations in physical variables

Choose Cartesian coordinates (X, Y) with the Y-axis pointing vertically upward, the X-axis being the direction of wave propagation, and with the origin located on the mean water level $Y = 0$. Let $(U(X,Y,T), V(X,Y,T))$ be the velocity field of the two-dimensional flow propagating in the X-direction over the flat bed $Y = -h_0$, and let $Y = H(X,T)$ be the water's free surface with mean water level $Y = 0$. According to the discussion in Chapter 2, the governing equations are given by the equation of mass conservation

$$U_X + V_Y = 0, \tag{6.1}$$

coupled with Euler's equation

$$\begin{cases} U_T + UU_X + VU_Y = -\dfrac{1}{\rho} P_X, \\ V_T + UV_X + VV_Y = -\dfrac{1}{\rho} P_Y - g, \end{cases} \tag{6.2}$$

where P is the pressure, g is the constant acceleration of gravity, and ρ is the constant density of water. The boundary conditions are

$$P = P_{atm} \qquad \text{on} \quad Y = H(X,T), \tag{6.3}$$

where P_{atm} is the (constant) atmospheric pressure at the water's free surface, with

$$V = H_T + UH_X \qquad \text{on} \quad Y = H(X,T) \tag{6.4}$$

and

$$V = 0 \qquad \text{on} \quad Y = -h_0. \tag{6.5}$$

For irrotational flows we also have

$$U_Y - V_X = 0 \tag{6.6}$$

throughout the flow.

Nondimensionalization

Since h_0 is the average depth of the water, the nondimensionalization of Y is chosen as

$$Y = h_0\, y, \tag{6.7}$$

which is to be understood as replacing the dimensional, physical variable Y by $h_0\, y$, where y is now a nondimensional version of the original Y. For the horizontal spatial variable, the nondimensionalization is also clear: if λ is some average or typical wavelength of the wave, we set

$$X = \lambda x. \tag{6.8}$$

The corresponding nondimensionalization of time is

$$T = \frac{\lambda}{\sqrt{gh_0}}\, t \tag{6.9}$$

(see the next subsection for an explanation of this particular choice). The governing equations for irrotational two-dimensional water waves read, in nondimensional form,

$$\begin{cases} \delta^2 \Phi_{xx} + \Phi_{yy} = 0 & \text{in} \quad \Omega(t), \\ \Phi_y = 0 & \text{on} \quad y = -1, \\ \zeta_t + \zeta_x \Phi_x - \dfrac{1}{\delta^2} \Phi_y = 0 & \text{on} \quad y = \varepsilon\,\zeta(x,t), \\ \Phi_t + \dfrac{\varepsilon}{2} \Phi_x^2 + \dfrac{\varepsilon}{2\delta^2} \Phi_y^2 + \zeta = 0 & \text{on} \quad y = \varepsilon\,\zeta(x,t), \end{cases} \tag{6.10}$$

where $x \mapsto \varepsilon\,\zeta(x,t)$ is a parametrization of the free surface at time t,

$$\Omega(t) = \{(x,y), -1 < y < \varepsilon\,\zeta(x,t)\}$$

is the fluid domain delimited above by the free surface and below by the flat bed $y = -1$, and $\Phi(\cdot,\cdot,t) : \Omega(t) \to \mathbb{R}$ is the velocity potential associated with the flow, so that the two-dimensional velocity field is given by $(u,v) = (\Phi_x, \Phi_y)$. Finally, ε and δ are two dimensionless parameters (the amplitude parameter and the shallowness parameter, respectively), defined as

$$\varepsilon = \frac{a}{h_0}, \qquad \delta = \frac{h_0}{\lambda}. \tag{6.11}$$

Making assumptions on the respective size of ε and δ leads to (simpler) asymptotic models from (6.10).

Long waves of small amplitude

In the **shallow-water regime**, defined by $\delta \ll 1$, one can derive the Green–Naghdi equations [131] without any assumption on ε, that is, for $\varepsilon = O(1)$. These equations couple the free surface elevation, ζ, to the vertically averaged horizontal component of the velocity,

$$u(x,t) = \frac{1}{1+\varepsilon\zeta(x,t)} \int_{-1}^{\varepsilon\zeta(x,t)} \partial_x \Phi(x,z,t)dz \tag{6.12}$$

and can be written as

$$\begin{cases} \zeta_t + \left[(1+\varepsilon\zeta)u\right]_x = 0, \\ u_t + \zeta_x + \varepsilon u u_x = \dfrac{\delta^2}{3} \dfrac{1}{1+\varepsilon\zeta} \left[(1+\varepsilon\zeta)^3 (u_{xt} + \varepsilon u u_{xx} - \varepsilon u_x^2)\right]_x, \end{cases} \tag{6.13}$$

where $O(\delta^4)$ terms have been discarded. Under the *additional* assumption that $\varepsilon \ll 1$ (this characterizes the regime of **waves of small amplitude**), the above system reduces, at first order, to the linear wave equation

$$\zeta_{tt} - \zeta_{xx} = 0 \quad \text{on} \quad y = 0, \tag{6.14}$$

with the general solution

$$\zeta(x,t) = f_+(x-t) + f_-(x+t),$$

where the $\pm$ sign refers to a wave of profile $f_\pm$ moving with unchanged shape to the right/left at constant unit speed. By virtue of the nondimensionalization that we have used, the corresponding dimensional speed is $\sqrt{gh_0}$, which explains our choice (6.9). Consequently, any perturbation of the surface splits up into two components moving in opposite directions. A natural issue is therefore to describe more accurately the motion of these two "unidirectional" waves.

The so-called **long-wave regime**

$$\delta \ll 1, \qquad \varepsilon = O(\delta^2), \tag{6.15}$$

emerges naturally since we obtain, instead of (6.13), a problem involving only one small parameter, ε. Korteweg and de Vries [184] found that in the regime (6.15) the right-going wave should satisfy the KdV equation

$$\zeta_t + \zeta_x + \varepsilon \frac{3}{2} \zeta \zeta_x + \delta^2 \frac{1}{6} \zeta_{xxx} = 0 \tag{6.16}$$

(see the Appendix to Chapter 7 for a formal derivation of KdV). In this context, let us point out that the evolution in time of $u = \zeta + O(\varepsilon)$ is also governed by the KdV equation. Notice that, at leading order, (6.16) reduces to the expected transport equation $\zeta_t + \zeta_x = 0$. Let us specify in what sense solutions to KdV approximate (6.10): given $\varepsilon_0 > 0$, there exists $T_0 > 0$ (independent of ε) such that, if (6.15) holds, then defining

$$\zeta^\varepsilon(x,t) = \zeta^+(\tau, x-t),$$

where $\tau = t\varepsilon$ and $\zeta^+(\tau,\xi)$ solves the KdV equation (6.16), for some $C > 0$ independent of $\varepsilon \in (0,\varepsilon_0)$, we have

$$|\zeta(x,t) - \zeta^\varepsilon(x,t)| \le C\,\varepsilon^2 t, \qquad t \in \left[0, \frac{T_0}{\varepsilon}\right],$$

for the solution to (6.10) with the same initial data (and similar estimates hold for the velocity field); cf. [8].

It has been noticed by Benjamin, Bona, and Mahoney [20] that the KdV equation belongs to a wider class of equations (the BBM equations):

$$\zeta_t + \zeta_x + \frac{3}{2}\varepsilon\,\zeta\zeta_x + \delta^2\left(\beta + \frac{1}{6}\right)\zeta_{xxx} + \delta^2 \beta\,\zeta_{xxt} = 0, \tag{6.17}$$

where[162] $\beta \le 0$. Equations (6.17) provide an approximation to the governing equations (6.10) of the same accuracy as KdV (cf. [8]) and exhibit, just like the particular case of KdV, a balance between the nonlinear effects (the $\zeta\zeta_x$ term) and the dispersive effects (the ζ_{xxx} and ζ_{xxt} terms) under the scaling (6.15). However, these equations do not account correctly for large-amplitude waves, whose behavior is more nonlinear than dispersive. In particular, one can show (cf. [295]) that for any initial wave profile $\zeta(\cdot,0) \in H^{k+1}(\mathbb{R}) \subset C^k(\mathbb{R})$ with[163] the integer $k \ge 3$, there exists a unique solution

$$\zeta \in C([0,\infty); H^k(\mathbb{R})) \cap C^1([0,\infty); H^{k-1}(\mathbb{R}))$$

[162] The condition $\beta \le 0$ is imposed to prevent ill-posedness issues: for $\beta > 0$, we see, by passing to Fourier transforms, that the linear part is not well-posed.

[163] There are results of this type requiring much lower regularity on the initial data (see the discussion in [295]), but for modelling purposes the setting of classical solutions is more appropriate.

to (6.17). Consequently, no equation from the BBM family (e.g., KdV) is appropriate as a model for breaking waves.

6.2 Wave breaking for long waves of moderate amplitude

For waves of moderate amplitude, characterized by larger values of ε, it is natural to investigate the following scaling:

$$\delta \ll 1, \qquad \varepsilon = O(\delta). \tag{6.18}$$

We are now going to show that within the regime (6.18), as it was the case in the regime (6.15), it is possible to derive an evolution equation for u that is consistent with the system (6.13), providing thus a **good approximation to the governing equations for water waves**. More precisely, given $\delta_0 > 0$, $M_1 > 0$, and $M_2 > 0$, there exist $d > 3$ and $t_0 > 0$ such that for every $u_0 \in H^{3+d}(\mathbb{R})$ with $\|u_0\|_{H^{3+d}(\mathbb{R})} \leq M_2$ and for every pair $(\varepsilon,\delta) \in \mathbb{R}^2$ satisfying

$$0 < \varepsilon \leq M_1\delta, \quad 0 < \delta \leq \delta_0, \tag{6.19}$$

the **Johnson equation**[164]

$$u_t + u_x + \frac{3}{2}\varepsilon u u_x + \frac{\delta^2}{12} u_{xxx} - \frac{\delta^2}{12} u_{xxt} + \frac{7\varepsilon\delta^2}{24} u u_{xxx} + \frac{4\varepsilon\delta^2}{3} u_x u_{xx} = 0, \tag{6.20}$$

with initial data u_0, has a unique solution

$$u^{\varepsilon,\delta} \in C^1\left(\left[0,\frac{t_0}{\varepsilon}\right]; H^d(\mathbb{R})\right) \cap C\left(\left[0,\frac{t_0}{\varepsilon}\right]; H^{d+1}(\mathbb{R})\right).$$

Setting

$$\zeta^{\varepsilon,\delta} = u^{\varepsilon,\delta} + \frac{\varepsilon}{4}[u^{\varepsilon,\delta}]^2 - \frac{\delta^2}{6} u_{xx}^{\varepsilon,\delta} - \frac{5\varepsilon\delta^2}{12} u^{\varepsilon,\delta} u_{xx}^{\varepsilon,\delta} - \frac{17\varepsilon\delta^2}{48}[u_x^{\varepsilon,\delta}]^2 \tag{6.21}$$

up to terms of order $O(\delta^4)$, for every pair (ε,δ) subject to (6.19) there exists a solution to (6.10) with initial average horizontal fluid velocity u_0 and initial wave profile ζ_0 obtained by specifying $u^{\varepsilon,\delta} = u_0$ on the right side of (6.21), such that the subsequent wave profile and the horizontal fluid velocity are approximated by $(\zeta^{\varepsilon,\delta}, u^{\varepsilon,\delta})$ with a precision $O(\delta^4 t)$ in $L^\infty([0,t] \times \mathbb{R})$ for all $t \in [0, \frac{t_0}{\varepsilon}]$. In view of the normalization (6.9), a time scale $O(\frac{1}{\varepsilon})$ for the nondimensionalized system (6.10) corresponds to the time scale $O(1)$ in the physical variables as $t = \frac{t_0}{\varepsilon}$ and $\varepsilon \leq M_1\delta$ yield $T \geq \frac{t_0}{M_1}\frac{\lambda^2}{h_0\sqrt{gh_0}}$. Notice that a nondimensional unit time corresponds in the physical variables to the time the wave needs to travel one wavelength.

Derivation of the Johnson equation

In our presentation we follow to some extent the (formal) asymptotic procedures due to Johnson [164] and refer the reader to [66] for a rigorous justification of these considerations. Henceforth, for the sake of simplicity, we write u and ζ instead of $u^{\varepsilon,\delta}$ and $\zeta^{\varepsilon,\delta}$, respectively, and we will use the notation $O(\delta)$, $O(\delta^2)$, etc., without explicitly mentioning the normed

[164]The first considerations of this type are due to Johnson [164] and were subsequently put on a firm and mathematically rigorous basis by Constantin and Lannes [66].

spaces to which we refer. Our aim is to show that by solving (6.20) and defining ζ by means of (6.21), we obtain functions that satisfy the system

$$\begin{cases} \zeta_t + \big[(1+\varepsilon\zeta)u\big]_x = \delta^4 r_1, \\ \\ u_t + \zeta_x + \varepsilon u u_x - \dfrac{\delta^2}{3}\dfrac{1}{1+\varepsilon\zeta}\Big[(1+\varepsilon\zeta)^3(u_{xt} + \varepsilon u u_{xx} - \varepsilon u_x^2)\Big]_x = \delta^4 r_2, \end{cases}$$

on the time interval $[0, \frac{t_0}{\varepsilon}]$, with $r_1, r_2 \in L^\infty([0, \frac{t_0}{\varepsilon}] \times \mathbb{R}; \mathbb{R})$. This means that our solutions are good approximations to the Green–Naghdi system (6.13), which in turn approximates well (in the above sense) the governing equations (6.10); cf. the considerations in [8].

Applying ∂_x^2 to (6.20) we get

$$u_{xxx} = -u_{xxt} - \frac{3\varepsilon}{2}(u u_x)_{xx} + O(\delta^2), \tag{6.22}$$

so that we can replace u_{xxx} in (6.20) by this expression truncated at $O(\delta^2)$ to get

$$u_t + u_x + \frac{3}{2}\varepsilon u u_x - \frac{\delta^2}{6}u_{xxt} + \frac{\varepsilon\delta^2}{6}u u_{xxx} + \frac{23\varepsilon\delta^2}{24}u_x u_{xx} = O(\delta^4). \tag{6.23}$$

We now seek w such that if $\zeta = u + \varepsilon w$, and u solves (6.20), then the second equation in (6.13) is satisfied up to a term of order $O(\delta^4)$. Since $\varepsilon = O(\delta)$, we have

$$\frac{1}{1+\varepsilon\zeta} = 1 - \varepsilon\zeta + O(\delta^2),$$

and within the regime (6.19), we need

$$\begin{aligned} u_t + [u+\varepsilon w]_x + \varepsilon u u_x &= \frac{\delta^2}{3}\Big\{u_{xt} + \varepsilon u u_{xx} - \varepsilon u_x^2 + 3\varepsilon[u+\varepsilon w]u_{xt}\Big\}_x \\ &\qquad - \frac{\varepsilon\delta^2}{3}[u+\varepsilon w]u_{xxt} + O(\delta^4) \\ &= \frac{\delta^2}{3}u_{xxt} + \varepsilon\delta^2\Big\{u_x u_{xt} + \frac{2}{3}u u_{xxt} + \frac{1}{3}(u u_{xx} - u_x^2)_x\Big\} + O(\delta^4). \end{aligned}$$

Adding to each side the expression

$$\frac{\varepsilon}{2}u u_x - \frac{\delta^2}{6}u_{xxt} + \varepsilon\delta^2\Big(\frac{1}{6}u u_{xx} + \frac{19}{48}u_x^2\Big)_x,$$

we can recast the previous relation as

$$\begin{aligned} &\varepsilon w_x + \Big[u_t + u_x + \frac{3\varepsilon}{2}u u_x - \frac{\delta^2}{6}u_{xxt} + \varepsilon\delta^2\Big(\frac{1}{6}u u_{xx} + \frac{19}{48}u_x^2\Big)_x\Big] \\ &= \frac{\varepsilon}{2}u u_x + \frac{\delta^2}{6}u_{xxt} + \varepsilon\delta^2\Big\{u_x u_{xt} + \frac{2}{3}u u_{xxt} + \Big(\frac{1}{2}u u_{xx} + \frac{1}{16}u_x^2\Big)_x\Big\} + O(\delta^4). \end{aligned}$$

From (6.23) we know that the term between the brackets on the left is $O(\delta^4)$, so that the second equation in (6.13) is satisfied up to $O(\delta^4)$ terms if

$$\varepsilon w_x = \frac{\varepsilon}{2} u u_x + \frac{\delta^2}{6} u_{xxt} + \varepsilon\delta^2 \left\{ u_x u_{xt} + \frac{2}{3} u u_{xxt} + \left(\frac{1}{2} u u_{xx} + \frac{1}{16} u_x^2 \right)_x \right\} + O(\delta^4).$$

Applying ∂_x and ∂_x^2 to (6.23) in turn, we find

$$u_{xt} = -u_{xx} + O(\delta), \tag{6.24}$$

$$u_{xxt} = -u_{xxx} - \frac{3\varepsilon}{2} (u u_x)_{xx} + O(\delta^2), \tag{6.25}$$

and hence

$$u_x u_{xt} + \frac{2}{3} u u_{xxt} = -\left(\frac{2}{3} u u_{xx} + \frac{1}{6} u_x^2 \right)_x + O(\delta),$$

allowing us to transform the previous requirement into

$$\varepsilon w_x = \frac{\varepsilon}{2} u u_x - \frac{\delta^2}{6} u_{xxx} - \varepsilon\delta^2 \left\{ \left(\frac{5}{12} u u_{xx} + \frac{17}{48} u_x^2 \right)_x \right\} + O(\delta^4). \tag{6.26}$$

This corresponds to the choice (6.21). It remains to check that with this choice the first equation in (6.13) is also satisfied up to $O(\delta^4)$. This amounts to checking that

$$u_t + \varepsilon w_t + u_x + 2\varepsilon u u_x + \varepsilon^2 w u_x + \varepsilon^2 w_x u = O(\delta^4). \tag{6.27}$$

But (6.21) yields

$$\begin{aligned} \varepsilon w_t &= \frac{\varepsilon}{2} u u_t - \frac{\delta^2}{6} u_{xxt} - \frac{5\varepsilon\delta^2}{12} u_t u_{xx} - \frac{5\varepsilon\delta^2}{12} u u_{xxt} - \frac{17\varepsilon\delta^2}{24} u_x u_{xt} \\ &= -\frac{\varepsilon}{2} u u_x - \frac{\delta^2}{6} u_{xxt} - \frac{3\varepsilon^2}{4} u^2 u_x + \frac{\varepsilon\delta^2}{3} u u_{xxx} + \frac{27\varepsilon\delta^2}{24} u_x u_{xx} + O(\delta^4) \end{aligned}$$

if we use, in the second equality, (6.24) and (6.25), and notice that

$$u_t = -u_x - \frac{3\varepsilon}{2} u u_x + \frac{\delta^2}{6} u_{xxt} + O(\delta^3)$$

in view of (6.23). On the other hand, from (6.26) and (6.21), we get

$$\varepsilon^2 w u_x + \varepsilon^2 w_x u = \frac{3\varepsilon^2}{4} u^2 u_x - \frac{\varepsilon\delta^2}{6} u_x u_{xx} - \frac{\varepsilon\delta^2}{6} u u_{xxx}.$$

Therefore, (6.27) holds in view of (6.25) and (6.23).

The evolution equation for the free surface waves

Unlike the case of the BBM equations (including KdV) in the regime (6.15), in the present setting the equation for the evolution of the free surface is not identical with the equation (6.20) for the average horizontal fluid velocity, being instead given by

$$\begin{aligned} &\zeta_t + \zeta_x + \frac{3}{2}\varepsilon\zeta\zeta_x - \frac{3}{8}\varepsilon^2\zeta^2\zeta_x + \frac{3}{16}\varepsilon^3\zeta^3\zeta_x \\ &\quad + \frac{\delta^2}{12}\zeta_{xxx} - \frac{\delta^2}{12}\zeta_{xxt} + \frac{7\varepsilon\delta^2}{24}(\zeta\zeta_{xxx} + 2\zeta_x\zeta_{xx}) = 0. \end{aligned} \tag{6.28}$$

To obtain this, we proceed as follows. First, notice that in view of (6.21)

$$\partial_x^k u = \partial_x^k \zeta + O(\delta), \qquad k = 0,1,2,3, \tag{6.29}$$

so that in terms of order $O(\delta^3)$ we may replace $\partial_x^k u$ by $\partial_x^k \zeta$ at the expense of an error term of order $O(\delta^4)$. Differentiate (6.21) with respect to time to get

$$\zeta_t = u_t + \frac{\varepsilon}{2} u u_t - \frac{\delta^2}{6} u_{xxt} - \frac{5\varepsilon\delta^2}{12} u_t u_{xx} - \frac{5\varepsilon\delta^2}{12} u u_{xxt} - \frac{17\varepsilon\delta^2}{24} u_x u_{xt}.$$

On the other hand, from (6.20) we get

$$u_{xxt} = -u_{xxx} - \frac{3\varepsilon}{2}(u u_x)_{xx} + O(\delta^2) \tag{6.30}$$

as well as

$$\partial_x^k u_t = -\partial_x^k u_x + O(\delta), \qquad k = 0,1,2.$$

The last four relations lead to

$$\zeta_t = u_t + \frac{\varepsilon}{2} u u_t + \frac{\delta^2}{6} u_{xxx} + \frac{15\varepsilon\delta^2}{8} \zeta_x \zeta_{xx} + \frac{2\varepsilon\delta^2}{3} \zeta \zeta_{xxx} + O(\delta^4).$$

Using (6.20) to express u_t, we transform this equation into

$$\begin{aligned}
\zeta_t &= -\left(1 + \frac{\varepsilon}{2} u\right)\left(u_x + \frac{3\varepsilon}{2} u u_x + \frac{\delta^2}{12} u_{xxx} - \frac{\delta^2}{12} u_{xxt} + \frac{7\varepsilon\delta^2}{24} u u_{xxx} + \frac{4\varepsilon\delta^2}{3} u_x u_{xx}\right) \\
&\quad + \frac{\delta^2}{6} u_{xxx} + \frac{15\varepsilon\delta^2}{8} \zeta_x \zeta_{xx} + \frac{2\varepsilon\delta^2}{3} \zeta \zeta_{xxx} + O(\delta^4) \\
&= -u_x - 2\varepsilon u u_x - \frac{3\varepsilon^2}{4} u^2 u_x + \frac{\delta^2}{12} u_{xxx} + \frac{\delta^2}{12} u_{xxt} \\
&\quad + \frac{13\varepsilon\delta^2}{24} \zeta_x \zeta_{xx} + \frac{7\varepsilon\delta^2}{24} \zeta \zeta_{xxx} + O(\delta^4)
\end{aligned}$$

when we recall (6.29) and (6.30). Applying ∂_x to (6.21) to obtain an expression for u_x, this becomes

$$\begin{aligned}
\zeta_t = -\zeta_x - \frac{3\varepsilon}{2} u u_x - \frac{\delta^2}{12} u_{xxx} - \frac{3\varepsilon^2}{4} u^2 u_x + \frac{\delta^2}{12} u_{xxt} \\
- \frac{7\varepsilon\delta^2}{12} \zeta_x \zeta_{xx} - \frac{\varepsilon\delta^2}{8} \zeta \zeta_{xxx} + O(\delta^4)
\end{aligned} \tag{6.31}$$

again using (6.29). On the other hand, (6.21) yields

$$u = \zeta - \frac{\varepsilon}{4} u^2 + O(\delta^2) = \zeta - \frac{\varepsilon}{4}\left(\zeta - \frac{\varepsilon}{4} u^2\right)^2 + O(\delta^2) = \zeta - \frac{\varepsilon}{4} \zeta^2 + O(\delta^2),$$

so that

$$u^2 = \zeta^2 - \frac{\varepsilon}{2} \zeta^3 + O(\delta^2) \tag{6.32}$$

and

$$u_{xx} = \zeta_{xx} - \frac{\varepsilon}{2} \zeta_x^2 - \frac{\varepsilon}{2} \zeta \zeta_{xx} + O(\delta^2). \tag{6.33}$$

Since

$$u = \zeta - \frac{\varepsilon}{4} u^2 + \frac{\delta^2}{6} u_{xx} + O(\delta^3)$$

by (6.21), the last three relations yield

$$u = \zeta - \frac{\varepsilon}{4} \zeta^2 + \frac{\varepsilon^2}{8} \zeta^3 + \frac{\delta^2}{6} \zeta_{xx} + O(\delta^3). \tag{6.34}$$

From (6.34) we then get

$$u_x = \zeta_x - \frac{\varepsilon}{2} \zeta \zeta_x + \frac{3\varepsilon^2}{8} \zeta^2 \zeta_x + \frac{\delta^2}{6} \zeta_{xxx} + O(\delta^3), \tag{6.35}$$

while (6.33) yields

$$\begin{cases} u_{xxx} = \zeta_{xxx} - \dfrac{3\varepsilon}{2} \zeta_x \zeta_{xx} - \dfrac{\varepsilon}{2} \zeta \zeta_{xxx} + O(\delta^2), \\ u_{xxt} = \zeta_{xxt} + \dfrac{3\varepsilon}{2} \zeta_x \zeta_{xx} + \dfrac{\varepsilon}{2} \zeta \zeta_{xxx} + O(\delta^2), \end{cases} \tag{6.36}$$

where in the last relation we took advantage of the fact that $\zeta_t = -\zeta_x + O(\delta)$ according to (6.31), and therefore

$$\partial_x^k \zeta_t = -\partial_x^{k+1} \zeta + O(\delta) \quad \text{for} \quad k = 0, 1, 2.$$

From (6.31)–(6.36) we obtain precisely (6.28) by neglecting the terms of order $O(\delta^4)$.

The approach used before can also provide us, in analogy to the way the BBM equations (6.17) arise as generalizations of the KdV equation (6.16), with a one-parameter family of equations

$$\begin{aligned} & u_t + u_x + \frac{3}{2} \varepsilon u u_x + \left(\beta + \frac{1}{6}\right) \delta^2 u_{xxx} + \beta \delta^2 u_{xxt} \\ & \quad + \left(\frac{3}{2}\beta + \frac{5}{12}\right) \varepsilon \delta^2 u u_{xxx} + \left(\frac{9}{2}\beta + \frac{41}{24}\right) \varepsilon \delta^2 u_x u_{xx} = 0 \end{aligned} \tag{6.37}$$

for the average horizontal velocity u, coupled with the equation

$$\begin{aligned} & \zeta_t + \zeta_x + \frac{3}{2} \varepsilon \zeta \zeta_x - \frac{3}{8} \varepsilon^2 \zeta^2 \zeta_x + \frac{3}{16} \varepsilon^3 \zeta^3 \zeta_x + \left(\beta + \frac{1}{6}\right) \delta^2 \zeta_{xxx} + \beta \delta^2 \zeta_{xxt} \\ & \quad + \left(\frac{3}{2}\beta + \frac{5}{12}\right) \varepsilon \delta^2 u u_{xxx} + \left(\frac{9}{2}\beta + \frac{23}{24}\right) \varepsilon \delta^2 u_x u_{xx} = 0 \end{aligned} \tag{6.38}$$

describing the evolution of the free surface, ζ. In (6.37) and in (6.38) we require $\beta < 0$ since for $\beta > 0$ ill-posedness issues arise at the linear level (as seen if one passes to Fourier transforms), while the limiting case $\beta = 0$ is avoided since it changes the type of the equation.

Equations (6.37) and (6.38) provide an approximation to the governing equations at the same order as the BBM equations, (6.17), namely, $O(\delta^4)$; cf. [66]. The only difference in the derivation of these equations lies in the different scalings (6.15) and (6.18). Since $O(\delta^2) = O(\delta)$ when δ is small, *the long-wave scaling* (6.15) *is contained in the scaling* (6.18), and, consequently, the BBM equations (6.17) can be recovered as a special case of

(6.37) and (6.38) when $\varepsilon = O(\delta^2)$: with $\varepsilon = O(\delta^2)$, by neglecting terms of order $O(\delta^4)$, both (6.37) and (6.38) reduce to (6.17). Notice that, with the scaling (6.18), we necessarily have $\varepsilon \ll 1$, and thus, in the regime (6.15), there is still the reduction to a simple wave equation at leading order; however, the dimensionless parameter ε is larger here, and so the nonlinear effects are stronger.

Remark. As mentioned in the Appendix to Chapter 5, in deriving models it is possible to obtain different equations in a given regime (e.g., in our setting there are trade-offs of type (6.30) that could be performed), and it is natural to choose from among them the one with better structural properties. Notice that (6.20) and (6.28) are obtained by setting $\beta = -\frac{1}{12}$, this being the only case when the coefficients of the terms of highest order in (6.38) are in a ratio 1 : 2. Unlike the case of the BBM family (6.17), where the choice $\beta = -\frac{1}{12}$ produced the integrable KdV equation (cf. [66, 160]), there is no value of the parameter β in (6.38) that yields an integrable equation (see the Notes to this chapter for more on this theme). However, we will see below that for $\beta = -\frac{1}{12}$ an in-depth qualitative study of (6.38) is possible.

Existence of breaking waves

The importance of (6.28) lies in its **relevance to the modelling of breaking waves**. Notice first that, for any initial data $\zeta_0 \in H^3(\mathbb{R})$, using Kato's semigroup approach for quasi-linear hyperbolic equations (see the Appendix to this chapter), one can show that there exists a unique solution

$$\zeta \in C([0,\tau);\, H^3(\mathbb{R})) \cap C^1([0,\tau);\, H^2(\mathbb{R}))$$

of (6.28) with initial data ζ_0, defined for a maximal time $\tau > 0$. Moreover, if $\tau < \infty$, then $\limsup_{t\uparrow\tau} \|\zeta(t,\cdot)\|_{H^3(\mathbb{R})} = \infty$. Let us now show that the only way singularities can develop in finite time for solutions of (6.28) is in the form of breaking waves.

Theorem 6.2. *If for some initial data $\zeta_0 \in H^3(\mathbb{R})$ the maximal existence time $\tau > 0$ of the solution to* (6.28) *is finite, then*

$$\sup_{t\in[0,\tau),\, x\in\mathbb{R}} |\zeta(t,x)| < \infty, \tag{6.39}$$

while

$$\limsup_{t\uparrow\tau} \{\zeta_x(t,x)\} = \infty. \tag{6.40}$$

Proof. Since $\tau < \infty$ means that $\limsup_{t\uparrow\tau} \|\zeta(t,\cdot)\|_{H^3(\mathbb{R})} = \infty$, it suffices to prove (6.39) and to show that the existence of $M > 0$ with

$$\zeta_x(t,x) \le M, \qquad t \in [0,\tau), \quad x \in \mathbb{R}, \tag{6.41}$$

ensures the boundedness of $\|\zeta(t,\cdot)\|_{H^3(\mathbb{R})}$ on finite time intervals.

The embedding $L^\infty(\mathbb{R}) \subset H^1(\mathbb{R})$ proves (6.39) since multiplying (6.28) by ζ and integrating over $\mathbb{R}$ yields

$$\partial_t \left\{ \int_{\mathbb{R}} \zeta^2\, dx + \frac{\delta^2}{12} \int_{\mathbb{R}} \zeta_x^2\, dx \right\} = 0. \tag{6.42}$$

Now choose $M_0 > 0$ such that

$$|\zeta(t,x)| \le M_0, \qquad t \in [0,\tau), \quad x \in \mathbb{R}. \tag{6.43}$$

If the initial data ζ_0 is more regular, e.g., if $\zeta_0 \in H^6(\mathbb{R})$, then Kato's theory would ensure $C^1([0,\tau); H^5(\mathbb{R}))$; by differentiating (6.28) twice with respect to the spatial variable, multiplying the outcome by ζ_{xx} and integrating, and taking advantage of some integration by parts, we obtain

$$\begin{aligned}\partial_t \left\{ \int_{\mathbb{R}} \zeta_{xx}^2\,dx + \frac{\delta^2}{12}\int_{\mathbb{R}} \zeta_{xxx}^2\,dx \right\} &= 15\varepsilon \int_{\mathbb{R}} \zeta\zeta_{xx}\zeta_{xxx}\,dx - \frac{15}{4}\varepsilon^2 \int_{\mathbb{R}} \zeta^2\zeta_{xx}\zeta_{xxx}\,dx \\ &\quad + \frac{9\varepsilon^3}{16}\int_{\mathbb{R}} \zeta_x^5\,dx + \frac{15\varepsilon^3}{8}\int_{\mathbb{R}} \zeta^3\zeta_{xx}\zeta_{xxx}\,dx + \frac{35\,\varepsilon\delta^2}{24}\int_{\mathbb{R}} \zeta_x\zeta_{xxx}^2\,dx.\end{aligned}$$

Assuming the validity of (6.41), then by using (6.43) and the Cauchy–Schwarz inequality, we infer from the previous relation and (6.42) that

$$\partial_t E(t) \le K\,E(t), \qquad t \in (0,\tau), \tag{6.44}$$

where

$$E(t) = \int_{\mathbb{R}} \zeta^2\,dx + \frac{\delta^2}{12}\int_{\mathbb{R}} \zeta_x^2\,dx + \int_{\mathbb{R}} \zeta_{xx}^2\,dx + \frac{\delta^2}{12}\int_{\mathbb{R}} \zeta_{xxx}^2\,dx$$

and

$$K = \frac{30M_0\varepsilon}{\delta} + \frac{15M_0^2\varepsilon^2}{2\delta} + \frac{27M^3\varepsilon^3}{4\delta^2} + \frac{15M_0^3\varepsilon^3}{4\delta} + \frac{35M\varepsilon}{2}.$$

Indeed, we have

$$\begin{aligned}\int_{\mathbb{R}} \zeta\zeta_{xx}\zeta_{xxx}\,dx \le M_0\int_{\mathbb{R}} |\zeta_{xx}\zeta_{xxx}|\,dx &\le \frac{M_0\sqrt{3}}{\delta}\int_{\mathbb{R}} \left(\zeta_{xx}^2\,dx + \frac{\delta^2}{12}\zeta_{xxx}^2\right)dx \\ &\le \frac{2M_0}{\delta}E(t),\end{aligned}$$

and, similarly,

$$\begin{aligned}-\int_{\mathbb{R}} \zeta^2\zeta_{xx}\zeta_{xxx}\,dx &\le \frac{M_0^2\sqrt{3}}{\delta}\int_{\mathbb{R}} \left(\zeta_{xx}^2\,dx + \frac{\delta^2}{12}\zeta_{xxx}^2\right)dx \le \frac{2M_0^2}{\delta}E(t), \\ \int_{\mathbb{R}} \zeta_x^5\,dx &\le M^3\int_{\mathbb{R}} \zeta_x^2\,dx \le \frac{12M^3}{\delta^2}E(t), \\ \int_{\mathbb{R}} \zeta^3\zeta_{xx}\zeta_{xxx}\,dx &\le \frac{M_0^3\sqrt{3}}{\delta}\int_{\mathbb{R}} \left(\zeta_{xx}^2\,dx + \frac{\delta^2}{12}\zeta_{xxx}^2\right)dx \le \frac{2M_0^3}{\delta}E(t), \\ \int_{\mathbb{R}} \zeta_x\zeta_{xxx}^2\,dx &\le M\int_{\mathbb{R}} \zeta_{xxx}^2\,dx \le \frac{12M}{\delta^2}E(t).\end{aligned}$$

From (6.44) we infer that $E(t) \le E(0)e^{Kt}$ for $t \in [0,\tau)$. While this estimate was derived for smoother initial data, by continuous dependence its validity is ensured for $\zeta_0 \in H^3(\mathbb{R})$. This completes the proof. □

Our next aim is to show that there are solutions to (6.28) that blow up in finite time in the form of breaking waves. We will prove this by analyzing the equation that describes the evolution of

$$M(t) = \sup_{x\in\mathbb{R}} \{\zeta_x(x,t)\}. \tag{6.45}$$

For this we have to write (6.28) in an equivalent form that is better suited for our purposes. Notice that

$$\left(1 - \frac{\delta^2}{12}\partial_x^2\right)^{-1} f = P * f, \qquad f \in L^2(\mathbb{R}), \tag{6.46}$$

where

$$P(x) = \sqrt{\frac{3}{\delta^2}}\, e^{-\sqrt{\frac{12}{\delta^2}}\,|x|}, \qquad x \in \mathbb{R},$$

with

$$\|P\|_{L^\infty(\mathbb{R})} = \sqrt{\frac{3}{\delta^2}} < \frac{2}{\delta} \tag{6.47}$$

and

$$\|P_x\|_{L^2(\mathbb{R})} = \sqrt{2}\left(\frac{3}{\delta^2}\right)^{\frac{3}{4}} < 4\delta^{-\frac{3}{2}}. \tag{6.48}$$

Applying $(1 - \frac{\delta^2}{12}\partial_x^2)^{-1}$ to (6.28), we obtain the equation

$$\begin{aligned}
&\zeta_t + P_x * \zeta + \frac{3\varepsilon}{4} P_x * \zeta^2 - \frac{\varepsilon^2}{8} P_x * \zeta^3 + \frac{3\varepsilon^3}{64} P_x * \zeta^4 \\
&+ \frac{\delta^2}{12}\partial_x^3 P * \zeta + \frac{7\varepsilon\delta^2}{24} P_x * \zeta_x^2 + \frac{7\varepsilon\delta^2}{24} P * (\zeta\zeta_{xxx}) = 0.
\end{aligned} \tag{6.49}$$

Differentiating this equation with respect to the spatial variable, we get

$$\begin{aligned}
&\zeta_{tx} + \partial_x^2 P * \zeta + \frac{3\varepsilon}{4}\partial_x^2 P * \zeta^2 - \frac{\varepsilon^2}{8}\partial_x^2 P * \zeta^3 + \frac{3\varepsilon^3}{64}\partial_x^2 P * \zeta^4 \\
&+ \frac{\delta^2}{12}\partial_x^4 P * \zeta + \frac{7\varepsilon\delta^2}{24}\partial_x^2 P * \zeta_x^2 + \frac{7\varepsilon\delta^2}{24} P_x * (\zeta\zeta_{xxx}) = 0.
\end{aligned}$$

Since $\zeta\zeta_{xxx} = \partial_x^2(\zeta\zeta_x) - 3\zeta_x\zeta_{xx}$, and taking advantage of the identity

$$\partial_x^2 P * f = P_x * \zeta_x = \frac{12}{\delta^2} P * f - \frac{12}{\delta^2} f, \qquad f \in L^2(\mathbb{R}), \tag{6.50}$$

in re-expressing the last three terms, we deduce that

$$\begin{aligned}
&\zeta_{tx} + 2P_x * \zeta_x - \frac{3}{8}\varepsilon^2 P_x * (\zeta^2\zeta_x) + \frac{3\varepsilon^3}{16} P_x * (\zeta^3\zeta_x) \\
&- \zeta_{xx} - \frac{7\varepsilon}{4} P * \zeta_x^2 - \frac{7\varepsilon}{4}\zeta_x^2 - \frac{7\varepsilon}{2}\zeta\zeta_{xx} + 5\varepsilon P_x * (\zeta\zeta_x) = 0.
\end{aligned} \tag{6.51}$$

Theorem 6.3. *If the initial wave profile* $\zeta_0 \in H^3(\mathbb{R})$ *satisfies*

$$\left|\inf_{x\in\mathbb{R}}\{\partial_x \zeta_0(x)\}\right|^2 > \frac{144}{7\varepsilon\delta^2}C_0 + \frac{549}{7\delta^2}C_0^2 + \frac{27\varepsilon}{7\delta^2}C_0^3 + \frac{27\varepsilon^2}{14\delta^2}C_0^4,$$

where

$$C_0 = \left\{\frac{\sqrt{3}}{\delta}\int_{\mathbb{R}}\left(\zeta_0^2 + \frac{\delta^2}{12}[\partial_x\zeta_0]^2\right)dx\right\}^{1/2} > 0,$$

then wave breaking occurs for the solution of (6.28) *in finite time* $\tau = O(\frac{1}{\varepsilon})$.

Proof. First, notice that for any $x_0 \in \mathbb{R}$ and $t \in [0,\tau)$ we have

$$\zeta^2(x_0,t) = \int_{-\infty}^{x_0}\zeta\zeta_x\,dx - \int_{x_0}^{\infty}\zeta\zeta_x\,dx \le \frac{\sqrt{3}}{\delta}\int_{\mathbb{R}}\left(\zeta^2 + \frac{\delta^2}{12}\zeta_x^2\right)dx = C_0^2,$$

using the invariant (6.42) in the last step. The invariant (6.42) also yields

$$\int_{\mathbb{R}}\zeta_x^2\,dx \le \frac{4\sqrt{3}}{\delta}\left\{\frac{\sqrt{3}}{\delta}\int_{\mathbb{R}}\left(\zeta^2 + \frac{\delta^2}{12}\zeta_x^2\right)dx\right\} = \frac{4\sqrt{3}}{\delta}C_0^2 < \frac{9}{\delta}C_0^2, \qquad t \in [0,\tau).$$

Therefore, using Young's inequality (see [33]) and (6.47)–(6.48), we obtain the set of inequalities

$$\begin{aligned}
&\|P_x * \zeta_x\|_{L^\infty} \le \|P_x\|_{L^2}\|\zeta_x\|_{L^2} < 12\delta^{-2}C_0,\\
&\|P * \zeta_x^2\|_{L^\infty} \le \|P\|_{L^\infty}\|\zeta_x^2\|_{L^1} = \|P\|_{L^\infty}\|\zeta_x\|_{L^2}^2 < 18\delta^{-2}C_0^2,\\
&\|P_x * (\zeta\zeta_x)\|_{L^\infty} \le \|P_x\|_{L^2}\|\zeta\zeta_x\|_{L^2} \le \|P_x\|_{L^2}\|\zeta\|_{L^\infty}\|\zeta_x\|_{L^2} < 12\delta^{-2}C_0^2,\\
&\|P_x * (\zeta^2\zeta_x)\|_{L^\infty} \le \|P_x\|_{L^2}\|\zeta^2\zeta_x\|_{L^2} \le \|P_x\|_{L^2}\|\zeta\|_{L^\infty}^2\|\zeta_x\|_{L^2} < 12\delta^{-2}C_0^3,\\
&\|P_x * (\zeta^3\zeta_x)\|_{L^\infty} \le \|P_x\|_{L^2}\|\zeta^3\zeta_x\|_{L^2} \le \|P_x\|_{L^2}\|\zeta\|_{L^\infty}^3\|\zeta_x\|_{L^2} < 12\delta^{-2}C_0^4
\end{aligned}$$

at every instant $t \in [0,\tau)$. Since (6.51) is at any fixed time an equality in the space of continuous functions, we can evaluate both sides at some fixed time t at a point $\xi(t) \in \mathbb{R}$ where

$$M(t) = \zeta_x(t,\xi(t)),$$

with $M(t)$ defined by (6.45). Since $\zeta_{xx}(t,\xi(t)) = 0$ as ζ is C^2 in the spatial variable (in view of the fact that $H^3(\mathbb{R}) \subset C^2(\mathbb{R})$), and using the result on the evolution of extrema proved in the Appendix to this chapter, (6.51) and the previous estimates enable us to derive the following differential inequalities for the locally Lipschitz function $M(t)$: for a.e. $t \in (0,\tau)$,

$$M'(t) \le \frac{7\varepsilon}{4}M^2(t) + \frac{7\varepsilon}{6}\left\{\frac{144}{7\varepsilon\delta^2}C_0 + \frac{549}{7\delta^2}C_0^2 + \frac{27\varepsilon}{7\delta^2}C_0^3 + \frac{27\varepsilon^2}{14\delta^2}C_0^4\right\} \tag{6.52}$$

and

$$M'(t) \ge \frac{7\varepsilon}{4}M^2(t) - \frac{7\varepsilon}{6}\left\{\frac{144}{7\varepsilon\delta^2}C_0 + \frac{549}{7\delta^2}C_0^2 + \frac{27\varepsilon}{7\delta^2}C_0^3 + \frac{27\varepsilon^2}{14\delta^2}C_0^4\right\}. \tag{6.53}$$

Notice that $\zeta_0 \not\equiv 0$ ensures $M(0) > 0$. At $t = 0$, the right side of (6.53) is, by our assumption on the initial wave profile, strictly positive. We infer that, up to the maximal existence time $\tau > 0$ of the solution, the function $M(t)$ must be strictly increasing and, moreover,

$$M'(t) \geq \frac{7\varepsilon}{12} M^2(t) \quad \text{for a.e.} \quad t \in (0, \tau).$$

Dividing by

$$M^2(t) \geq M^2(0) > 0, \qquad t \in [0, \tau),$$

and integrating, we get

$$\frac{1}{M(t)} \leq \frac{1}{M(0)} - \frac{7\varepsilon}{12} t, \qquad t \in (0, \tau).$$

Since $M(t) > 0$ we must have

$$\lim_{t \uparrow \tau} M(t) = \infty$$

and

$$\tau \leq \frac{12}{7\varepsilon\, M(0)}. \tag{6.54}$$

On the other hand, knowing that $M(t) \geq M(0) > 0$ for all $t \in [0, \tau)$, our assumption on $M(0)$, together with (6.52), yield

$$M'(t) \leq \frac{35\varepsilon}{12} M^2(t) \quad \text{for a.e.} \quad t \in (0, \tau),$$

so that

$$\frac{1}{M(0)} - \frac{1}{M(t)} \leq \frac{35\varepsilon}{12} t, \qquad t \in (0, \tau).$$

Since $\lim_{t \uparrow \tau} M(t) = \infty$ we deduce that

$$\tau \geq \frac{12}{35\varepsilon\, M(0)}. \tag{6.55}$$

From the estimates (6.54) and (6.55) we infer that the finite maximal existence time $\tau > 0$ is of order $O(\frac{1}{\varepsilon})$. $\square$

Remark. Since we consider waves propagating over a flat bed, the breaking cannot be caused by the changes in the bottom topography. Intuitively the result of Theorem 6.3 can be interpreted as follows: the wave with a sufficiently steep initial profile will develop a verticality in its slope in a nondimensional time of order $O(\frac{1}{\varepsilon})$, that is, in real time of order $O(1)$.

6.3 Appendix for Chapter 6

6.3.1 Semigroup theory approach toward well-posedness

Semigroup theory uses abstract methods of operator theory to investigate problems (mostly existence and uniqueness of solutions, but also qualitative aspects like stability) concerning the evolution of a system in time, described by means of an equation of the type

$$\frac{d}{dt}u = \mathcal{A}(x,t;u)u + f(x,t;u), \tag{6.56}$$

where u is the unknown function characterizing the state of the system, f is given, the real variable t is interpreted as time, and $\mathcal{A}$ is an operator, with the set of space variables $x \in \mathbb{R}^n$. Equation (6.56) is *linear* if the functions f and $\mathcal{A}$ depend only on the independent variables (x,t), it is termed *semilinear* if only f depends on u, and *quasi-linear* if $\mathcal{A}$ depends on u. We briefly outline some basic aspects and a key result that was used in Section 6.1; we refer the reader to [90] for a thorough introduction and to [248] for a systematic exposition of semigroup theory.

A C_0 *(or strongly continuous) one-parameter semigroup* on a complex Banach space X is a family of bounded linear operators $T(t): X \to X$ parametrized by $t \geq 0$, subject to the following conditions:

(i) $T(0)$ is the identity I;

(ii) for $t, s \geq 0$ we have $T(t+s) = T(t)T(s)$;

(iii) the map $(t,x) \mapsto T(t)x$ from $[0,\infty) \times X$ to X is continuous.[165]

This definition embodies the idea that a linear evolution equation

$$\frac{d}{dt}u(t) = Au(t), \qquad t > 0, \tag{6.57}$$

where A is an operator in the Banach space X, possesses for initial data $u(0) = u_0$ in the domain of A, $\mathcal{D}(A)$, the unique solution $u(t) = T(t)u_0$ for $t \geq 0$. This solution varies continuously in X as the initial data and t vary, with the property (ii) expressing the fact that starting at u_0, the state $T(t+s)u_0$ at time $t+s$ agrees with the state at time t, starting at $T(s)u_0$. Semigroup theory generalizes the theory of ordinary differential equations to the case of certain partial differential equations: to every C_0-semigroup of bounded linear operators in X there corresponds a linear operator A in X, defined by $Au = \lim_{t \downarrow 0} \frac{T(t)u - u}{t}$ with domain the set $\mathcal{D}(A)$ of all $u \in X$ for which this limit exists, called the *infinitesimal generator* of the semigroup, and the solution to (6.57) with initial data $u_0 \in \mathcal{D}(A)$ is precisely $u(t) = T(t)u_0$ for $t \geq 0$. In general, A is a densely defined unbounded closed linear operator,[166] but if it is bounded, then $T(t) = e^{At}$ for $t \geq 0$ with $T(t) \to T(t_0)$ in the operator norm as $t \to t_0$ (in which case we have a uniformly continuous semigroup), and we are within the framework of the theory of ordinary differential equations (in the Banach space X). Conversely, if we have a uniformly continuous semigroup, then its generator is bounded. This explains why we do not require uniform convergence instead of strong convergence in (iii). Notice that (i), (ii), and the milder requirement of $T(t)u \rightharpoonup u$ (weak convergence) as $t \downarrow 0$ imply the validity of (iii) (see [90]), so that a generalization of the notion of one-parameter semigroup is ruled out. We would like to emphasize that in order to use semigroup theory to investigate an evolution equation, the identification of a suitable Banach space is not a minor technical detail: a relaxed attitude toward such technical details can lead to serious errors. For example, on $L^2(\mathbb{R})$ the C_0-semigroup $T(t)f = \mathcal{F}^{-1}(e^{-i\xi^2 t}\mathcal{F}(f))$, where $\mathcal{F}(f)$ is the Fourier transform of $f \in L^2(\mathbb{R})$, is generated by the operator $A = i\partial_x^2$ with $\mathcal{D}(A) = H^2(\mathbb{R})$, but the "same" operators are not bounded on $L^p(\mathbb{R})$ for any $1 \leq p < 2$ (unless $t = 0$); cf. [150].

[165] This requirement can be shown to hold if (i) and (ii) are granted, and for every fixed $x \in X$ the map $t \mapsto T(t)x$ is continuous at $t = 0$ (cf. [248, Chapter 1]).

[166] A closed linear operator is one whose graph is closed in $X \times X$. Notice (cf. [262]) that not all unbounded operators are closable (in the sense that we can suitably extend their domain of definition to obtain a closed extension).

If $\{T(t)\}_{t\geq 0}$ is a C_0-semigroup, one can show (see [248]) the existence of constants $M \geq 1$ and $\omega \geq 0$ such that

$$\|T(t)\| \leq M e^{\omega t}, \qquad t \geq 0. \tag{6.58}$$

C_0-semigroups satisfying (6.58) with $M = 1$ and $\omega = 0$ are called *contraction semigroups*, and the characterization of their generators is provided by the Lumer–Phillips theorem, which we now present in a form suitable for applications (see [90]). These generators are the closed operators A with a domain $\mathcal{D}(A)$ dense in X, such that $\lambda I - A$ is onto X for some $\lambda > 0$, and such that[167] for all $u \in \mathcal{D}(A)$, there exists $f \in X^*$ with $\|f\|_{X^*} = 1$, $f(u) = \|u\|$ (the existence of f being ensured by the Hahn–Banach theorem, with f generally not unique, the most notable exception being if X is a Hilbert space), and $\Re e\{f(Au)\} \leq 0$. In this context, let us also mention that Stone's theorem [248] establishes a one-to-one correspondence between skew-adjoint operators on a Hilbert space X and one-parameter families of unitary operators[168] on X (each family being a group of isometries on the Hilbert space X).

For generators A of semigroups satisfying (6.58) with $M = 1$, the operators $-A$ are called *quasi-m-accretive*. Semigroups of this type are very useful in evolution problems where the operator A also depends on time, i.e., we have a family $\{A(t)\}_{t\geq 0}$ generating C_0-semigroups $\{S(t,s)\}_{s\geq 0}$, since (cf. [248]) the validity of (6.58) with $M = 1$ and some fixed $\omega \geq 0$ for all t ensures a certain stability property that plays an important role in the existence theory of corresponding evolution systems (see below). For $\alpha \in \mathbb{R}$, notice that A generates the C_0-semigroup $\{T(t)\}_{t\geq 0}$ if and only if $A - \alpha I$ generates the semigroup $\{e^{-\alpha t}T(t)\}_{t\geq 0}$ so that, by the Lumer–Phillips theorem, quasi-m-accretive operators A are precisely those for which $A + \omega I$ are m-accretive. For applications, it is worth noticing the following useful perturbation result [171]: if A and B are both quasi-m-accretive on a Hilbert space X, with corresponding constants ω_A and ω_B in (6.58), and if B is relatively bounded with respect to A with A-bound $b < 1$, that is, $\mathcal{D}(A) \subset \mathcal{D}(B)$ and

$$\|Bu\| \leq a\,\|u\| + b\,\|Au\|, \qquad u \in \mathcal{D}(A),$$

for some $a \geq 0$, then[169] $A + B$ with domain $\mathcal{D}(A)$ is quasi-m-accretive, with corresponding $\omega = \omega_A + \omega_B$. Examples of unbounded generators of hyperbolic type[170] are as follows:

(E1) Given $a \in C^1(\mathbb{R}) \cap L^\infty(\mathbb{R})$ with $a' \in L^\infty(\mathbb{R})$ and given $b \in L^\infty(\mathbb{R})$, the operator $A_1 = a(x)\,\partial_x + b(x)$ with

$$\mathcal{D}(A_1) = \{u \in L^2(\mathbb{R}) : au \in H^1(\mathbb{R})\} \subset X = L^2(\mathbb{R})$$

is quasi-m-accretive with the corresponding $\omega = \frac{1}{2}\sup_{x\in\mathbb{R}}\{|a'(x)|\} + \sup_{x\in\mathbb{R}}\{|b(x)|\}$ in (6.58). Indeed, defining $\tilde{A}_1$ rigorously by $\tilde{A}_1 u = (au)_x - a_x u$ for $u \in \mathcal{D}(\tilde{A}_1) = \mathcal{D}(A_1)$, as in [52], one shows that the operator $A_0 = \tilde{A}_1 + \frac{1}{2}a'$ with $\mathcal{D}(A_0) = \mathcal{D}(A_1)$ is skew adjoint and thus will generate a C_0-semigroup of contractions. Since $a', b \in L^\infty(\mathbb{R})$, the operator $A_1 - A_0 = b(x) - \frac{1}{2}a'(x)$ is bounded, and the above perturbation result yields the claim. Notice that in this example, the function a could vanish at certain points, in which case A_1 is a degenerate linear differential operator whose domain is not $H^1(\mathbb{R})$.

[167] This next condition defines *dissipative* operators A, or *accretive* operators $-A$. Together with the previous condition we have the *m-dissipative* operators A, and the *m-accretive* operators $-A$, respectively.

[168] Bounded linear operators $U : X \to X$ satisfying $U^*U = UU^* = I$, where I is the identity operator and U^* is the adjoint of U. Such operators preserve the inner product $\langle\cdot,\cdot\rangle$ on the Hilbert space X, that is, $\langle Ux, Uy\rangle = \langle x, y\rangle$ for all $x, y \in X$. Notice that if U preserves the inner product, then U is an isometry. Conversely, isometric surjective linear operators on a Hilbert space are unitary operators (cf. [171, Chapter v]).

[169] This might fail if $b < 1$ is replaced by $b = 1$ (see [248]).

[170] The parabolic type is not relevant for our purposes.

(E2) For any real constant $\alpha \neq 0$ the operator $A_3 = \alpha\,\partial_x^3$ with $\mathcal{D}(A_3) = H^3(\mathbb{R}) \subset X = L^2(\mathbb{R})$ is skew adjoint (see [248]) and will therefore generate a semigroup of contractions in X.

(E3) The previous two examples can be combined using the above perturbation result: since for $a \in H^3(\mathbb{R})$ the operator A_1 is relatively bounded with respect to A_3 with A_3-bound ε for every $\varepsilon > 0$ (see [248]), the operator $A = \alpha\,\partial_x^3 + a(x)\,\partial_x$ with $\mathcal{D}(A) = H^3(\mathbb{R})$ is quasi-m-accretive, with corresponding $\omega = \frac{1}{2}\sup_{x\in\mathbb{R}}\{|a'(x)|\}$.

The main idea behind the semigroup approach to the semilinear evolution equation

$$\frac{d}{dt}u(t) = Au(t) + f(t,u), \qquad t > 0, \tag{6.59}$$

where A generates the C_0-semigroup $\{T(t)\}_{t\geq 0}$ and $f : [0,\infty) \times X \to X$ is continuous in t and locally Lipschitz continuous in u, uniformly in t on bounded intervals, is to associate to (6.59) the integral equation

$$u(t) = T(t)u_0 + \int_0^t T(t-s)f(s,u(s))\,ds \tag{6.60}$$

obtained formally by means of a variation of constants formula. Solutions $u \in C([0,t_0);X)$ with $t_0 > 0$ to (6.60) are called "mild solutions" of (6.59). Notice that, in contrast to (6.57), the nonlinear term f might induce the new phenomenon that these solutions may not be defined for all times $t \geq 0$ (called "blowup"; see [248]): for every $u_0 \in X$, (6.59) with data $u(0) = u_0$ has a unique mild solution defined on some maximal interval $[0,t_{max})$ with $t_{max} > 0$, and if $t_{max} < \infty$, then $\limsup_{t\uparrow t_{max}} \|u(t)\| = \infty$. Moreover, if f is continuously differentiable, then the mild solution with $u_0 \in \mathcal{D}(A)$ is a classical solution, that is, it is continuously differentiable for $t \in (0,t_{max})$, with $u(t) \in \mathcal{D}(A)$ and such that (6.59) is satisfied.

The passage from the autonomous linear equation (6.57) to that of the linear evolution equation

$$\frac{d}{dt}u(t) = A(t)u(t), \qquad t > 0, \tag{6.61}$$

is quite intricate. In the simplest setting of ordinary differential equations in the Banach space X, when $A(t)$ are bounded linear operators with $t \mapsto A(t)$ continuous in the uniform operator topology, the solution operator for the initial data $u_0 \in X$ is given by $u(t) = U(t,0)u_0$, where the two-parameter family $\{U(t,s)\}_{0\leq s\leq t}$ of bounded linear operators, called an *evolution system*, is such that $U(t,t) = I$, $U(t,r)U(r,s) = U(t,s)$ for $0 \leq s \leq r \leq t$, and the map $(t,s) \mapsto U(t,s)$ is continuous in the uniform operator topology for $0 \leq s \leq t$. Notice that if $A(t)$ is time independent, $A(t) = A$ with A a bounded linear operator, then $U(t,s) = e^{A(t-s)}$, but even for time-dependent bounded linear operators on $X = \mathbb{R}^2$, explicit formulas for the evolution system are not available. In the case when the operators $A(t)$ are unbounded, there is a strong distinction between the hyperbolic case and the parabolic case. Since we neglect dissipation, the hyperbolic case is the one relevant to us, and we refrain from discussing parabolic aspects. The lack of a strong regularization effect in time is typical for hyperbolic problems; if present, it is very modest, and one has to trade regularity in time for that in space. Let Y be a Banach space with norm $\|\cdot\|_Y$ that is continuously embedded and dense in X (i.e., Y is a dense subspace of X and there is a constant $C > 0$ with $\|y\| \leq C\,\|y\|_Y$ for all $y \in Y$). Assume there is $t_0 > 0$ such that

(H1) there is a family $\{Q(t)\}_{t\in[0,t_0]}$ of isomorphisms of Y onto X such that $t \mapsto Q(t)v$ is continuously differentiable in X on $[0,t_0]$ and

$$Q(t)\,A(t)\,[Q(t)]^{-1} = A(t) + B(t), \tag{6.62}$$

where $\{B(t)\}_{t\in[0,t_0]}$ is a family of bounded linear operators on X with $(t,u) \mapsto B(t)u$ continuous from $[0,t_0] \times X$ to X;

(H2) there exists $\omega > 0$ such that, for each $t \in [0,t_0]$, the operator $A(t)$ generates a C_0-semigroup $\{S(t,s)\}_{s\geq 0}$ on the Banach space X with $\|S(t,s)\| \leq e^{\omega s}$ for $s \geq 0$;

(H3) for each $t \in [0,t_0]$ we have $Y \subset \mathcal{D}(A(t))$, and $A(t)$ is a bounded operator from Y into X, depending continuously on t.

Then (see [248]) there is an evolution system $\{U(t,s)\}_{0\leq s\leq t\leq t_0}$ in X satisfying

- $\|U(t,s)\| \leq e^{\omega(t-s)}$ for $0 \leq s \leq t \leq t_0$;
- $U(t,s)Y \subset Y$ and whenever $v \in Y$, $U(t,s)v$ is continuous in Y for $0 \leq s \leq t \leq t_0$;
- if $v \in Y$, then $\lim_{h\downarrow 0} \frac{U(t+h,s)v-U(t,s)v}{h}\Big|_{t=s} = A(t)v$ for $0 \leq s \leq t \leq t_0$;
- if $v \in Y$, then $\frac{\partial}{\partial s}U(t,s)v = -U(t,s)A(s)v$ for $0 \leq s \leq t \leq t_0$;

and for every $y \in Y$ the problem (6.61) with data $u(0) = v$ has a unique classical solution

$$u \in C([0,t_0];Y) \cap C^1([0,t_0];X) \tag{6.63}$$

given by $u(t) = U(t,0)v$ for $t \in [0,t_0]$. Moreover, the inhomogeneous linear evolution equation

$$\frac{d}{dt}u(t) = A(t)u(t) + f(t), \qquad t \in (0,t_0], \tag{6.64}$$

with $f \in C([0,t_0];Y)$, has, for any initial data $v \in Y$, a unique solution u with the regularity specified by (6.63), given by

$$u(t) = U(t,0)v + \int_0^t U(t,s)f(s)ds, \qquad 0 \leq t \leq t_0.$$

On the basis of these considerations it is possible to investigate the quasi-linear evolution equation (6.56): with the function-space (with respect to the x-variable) notation $A(t,u) = \mathcal{A}(x,t;u)$ and $F(t,u) = f(x,t;u)$, one specifies appropriate classes of functions w to which a prospective solution of (6.56) is likely to belong, and, subsequently, denoting by $u(w)$ the unique solution of

$$\begin{cases} \dfrac{d}{dt}u = A(t,w(t))u + F(t,w(t)), & t \in (0,t_0], \\ u(0) = u_0, \end{cases}$$

the existence of a solution on a possibly smaller time interval $[0,t_0^*]$, with $t_0^* \in (0,t_0]$, is established by a fixed point argument for the map $w \mapsto u(w)$. There are many technical difficulties in realizing the preceding oversimplified outline. We now present a far-reaching result in this direction, due to Kato [168], in a simple form sufficient for our purposes. Consider the Cauchy problem for the quasi-linear equation of evolution

$$\begin{cases} \dfrac{d}{dt}u = A(u)u + F(u), & t \in (0,t_0], \\ u(0) = u_0. \end{cases} \tag{6.65}$$

Let X and Y be reflexive Banach spaces with Y continuously and densely embedded in X. Let S be an isomorphism (bicontinuous linear map) of Y onto X and assume that

(K1) given $C > 0$, for every $y \in Y$ with $\|y\|_Y \leq C$, $A(y)$ generates a C_0-semigroup in X subject to (6.58) with $M = 1$ and ω depending only on C;

(K2) for every $y \in Y$, $A(y)$ is a bounded linear operator from Y to X and

$$\|[A(y) - A(z)]w\|_X \leq \mu_A \|y - z\|_X \|w\|_Y, \qquad y, z, w \in Y,$$

for some constant μ_A depending only on $\max\{\|y\|_Y, \|z\|_Y\}$;

(K3) for every $C > 0$ there is a constant $\mu_1(C) > 0$ such that $S\,A(y)\,S^{-1} - A(y) = B(y)$ for some bounded linear operator $B(y)$ on X satisfying

$$\|[B(y) - B(z)]\,w\|_X \le \mu_1(C)\,\|w\|_X, \qquad w \in X,$$

for all $y, z \in Y$ with $\|y\|_Y, \|z\|_Y \le C$;

(K4) for each $C > 0$ we have

$$\|f(y) - f(z)\|_X \le \mu_2\,\|y - z\|_X, \qquad y, z \in X,$$

and

$$\|f(y) - f(z)\|_Y \le \mu_3\,\|y - z\|_Y, \qquad y, z \in Y,$$

for some constants μ_2 and μ_3 depending only on $\max\{\|y\|_X, \|z\|_X\}$ and $\max\{\|y\|_Y, \|z\|_Y\}$, respectively.

Then for any $u_0 \in Y$ there is a maximal time $t_0 \in (0, \infty]$ such that (6.65) has a unique solution $u \in C([0,t_0);Y) \cap C^1([0,t_0);X)$ depending continuously on the initial data, that is, for any $\tau \in [0,t_0)$ and any sequence $u_0^n \to u_0$ in Y there exists an integer $N \ge 1$ such that for $n \ge N$ there are unique solutions $u^n \in C([0,\tau];Y) \cap C^1([0,\tau];X)$ to (6.65) with $u^n(0) = u_0^n$, and $u^n(t) \to u(t)$ in Y, uniformly in $t \in [0,\tau]$. Moreover (cf. [169, 170]), if $t_0 < \infty$, then $\limsup_{t \uparrow t_0} \|u(t)\|_Y = \infty$.

We now briefly indicate the applicability of this approach to the well-posedness of the model equations discussed in this chapter.

Example 6.4 (the KdV equation). Write (6.16) in the form (6.65) with $F \equiv 0$ and

$$A(u) = -\frac{\delta^2}{6}\,\partial_x^3 - \Big(1 + \frac{3\varepsilon}{2}\,u\Big)\,\partial_x, \qquad \mathcal{D}(A(u)) = H^s(\mathbb{R}) \subset X = L^2(\mathbb{R})$$

for $u \in Y = H^s(\mathbb{R})$ with $s \ge 3$. The condition (K4) is trivially satisfied, (K2) is easily checked, and (K1) holds in view of (E3). We now consider (K3) with $S = (1 - \partial_x^2)^{s/2}$, given explicitly as the pseudodifferential operator

$$Su = \mathcal{F}^{-1}\Big((1 + \xi^2)^{s/2}\,\mathcal{F}(u)\Big), \qquad u \in Y,$$

in terms of the Fourier transform $\mathcal{F}$. Since ∂_x^3 and ∂_x commute with S, an explicit calculation yields that, given $y \in Y$, for $B(y) = SA(y)S^{-1} - A(y)$, we have

$$B(y)\,u = -\frac{3\varepsilon}{2}\Big(S(y w_x) - y\,(S w_x)\Big) = -\frac{3\varepsilon}{2}\,[[S, M_y]]\,w_x, \qquad u \in Y,$$

with $w = S^{-1}u \in H^{2s}(\mathbb{R})$, where

$$[[S, M_y]] = S M_y - M_y S$$

denotes the commutator of S and the multiplication operator M_y with $y \in Y$. It is known that for $y \in H^s(\mathbb{R})$ the operator $[[S, M_y]]$ is a bounded linear operator from $H^{s-1}(\mathbb{R})$ to X, with norm estimated by a universal constant times $\|y\|_{H^s(\mathbb{R})}$; cf. [168]. Thus

$$\|B(y)u\|_X \le c\,\|y\|_Y\,\|S^{-1}u_x\|_{H^{s-1}(\mathbb{R})} \le \tilde{c}\,\|y\|_Y\,\|u\|_X, \qquad u \in Y,$$

for some $c, \tilde{c} > 0$ independent of $u, y \in Y$. Since Y is dense in X, we deduce that $B(y)$ is a bounded linear operator on X with norm at most $\tilde{c}\,\|y\|_Y$. (K3) follows at once since $B(y)$ depends linearly on y. We conclude that for any $\zeta_0 \in H^s(\mathbb{R})$, there is a maximal time $t_s \in (0,\infty]$ such that the KdV equation (6.16) with initial data $\zeta(0) = \zeta_0$ has a unique solution $\zeta \in C([0,t_s); H^s(\mathbb{R})) \cap C^1([0,t_s); L^2(\mathbb{R}))$, depending continuously on the initial data. Notice that this regularity shows that $\frac{d\zeta}{dt} \in C([0,t_s); H^{s-3}(\mathbb{R}))$, by expressing it using the equation, so that actually $\zeta \in C([0,t_s); H^s(\mathbb{R})) \cap C^1([0,t_s); H^{s-3}(\mathbb{R}))$. ∎

Example 6.5 (BBM family). In the case $\beta = 0$ we have that (6.17) becomes the KdV equation. Let $\beta < 0$; for simplicity we choose $\beta = -\delta^{-2}$ so that (6.17) becomes

$$\zeta_t - \zeta_{xxt} + \zeta_x + \frac{3\varepsilon}{2}\zeta\zeta_x + \theta\,\zeta_{xxx} = 0 \tag{6.66}$$

with $\theta = \frac{\delta^2}{6} - 1$. It is convenient to rewrite (6.66) as

$$\frac{dm}{dt} = \theta\, m_x - (1+\theta)(Q^{-2}m)_x - \frac{3\varepsilon}{2}(Q^{-2}m)(Q^{-2}m)_x \tag{6.67}$$

in terms of $m = Q^2\zeta$, where the pseudodifferential operator $Q = (1-\partial_x^2)^{1/2}$ is an isomorphism from $H^s(\mathbb{R})$ to $H^{s-1}(\mathbb{R})$ for any $s \in \mathbb{R}$, defined analogously to S from the previous example by using the Fourier transform. We apply Kato's semigroup approach in the setting $X = L^2(\mathbb{R})$ and $Y = H^s(\mathbb{R})$ with $s \geq 1$ an integer, by setting (for $m \in Y$ fixed)

$$A(m) = \theta\,\partial_x \quad \text{with} \quad \mathcal{D}(A) = H^1(\mathbb{R}), \quad F(m) = -(1+\theta)(Q^{-2}m)_x - \frac{3\varepsilon}{2}(Q^{-2}m)(Q^{-2}m)_x$$

and choosing $S = Q^s$. Hypothesis (K1) holds in view of (E1), and since (6.67) is actually semilinear, (K2)–(K4) are easy to check with S and $A(m)$ commuting, as one can see from their explicit representations in terms of the Fourier transform. We conclude, given $s \geq 1$, that for any $m_0 \in H^s(\mathbb{R})$ there is a maximal time $t_s \in (0,\infty]$ such that (6.67) with initial data $m(0) = m_0$ has a unique solution $m \in C([0,t_s); H^s(\mathbb{R})) \cap C^1([0,t_s); L^2(\mathbb{R}))$, depending continuously on the initial data. Expressing $\frac{dm}{dt} \in C([0,t_s); H^{s-1}(\mathbb{R}))$ from (6.67), we get $m \in C([0,t_s); H^s(\mathbb{R})) \cap C^1([0,t_s); H^{s-1}(\mathbb{R}))$. Since $m = Q^2\zeta$ and Q^2 is an isomorphism from $H^r(\mathbb{R})$ to $H^{r-2}(\mathbb{R})$ for any $r \in \mathbb{R}$, we can formulate our conclusion in terms of ζ: given the integer $k \geq 3$, for any $\zeta_0 \in H^k(\mathbb{R})$, there is a maximal time $t_k \in (0,\infty]$ such that (6.66), with initial data $\zeta(0) = \zeta_0$, has a unique solution $\zeta \in C([0,t_k); H^k(\mathbb{R})) \cap C^1([0,t_k); H^{k-1}(\mathbb{R}))$, depending continuously on the data. ∎

Example 6.6 (the equations for waves of moderate amplitude). Since the analysis of the well-posedness for (6.20) and (6.28) is similar, we discuss only the case of (6.28). Moreover, as the exact values of $\delta, \varepsilon > 0$ are irrelevant in this context, for simplicity we set $\varepsilon = 4$ and $\delta = \sqrt{12}$ (for convenience) obtaining the equation

$$\zeta_t + \zeta_x + 6\zeta\zeta_x - 6\zeta^2\zeta_x + 12\zeta^3\zeta_x + \zeta_{xxx} - \zeta_{xxt} + 14\zeta\zeta_{xxx} + 28\zeta_x\zeta_{xx} = 0.$$

As in Example 6.5, we set $m = Q^2\zeta$ with $Q = (1-\partial_x^2)^{1/2}$ and write the above equation equivalently as

$$\begin{aligned} m_t - m_x - 14m_x(Q^{-2}m) - 28m(Q^{-2}m)_x + 48(Q^{-2}m)(Q^{-2}m)_x \\ -6(Q^{-2}m)^2(Q^{-2}m)_x + 12(Q^{-2}m)^3(Q^{-2}m)_x + 2(Q^{-2}m)_x = 0. \end{aligned} \tag{6.68}$$

We apply Kato's semigroup approach in the setting $X = L^2(\mathbb{R})$ and $Y = H^1(\mathbb{R})$ with $S = Q$ by choosing

$$F(m) = -48(Q^{-2}m)(Q^{-2}m)_x + 6(Q^{-2}m)^2(Q^{-2}m)_x - 12(Q^{-2}m)^3(Q^{-2}m)_x - 2(Q^{-2}m)_x$$

and

$$[A(m)]\,w = \partial_x\Big([1 + 14(Q^{-2}m)]\,w\Big) + 14(Q^{-2}m)_x\,w$$

with $\mathcal{D}(A(m)) = \{w \in X : [1 + 14(Q^{-2}m)]\,w \in H^1(\mathbb{R})\}$. Since $m \in L^2(\mathbb{R})$ if and only if $Q^{-2}m \in H^2(\mathbb{R})$, (K1) holds true in view of (E1), while (K2) and (K4) are easily checked. As for (K3), since ∂_x and Q commute, denoting $a = 1 + 14(Q^{-2}y) \in H^3(\mathbb{R})$ for $y \in H^1(\mathbb{R})$, a direct computation yields

$$\Big(QA(y)Q^{-1} - A(y)\Big)\,w = \Big([[Q, M_a]]\,\partial_x Q^{-1} + 2\,[[Q, M_{a_x}]]\,Q^{-1}\Big)\,w, \qquad w \in H^1(\mathbb{R}),$$

with the notation from Example 6.4 for the commutator and for the multiplication operator. An explicit calculation confirms at once that both $\partial_x Q^{-1}$ and $[[Q, M_{a_x}]]\,Q^{-1}$ extend to bounded linear operators on X, with the norm of the first a constant and that of the second estimated from above by a universal constant times $\|\partial_x^2\, Q^{-2}y\|_{L^\infty} + \|\partial_x\, Q^{-2}y\|_{L^\infty}$. On the other hand, since Q is a first-order pseudodifferential operator, a remarkable result from harmonic analysis (see [283, Section VII.3.5]) ensures that for $f \in H^\alpha(\mathbb{R})$ with $\alpha > 3/2$, $[[Q, M_f]]$ extends to a bounded linear operator on X, whose norm is estimated by a universal constant times $\|\partial_x f\|_{L^\infty}$. This proves that (K3) holds true, and we deduce that for any $m_0 \in H^1(\mathbb{R})$ there is a maximal time $t_0 \in (0,\infty]$ such that (6.68), with initial data $m(0) = m_0$, has a unique solution $m \in C([0,t_0); H^1(\mathbb{R})) \cap C^1([0,t_0); L^2(\mathbb{R}))$, depending continuously on the initial data. Expressing $\frac{dm}{dt} \in C([0,t_0); L^2(\mathbb{R}))$ from (6.68) we see that $m \in C([0,t_0); H^1(\mathbb{R})) \cap C^1([0,t_0); L^2(\mathbb{R}))$. Since $m = Q^2\zeta$, and Q^2 is an isomorphism from $H^r(\mathbb{R})$ to $H^{r-2}(\mathbb{R})$ for any $r \in \mathbb{R}$, we proved that, for any $\zeta_0 \in H^3(\mathbb{R})$, there is a maximal time $t_0 \in (0,\infty]$ such that (6.28) with initial data $\zeta(0) = \zeta_0$ has a unique solution $\zeta \in C([0,t_0); H^3(\mathbb{R})) \cap C^1([0,t_0); H^2(\mathbb{R}))$, depending continuously on the initial data. It is of interest to note that, if $\zeta_0 \in H^k(\mathbb{R})$ for some $k \geq 4$, then $\zeta \in C([0,t_0); H^k(\mathbb{R})) \cap C^1([0,t_0); H^{k-1}(\mathbb{R}))$ with the same maximal existence time $t_0 > 0$ as for $k = 3$. We only sketch the proof, as it follows the reasoning of Kato's similar result for KdV [170]. For example, let $k = 4$ and first write (6.68) in the form

$$m_t = am_x + 2a_x m + F(m),$$

as above, with $a = 1 + 14(Q^{-2}m)$. Subsequently, apply the operator Q to get

$$w_t = Q\Big(a\,(Q^{-1}w)_x\Big) + 2Q\Big(a_x\,(Q^{-1}w)\Big) + QF(Q^{-1}w) \tag{6.69}$$

for $w = Qm$. Notice now that F is locally Lipschitz from $H^s(\mathbb{R})$ to $H^{s+1}(\mathbb{R})$ for any $s \geq 0$. Also $a - 1 \in H^{s+2}(\mathbb{R})$ if $m \in H^s(\mathbb{R})$, so that the previously mentioned remarkable commutator property ensures that $w \mapsto Q(a\,(Q^{-1}w)_x) - a\,w_x$ as well as $w \mapsto Q(a_x\,(Q^{-1}w)) - a_x w$ are bounded linear operators on $L^2(\mathbb{R})$. Viewing $m \in C([0,t_0); H^1(\mathbb{R}))$ as a known function, we can therefore write (6.69) on $(0,t_0)$ as

$$w_t = p(t)\,w_x + q(t)\,w + f(t)$$

with $p, q \in C([0,t_0); L^2(\mathbb{R}))$ and $f \in C([0,t_0); H^1(\mathbb{R}))$. This linear equation of the form (6.64) will have a unique solution $w \in C([0,t_0); H^1(\mathbb{R})) \cap C^1([0,t_0); L^2(\mathbb{R}))$. But it already has the solution $w = Qm$. Consequently, $m \in C([0,t_0); H^2(\mathbb{R})) \cap C^1([0,t_0); H^1(\mathbb{R}))$ if $m_0 \in H^1(\mathbb{R})$. We can repeat the procedure to deal with $k = 5$, and so on. ∎

Remark. For (6.16) and (6.17) one can actually show, using the structure of the equations, that the $H^s(\mathbb{R})$-norms do not blow up in finite time (see, e.g., [295]) so that $t_s = \infty$ for any $s \geq 3$, whereas the existence of a choice of parameters and some suitable initial data for (6.28) results in solutions with $t_3 < \infty$, as proved in Theorem 6.3. This shows that, while abstract considerations are very useful in the investigation of specific nonlinear partial differential equations, the abstractions alone rarely provide sufficient detailed knowledge.

6.3.2 The evolution of extrema

In mathematical physics one often studies the evolution in time of certain quantities. It is tempting to argue formally as follows. Let $F(t,x)$ be a smooth function defined for $t \geq 0$ and $x \in \mathbb{R}$, and such that at every fixed $t \geq 0$ the function $x \mapsto F(t,x)$ decays to zero as $|x| \to \infty$. Define

$$f(t) = \inf_{x \in \mathbb{R}} \{F(t,x)\}.$$

The smoothness of F and its decay properties ensure the existence of $\xi(t) \in \mathbb{R}$, where $F(t,\xi(t)) = f(t)$. Then

$$f'(t) = F_t(t,\xi(t)) + \xi'(t)\, F_x(t,\xi(t)) = F_t(t,\xi(t))$$

since at $x = \xi(t)$ the function $x \mapsto F(t,x)$ attains its minimum. The above relation could be useful to derive differential inequalities for the time evolution of the minimum. However, this formal approach is faulty since the function $t \mapsto \xi(t)$ is generally multivalued and, even if it were single-valued, it might be discontinuous. Both facts can easily be seen by analyzing a double-well graph and modifying its shape smoothly. Nevertheless, the above (faulty) formal argument is surprisingly close to being correct. We formulate a version of the correct result (see [51]) in a setting suitable for our study of breaking waves.

Theorem A 6.1. *For $T > 0$ and $v \in C^1([0,T); H^2(\mathbb{R}))$ there exists $\xi(t) \in \mathbb{R}$ such that*

$$m(t) := \inf_{x \in \mathbb{R}} [v_x(t,x)] = v_x(t,\xi(t)), \qquad t \in [0,T).$$

The function m is a.e. differentiable on $(0,T)$ with

$$\frac{dm}{dt}(t) = v_{tx}(t,\xi(t)) \quad a.e. \quad on \quad (0,T). \tag{6.70}$$

Proof. Fix $t \in [0,T)$. If $m(t) \geq 0$, then $v(t,\cdot)$ is nondecreasing on $\mathbb{R}$ and since $v(t,\cdot) \in L^2(\mathbb{R})$ this implies $v(t,\cdot) \equiv 0$. Therefore, we can assume $m(t) < 0$. As $v_x(t,\cdot) \in H^1(\mathbb{R})$ we have $\lim_{|x|\to\infty} v_x(t,x) = 0$ so that there exists $\xi(t) \in \mathbb{R}$ with $m(t) = v_x(t,\xi(t))$. For $t, s \in [0,T)$ with $m(t) \leq m(s)$, we have

$$0 \leq m(s) - m(t) = \inf_{x \in \mathbb{R}} [v_x(s,x)] - v_x(t,\xi(t)) \leq v_x(s,\xi(t)) - v_x(t,\xi(t)),$$

and from the Sobolev embedding $H^1(\mathbb{R}) \subset L^\infty(\mathbb{R})$ we get

$$|m(s) - m(t)| \leq ||v_x(t,\cdot) - v_x(s,\cdot)||_{L^\infty(\mathbb{R})} \leq c\,||v_x(t,\cdot) - v_x(s,\cdot)||_{H^1(\mathbb{R})}$$

for some constant $c > 0$ independent of time $t \geq 0$. The mean-value theorem for Banach spaces (cf. [96]) yields

$$|m(t) - m(s)| \leq c\,|t-s| \sup\{||v_{tx}(\tau,\cdot)||_{H^1(\mathbb{R})} : \tau \in (\min\{s,t\}, \max\{s,t\})\}.$$

Since $v_{tx} \in C\left([0,T); H^1(\mathbb{R})\right)$, the above relation shows that m is locally Lipschitz on $[0,T)$ and therefore a.e. differentiable on $(0,T)$ by Rademacher's theorem (cf. [112]). From the definition of m,

$$m(t+h) = v_x(t+h,\xi(t+h)) \leq v_x(t+h,\xi(t)), \qquad h > 0,$$

and so

$$\frac{m(t+h) - m(t)}{h} \leq \frac{v_x(t+h,\xi(t)) - v_x(t,\xi(t))}{h}.$$

Letting $h \to 0^+$, we get

$$\limsup_{h\to 0^+} \frac{m(t+h)-m(t)}{h} \le v_{tx}(t,\xi(t)), \tag{6.71}$$

since $v_x \in C^1([0,T);H^1(\mathbb{R}))$ means that

$$\lim_{h\to 0^+} \left\| \frac{v_x(t+h,\cdot)-v_x(t,\cdot)}{h} - v_{tx}(t,\cdot)\right\|_{H^1(\mathbb{R})} = 0$$

and, as $L^\infty(\mathbb{R}) \subset H^1(\mathbb{R})$, we infer that

$$\lim_{h\to 0^+} \left| \frac{v_x(t+h,\xi(t))-v_x(t,\xi(t))}{h} - v_{tx}(t,\xi(t))\right| = 0.$$

Similarly, for $h > 0$ we get

$$\frac{m(t)-m(t-h)}{h} = \frac{m(t)-v_x(t-h,\xi(t-h))}{h} \ge \frac{v_x(t,\xi(t))-v_x(t-h,\xi(t))}{h},$$

so that

$$\limsup_{h\to 0^+} \frac{m(t)-m(t-h)}{h} \ge v_{tx}(t,\xi(t)). \tag{6.72}$$

From (6.71) and (6.72) we infer the validity of (6.70) at all points where m is differentiable; this completes the proof. □

6.4 Notes for Chapter 6

The point of view set out in this chapter is rooted in the papers [164] and [66]. We now present two directions in which these considerations can be further developed.

Relevance of integrable systems

The first direction concerns the possibility of using integrable systems. Since breaking waves can only arise for nonlinear equations—model or exact—the most one can hope for is to be able to use the extraordinary mathematical structure of integrable systems to pursue in-depth qualitative studies. But, as pointed out earlier, none of the model equations (6.38) is integrable. If, however, instead of the vertically averaged velocity $u(x,t)$ given by (6.12), following [164] we consider for $\rho \in [0,1]$ the horizontal fluid velocity $u_\rho(x,t)$ evaluated at the level line ρ of the fluid domain,[171]

$$u_\rho(x,t) = \partial_x \Phi(x,z,t)\Big|_{z=(1+\varepsilon\zeta)\rho-1},$$

then (cf. [66]) one can obtain an approximation to the governing equations consistent with (6.13) built on a two-parameter family of equations of the form

$$u_t + u_x + \frac{3}{2}\varepsilon u u_x + \delta^2(p\, u_{xxx} + q u_{xxt}) = \varepsilon\delta^2(\gamma u u_{xxx} + \theta u_x u_{xx}), \tag{6.73}$$

[171] So that $\rho = 0$ and $\rho = 1$ correspond to the bottom and surface, respectively.

where, for simplicity, we write u instead of u_ρ. The family of equations (6.73) depends on the parameter ρ and on an additional parameter[172] $p < \frac{1}{6}$, as follows:

$$q = p - \frac{1}{6}, \quad \gamma = -\frac{3}{2}p - \frac{1}{6}, \quad \theta = -\frac{9}{2}p + \frac{3}{2}\rho^2 - \frac{35}{24}.$$

The free surface ζ is then described, up to $O(\delta^4)$, in terms of the horizontal fluid velocity at a given depth, by

$$\zeta = u + \frac{\varepsilon}{4}u^2 - \frac{\delta^2}{6}u_{xx} - \frac{\varepsilon\delta^2}{48}[20uu_{xx} + 17uu_{xxx}]; \tag{6.74}$$

cf. [66]. Notice that, since the flow is irrotational, the specification of the horizontal velocity u at a certain depth determines uniquely the velocity field throughout the fluid. Indeed, from (6.5), we infer that $u_y = 0$ on the flat bed, and the corresponding boundary value problem for the harmonic function u in the strip beneath the given depth determines u uniquely in that strip. As for the vertical velocity component v, being the harmonic conjugate function with $v + iu$ analytic, in that strip, it is determined up to a constant and (6.5) shows that there is no degree of freedom. The uniqueness of the velocity field (u, v) throughout the fluid follows by analytic continuation.

Among the equations (6.73) there are precisely two integrable equations (see [160]):

- at the depth $\rho = \frac{1}{\sqrt{2}}$ and with $p = -\frac{1}{4}$ we find the **Camassa–Holm equation**

$$U_t + \widehat{\kappa}U_x + 3UU_x - U_{txx} = 2U_xU_{xx} + UU_{xxx} \tag{6.75}$$

  with $\widehat{\kappa} = \frac{4\sqrt{3}}{5\delta\sqrt{5}}$ by setting $u(x,t) = \frac{\delta\sqrt{5}}{\varepsilon\sqrt{3}}U(\frac{2\sqrt{3}}{\delta\sqrt{5}}(x - \frac{3}{5}t), t)$;

- at the depth $\rho = \frac{\sqrt{23}}{6}$ and with $p = -\frac{1}{18}$ we find the **Degasperis–Procesi equation**

$$U_t + \widehat{\kappa}U_x + 4UU_x - U_{txx} = 3U_xU_{xx} + UU_{xxx} \tag{6.76}$$

  with $\widehat{\kappa} = \frac{1}{2\delta\sqrt{2}}$ by setting $u(x,t) = \frac{4\delta\sqrt{2}}{\varepsilon}U(\frac{\sqrt{2}}{3\delta}x - \frac{1}{4}t, t)$.

Equation (6.75) was first derived abstractly as a bi-Hamiltonian system by Fokas and Fuchssteiner [120] but gained prominence after Camassa and Holm [41] independently rederived it as an approximation to the Euler equations of hydrodynamics and also discovered a number of the intriguing properties of this equation. Camassa and Holm [41] found a Lax pair formulation for (6.75), and this lies at the core of showing, via direct and inverse scattering [58, 239] (see also [67] for the spatially periodic case), that the equation is a completely integrable Hamiltonian system. The Camassa–Holm equation shares, with KdV, this integrability property, as well as possessing solitary-wave solutions that are solitons [70, 247]. As for (6.76), first derived by Degasperis and Procesi [92], it has a Lax pair formulation that leads to an inverse scattering approach [60], and its solitary waves are also solitons [220]. Just like the KdV equation, (6.75) and (6.76) have infinitely many integrals of motion.

A further common structural property of these three model equations is that, in the spatially periodic case,[173] each is a re-expression of geodesic flow: KdV and Camassa–Holm for the L^2 and the H^1 right-invariant metric on the Bott–Virasoro group, respectively (cf. [246, 234, 64]), while (6.76) arises as the geodesic flow of a right-invariant symmetric linear connection that is nonmetric (meaning that there does not exist a Riemannian metric which is preserved by this connection) on the diffeomorphism group; cf. [109]. For example, denote by $\mathcal{D}$ the group of C^∞-smooth

[172] The restriction $p < \frac{1}{6}$ is again imposed to avoid ill-posedness and change-of-type issues.

[173] The geometric interpretation is more elegant for spatially periodic solutions since in this setting it suffices to consider smooth functions as elements of the group, whereas on the line the behavior at infinity has to be controlled (see [44, 142]).

orientation-preserving diffeomorphisms of the circle $\mathbb{S}^1 = \mathbb{R}/\mathbb{Z}$. The topology on $\mathcal{D}$ is induced from the standard Fréchet topology generated by the countable system of Sobolev norms

$$\|\varphi\|_n = \sum_{k=0}^{n} \int_0^1 (\partial_x^k \varphi)^2 \, dx, \qquad n \geq 0.$$

The Fréchet manifold $\mathcal{D}$ is a Lie group with multiplication given by the composition of diffeomorphisms (cf. [64]), in the sense that it is a group such that both the group operation and the inversion are smooth maps. The *Bott–Virasoro group* Vir is the Fréchet manifold $\mathcal{D} \times \mathbb{R}$ with multiplication

$$(\varphi_1, k_1) \circ (\varphi_2, k_2) = \Big(\varphi_1 \circ \varphi_2, k_1 + k_2 - \frac{1}{2} \int_0^1 \log(\varphi_1(\varphi_2(x)))_x \, d \log \varphi_2'(x)\Big),$$

with the map $(\varphi_1, \varphi_2) \mapsto -\frac{1}{2} \int_0^1 \log(\varphi_1 \circ \varphi_2)_x \, d \log \varphi'$ referred to as the *Bott cocycle*. Passing to the universal cover $\mathbb{R} \mapsto \mathbb{R}/\mathbb{Z}$ of the circle $\mathbb{S}^1 = \mathbb{R}/\mathbb{Z}$, we identify a diffeomorphism $\varphi \in \mathcal{D}$ with a lift $\tilde{\varphi} : \mathbb{R} \to \mathbb{R}$ of the form $\tilde{\varphi}(x) = x + \nu(x)$, with $0 \leq \nu(0) < 1$ and $\nu : \mathbb{R} \to \mathbb{R}$ a smooth 1-periodic function. Vir is a Lie group whose Lie algebra vir can be identified with the Fréchet space $C^\infty(\mathbb{S}^1) \times \mathbb{R}$ with Lie bracket

$$[(u,a),(v,b)] = \Big(u_x v - v_x u, \int_0^1 u(x) v_{xxx}(x) \, dx\Big),$$

where the map $(u,v) \mapsto \int_0^1 u v_{xxx} \, dx$ is referred to as the *Gelfand–Fuchs 2-cocycle*. Since vir is not a Hilbert space, we can consider for a given $k \geq 0$ the Sobolev-type scalar product

$$\langle (u,a),(v,b) \rangle_k = \sum_{j=0}^{k} \int_0^1 \partial_x^j u \cdot \partial_x^j v \, dx + ab, \qquad (u,a),(v,b) \in vir.$$

This scalar product induces a right-invariant (weak[174]) Riemannian metric μ_k on Vir if we transport the inner product from vir to any tangent space by right translation.[175] A *geodesic* $\gamma : [0,t_0] \to Vir$, $t_0 > 0$, with respect to the smooth (weak) Riemannian metric μ_k on Vir is defined as a stationary point[176] of the action functional

$$\mathcal{A}_k(\gamma(s,\cdot)) = \frac{1}{2} \int_0^{t_0} \mu_k(\gamma_t(s,t), \gamma_t(s,t)) \, dt.$$

Alternatively, one can show the existence of a Levi-Civita connection on Vir, which allows us to define the geodesics as the curves whose tangent vectors are parallel with respect to the connection (cf. [64]). It turns out that the curve $t \mapsto \gamma(t) = (\varphi(t), \alpha(t)) \in Vir$, with $(\varphi(0), \alpha(0)) = (Id, 0) \in Vir$ and $(\varphi_t(0), \alpha'(0)) = (u_0, a_0) \in vir$, is a geodesic with respect to the metric μ_0 if and only if

$$\alpha'(t) \equiv a_0 - \frac{1}{2} \int_0^1 u_x(t, \varphi(t,x)) \, d \log \varphi_x(t,x) \tag{6.77}$$

---

[174]The topology induced by this metric is not identical to the topology defined on Vir.

[175]It is natural to build in an invariance since this leads to structural properties. Starting with a metric on the Lie algebra (that is, on the tangent space at the neutral element), the choice of right invariance is suggested by the following property. If we think of the diffeomorphism $\varphi(t)$ as giving the configuration of some particles at time t, the material velocity field is defined by $(t,x) \mapsto \varphi_t(t,x)$, while the spatial velocity field is given by $u(t,y) = \varphi_t(t,x)$ with $y = \varphi(t,x)$, that is, $u = \varphi_t \circ \varphi^{-1}$. In terms of u we have the Eulerian description and in terms of (φ, φ_t) the Lagrangian description. Notice that if we replace the path $t \mapsto \varphi(t)$ in $\mathcal{D}$ by $t \mapsto \varphi(t) \circ \eta$ with a fixed $\eta \in \mathcal{D}$, then the spatial velocity u is unchanged.

[176]In the sense that $\frac{d}{ds}\Big|_{s=0} \mathcal{A}_k(\gamma(s,\cdot)) = 0$ for any smooth variation $\gamma(s,t) : (-\varepsilon, \varepsilon) \times [0,t_0] \to Vir$ of the curve γ, with $\gamma(0,t) = \gamma(t)$, and such that $\gamma(s,0) = \gamma(0)$ and $\gamma(s,t_0) = \gamma(t_0)$ for any $s \in (-\varepsilon, \varepsilon)$.

and

$$\varphi_t = u(t,\varphi(t)), \tag{6.78}$$

with $\varphi(0,x) = x$, where u satisfies the so-called *Euler equation*

$$A_k u_t = -(2u_x A_k u + u A_k u_x) + a_0 u_{xxx} \tag{6.79}$$

with

$$A_k = \sum_{j=0}^{k} (-1)^j \partial_x^{2j}$$

and initial data $u(0,x) = u_0(x)$. For $k = 0$ the Euler equation (6.79) is precisely KdV in the form (6.16) if we set $\zeta(x,t) = A\,u(Bx,t) + C$ with $A = -(\frac{4\delta^2}{3a_0})^{1/3}$, $B = -(\frac{6a_0}{\delta^2})^{1/3}$, and $C = -\frac{2}{3\varepsilon}$, while for $k = 1$ the Euler equation (6.79) becomes the Camassa–Holm equation (6.75) with $\hat{k} = -3a_0$ if we set $U(t,x) = u(t,x) + a_0$. While this geometric interpretation is interesting by itself, and the existence of geodesics on Vir can be proved,[177] the geodesic flow for $k \in \{0,1\}$ is not that interesting from a geometric point of view since the Riemannian exponential map (that is, the time-one map of the geodesic flow) is not a C^1-diffeomorphism from a neighborhood of $0 \in vir$ onto a neighborhood of $(Id,0)$ in Vir. This deficiency can be corrected for $k = 1$ as follows. Setting

$$u(x,t) = U\left(x - \frac{\hat{k}}{2}t, t\right) + \frac{\hat{k}}{2},$$

we transform (6.75) into the equivalent equation

$$u_t - u_{xxt} + 3uu_x = 2u_x u_{xx} + uu_{xxx}, \tag{6.80}$$

while preserving the periodicity in the spatial variable. Just like (6.75) was the Euler equation for the geodesic flow induced by the H^1 right-invariant metric on Vir, it turns out that (6.80) is the Euler equation for the H^1 right-invariant metric on $\mathcal{D}$. Here $\mathcal{D}$ is the Lie group of smooth orientation-preserving diffeomorphisms of the circle, with the Lie algebra $T_{Id}\mathcal{D} \equiv C^\infty(\mathbb{S}^1)$ (the smooth 1-periodic functions on $\mathbb{R}$), and the Lie bracket

$$[u,v] = u_x v - u v_x, \qquad u, v \in C^\infty(\mathbb{S}^1).$$

In this setting, however, the geodesic exponential map is a C^1-diffeomorphism from a small neighborhood of $0 \in C^\infty(\mathbb{S}^1)$ to a small neighborhood of the identity Id in $\mathcal{D}$ (cf. [65]) and can be used as a local chart. This fixes a certain built-in deficiency of $\mathcal{D}$: the Lie group exponential map[178] can be shown to be a smooth map from the Lie algebra $C^\infty(\mathbb{S}^1)$ to the Lie group $\mathcal{D}$, but it fails to be locally surjective (cf. [231]) although its Fréchet derivative at 0 is the identity.[179] Thus $\exp_L$ does

[177] The technical intricacy is due to the fact that although (6.78) is an ordinary differential equation with a smooth right-hand side, in Fréchet spaces existence and uniqueness of solutions are generally not guaranteed (cf. [137]) so that the proof depends on structural properties (see [64]).

[178] Defined for $u \in C^\infty(\mathbb{S}^1)$ by $[\exp_L(u)](x_0) = \eta(1,x_0) \in \mathcal{D}$, where $\eta(t,x_0)$ is the unique global solution of the ordinary differential equation $\eta'(t) = u(\eta(t))$ with $\eta(0) = x_0$.

[179] Such anomalies can never occur for Banach manifolds by the inverse function theorem (cf. [137]), but $\mathcal{D}$ is a Fréchet manifold. One cannot lower the regularity requirements (for example, by requiring some Sobolev-type regularity instead of C^∞-smoothness) without losing the Lie group property, the price to pay being that analysis in Fréchet space offers additional intricacies of this type. Notice the contrast with the case of finite-dimensional Lie groups, where the map $\exp_L$ is always a local diffeomorphism from the Lie algebra to the Lie group [16].

not provide us with a chart near the identity $Id \in \mathcal{D}$. Not only does the H^1 right-invariant metric endow $\mathcal{D}$ with a nice geometric structure, but the connection to (6.75) provides us with considerable insight into the structure of the latter equation. More precisely, the invariance properties built into the geodesic flow translate into strong invariance properties in terms of the solutions to (6.75); see, e.g., relation (6.83) below. While at first it might seem surprising that the mere elimination of the constant in (6.75) has such a strong effect in geometric terms, note, however, that the natural inclusion $\mathcal{D} \subset Vir$, whereby we associate to $\varphi \in \mathcal{D}$ the element $(\varphi,0) \in Vir$, does not define a subgroup of Vir, and it turns out (cf. [64]) that the geodesic exponential map on Vir for the H^1 right-invariant metric, when projected to $\mathcal{D}$, is different from the geodesic exponential map on $\mathcal{D}$ for the H^1 right-invariant metric. Notice also that this feature of improved regularity in terms of the geodesic exponential map cannot be replicated in the case $k=0$: the inviscid Burgers equation $u_t + 3uu_x = 0$ is in this case the Euler equation, and the geodesic exponential map on $\mathcal{D}$ for the L^2 right-invariant metric is not a local C^1-diffeomorphism (cf. [65]). Moreover, in finite-dimensional Riemannian geometry, the geodesic pseudodistance $d(x,y)$, defined as the infimum of the lengths of all piecewise C^1-curves joining x to y, is a distance, that is, $d(x,y) > 0$ if $x \neq y$. On $\mathcal{D}$, this is the case for the H^1 right-invariant metric (cf. [65]), but the geodesic pseudodistance vanishes identically for the L^2 right-invariant metric (cf. [229]).

The structural properties that single out these equations permit an in-depth qualitative study of (6.75) and (6.76). For example, in the case of (6.75), studies of the long-time asymptotics of the solutions for a large class of initial data are available. More precisely, the solitons are orbitally stable in the sense that a small perturbation will result in waves with a similar profile; cf. [70]. Moreover, while there are smooth initial profiles with compact support, such that the resulting solution to (6.75) develops singularities in finite time (see, e.g., [51] for blowup results and [30, 31, 149] for alternative theories of continuation of the solutions after blowup), any classical solution such that its initial data u_0 decays rapidly and initially satisfies

$$U - U_{xx} + \frac{\widehat{\kappa}}{2} > 0 \tag{6.81}$$

will not develop singularities in finite time.[180] To see this one first proves (cf. [45]) a result analogous to Theorem 6.2, characterizing blowup in finite time $T > 0$ of the solution $U \in C^1([0,T); H^2(\mathbb{R})) \cap C([0,T); H^3(\mathbb{R}))$ by

$$\liminf_{t \to T} \left[\min_{x \in \mathbb{R}} \{U_x(x,t)\} \right] = -\infty, \tag{6.82}$$

while

$$\partial_t \int_{\mathbb{R}} (U^2 + U_x^2)\, dx = 0, \qquad t \in (0,T),$$

ensures that u is uniformly bounded as long as it exists. The preservation of (6.81) for all $t \in (0,T)$ can be inferred, either from the isospectral problem associated to (6.75), cf. [45], or by noticing that, if we set

$$m = U - U_{xx} + \frac{\widehat{\kappa}}{2}$$

and we define for $t \in [0,T)$ the diffeomorphisms $\nu(\cdot,t) : \mathbb{R} \to \mathbb{R}$ by

$$\begin{cases} \nu_t = U(\nu,t), & t \in (0,T), \\ \nu(x,0) = x, & x \in \mathbb{R}, \end{cases}$$

[180]The same result holds even for weak solutions if (6.81) holds in the sense of measures, as one can see by adapting the approach pursued in [68].

then

$$m(v(x,t),t)\cdot v_x^2(x,t)=m_0(x), \qquad (x,t)\in\mathbb{R}\times[0,T). \tag{6.83}$$

This remarkable identity can be checked directly, its rationale being explained using the geometric interpretation in the spatially periodic case by observing that the right invariance of the metric ensures the conservation of the left angular momentum (see [64, 182]). But the Green's formula representation

$$\frac{\hat{\kappa}}{2}+U(x,t)=\frac{1}{2}e^{-x}\int_{-\infty}^{x}e^{y}\,m(y,t)\,dy+\frac{1}{2}e^{x}\int_{x}^{\infty}e^{-y}m(y,t)\,dy, \qquad (x,t)\in\mathbb{R}\times[0,T),$$

yields

$$U_x(x,t)=-\frac{1}{2}e^{-x}\int_{-\infty}^{x}e^{y}\,m(y,t)\,dy+\frac{1}{2}e^{x}\int_{x}^{\infty}e^{-y}m(y,t)\,dy, \qquad (x,t)\in\mathbb{R}\times[0,T),$$

so that the positivity of m ensures

$$U_x(x,t)\geq -U(x,t)-\frac{\hat{\kappa}}{2}, \qquad (x,t)\in\mathbb{R}\times[0,T),$$

preventing blowup in finite time in view of (6.82) because the solution u is known to be uniformly bounded as long as it exists. Notice that the inverse scattering approach used in [29] permits an exact description of the long-time behavior of the solution u with localized smooth initial data satisfying (6.81).

The utility of these qualitative studies of the horizontal fluid velocity at a certain depth lies in their providing information about the surface wave profile ζ by means of (6.74) when the solution of (6.75) does not develop singularities in finite time. Indeed, for a global solution of (6.75) with smooth data, all the derivatives remain bounded at later times so that, by (6.74), the surface wave profile ζ enjoys the same property. The fact that we did not mention the derivation of an evolution equation for ζ in this case is not accidental: one would obtain an equation similar to (6.28) which could be used to infer the existence of breaking waves, but, for (6.28), we are not aware of any global existence results for a sufficiently wide class of initial data. Herein lies the big structural advantage of (6.75). Similar considerations are valid for (6.76), although the qualitative study of this equation is not as advanced as that of (6.75); see, however, [110] for global existence and blowup results, [60] for the inverse scattering approach, and [220] for the soliton interaction property of its solitary waves.

Variable depth and underlying currents

The equations presented in this chapter appear to arise in very special circumstances (irrotational water flow over a flat bed). These models can offer an accurate and comprehensive description of real phenomena only if they are fairly robust to changes in the ambient physical conditions. Within the confines of inviscid two-dimensional theory (so that variations in one horizontal direction are negligible) one has to address the issues of variable depth and underlying currents.

Let us first discuss the case of variable depth $Y=-h_0+B(X)$ with h_0 being the reference depth. The description of the variations of the bed necessitates the introduction of two additional parameters: if b_0 is the typical amplitude of the variations of the bottom topography and λ_0 is the wavelength of the bottom variations, set

$$\alpha=\frac{\lambda}{\lambda_0}, \qquad \beta=\frac{b_0}{h_0}.$$

If the variations of the bottom topography satisfy

$$\alpha\beta = O(\delta^4), \tag{6.84}$$

then the evolution of the free surface waves in the long-wave regime (6.15) is described by the KdV-type equation[181]

$$\zeta_t + \kappa\zeta_x + \frac{1}{2}\kappa_x\zeta + \frac{3\varepsilon}{2\kappa}\zeta\zeta_x + \frac{\delta^2}{6}\kappa^5\zeta_{xxx} = 0, \tag{6.85}$$

where

$$\kappa(x) = \sqrt{1-\beta B(\alpha x)}. \tag{6.86}$$

Notice that, in the case of a flat bed, we have $B \equiv 0$ so that $\kappa \equiv 1$ and (6.85) particularizes to the KdV equation (6.16). One can prove rigorously that (6.85) provides an approximation to the governing equations up to order $O(\delta^4)$; cf. [157]. While integrability for (6.85) in the full generality of (6.84) is not granted and we are not aware of a study of global existence and blowup, notice that, in the special case

$$\alpha = O(\varepsilon), \qquad \beta = O(\varepsilon), \tag{6.87}$$

which arises when the scale of the depth variation is the same as the scale on which the wave will naturally evolve over constant depth, we have

$$\kappa(x) = 1 + \frac{1}{2}\beta B(\alpha x) - \frac{1}{8}[\beta B(\alpha x)]^2 + \cdots = 1 + O(\varepsilon),$$
$$\kappa_x(x) = -\frac{\beta\alpha B'(\alpha x)}{2\kappa(x)} = O(\varepsilon^2), \qquad \kappa_{xx} = O(\varepsilon^2),$$

so that the $O(\delta^4)$-accurate approximation to the governing equations is preserved if we replace (6.85) by

$$\zeta_t + \kappa\,\zeta_x + \frac{3\varepsilon}{2}\kappa\,\zeta\zeta_x + \frac{\delta^2}{6}\kappa^3\zeta_{xxx} = 0. \tag{6.88}$$

Moreover, setting $\eta(r,t) = \zeta(x,t)$ with $r = \int_0^x \frac{ds}{\kappa(s)}$ we have

$$\zeta_x = \frac{\eta_r}{\kappa}, \quad \zeta_t = \eta_t, \quad \zeta_{xx} = \frac{\eta_{rr}}{\kappa^2} - \frac{\kappa_x}{\kappa^2}\eta_r = \frac{\eta_{rr}}{\kappa^2} + O(\varepsilon^2),$$
$$\zeta_{xxx} = \frac{\eta_{rrr}}{\kappa^3} - 3\frac{\kappa_x}{\kappa^3}\eta_{rr} - \frac{\kappa_{xx}}{\kappa^2}\eta_r + 2\frac{\kappa_x^2}{\kappa^3}\eta_r = \frac{\eta_{rrr}}{\kappa^3} + O(\varepsilon^2),$$

so that (6.88) is transformed into the classical KdV equation (6.16) in the form

$$\eta_t + \eta_r + \frac{3\varepsilon}{3}\eta\eta_r + \frac{\delta^2}{6}\eta_{rrr} = 0$$

by neglecting terms of order $O(\delta^4)$. Consequently, for topographical variations of the bed of type (6.87) the regime (6.15) of long waves of small amplitude is not appropriate for studying breaking waves.

For topographical variations of type (6.84) one can also investigate the unidirectional propagation of long waves of moderate amplitude in the regime (6.18). An approach similar to the one

[181] This is a KdV-type equation for variable depth. Notice that there are several equations of this type, obtained from each other either directly by a change of variables or by neglecting terms of order $O(\delta^4)$ after changing variables; we illustrate both procedures.

presented above for the case of a flat bed was developed in [157], obtaining, instead of (6.28), the following equation describing the evolution of the surface water waves:

$$\begin{aligned}\zeta_t + \kappa\,\zeta_x + \frac{1}{2}\kappa_x\,\zeta + \frac{3}{2\kappa}\varepsilon\,\zeta\,\zeta_x - \frac{3}{8\kappa^3}\varepsilon^2\zeta^2\zeta_x + \frac{3}{16\kappa^5}\varepsilon^3\zeta^3\zeta_x + \frac{\delta^2}{12}(2\kappa^5-\kappa)\,\zeta_{xxx}\\ -\frac{\delta^2}{12}\zeta_{xxt} + \varepsilon\delta^2 E\,\zeta\,\zeta_{xxx} + \frac{\varepsilon\delta^2}{2}\Big(\zeta_{xx}\,\partial_x(F\zeta) + \zeta_x\,\partial_x^2(F\zeta)\Big) = 0,\end{aligned} \tag{6.89}$$

where κ is defined in (6.86) and

$$E = \frac{1}{6}\kappa^4 + \frac{1}{8}, \qquad F = \frac{5}{24}\kappa^4 + \frac{3}{8}.$$

For a flat bed we have $\kappa \equiv 1$ and (6.89) is precisely (6.28). It turns out that (6.89) provides an approximation to the governing equations for water waves over variable depth with accuracy $O(\delta^4)$; cf. [157]. Moreover, the proof of Theorems 6.2 and 6.3 can be adapted to the present setting (see [157]), showing that (6.89) models breaking waves. The derivation and investigation of evolution equations for the horizontal fluid velocity at certain depths, where one expects perturbations of the Camassa–Holm and Degasperis–Procesi model equations (6.75) and (6.76), remain to be explored. As pointed out in the above discussion of the Camassa–Holm equation (6.75), global existence results for such equations provide information about the long-time behavior of the surface waves by means of a relation analogous to (6.74).

Compared with the case of variable depth, the situation for underlying currents in water over a flat bed gives rise to a more satisfactory outcome in terms of integrable equations. For the KdV equation, no matter the form of the flow without stagnation points below the surface, we can recover its classical form after suitable scaling (see [117] and the recent discussion in [166]). The derivation of evolution equations of type (6.28) for waves at the surface of a flow with vorticity, as well as that of equations of type (6.73) for the horizontal fluid velocity at certain depth, remains to be investigated. A promising result in this direction is that for underlying currents of positive constant vorticity $\omega > 0$ one can (after scaling) recover the Camassa–Holm equation (6.75), with only minor changes: the value of $\widehat{\kappa}$ in (6.75) and the specific depth at which this equation is obtained both depend on ω (see [165]). The presence of the underlying current of constant vorticity introduces a new feature. Indeed, in contrast to the case of a uniform flow (within the irrotational setting), where the direction in which the waves propagate (left to right or in the opposite direction) is irrelevant, when the ambient flow has a nonzero vorticity, differences are immediately apparent (cf. [165]): if the underlying current is given by the velocity field $(\omega(y+h_0), 0)$, then in the regime (6.18) of long waves of moderate amplitude propagating in the direction of the current (i.e., downstream, or from left to right), at a certain depth, the horizontal velocity component of the wave-current interaction will satisfy a Camassa–Holm equation if we discard terms of order $O(\delta^4)$, while if the waves propagate upstream (i.e., from right to left), the derivation of a Camassa–Holm equation is not physically realizable at any depth level.

Finally, the case of underlying currents in a region whose bottom topography slowly changes is considerably more complicated. Superposing on this configuration a nonlinear gravity wave which evolves on the same scale as the bottom variation, one can show (cf. [161]) that, in the regime (6.15) of long waves with a small amplitude, the surface wave satisfies a new variable coefficient KdV equation, which reduces to the classical form in the absence of either the shear flow or any depth variation. There are no studies of this type in the regime (6.18) of long waves of moderate amplitude.

Chapter 7

Modelling tsunamis

"There is nothing like looking, if you want to find something.
You usually find something, if you look,
but it is not always quite the something you were after."
J. R. R. Tolkien

The word "tsunami" means "harbor wave" in Japanese. The name captures the essential feature of these waves: they are hard to detect at sea due to their deceptively small amplitude (usually with waves rising almost uniformly over a wavelength of several hundreds of kilometres to less than $1\,m$ above the usual sea level), but as they approach the shore tsunami waves sometimes rise to large amplitudes (up to 10–15 m). Earthquakes, volcanic eruptions, great landslides, and even more localized impulses—such as meteorite strikes into the sea—sometimes cause a large volume of water to be rapidly displaced from its original state with an almost flat surface. A new pattern featuring more pronounced rises and/or depressions is generated, with the resulting wave moving away from the location of its creation in the form of a tsunami wave.

To accurately predict the appearance of a tsunami, it is of paramount importance to understand the motion of these exceptional waves. We are especially interested in whether a balance between dispersion and nonlinearity is likely to occur, thus providing a setting appropriate for the application of soliton theory to the modelling of the tsunamis. The long distances involved in some tsunamis could be an indication of the relevance of soliton theory to the modelling of tsunamis. If so, a detailed study of the interaction of such waves is possible via inverse scattering theory. We will therefore address the following issue: *Is soliton theory, or more precisely the KdV equation, relevant to the modelling of tsunamis?* We discuss this in the context of the two largest tsunamis for which records are available—the December 2004 and the May 1960 tsunamis. The tsunami of December 2004 was one of the deadliest natural disasters in history, devastating many coastal communities across the Indian Ocean/Bay of Bengal, killing more than 275,000 people in 11 countries, and causing enormous material damage. The 1960 Chile tsunami is considered as the most likely candidate to illustrate the relevance of soliton theory to tsunamis since it involved the largest ever recorded earthquake, and the tsunami waves propagated all the way across the Pacific Ocean in water of practically constant depth, over distances in excess of 10,000 km. Our contention is that, in both cases, the scales involved do not permit a balancing effect of dispersion and nonlinearity, and therefore **KdV theory is not applicable**.

7.1 General considerations

A tsunami is a water wave of very long wavelength, generated by a sudden displacement of the seafloor (e.g., due to an earthquake) or the disruption of a large body of water (e.g., by landslides, rockslides, or, more rarely, by meteor impact). In deep water, these waves are generally of relatively small amplitude (less than $1\,m$; cf. [36]) so that linear theory captures well their main characteristics. Typically, regions that suffer earthquake activity are elliptical in shape, with a major axis several hundreds of km long corresponding to the active part of the fault, and most of the energy is transmitted at right angles to the major axis (cf. [36]), so that it is reasonable to assume that tsunamis generated by earthquakes are two-dimensional gravity water waves. According to the considerations made in Section 3.1.3, for practical purposes the appropriate range for shallow-water waves is $\delta = \frac{d}{\lambda} < 0.07$, where d is the average water depth and λ denotes the wavelength. The deepest point in the ocean being in the Marianas Trench of the Western Pacific Ocean at approximately $11021\,m$, we conclude that all tsunamis with wavelengths λ in excess of $158\,km$ qualify as shallow-water waves, even if they occur in the deepest parts of the ocean. In particular, the propagation speed of these shallow-water waves of small amplitude (for which linear theory is adequate) will be roughly $c = \sqrt{gd}$, where $g = 9.8\,m/s^2$ is the approximate gravitational acceleration at sea level. As long as the water depth remains approximately constant, these waves travel rapidly with almost no change of form. Nevertheless, tsunamis are barely noticeable in the open ocean. To illustrate this, we borrow an example from [271]: how would an observer sitting in a boat in the middle of the Indian Ocean experience the 2004 Boxing Day tsunami? The tsunami wave that traveled to the west was about $100\,km$ long and $1\,m$ high. The depth of the Bay of Bengal/Indian Ocean between Sumatra and Sri Lanka having a relatively flat bed of average depth $d = 4\,km$, we can estimate the speed of the tsunami to be $c = \sqrt{gd} \approx 720\,km/h$. For the observer sitting in the boat, two consecutive crests will be experienced 8–9 minutes apart, a time in which the boat would move up by about $1\,m$, and then back down by $1\,m$. It is therefore unlikely that the observer would notice that the tsunami wave had gone by. In contrast to this behavior in water of constant depth, as the tsunami enters the coastal waters, it undergoes a transformation. The rate at which a wave loses its energy is roughly inversely related to its wavelength (cf. [162]). Therefore, since a tsunami has a very large wavelength, it will lose little energy as it propagates. On the other hand, the speed of the tsunami being directly related to the water depth, as the depth of the water decreases, the velocity of the tsunami decreases. The time period of the wave remains the same, so the wavelength decreases. To argue that thus more water is forced between the wave crests, causing the height of the wave to increase due to the conservation of the water mass (that is, smaller wavelength makes the vertical dimension higher if we are to keep the same volume of water) is not appropriate since, as seen in Chapter 4, the wave motion and the motion of the water are different: the wave is an energy pattern. The proper explanation for the increase in wave height as the tsunami reaches the coastal waters is that the energy flow associated with the wave must remain roughly constant. Since (cf. [206]) within the framework of linear theory the energy flow is proportional to $a^2 c \approx a^2 \sqrt{gd}$, where a is the typical wave amplitude, if d decreases, a must increase. Because of this "shoaling" effect, a tsunami that was imperceptible in deep water may grow to have wave heights of several meters or more.[182] If the trough of the tsunami wave precedes the crest,

[182]For example, a tsunami wave with amplitude $a = 2\,m$ near the shore, say for $d = 2\,m$, would have in a deep-sea area with $d = 4\,km$ an amplitude of about $0.3\,m$.

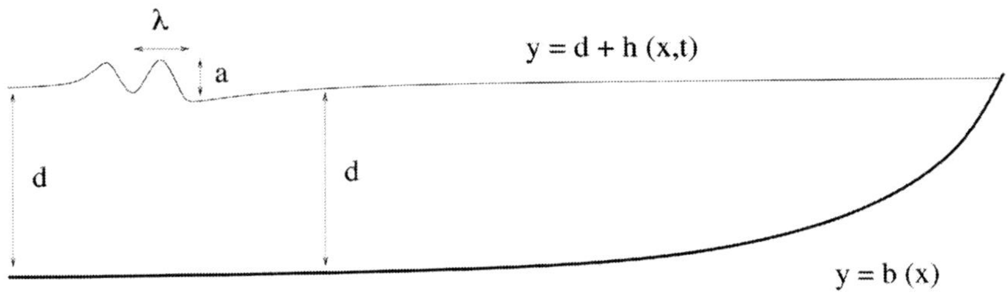

Figure 7.1. *The main parameters used in describing the propagation of the tsunami.*

it reaches the coast first, causing a phenomenon called "drawdown," where the sea appears to retreat from the shore. Drawdown is followed immediately by the crest of the tsunami wave which can catch people observing the drawdown off guard.

The problem of the modelling of tsunamis presents three major aspects:

(1) *tsunami wave generation* in the event of a major earthquake, landslide, etc.;

(2) *tsunami wave propagation* across the open ocean;

(3) *tsunami wave impact* upon coastal regions.

The second aspect here is within the framework of earlier considerations, and we will cover it in some detail. Before proceeding with this, we provide some background information on the other two aspects.

For tsunamis generated by an earthquake, not only the magnitude of the earthquake is of relevance, but also its type. Indeed, most earthquakes occur at the boundary of tectonic plates, where, depending on the relative motion of adjacent plates, we distinguish three types of faults: one tectonic plate can slide over, under, or past another, producing a thrust fault, a normal fault, or a strike-slip fault, respectively. Strike-slip faults do not generate tsunamis since they do not produce a significant vertical movement of the ocean floor. In contrast to this, a thrust fault raises the floor of the ocean, and also the water above, creating a water wave of elevation in the region above the fault, while a normal fault lowers the floor of the ocean and thus creates a depression in the water surface above the fault. These basic considerations are used in identifying the possible location of the source of a tsunami, since the history of any particular earthquake zone indicates the type of fault that is most likely to occur.

The dynamics of the tsunami near the shore is, perhaps, the most significant aspect of these waves. The simple argumentation used above indicates that, as the tsunami approaches the shore, its wavelength shortens, and (generally) its amplitude grows. Due to variations in the topography of the sea bed, this process is not uniform: often tsunamis completely destroy some coastal locations, while sparing locations only a few kilometres away. In contrast to the case of the propagation in the open ocean, in order to predict what happens on a particular coastline, we must know the bottom topography there.

Let us now discuss the tsunami wave propagation across the open ocean, in water with an almost flat bed. Let x be the direction of wave propagation, with the fluid assumed to extend to $-\infty$, the shoreline in the absence of waves being at $x = 0$, and we let y be the vertical direction, the free surface being $y = d + h(x,t)$. The impermeable bed has the equation $y = b(x)$ with $b(0) = d$, and for a beach we require that $b'(d) > 0$, u is the fluid velocity component in the x-direction, v is the fluid velocity component in the y-direction, and P is the pressure. Moreover, let λ be the typical wavelength, let d be the average depth of the water far from the shoreline, and let a be the maximum amplitude of the wave in the open ocean. The governing equations are Euler's equation

$$\begin{cases} u_t + uu_x + vu_y = -P_x, \\ v_t + uv_x + vv_y = -P_y - g \end{cases}$$

coupled with the equation $u_x + v_y = 0$, expressing mass conservation, and with the condition $u_y = v_x$ for irrotational flow (assuming the absence of strong currents). The dynamic boundary condition $P = P_{atm}$ (constant atmospheric pressure) on $y = d + h(x,t)$ and the kinematic boundary conditions $v = ub_x$ on $y = b(x)$ and $v = h_t + uh_x$ on $y = d + h(x,t)$ complete the formulation of the governing equations. We assume that only close to the shoreline will there be a significant bottom variation, while far out the bed is almost flat. As discussed in Chapter 6, two fundamental parameters appear:

- $\varepsilon = a/d$, the wave amplitude parameter;
- $\delta = d/\lambda$, the long wave or shallowness parameter.

Investigating the governing equations for data representing a disturbance occurring far from the shore, one can rigorously analyze the solution in the region where the bed is practically flat (with variations of order ε), obtaining the validity ranges for the main physical regimes encountered in the modelling of two-dimensional shallow-water waves [8]:

1. shallow-water, large-amplitude regime ($\delta \ll 1$, $\varepsilon \sim 1$) leading at first order to the shallow-water equations [162] and at second order to the Green–Naghdi model [131], which takes into account the dispersive effects neglected by the shallow-water equations;

2. shallow-water, moderate-amplitude regime ($\delta \ll 1$, $\varepsilon \sim \delta$) leading to the Camassa–Holm equation [41] (see [66]);

3. shallow-water, small-amplitude or long-wave regime ($\delta \ll 1$, $\varepsilon \sim \delta^2$) leading at first order to the linear wave equation

$$h_{tt} - h_{xx} = 0 \tag{7.1}$$

with the general solution

$$h(x,t) = h_+(x-t) + h_-(x+t), \tag{7.2}$$

where the sign $\pm$ refers to a wave of profile $h_\pm$ moving with unchanged shape to the right/left at constant unit speed by virtue of the nondimensionalization process, the corresponding dimensional speed being $\sqrt{gd}$. The small effects that were ignored at first order (small amplitude, long wave) build up on longer time/spatial scales to have a significant cumulative nonlinear effect, so that on a longer time scale each of the waves that make up the solution (7.2) to (7.1) satisfies the KdV equation [272].

Therefore, KdV arises in the regime

$$\varepsilon = O(\delta^2). \tag{7.3}$$

To answer the question of whether KdV theory can be used effectively to predict tsunamis, we will first present some general considerations showing that KdV dynamics occur only if the propagation distance is long enough. In subsequent sections we will discuss the two major tsunamis of the 20th century.

In the derivation of the KdV equation from the governing equations (see the Appendix), one can see that, after the nondimensionalization $x \mapsto \lambda x$, $t \mapsto \frac{\lambda}{\sqrt{gd}}\, t$, the region where we expect a KdV-type balance to occur is given by

$$x - t = O(1), \quad t\varepsilon = O(1),$$

in nondimensional variables, that is,

$$\frac{x - t\sqrt{gd}}{\lambda} = O(1), \quad \frac{\varepsilon t\sqrt{gd}}{\lambda} = O(1)$$

in the original physical variables. The second relation yields $\frac{t\sqrt{gd}}{\lambda} = O(\varepsilon^{-1})$, which implemented in the first leads us to $\frac{x}{\lambda} = O(\varepsilon^{-1})$ so that *the length scale on which one sees KdV-type dynamics is*

$$x = O(\varepsilon^{-1}\lambda). \tag{7.4}$$

This corrects[183] the classical scale for the evolution of KdV, believed to be $x = O(\varepsilon^{-1} d)$; cf. the recent surveys [271, 272] and the classical paper [138]. The change of scale is considerable since $d/\lambda \ll 1$ for tsunami waves due to wavelengths of several hundreds of kilometers (the deepest part of the world's seas is about 11 *km*). Since in the KdV regime $\varepsilon \approx \delta^2$, *the length scale on which one sees KdV-type dynamics can also be expressed as*

$$\varepsilon^{-1}\lambda \approx d^{\frac{5}{2}} a^{-\frac{3}{2}}, \tag{7.5}$$

which is independent of λ. We would like to emphasize that checking that the wave propagation occurs within the regime $\varepsilon \approx \delta^2$ does not automatically mean that KdV dynamics occurs; the propagation distance also has to be long enough. It is instructive to present some estimates of the distance that waves will have to travel before the nonlinear dispersion balance, epitomized by the KdV equation, can occur. The results are given in the table below (but note that, in the cases when $\varepsilon = a/h_0 \geq 1$, the estimate is not applicable).

Amplitude	0.1 m	0.5 m	1 m	5 m	10 m
Depth 5 *m*	1.77 *km*	0.16 *km*	0.056 *km*	N/A	N/A
Depth 10 *m*	10 *km*	0.89 *km*	0.32 *km*	0.028 *km*	N/A
Depth 100 *m*	3162 *km*	283 *km*	100 *km*	8.9 *km*	3.2 *km*
Depth 1 *km*	10^6 *km*	89442 *km*	31622 *km*	2828 *km*	1000 *km*
Depth 4 *km*	$32 \cdot 10^6$ *km*	$28 \cdot 10^6$ *km*	10^6 *km*	$9 \cdot 10^4$ *km*	$32 \cdot 10^3$ *km*

[183] A fact confirmed by H. Segur (private communication).

These figures demonstrate that, as John Scott Russell so eloquently described in his 1844 "Report on Waves," a solitary wave of moderate amplitude in a river could evolve and be followed on horseback; the distances needed for a KdV balance are no more than a few kilometres. On the other hand, a wave generated in deep water, say, $4\,km$, would have to travel hundreds or even thousands of kilometers before the KdV balance would be achieved.

The previous considerations enable us to reach the following conclusions concerning the propagation of tsunamis in the open sea, in water of almost constant depth. If the propagation is one-directional and occurs within the shallow-water, small-amplitude regime (7.3), as is often the case, linear theory applies, and from (7.1) and (7.2) we infer that the wave profile is given by

$$h(x,t) = h_0(x - \sqrt{gd}\,t) + O(\varepsilon),$$

where $x \mapsto h_0(x)$ is the initial profile of the disturbance in the deep water, generated by the seismic activity. Consequently, to leading order, the wave profile does not change its shape as it propagates. Therefore, the initial profile of the disturbance is the single most important ingredient in the formation of tsunami waves and will determine if a wave of depression or a wave of elevation is the leading wave. The observed phenomena of either one or several large tsunami waves crashing on the shore, with large wave fronts sometimes tens of minutes apart, as well as the occurrence of drawdown, is also explained by the particular form of the initial disturbance.[184]

The role of dispersion

Our contention is that dispersion plays a minor role and can be ignored. Even if the initial profile comprises waves with short wavelengths, after some time the thrust of the tsunami propagation problem is confined to long waves. Indeed, since out in the open sea the tsunami waves are of small amplitude, we can use linear theory. According to the discussion in Section 3.1.3, the dispersion relation for waves with wavelength λ in water of constant depth d can be written as a formula expressing the propagation speed c_λ of these waves in the form

$$c_\lambda = \sqrt{\frac{g\lambda}{2\pi}\tanh\left(\frac{2\pi d}{\lambda}\right)}. \tag{7.6}$$

The right side is an increasing function of $\lambda \in (0,\infty)$, with the values of c_λ confined to the interval $(0,\sqrt{gd})$. The dispersive effect of linear water waves (longer waves travel faster) shows that, after initiation, provided that the water depth remains roughly constant, the longer waves move to the front and the short waves are then confined to the back of the tsunami profile. Behind the leading long waves a dispersive effect might be noticeable, but dispersion cannot be essential for the leading waves in a tsunami.

We conclude our general considerations by discussing, within the framework of linear theory, how the large-time response to a localized disturbance depends on whether we are

[184]The number of crests and troughs in the initial disturbance will prescribe the number of fronts that will steepen as the tsunami wave enters the coastal regions. While we do not discuss the steepening effect that a decreasing depth has on these fronts, note that the number of crests that reach the shore might be smaller because some of the fronts may coalesce, as the faster ones catch up to the slower ones (but cannot overtake them) in a manifestation of the "confluence of shocks" (see [314]).

in the shallow-water or deep-water regime. Consider the disturbance

$$h(x,t) = \int_0^\infty a(k)\cos(kx - \omega t)\,dk \tag{7.7}$$

representing a superposition of two-dimensional harmonic waves $a\cos(kx-\omega t)$ traveling to the right. In (7.7) the dispersion relation $\omega = \sqrt{gk\tanh(kd)}$ for the wavenumber $k = 2\pi/\lambda > 0$, derived in Section 3.1.3, is implicitly assumed, and the amplitude $a(k)$ depends on the wavelength λ. Setting $a(-k) = a(k)$ for $k > 0$, we can express (7.7) in the more convenient form

$$h(x,t) = \frac{1}{2}\int_{\mathbb{R}} a(k)e^{i(kx-\omega t)}\,dk. \tag{7.8}$$

We are interested in the large-time response to the localized smooth initial disturbance

$$h_0(x) = \int_0^\infty a(k)\cos(kx)\,dk = \frac{1}{2}\int_{\mathbb{R}} a(k)e^{ikx}\,dk. \tag{7.9}$$

In the shallow-water regime, the dispersion relation becomes $\omega = k\sqrt{gd}$ as $\tanh(kd) \approx kd$ for $kd = 2\pi\delta \to 0$, so that

$$h(x,t) = \frac{1}{2}\int_{\mathbb{R}} a(k)e^{ik(x-ct)}\,dk = h_0(x-ct)$$

with $c = \sqrt{gd}$. This shows that in linear shallow-water theory the initial profile propagates with speed c and **without change of shape**. On the other hand, in the deep-water regime, the dispersion relation becomes $\omega = \sqrt{gk}$ as $\tanh(kd) \approx 1$ for $kd = 2\pi\delta \to \infty$, so that (7.8) now takes the form

$$h(x,t) = \frac{1}{2}\int_{\mathbb{R}} a(k)e^{i(kx-\sqrt{gk}\,t)}\,dk. \tag{7.10}$$

A localized smooth initial disturbance implies that the function $x \mapsto h_0(x)$ is smooth and rapidly decreasing (in the sense that the function and all its derivatives decay faster than polynomially as $|x| \to \infty$). Since (by (7.9)) the function $a(k)$ is twice the Fourier transform of h_0, we deduce from the Paley–Wiener theorem (see [290, Section 7.2]) that the function $k \mapsto a(k)$ is smooth and rapidly decreasing.[185]

Let us first fix a location $x_0 \in \mathbb{R}$ and describe the character of the evolution of the wave packet (7.10) at that precise location as time passes by. Taking into account the properties of the function a, we have

$$h(x_0,t) = \frac{1}{2}\int_{\mathbb{R}} a(k)e^{i(kx_0-\sqrt{gk}\,t)}\,dk = O(t^{-n}) \quad \text{as} \quad t \to \infty$$

for any integer $n \geq 1$. To discuss in more detail the asymptotics of (7.10) as $t \to \infty$, we seek its behavior as t becomes large for a fixed value c_0 of x/t: this corresponds to the

[185]However, unless $h_0 \equiv 0$, a theorem of Hardy (cf. [103]) ensures that a and h_0 cannot both be compactly supported. It is more realistic to assume that a has compact support since we deal with gravity water waves so that (see the discussion of surface tension in the Appendix to Chapter 1) the wavelength λ should exceed $1.7\,cm$, and consequently the values of $k = 2\pi/\lambda > 0$ for which $a(k) \neq 0$ are confined to a bounded set.

idea of the observer's gaze moving at a fixed speed c_0. The method of stationary phase (see [283, Chapter VIII]) yields

$$h(c_0 t,t)=\frac{1}{2}\int_{\mathbb{R}} a(k)e^{it(c_0k-\sqrt{gk})}dk=\frac{a(k_0)}{2\sqrt{t}}\sqrt{\frac{\pi g}{c_0^3}}\cos\left(\frac{gt}{4c_0}+\frac{\pi}{4}\right)+O(t^{-1})$$

as $t\to\infty$, where $k_0=\frac{g}{4c_o^2}$ is the stationary point of the phase function $\theta(k)=c_0k-\sqrt{gk}$. Since $x/t=c_0$, we may write the above expression in the more convenient form

$$h(x,t)=\frac{a(k_0)}{2\sqrt{t}}\sqrt{\frac{\pi g}{c_0^3}}\cos\Big(k_0x-t\sqrt{gk_0}-\frac{\pi}{4}\Big)+O(t^{-1})\quad\text{as}\quad t\to\infty. \tag{7.11}$$

The asymptotic expression (7.11) brings to light several interesting aspects. The dominant part of the wave packet (7.10) is the harmonic wave $\cos\big(k_0x-\sqrt{gk_0}\,t\big)$, in spite of the fact that the initial state of the wave h_0 was not necessarily harmonic. This wave, noticed by the observer with a phase shift equal to $\pi/4$ and a decaying amplitude $\frac{a(k_0)}{2\sqrt{t}}\sqrt{\frac{\pi g}{c_0^3}}$ as time progresses, is singled out among the large number of harmonic-wave components of the initial disturbance $h_0(x)$ by the fact that the observer's gaze moves precisely at the group speed $c_g=\omega'(k_0)=\frac{\sqrt{g}}{2\sqrt{k_0}}$. In other words, an observer moving with the group speed c_g will continually see the main disturbance of the water's surface as a sinusoidal wave of constant length (the wavelength is determined by the constant speed at which the observer is moving), albeit subject to gradual amplitude attenuation. Note that, according to the discussion in Section 3.1.3, the speed of a single deep-water wave of wavelength λ is $c_\lambda=\sqrt{\frac{g\lambda}{2\pi}}$ but, with $c_g=\frac{1}{2}c_\lambda$, after time t, waves of the specific wavelength λ in the wave group are observed at the distance $\frac{1}{2}c_\lambda t$ and not at the distance $c_\lambda t$. In other words, in the aftermath of some complicated but localized initial disturbance, the wave pattern will be locally sinusoidal, with wavenumber k, in the neighborhood of a distance $x=\frac{1}{2}\sqrt{\frac{g}{k}}\,t$ from the initial disturbance region. To illustrate this, consider the simple case of a packet of two progressive harmonic waves, moving in the same direction and having the same amplitude a but slightly different wavelengths. The elevation about the still water surface $y=d$ is given by

$$h(x,t)=a\,\cos(kx-\omega t)+a\,\cos([k+\Delta k]x-[\omega+\Delta\omega]t)$$

with appropriate increments Δk, $\Delta\omega$, corresponding to a slight change in the wavelength. This can be rewritten as

$$h(x,t)=A(x,t)\cos\left(\left[k+\frac{\Delta k}{2}\right]x-\left[\omega+\frac{\Delta\omega}{2}\right]t\right),$$

where

$$A(x,t)=2a\cos\left(\frac{\Delta k}{2}x-\frac{\Delta\omega}{2}t\right).$$

We can look upon this as a progressive wave, modified by having the variable amplitude A. Since Δk and $\Delta\omega$ are infinitesimal increments, A varies very slowly with x for a fixed value of t, and so there are considerable distances over which A is nearly zero or nearly $2a$. At any instant the waveform is as shown in Figure 7.2. The variation in amplitude depends on

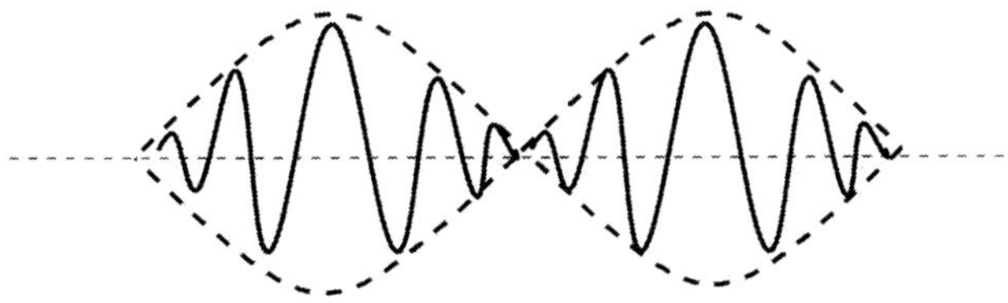

Figure 7.2. *The modulating wave describes the variation in amplitude.*

the **modulating wave** $\cos\left(\frac{\Delta k}{2}x - \frac{\Delta\omega}{2}t\right)$, whose outline is shown touching all the individual crests and troughs. The profile of the resulting wave is a series of wave groups separated by regions of almost calm water (almost free from waves). The speed of the envelope of the wave packet is called the **group speed**, $c_g = \frac{\Delta\omega}{\Delta k}$, and is less than the speed c_λ at which the individual waves with wavelength λ move, with equality of speeds arising only for vanishingly small depth or wave number; see below. We are interested in the limit Δk, $\Delta\omega \to 0$, corresponding to waves of nearly equal wavelengths, so that the group speed of waves having wavelength approximately k_0 is given by

$$c_g = \frac{\partial\omega}{\partial k}(k_0). \tag{7.12}$$

The individual wave crests move with speed $c = \lim_{\Delta k,\,\Delta\omega\to 0}\frac{\omega+\Delta\omega/2}{k+\Delta k/2}$, so that the speed of any particular wave crest is $c_\lambda = \frac{\omega}{k}$. Since $\omega^2 = gk\tanh(kd)$ and $kd = 2\pi\delta$, we see that

$$\frac{\partial\omega}{\partial k} = \frac{1}{2}\sqrt{\frac{g}{k}}\tanh(kd) + \frac{1}{2}\sqrt{gk}\,\frac{d}{\cosh^2(kd)}\,\frac{1}{\sqrt{\tanh(kd)}}.$$

For shallow-water waves ($\delta \to 0$ at d fixed) we obtain the group speed[186]

$$c_g = \sqrt{gd} = c, \tag{7.13}$$

while in deep water ($\delta \to \infty$ at k fixed) we obtain

$$c_g \approx \frac{1}{2}\sqrt{\frac{g}{k}} = \frac{1}{2}c_\lambda. \tag{7.14}$$

Relation (7.14) shows that a wave in deep water has no continuing individuality: the individual wave crests vary in amplitude as they pass through the group moving at a speed $c = 2c_g$, rising at the rear, then increasing in height and gradually diminishing as they approach the front of the group. If we fix our eyes on a crest and follow it, it will soon lose its identity as the individual wavelets in a wave group diminish toward the end of the group while the speed of passage of the undulation is twice the group speed (they "pass through" the group). The group as an entity can be followed, whilst the individual wave is evanescent; the behavior will be more comprehensible if we realize that the tops of the wave are not individual entities but local manifestations of the general movement in a group of waves. To understand the phenomenon it is useful to perform the following experiment: float a piece of wood on the surface of the water at one end of a bathtub, and then tap the top of the

[186]Confirming that shallow-water waves are nondispersive.

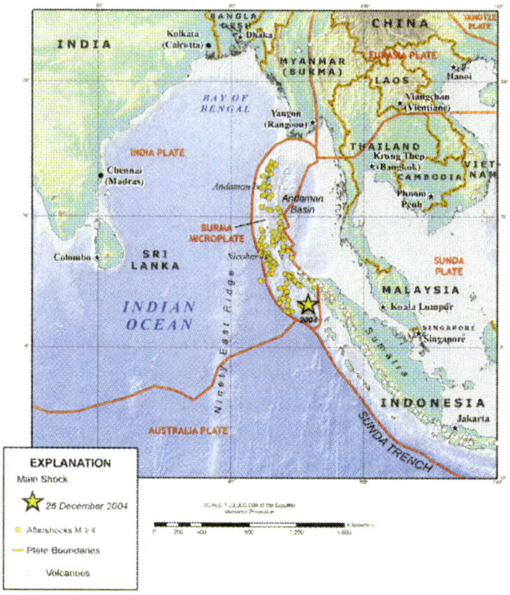

Figure 7.3. *The tectonic plates in the region where the 2004 tsunami occurred.* [Courtesy of the U.S. Geological Survey]

wood, causing small oscillations which will produce little wavelets that are grouped in wave packets. One will see these rather fast wavelets grow from zero at the rear of the packet, travel through the packet, and disappear at the front. As an alternative experiment, one could also watch waves on a lake when a fairly strong breeze produces waves. These short waves can be watched satisfactorily from the bank because, although the water nearby is shallow, their wavelength is short enough for their behavior to correspond to that of waves on deep water. The phenomena associated with the propagation of wave groups can be clearly seen after a short time. The waves produced by the wind travel in well-defined groups, typically six or seven waves apiece.[187] If we pick out, first, a large wave in a particular group, then in a short time it will be the following wave which will take over the office of largest of the group and immediately afterward the ware following that, and so on. Alternatively, by watching an individual wave instead of a group, one can follow the wave as it grows when approaching the center of a group; then the ware diminishes after this point until it becomes quite small as it passes out of the front of the group. Continuing to follow it as it enters the new group in front of the original one, we see that it grows again only to diminish once more as it gains the front of that group.

In view of the dispersive effect of linear deep-water waves—the fundamental feature being that longer waves travel faster—after the initiation of the tsunami, provided that the typical water depth remains roughly constant, the longer waves move to the front and

[187]One can make reference to the ninth wave being the highest, as mentioned in Tennyson's poem about King Arthur:
"Wave after wave, each mightier than the last,
Till last, a ninth one, ..."

the short wavelengths are then confined to the back of the tsunami profile. Thus, even if the initial wave profile comprises waves with short wavelength, after some time the thrust of the tsunami problem is confined to shallow-water waves. Behind these leading long waves, a dispersive effect might be noticeable, but dispersion cannot be essential for the leading waves in a tsunami. As for a possible balance between dispersion and the accumulation of small nonlinear effects, the distances needed for this balance are too large for such a mechanism to take place even across our oceans. Our contention is that, up to the location where the change of depth occurs on a scale larger than that controlled by the amplitude parameter ε, all nonlinear effects can be neglected. More precisely, in an open ocean with an almost flat bed, the shallow-water, linear-wave regime is an appropriate model for the propagation of a tsunami.

7.2 The December 2004 tsunami

The December 2004 tsunami was the first major tsunami detected by satellite altimetry: the Jason satellite, operated jointly by NASA and the French space agency (CNES), accurately measured the tsunami wave height along more than $2000\,km$ (north to south) in the Indian Ocean, about two hours after the main earthquake shock. The along-track sea level profile in Figure 7.4 was obtained by subtracting the background profile (measured during the previous cycle, 10 days earlier) from the original along-track record of the satellite. These satellite measurements suggest $a \approx 1\,m$ and $\lambda \approx 100\,km$. Since the Indian Ocean between Sumatra and Sri Lanka has a relatively flat bed with $d \approx 4\,km$, we can compute

$$\varepsilon = a/d \approx 25 \times 10^{-5}, \quad \delta = d/\lambda \approx 4 \times 10^{-2},$$

so that the tsunami propagation across the ocean occurred in KdV regime $\varepsilon \approx \delta^2$. Several researchers (see [82, 95] and references therein) have concluded, on this basis, that KdV dynamics had considerable relevance. However,[188] the length scale on which KdV dynamics becomes relevant is of order $\varepsilon^{-1}\lambda \approx 4 \times 10^5\,km$, and the distance to India and Sri Lanka is less than $1600\,km$. As for the waves traveling westward across the Andaman basin toward Thailand, a and λ are left unchanged, but $d \approx 1\,km$. This gives $\varepsilon = 10^{-3}$, $\delta = 10^{-2}$, and a length scale of about $10^5\,km$ needed for KdV dynamics, the available distance being only about $700\,km$.

Some theoretical considerations are validated by Photographs 11 and 12:

- the two-dimensional character of the wave motion is visible, the wave crest line being parallel to the shore;
- the leading tsunami wave was a wave of depression since the shoreline receded;[189]

[188] Segur has also argued in [271] that the propagation distances were too short for KdV dynamics to develop, but in [271] the incorrect scale $x = O(\varepsilon^{-1}d)$ was used. Even with this considerably shorter scale—the correct scale $x = O(\varepsilon^{-1}\lambda)$ (cf. the discussion in Section 7.1) being 25 times larger—the propagation distances were deemed too short.

[189] The generating earthquake raised the ocean floor a few meters over a distance of about $100\,km$ to the west of the epicenter and lowered it to the east (over $900\,km$) so that the initial wave pattern featured, westward, a wave of elevation followed by a wave of depression, while to the east of the epicenter the initial wave profile consisted of a depression followed by an elevation [271]. This, and the fact that for India and Sri Lanka, the first tsunami wave reaching the shore was a wave of elevation, is consistent with our conclusions in Section 7.1 that the leading waves do not change (at leading order) while they propagate across the open sea in water with a relatively flat bed.

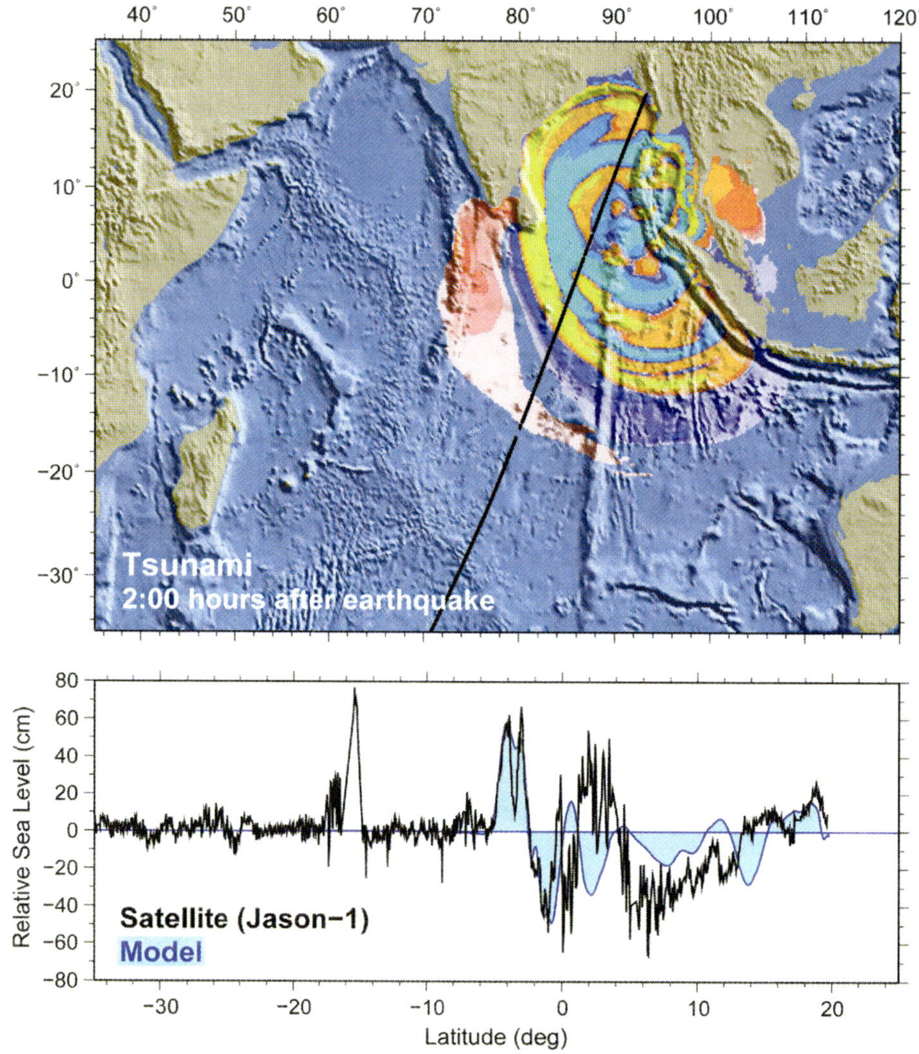

Figure 7.4. *Altimetry sea-level taken along track* 129 *of the Jason-1 satellite* [from the NOAA website]. Note that 1 degree latitude corresponds to about 111 km in this figure.

- the second wave of elevation is larger than the first, which is contrary to what soliton theory would predict (the larger solitons travel faster and move to the front).

As a further indication that linear shallow-water theory is appropriate in this context for the modelling of the propagation of the tsunami waves in the Indian Ocean, notice that, within this setting, the wave speed should be approximately $c = \sqrt{gd}$. With $g = 9.8 m/s^2$ and $d = 4 km$, we find the propagation speed of the tsunami to be approximately $c \approx 720 km/h$. Thus the approximate time needed to cover the distance of about $1550 km$ across the Indian Ocean/Bay of Begal, between the epicenter of the earthquake and the affected coast of Sri Lanka, is $2.2 h$, which is about right [61, 82].

Photograph 11. *The first tsunami wave of elevation approaching Hat Ray Leach beach (Thailand) in December 2004.* [Copyright APA, 2010]

Photograph 12. *The second tsunami wave of elevation, visible behind the first, approaching Hat Ray Leach beach (Thailand) in December 2004.* [Copyright APA, 2010]

7.3 The 1960 Chile tsunami

The tsunami of May 22, 1960 was caused by the largest earthquake ever recorded. Tsunami waves propagated in the northwest direction across the Pacific Ocean, reaching Hawaii with wave heights of $10m$ after 14.8 hours, at $10,000km$ from the epicenter of the earthquake; 7 hours later and about $7,000km$ further away, tsunami waves of $6m$ height hit the Japanese islands of Honsu and Hokkaido. Due to the extremely long distances of propagation it was speculated [190, 191, 272] that KdV dynamics was a relevant feature in this case. However, the ocean floor of the Central Pacific Basin between Chile and Hawaii is relatively uniform, with a mean depth $d = 4.3km$. Using (7.5), for KdV to be relevant within a distance

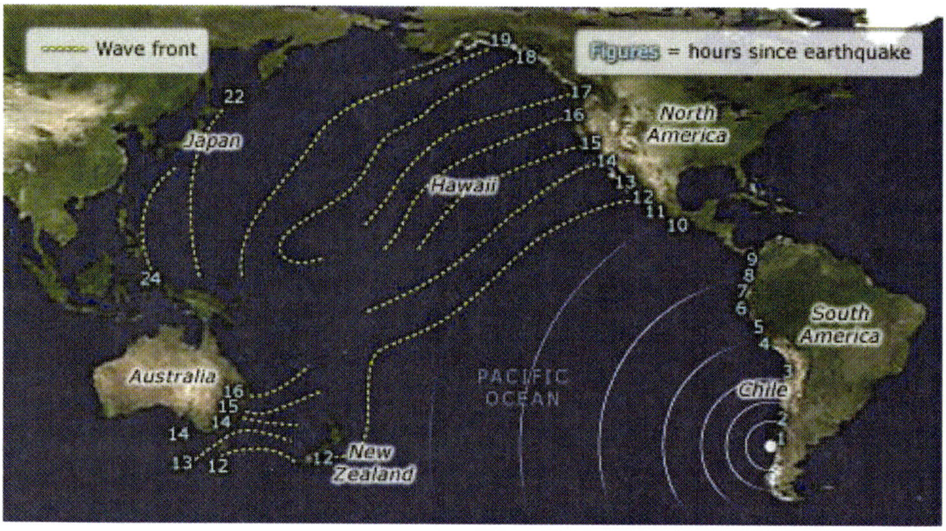

Figure 7.5. *The travel time for the* 1960 *Chilean tsunami* [from the NOAA website].

$l = 10{,}000\,km$, an initial wave amplitude

$$a = d^{5/3} l^{-2/3} > \frac{4^{5/3} \cdot 10^5}{10^{14/3}}\, m = 4 \cdot (160)^{1/3} \approx 22\,m$$

is required. However, the wave measurements[190] of the tsunami waves propagating southeast show very few waves in excess of $10\,m$ height, and one should take into account that the Chilean coast presents gradual bottom slopes extending for $150\,km$ from the coast to a deep sea trench about $3\,km$ deep along the coast [222], a feature which greatly amplifies the magnitudes of waves approaching the shore. A further indication that an initial wave amplitude in excess of $22\,m$ is out of the question is the fact that on smaller islands in the Pacific Ocean fronted by steep offshore slopes the recorded waves were less than $2\,m$ in height [36]. As for the tsunami waves propagating on this occasion from Hawaii to Japan, taking $d = 6\,km$ and $l = 7{,}000\,km$, the estimate (7.5) requires waves with amplitudes

$$a = d^{5/3} l^{-2/3} = 60 \cdot \left(\frac{6}{7}\right)^{2/3} m > 48\,m$$

near Hawaii.[191] We deduce that the 1960 tsunami cannot be regarded as a manifestation of soliton theory. This is further supported by the fact that the first two tsunami waves

[190]In 1960 the Pacific Tsunami Warning System was already in place and tide gauge stations throughout the Chilean coast allowed for accurate measurements of the tsunami amplitudes as they hit the Chilean coastal regions shortly after initiation—these being the waves propagating east/southeast in contrast to the ones propagating in the northwest direction towards Hawaii.

[191]These considerations are, however, less reliable than the estimates we derived above for the wave propagation from Chile to Hawaii, since, after encountering the Hawaiian Islands, the two-dimensional character of the tsunami waves was most likely lost due to diffraction, reflection, and wave breaking. Also, the bathymetry of this region of the Pacific Ocean presents variations on a relatively large scale: with depth ranging from $4.3\,km$ up to $6\,km$.

reaching Hilo in Hawaii were smaller than the third, most destructive, wave [293]. This could not occur if KdV dynamics were significant throughout the propagation of the tsunami across the Pacific Ocean since after some time the KdV solution evolves into an ordered set of solitons, in decreasing order of amplitude (the tallest being in front), followed by an oscillatory tail; cf. the discussion in the Appendix to Chapter 5.

We conclude by making, in support of the shallow-water, linear-theory model, the observation that the theoretical speed $c \approx \sqrt{gd}$ gives a speed of about $720 km/h$ for a depth of $d = 4.3 km$ (the appropriate average depth). This equates to a total time of travel from close to the Chilean shore to Hawaii of approximately $14h$, which is about right [59, 48].

7.4 Appendix for Chapter 7

7.4.1 Nondimensionalization and multiple scales approach

Asymptotic analysis is a useful tool that leads to analytical insight and—occasionally—to numerical estimates for solutions of complicated problems that are described within a mathematical framework but are not susceptible to the more usual techniques. This approach quite often provides the first detailed, useful insights into the structure of involved, physically interesting, nonlinear problems. Typically, such problems are defined by differential equations, which provide the basic model for the phenomenon under consideration, and which also contain one (or more) small parameters (after some suitable nondimensionalization). The process of nondimensionalization ensures the removal of scales ("size") associated with a specific problem. This leads to a standardization of all similar problems; e.g., the waves generated by a duck or aircraft carrier are essentially the same, as is the movement of water in a bath or the ocean, but in both cases the scales are very different. The procedure is to introduce nondimensional variables, whose original units are suitable for the given problem. For example, in describing the topography of geographic regions, a scale of $100m$ might be appropriate for a small village, but $10km$ (or even $100km$) for countries. When this "scaling" is performed,[192] by redefining all the variables, the resulting governing equations usually contain one (or more) nondimensional parameters. Let us suppose that there is only one parameter of interest, say, λ; an asymptotic expansion is then a formal power series expansion in a small parameter ε (with $\varepsilon = \lambda$ if $\lambda \to 0$ or $\varepsilon = 1/\lambda$ if $\lambda \to \infty$ being the limits of interest). The coefficient of each power of ε then yields—it is hoped—a simpler problem than the original one; typically, we work in detail with only the first few terms—perhaps the first two—in an asymptotic expansion. (The issues of asymptotic validity and of convergence are important and must be addressed because of the formality of this procedure; in many cases, the procedure for specific problems can be rigorously justified.) There is a small set of standard methods that come under the umbrella of asymptotic methods for differential equations, and of these perhaps the most successful and powerful technique is the *method of multiple scales*.

Multiple scales appear quite naturally in many physical scenarios. For example, seasons (a scale of about 4 months) are appropriate for the description of temperature variations over a year, while, on the other hand, hours are more suited to temperature variations over a few days: some weather events occur on a seasonal scale, and others on significantly shorter time scales. Consider, for example, the "El Nino"; this is a disruption of the ocean-atmosphere system (in the tropical Pacific) happening once every two to six years which has a strong influence on the precipitation

[192]Practical examples of nondimensionalization occurring in everyday life are measuring devices. Thermometers are calibrated relative to some known unit (e.g., grade Celsius $^\circ C$ in Europe and grade Fahrenheit $^\circ F$ in the US), and measurements are made relative to this standard. To recover the value of the measurement in familiar terms one has to scale back.

distribution in parts of the world, e.g., increased rain fall in Latin America and droughts in Australia. This event is naturally described on scales of many months, or even a year or two. On the other hand, the weather in Calgary (in Canada) is strongly influenced by a warm wind—the "Chinook"—which is generated by the Rocky Mountains blocking a strong Pacific atmospheric current [274]. This event has a time scale of just a few hours: it was reported that, on January 11, 1983, the temperature in Calgary rose from $-17°C$ to $13°C$ (a rise of $30°C$) in 4 hours and, on February 7, 1964, it rose $28°C$, while the humidity dropped by 43 percent! In this case, the monthly temperature data does not contain these weather events at all.

We now provide a brief description of the basic idea of the method of multiple scales. The first stage involves the introduction of suitable temporal and spatial scales, the choice of which will be guided by the nature of the problem under consideration; it might be years or hours (as indicated above), or *mm* or *km*, as appropriate. It is clear that the wrong scales may confuse and mislead, but these are not likely to be selected, if only on physical grounds. Certainly, the aim should be to use the smallest number of scales that reasonably represent the problem. The largest time or distance scale will certainly need to be included, because most macrophysical quantities will normally manifest themselves on such scales. On the other hand, there will probably be some underlying, small-scale structure that plays a role in the solution; this should also be accessible from the model. The usual guiding principle is to include each scale that is associated with each different physical contribution to the phenomenon, e.g., period of underlying oscillation, time scale of any damping processes, and time scale of any external forcing. Obviously, if there is any doubt, include all possibilities (in the theoretical sense); thus, if a time-dependent problem contains one small parameter, ε, in such a way that time scales based on ε, ε^2, ε^3, etc., are consistent with the governing equations, then we might start by introducing a set of time scales: $T_0 = t$, $T_1 = \varepsilon t$, $T_2 = \varepsilon^2 t$, and so on. Let the solution we seek be represented by $u(t;\varepsilon)$; we suppose that this solution can be written in the form

$$u(t;\varepsilon) = U(T_0, T_1, T_2, \ldots; \varepsilon) \approx \sum_{k=0}^{\infty} \varepsilon^k \, U_k(T_0, T_1, T_2, \ldots)$$

as $\varepsilon \to 0$; the solution is then described by its evolution on different scales. This might be thought of as seconds for the shortest times, then minutes, then hours, and so on.[193] Formally, the scale is interpreted by treating each T_n as fixed, as $\varepsilon \to 0$; thus we obtain, from above, the sequence of time scales $t = O(1)$, $t = O(\varepsilon^{-1})$, $t = O(\varepsilon^{-2})$, and so on. (We comment that the technique involves treating the function, u, as sitting, now, in a higher-dimensional space, the required solution being a choice, driven by the definition of the various scales involved, of a suitable subspace.) The use of this power series in the governing equations will generate, at leading order, a problem that may be nonlinear (if the underlying problem is indeed nonlinear), but every problem at higher order will certainly be linear; a nonlinear problem may become linear, at the first order. The nonlinearity is often evident by the interaction of some physical quantity, with itself, but on different (time) scales, even if, on any one time scale, the development is a purely linear process. The procedure involves solving at each order in ε, using the appropriate boundary and initial conditions consistent with the power series expansion, and by removing *secular terms*. The secular terms are those that will produce a nonuniformity in the expansion, e.g., terms of the type $\varepsilon t \sin(\varepsilon t)$ in a problem that is known to be periodic: terms of this type become unbounded in time despite of the fact that ε is very small. These are removed by making specific choices of the dependence on the longer scales in the problem; this choice, and that of the number of scales to be used, being sufficient to ensure a unique

[193] For example, if $\varepsilon = \frac{1}{60}$ and t represents seconds, variations on the scale T_0 are on a fast scale (seconds: $T_0 = 1$ means $t = 1$), variations on the scale T_1 are on a slower scale (minutes: $T_1 = 1$ means $t = 60$), and variations on the scale T_2 are on a much slower scale (hours: $T_2 = 1$ means $t = 3600$). We can imagine (cf. [242]) that we have a watch and we attempt to observe the behavior of u on the different scales of the watch. Thus, instead of determining u as a function of t, we view it as a function of $T_0, T_1, T_2, \ldots$.

solution. (The word "secular," based on the Latin *saeculum* (generation or age), was originally used in astronomy to denote phenomena which change too slowly to be observed, as compared with those that repeat, essentially periodically, on shorter time scales. They appear on times measured by εt, as $\varepsilon \to 0$, and so become important only for $t \to \infty$.) The solution, first expressed in terms of all the relevant time scales, is rewritten solely in the original t (which is equivalent to selecting a suitable subspace), thereby producing the required (asymptotic) solution. The validity of this final form of solution can be discussed and analyzed without the need to introduce the original scales used in its derivation.

These ideas, and the technique, will be employed in presenting a derivation of the KdV equation for water waves; for other examples, and more background information on asymptotic methods, we refer the reader to [163].

Example (derivation of the KdV model for shallow-water waves). According to the discussion in Chapter 2, the propagation of two-dimensional waves in irrotational flow at the surface of water of average depth d, and above the flat bed $y = 0$, is governed by Euler's equation

$$\begin{cases} u_t + uu_x + vu_y = -P_x, \\ v_t + uv_x + vv_x = -P_y - g \end{cases}$$

coupled with the equations, respectively, of mass conservation and irrotational flow

$$\begin{cases} u_x + v_y = 0, \\ u_y = v_x \end{cases}$$

and subject to the boundary conditions

$$\begin{cases} P = P_{atm} \quad \text{on} \quad y = d + \eta(x,t), \\ v = \eta_t + u\eta_x \quad \text{on} \quad y = d + \eta(x,t), \\ v = 0 \quad \text{on} \quad y = 0. \end{cases}$$

Introducing the velocity potential $\phi(x, y, t)$, defined up to an additive function of time, with $\phi_x = u$ and $\phi_y = v$, and taking into account Bernoulli's theorem which states that Euler's equation is equivalent to requiring that the expression $\phi_t + \frac{\phi_x^2+\phi_y^2}{2} + gy + P$ is constant throughout the fluid, the governing equations can be expressed in the form

$$\begin{cases} \phi_{xx} + \phi_{yy} = 0 \quad \text{for} \quad x \in \mathbb{R},\ 0 < y < d + \eta(x,t), \\ \phi_y = 0 \quad \text{on} \quad y = 0, \\ \phi_y = \eta_t + \phi_x \eta_x \quad \text{on} \quad y = d + \eta(x,t), \\ \phi_t + \frac{\phi_x^2+\phi_y^2}{2} + gy = 0 \quad \text{on} \quad y = d + \eta(x,t). \end{cases} \tag{7.15}$$

The first boundary condition was transformed into the last equation in (7.15) by taking advantage of Bernoulli's theorem and by choosing appropriately the additive function of time in the definition of ϕ.

The problem (7.15) is nondimensionalized using a typical wavelength λ and a typical amplitude of the wave a. That is, we define the set of nondimensional variables

$$x \mapsto \lambda x, \quad y \mapsto dy, \quad t \mapsto \frac{\lambda}{\sqrt{gd}}t, \quad u \mapsto u\sqrt{gd}, \quad v \mapsto v\frac{d\sqrt{gd}}{\lambda}, \quad \eta \mapsto a\eta;$$

it is understood that x is replaced by λx, so that afterward the symbol x stands for a nondimensional variable, and correspondingly for all the other variables. This convention avoids the need to introduce further and different notation. Setting the constant water density $\rho = 1$, the pressure in the new nondimensional variables is

$$P = P_0 + gd(1-y) + gdp,$$

with the nondimensional pressure variable p measuring the deviation from the hydrostatic pressure distribution. We obtain the following boundary value problem in nondimensional variables:

$$\begin{cases} \delta^2 \phi_{xx} + \phi_{yy} = 0 \quad \text{in} \quad 0 < y < 1+\varepsilon\eta, \\ \phi_y = 0 \quad \text{on} \quad y = 0, \\ \phi_y = \delta^2(\eta_t + \varepsilon\,\phi_x\,\eta_x) \quad \text{on} \quad y = 1+\varepsilon\eta, \\ \phi_t + \frac{\varepsilon}{2}(\phi_x^2 + \frac{1}{\delta^2}\phi_y^2) + \eta = 0 \quad \text{on} \quad y = 1+\varepsilon\eta, \end{cases} \tag{7.16}$$

where

$$\varepsilon = \frac{a}{d} \quad \text{and} \quad \delta = \frac{d}{\lambda}$$

are the amplitude parameter and the shallowness parameter, respectively.

We are interested in small-amplitude long waves, that is, in the limits as $\varepsilon \to 0$ and $\delta \to 0$. Noticing the way in which ε and δ appear in equations (7.16), the shallow-water, small-amplitude regime

$$\delta \ll 1, \quad \varepsilon = O(\delta^2)$$

arises naturally. Setting $\delta^2 = \alpha\varepsilon$ for some $\alpha > 0$ bounded, equations (7.16) become

$$\begin{cases} \alpha\varepsilon\,\phi_{xx} + \phi_{yy} = 0 \quad \text{in} \quad 0 < y < 1+\varepsilon\eta, \\ \phi_y = 0 \quad \text{on} \quad y = 0, \\ \phi_y = \alpha\varepsilon\,(\eta_t + \varepsilon\,\phi_x\,\eta_x) \quad \text{on} \quad y = 1+\varepsilon\eta, \\ \phi_t + \frac{\varepsilon}{2}\phi_x^2 + \frac{1}{2\alpha}\phi_y^2 + \eta = 0 \quad \text{on} \quad y = 1+\varepsilon\eta. \end{cases} \tag{7.17}$$

Let us first expand ϕ and η in powers of ε: we assume that (7.17) has a solution of the form

$$\phi(x,y,t) = \sum_{k=0}^{\infty} \varepsilon^k\,\phi_k(x,y,t), \qquad \eta(x,t) = \sum_{k=0}^{\infty} \varepsilon^k\,\eta_k(x,t).$$

The leading order approximation (ε^0) yields

$$\begin{cases} \partial_y^2\phi_0 = 0 \quad \text{in} \quad 0 < y < 1, \\ \partial_y\phi_0 = 0 \quad \text{on} \quad y = 0, \\ \partial_y\phi_0 = 0 \quad \text{on} \quad y = 1, \\ \partial_t\phi_0 + \eta_0 = 0 \quad \text{on} \quad y = 1. \end{cases} \tag{7.18}$$

The first three equations in (7.18) imply $\partial_y\phi_0 \equiv 0$ so that

$$\phi_0(x,y,t) = F(x,t) \qquad \text{in} \quad 0 < y < 1, \tag{7.19}$$

and, using this in the last equation in (7.18), we obtain

$$F_t + \eta_0 = 0 \quad \text{on} \quad y = 1. \tag{7.20}$$

At the next level of approximation (ε^1) we obtain, from (7.17), the system

$$\begin{cases} \alpha\,\partial_x^2\phi_0 + \partial_y^2\phi_1 = 0 \quad \text{in} \quad 0 < y < 1, \\ \partial_y\phi_1 = 0 \quad \text{on} \quad y = 0, \\ \partial_y\phi_1 = \alpha\,\partial_t\eta_0 \quad \text{on} \quad y = 1, \\ \partial_t\phi_1 + \eta_1 + \frac{1}{\alpha}(\partial_y\phi_1)(\partial_y\phi_0) + \frac{1}{2}(\partial_x\phi_0)^2 = 0 \quad \text{on} \quad y = 1. \end{cases} \tag{7.21}$$

From the first two equations, in combination with (7.19), we deduce that

$$\partial_y \phi_1 = -\alpha\, y\, F_{xx} \quad \text{for} \quad 0 \le y \le 1.$$

Therefore, the third equation in (7.21) becomes $F_{xx} = -\partial_t \eta_0$. In conjunction with (7.20), this yields the linear wave equation

$$F_{xx} - F_{tt} = 0$$

with the general solution

$$F(x,t) = f(x-t) + \tilde{f}(x+t),$$

where the signs $-/+$ refer to the wave profiles $f/\tilde{f}$, respectively, moving with unchanged shape to the right/left, each with constant speed 1. We note that the appearance of unit speed here is by virtue of the nondimensionalization, the corresponding dimensional speed being $\sqrt{gd}$. For definiteness, let us choose the wave moving to the right, that is,

$$F(x,t) = f(x-t). \tag{7.22}$$

It is significant to observe that we did not use the last equation in (7.21) in the development so far. Its purpose is to provide information about η_1, and, at this stage, we have restricted our attention to the leading order term η_0 in the asymptotic development of the free surface profile η.

The form of (7.22) suggests that, in some neighborhood of the (x,t)-space, it might be possible to obtain some precise information about the evolution of the water's free surface. In order to do this, we look at (7.17) in a frame of reference which is moving with a speed of unity to the right. In addition to the moving frame, we need suitable slow time scales; in this problem, the slow evolution due to both nonlinearity and dispersion is on the scale $O(\varepsilon^{-1})$, as is evident from a cursory examination of (7.17). Although other time scales could be included (i.e., $O(\varepsilon^{-2})$, $O(\varepsilon^{-3})$, etc.), they turn out to be redundant here. Thus we restrict the choice in this discussion to a frame moving with the right-going wave and the slow time scale $T = \varepsilon t$. Then, with

$$\begin{cases} \xi = x - t, \\ T = \varepsilon t, \end{cases}$$

we have $x = \xi + \frac{T}{\varepsilon}$ and $t = \frac{T}{\varepsilon}$, so that

$$\partial_x = \partial_\xi \frac{\partial \xi}{\partial x} + \partial_T \frac{\partial T}{\partial x} = \partial_\xi, \qquad \partial_t = \partial_\xi \frac{\partial \xi}{\partial t} + \partial_T \frac{\partial T}{\partial t} = -\partial_\xi + \varepsilon\, \partial_T.$$

The system (7.17) in the (ξ, T)-coordinates is

$$\begin{cases} \alpha\varepsilon\, \phi_{\xi\xi} + \phi_{yy} = 0 \quad \text{in} \quad 0 < y < 1 + \varepsilon\eta, \\ \phi_y = 0 \quad \text{on} \quad y = 0, \\ \phi_y = \alpha\varepsilon\,(-\eta_\xi + \varepsilon\, \eta_T + \varepsilon\, \phi_\xi\, \eta_\xi) \quad \text{on} \quad y = 1 + \varepsilon\eta, \\ \varepsilon\, \partial_T \phi - \partial_\xi \phi + \frac{\varepsilon}{2} \phi_\xi^2 + \frac{1}{2\alpha} \phi_y^2 + \eta = 0 \quad \text{on} \quad y = 1 + \varepsilon\eta. \end{cases} \tag{7.23}$$

Expanding ϕ and η in powers of ε,

$$\phi(x,y,t) = \sum_{k=0}^{\infty} \varepsilon^k \phi_k(\xi, y, T), \qquad \eta(x,t) = \sum_{k=0}^{\infty} \varepsilon^k \eta_k(\xi, T),$$

we obtain a leading order (ε^0) approximation of (7.23) of the form

$$\begin{cases} \partial_y^2 \phi_0 = 0 \quad \text{in} \quad 0 < y < 1, \\ \partial_y \phi_0 = 0 \quad \text{on} \quad y = 0, \\ \partial_y \phi_0 = 0 \quad \text{on} \quad y = 1, \\ -\partial_\xi \phi_0 + \frac{1}{2\alpha} (\partial_y \phi_0)^2 + \eta_0 = 0 \quad \text{on} \quad y = 1. \end{cases} \tag{7.24}$$

The first three equations in (7.24) yield $\partial_y\phi_0 \equiv 0$, so that

$$\phi_0 = f(\xi,T) \qquad \text{in} \quad 0 < y < 1, \tag{7.25}$$

and the last equation in (7.24) implies

$$\eta_0 = \partial_\xi f. \tag{7.26}$$

The next level of approximation (ε^1) to (7.23) gives

$$\begin{cases} \partial_y^2\phi_1 + \alpha\,\partial_\xi^2\phi_0 = 0 \quad \text{in} \quad 0 < y < 1, \\ \partial_y\phi_1 = 0 \quad \text{on} \quad y = 0, \\ \partial_y\phi_1 = -\alpha\,\partial_\xi\,\eta_0 \quad \text{on} \quad y = 1, \\ \partial_T\phi_0 - \partial_\xi\phi_1 + \frac{1}{\alpha}(\partial_y\phi_0)(\partial_y\phi_1) + \frac{1}{2}(\partial_\xi\phi_0)^2 + \eta_1 = 0 \quad \text{on} \quad y = 1. \end{cases} \tag{7.27}$$

From the first two equations in (7.27), in combination with (7.25), we deduce that

$$\partial_y\phi_1 = -\alpha\, y\, f_{\xi\xi}(\xi,T) \quad \text{in} \quad 0 \le y \le 1.$$

Now (7.26) and the third equation in (7.27) yield

$$\phi_1 = -\frac{\alpha}{2}\, y^2\, f_{\xi\xi}(\xi,T) + \beta(\xi,T) \quad \text{in} \quad 0 \le y \le 1 \tag{7.28}$$

for some function of integration β. From the last equation in (7.27) we now obtain

$$\eta_1 = -f_T - \frac{\alpha}{2}\, f_{\xi\xi\xi} + \beta_\xi - \frac{1}{2}\, f_\xi^2 \tag{7.29}$$

if we recall (7.25).

The third level of approximation (ε^2) to (7.23) is

$$\begin{cases} \partial_y^2\phi_2 + \alpha\,\partial_\xi^2\phi_1 = 0 \quad \text{in} \quad 0 < y < 1, \\ \partial_y\phi_2 = 0 \quad \text{on} \quad y = 0, \\ \partial_y\phi_2 = -\alpha\,\partial_\xi\,\eta_1 + \alpha\,\partial_T\eta_0 + \alpha\,(\partial_\xi\phi_0)(\partial_\xi\eta_0) \quad \text{on} \quad y = 1, \\ \partial_T\phi_1 - \partial_\xi\phi_2 + (\partial_\xi\phi_0)(\partial_\xi\phi_1) + \frac{1}{2\alpha}(\partial_y\phi_1)^2 \\ \qquad\qquad + \frac{1}{\alpha}(\partial_y\phi_0)(\partial_y\phi_2) + \eta_2 = 0 \quad \text{on} \quad y = 1. \end{cases} \tag{7.30}$$

From the first two equations in (7.30) we deduce, in view of (7.28), that

$$\phi_2 = \frac{\alpha^2}{24}\, y^4\partial_\xi^4\, f(\xi,T) - \frac{\alpha}{2}\, y^2\,\partial_\xi^2\beta(\xi,T) + \theta(\xi,T)$$

for a second function of integration, θ. Using this in combination with (7.25)–(7.26) and (7.29) in the third equation of (7.30), we obtain the following equation:

$$\frac{\alpha^2}{3}\, f_{\xi\xi\xi\xi} + 2\alpha\, f_\xi\, f_{\xi\xi} + 2\alpha\, f_{T\xi} = 0$$

for $f(\xi,T)$. In view of (7.26), since $\alpha > 0$, the above nonlinear partial differential equation is precisely

$$\frac{\alpha}{6}\,\partial_\xi^3\eta_0 + \eta_0\,\partial_\xi\eta_0 + \partial_T\eta_0 = 0. \tag{7.31}$$

The choice $\alpha = 6$ (or a suitable scaling) transforms (7.31) into the standard form of the KdV equation. This shows that, in this region of the (x,t)-space, the KdV equation describes the evolution of the

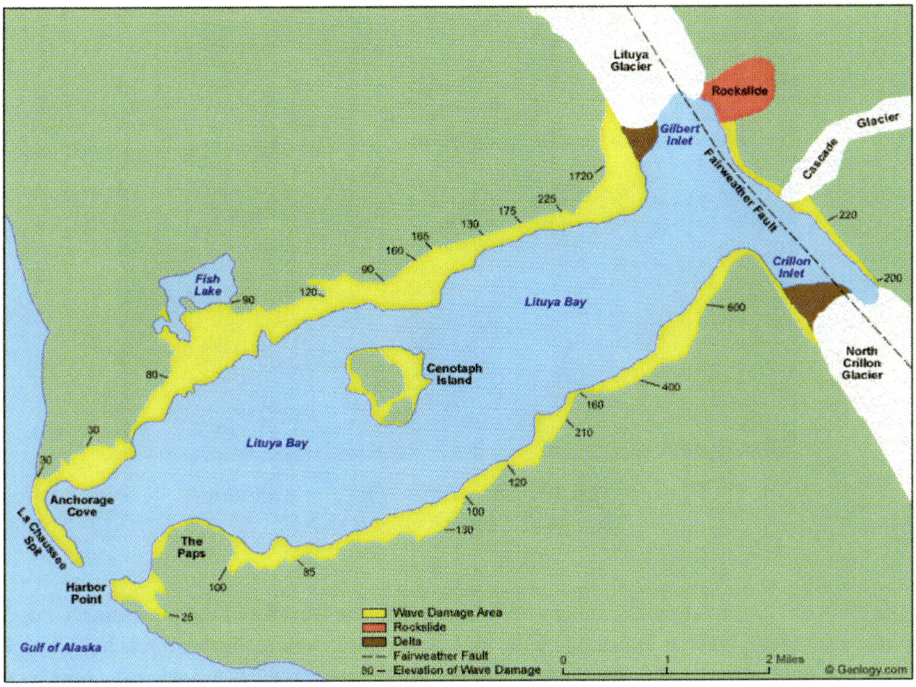

Figure 7.6. *Detail map of Lituya Bay, Alaska.*
[Courtesy of the U.S. Geological Survey]

leading order of the free surface. We observe that the region where the KdV equation is valid, at leading order, is given by $\xi = O(1)$ and $T = O(1)$; this is a neighborhood of the wave front (defined by $\xi = 0$) and for times $t = \varepsilon^{-1} T_0$, for some bounded constant, T_0, fixed as $\varepsilon \to 0$. Therefore, KdV models the propagation of small-amplitude waves in shallow water (recall that we worked in the long-wave regime $\delta \to 0$). These formal considerations can be rigorously justified (see, e.g., the discussion in [8]).

7.4.2 The highest wave ever measured

Lituya Bay is an inlet (see Photograph 13) on the northeast shore of the Gulf of Alaska, about $11.3km$ long and up to $3.2km$ wide, with a maximum depth of about $219m$ but with a sill of only $9.7m$ in depth separating it from the Gulf of Alaska. On the night of July 9, 1958, an earthquake loosened about 30.6 million m^3 of rock high above the northeastern shore of Lituya Bay. This mass of rock plunged from approximately $914m$ down into the waters of Gilbert Inlet; see the map in Figure 7.6, in which the numbers represent the approximate elevation of the wave (in feet) as it traveled through the bay. The impact generated a local tsunami—the tsunami was not generated by the earthquake—contained within the bay. The force of the wave removed all trees and vegetation from elevations as high as $524m$ above sea level: millions of trees were uprooted and swept away by the wave. There were three fishing boats anchored at the mouth of Lituya Bay on that day. Unfortunately, one of the boats was close to shore and the huge waves overtook it killing the two

Photograph 13. *Lituya Bay after the tsunami.*
[Image ID: cgs02082, NOAA]

people on board. Amazingly, the other two boats "rode" the waves and the crews survived. The height of the biggest wave is indicated by the high water mark (where the water reached its highest point, at the uppermost edge of the damaged area), yielding a wave $524\,m$ high—the biggest ever measured.[194]

7.4.3 The March 2011 Japan tsunami

While this book was in production the fourth largest earthquake in the world and the largest earthquake in Japan (since instrumental recordings began in 1900) generated a tsunami that was observed throughout the Pacific region. Tsunami waves 10–15 m high wreaked havoc in Japan along a stretch of coastline $670\,km$ long, resulting in a death toll in the tens of thousands [285]. In addition, the tsunami caused a number of dramatic accidents at the Fukushima nuclear power plant, and the associated evacuation zones affected hundreds of thousands of residents. The tsunami wave height was generally less than $1\,m$ as it crossed the Pacific but increased as it reached shallow coastal waters, with waves up to $2\,m$ observed in Hawaii and $3\,m$ high waves reaching the coast of Chile about $20\,h$ after the earthquake [285]. The impact on the US west coast was lessened as the arrival of the tsunami waves largely coincided with low tide. The earthquake occurred near the Japan Trench, where the ocean depth exceeds $7\,km$, as a consequence of a thrust fault due to subduction of the Pacific Plate beneath the tectonic plate underlying Japan's main island Honshu. The main earthquake was preceded by a number of large foreshocks, and hundreds of aftershocks were reported [280]. According to the Japanese Meteorological Agency, an upthrust of the seabed of up to $5\,m$ occurred over an area 300–400$\,km$ long by $150\,km$ wide, approximately $70\,km$ offshore. Reports indicate that the tsunami wave was about $3\,m$ high at initiation. Its two-dimensional character was visible in live coverage by Japan's national public broadcaster.

Concerning the mathematical modelling of the 2011 Japan tsunami, let us notice that, in contrast to the December 2004 and the 1960 Chile tsunami, the tsunami propagation in the open

[194]It is believed that larger tsunamis have occurred, generated by asteroids or meteors falling into the ocean, in the geologic past.

ocean is of secondary importance. Indeed, the tsunami waves propagating toward Japan reached its coast about 1 h after the main earthquake and were generated near the shore, evolving in a water region of practically constantly diminishing depth. The propagation of the tsunami waves toward Hawaii also occurred in a region with considerable changes in the ocean bathymetry. Indeed, about $1600\,km$ east of Japan lies the Shatsky Rise, an oceanic plateau with water depths less than $5\,km$; cf. [320]. This plateau of the size of Japan is surrounded to the west, east, and south by the Nothern Pacific Basin that has an almost flat bed about $6\,km$ deep. Tsunami waves are sensitive to water depth variations, and a seamount (submarine mountain) scatters or deflects tsunami energy. This explains why the tsunami was not a major damaging event in Hawaii.[195] From the point of view of modelling, the considerations presented throughout this chapter (whether at linear level or including weakly nonlinear, long wave effects) are valid under the assumption of a flat bed and cannot be applied to predict how the waves behave in locations that present large variations in the bottom topography. Consequently, to gain a better understanding of the March 2011 Japan tsunami, a much more complicated approach has to be implemented. The relevance of detailed bottom topography suggests that the wave dynamics should be analyzed separately in each particular region.

7.5 Notes for Chapter 7

The full complexity of tsunamis is beyond our ability to grasp, but we can break it up into smaller components which can be understood separately, and then combine them with the intention of producing a broader and more comprehensive picture. After all, in some sense the essence of physics is to identify simple systems whose behavior depends upon a very small number of relevant factors, each of which can be expressed in numerical terms. Once we understand these simple systems, combining them in ever more complex ways provides us with more insight. If the conclusions reached this way have overwhelming evidence in their support, it is reasonable to move away from the position of regarding them as having a provisional status. In this chapter, based mainly upon the considerations in [48, 59, 61, 62], we addressed some issues related to the propagation of the tsunamis across the deep oceans, reaching the conclusion that the balance of nonlinearity and dispersion, a characteristic of soliton theory, is not to be expected. Consequently, in deep water, linear shallow-water theory is appropriate for the modelling, while, near the shore, where the water depth rapidly diminishes, a nonlinear framework becomes necessary. Since the bottom topography is a main factor in the evolution of the wave form near the shore, the investigation of the tsunami as it approaches the shore is a study at the boundary between mathematical analysis and numerical simulations. These two approaches provide different views, the first being broader but fuzzier and the second sharp but narrow in scope, with the degree of overlap increasing the predictive powers of both approaches. Since, throughout these lecture notes, we have dealt exclusively with water-wave propagation over a flat bed, the important problem of the evolution of tsunamis over variable depth is beyond our scope; we refer the reader to [61] for a discussion of this aspect. A further issue, which is of practical interest, is the possibility of incorporating vorticity in the flow. This is motivated by the fact that, in the background state of the water, prior to the arrival of the tsunami, there could be strong localized currents. For some progress in this direction we refer the reader to [61] for results within the shallow-water regime, and to [63, 49, 127] for some more general results on the governing equations.

[195]Actually, the biggest tsunamis to hit Hawaii have not been generated from Japan, but, for example, from Chile in 1960.

Bibliography

[1] M. J. ABLOWITZ AND P. A. CLARKSON, *Solitons, Nonlinear Evolution Equations and Inverse Scattering*, Cambridge University Press, Cambridge, UK, 1991.

[2] S. AGMON, On the eigenfunctions and on the eigenvalues of general elliptic boundary value problems, Comm. Pure Appl. Math. 15 (1962), 119–147.

[3] S. AGMON, A. DOUGLIS, AND L. NIRENBERG, Estimates near the boundary for solutions of elliptic partial differential equations satisfying general boundary conditions. I, Comm. Pure Appl. Math. 12 (1959), 623–727.

[4] S. AGMON, A. DOUGLIS, AND L. NIRENBERG, Estimates near the boundary for solutions of elliptic partial differential equations satisfying general boundary conditions. II, Comm. Pure Appl. Math. 17 (1964), 35–92.

[5] L. V. AHLFORS, *Complex Analysis*, McGraw–Hill, New York, 1978.

[6] G. AIRY, Tides and waves, Encyclop. Metrop. 3 (1841), 396 pp.

[7] T. AKTOSUN, Inverse scattering transform, KdV, and solitons, in *Current Trends in Operator Theory and Its Applications* (Ed. J. A. Ball, J. W. Helton, M. Klaus, and L. Rodman), Oper. Theory Adv. Appl. 149, Birkhäuser, Basel, 2004, pp. 1–22.

[8] B. ALVAREZ-SAMANIEGO AND D. LANNES, Large time existence for 3D water-waves and asymptotics, Invent. Math. 171 (2008), 485–541.

[9] C. J. AMICK, Bounds for water waves, Arch. Ration. Mech. Anal. 99 (1987), 91–114.

[10] C. J. AMICK, L. E. FRAENKEL, AND J. F. TOLAND, On the Stokes conjecture for the wave of extreme form, Acta Math. 148 (1982), 193–214.

[11] C. J. AMICK AND J. F. TOLAND, On solitary water waves of finite amplitude, Arch. Ration. Mech. Anal. 76 (1981), 9–95.

[12] C. J. AMICK AND J. F. TOLAND, On periodic water-waves and their convergence to solitary waves in the long-wave limit, Philos. Trans. R. Soc. Lond. Ser. A Math. Phys. Eng. Sci. 303 (1981), 633–669.

[13] S. ANGENENT, Parabolic equations for curves on surfaces, Ann. of Math. (2) 132 (1990), 451–483.

[14] J. V. ARMITAGE AND W. F. EBERLEIN, *Elliptic Functions*, London Math. Soc. Student Texts, Cambridge University Press, Cambridge, UK, 2006.

[15] V. I. ARNOLD, Sur la géométrie différentielle des groupes de Lie de dimension infinie et ses applications à l'hydrodynamique des fluides parfaits, Ann. Inst. Fourier (Grenoble) 16 (1966), 319–361.

[16] V. I. ARNOLD, *Mathematical Methods of Classical Mechanics*, Springer-Verlag, New York, 1989.

[17] V. I. ARNOLD AND B. A. KHESIN, *Topological Methods in Hydrodynamics*, Springer-Verlag, New York, 1998.

[18] M. L. BANNER AND D. H. PEREGRINE, Wave breaking in deep water, Annu. Rev. Fluid Mech. 25 (1993), 373–397.

[19] J. T. BEALE, The existence of solitary water waves, Comm. Pure Appl. Math. 30 (1977), 373–389.

[20] T. B. BENJAMIN, J. L. BONA, AND J. J. MAHONEY, Model equations for long waves in nonlinear dispersive systems, Philos. Trans. R. Soc. Lond. Ser. A Math. Phys. Eng. Sci. 227 (1972), 47–78.

[21] H. BERESTYCKI, L. NIRENBERG, AND S. R. S. VARADHAN, The principal eigenvalue and maximum principle for second-order elliptic operators in general domains, Comm. Pure Appl. Math. 47 (1994), 47–92.

[22] M. S. BERGER, *Nonlinearity and Functional Analysis*, Academic Press, New York, London, 1977.

[23] R. P. BOAS, *Invitation to Complex Analysis*, Mathematical Association of America, Washington, DC, 2010.

[24] J. BOCHNAK, M. COSTE, AND M.-F. ROY, *Real Algebraic Geometry*, Springer-Verlag, Berlin, 1998.

[25] R. BÖHME, Die Lösung der Verzweigungsgleichungen für nichtlineare Eigenwertprobleme, Math. Z. 127 (1972), 105–126.

[26] J.-M. BONY, Principe du maximum, inégalite de Harnack et unicité du problème de Cauchy pour les opérateurs elliptiques dégénérés, Ann. Inst. Fourier (Grenoble) 19 (1969), 277–304.

[27] G. BORG, Eine Umkehrung der Sturm-Liouvilleschen Eigenwertaufgabe, Acta Math. 78 (1946), 1–96.

[28] J. V. BOUSSINESQ, Théorie de l'intumescence appelée 'onde solitaire' ou 'de translation', se propageant dans un canal rectangulaire, C. R. Acad. Sci. Paris 72 (1871), 755–759.

[29] A. BOUTET DE MONVEL, A. KOSTENKO, D. SHEPELSKY, AND G. TESCHL, Long-time asymptotics for the Camassa–Holm equation, SIAM J. Math. Anal. 41 (2009), 1559–1588.

[30] A. BRESSAN AND A. CONSTANTIN, Global conservative solutions of the Camassa-Holm equation, Arch. Ration. Mech. Anal. 183 (2007), 215–239.

[31] A. BRESSAN AND A. CONSTANTIN, Global dissipative solutions of the Camassa-Holm equation, Anal. Appl. (Singap.) 5 (2007), 1–27.

[32] H. BRÉZIS, On a characterization of flow invariant sets, Comm. Pure. Appl. Math. 23 (1970), 261–263.

[33] H. BRÉZIS, *Analyse Fonctionnelle*, Masson, Paris, 1983.

[34] H. BRÉZIS, *Functional Analysis, Sobolev Spaces and Partial Differential Equations*, Springer, New York, 2011.

[35] R. F. BROWN, *A Topological Introduction to Nonlinear Analysis*, Birkhäuser Boston, Boston, 2004.

[36] E. BRYANT, *Tsunami: The Underrated Hazard*, Springer Praxis Books, Springer, Berlin, Heidelberg, 2008.

[37] A. E. BRYSON, Waves in fluids, National Committee for Fluid Mechanics Films (Encyclopaedia Britannica Educational Corporation), Chicago, 60611.

[38] B. BUFFONI AND J. F. TOLAND, *Analytic Theory of Global Bifurcation*, Princeton University Press, Princeton, 2003.

[39] R. B. BURCKEL, *An Introduction to Classical Complex Analysis*, Academic Press, New York, London, 1979.

[40] J. C. BURNS, Long waves in running water, Proc. Cambridge Philos. Soc. 49 (1953), 695–706.

[41] R. CAMASSA AND D. HOLM, An integrable shallow water equation with peaked solitons, Phys. Rev. Lett. 71 (1993), 1661–1664.

[42] Y.-Y. CHEN, H.-C. HSU, AND G.-Y. CHEN, Lagrangian experiment and solution for irrotational finite-amplitude progressive gravity waves at uniform depth, Fluid Dynam. Res. 42 (2010), 045511.

[43] C. CHICONE, *Ordinary Differential Equations with Applications*, Springer-Verlag, New York, 1999.

[44] A. CONSTANTIN, Existence of permanent and breaking waves for a shallow water equation: A geometric approach, Ann. Inst. Fourier (Grenoble) 50 (2000), 321–362.

[45] A. CONSTANTIN, On the scattering problem for the Camassa-Holm equation, Proc. R. Soc. Lond. Ser. A Math. Phys. Eng. Sci. 457 (2001), 953–970.

[46] A. CONSTANTIN, On the deep water wave motion, J. Phys. A 34 (2001), 1405–1417.

[47] A. CONSTANTIN, The trajectories of particles in Stokes waves, Invent. Math. 166 (2006) 523–535.

[48] A. CONSTANTIN, On the relevance of soliton theory to tsunami modelling, Wave Motion 46 (2009), 420–426.

[49] A. CONSTANTIN, A dynamical systems approach towards isolated vorticity regions for tsunami background states, Arch. Ration. Mech. Anal. 200 (2011), 239–253.

[50] A. CONSTANTIN, M. EHRNSTRÖM, AND E. WAHLÉN, Symmetry of steady periodic gravity water waves with vorticity, Duke Math. J. 140 (2007), 591–603.

[51] A. CONSTANTIN AND J. ESCHER, Wave breaking for nonlinear nonlocal shallow water equations, Acta Math. 181 (1998), 229–243.

[52] A. CONSTANTIN AND J. ESCHER, Global existence and blow-up for a shallow water equation, Ann. Sc. Norm. Super. Pisa Cl. Sci. (4) 26 (1998), 303–328.

[53] A. CONSTANTIN AND J. ESCHER, Symmetry of steady periodic surface water waves with vorticity, J. Fluid Mech. 498 (2004), 171–181.

[54] A. CONSTANTIN AND J. ESCHER, Symmetry of steady deep-water waves with vorticity, European J. Appl. Math. 15 (2004), 755–768.

[55] A. CONSTANTIN AND J. ESCHER, Particle trajectories in solitary water waves, Bull. Amer. Math. Soc. 44 (2007), 423–431.

[56] A. CONSTANTIN AND J. ESCHER, Analyticity of periodic traveling free surface water waves with vorticity, Ann. of Math. (2) 173 (2011), 559–568.

[57] A. CONSTANTIN, J. ESCHER, AND H.-C. HSU, Pressure beneath a solitary water wave: Mathematical theory and experiments, Arch. Ration. Mech. Anal. 201 (2011), 251–269.

[58] A. CONSTANTIN, V. GERDJIKOV, AND R. IVANOV, Inverse scattering transform for the Camassa-Holm equation, Inverse Problems 22 (2006), 2197–2207.

[59] A. CONSTANTIN AND D. HENRY, Solitons and tsunamis, Z. Naturforsch. 64a (2009), 65–68.

[60] A. CONSTANTIN, R. IVANOV, AND J. LENELLS, Inverse scattering transform for the Degasperis-Procesi equation, Nonlinearity 23 (2010), 2559–2575.

[61] A. CONSTANTIN AND R. S. JOHNSON, Propagation of very long water waves, with vorticity, over variable depth, with applications to tsunamis, Fluid Dynam. Res. 40 (2008), 175–211.

[62] A. CONSTANTIN AND R. S. JOHNSON, On the non-dimensionalisation, scaling and resulting interpretation of the classical governing equations for water waves, J. Nonlinear Math. Phys. 15 (2008), 58–73.

[63] A. CONSTANTIN AND R. S. JOHNSON, Addendum: Propagation of very long water waves, with vorticity, over variable depth, with applications to tsunamis, Fluid Dynam. Res. 42 (2010), Art. No. 038901.

[64] A. CONSTANTIN, T. KAPPELER, B. KOLEV, AND P. TOPALOV, On geodesic exponential maps of the Virasoro group, Ann. Global Anal. Geom. 31 (2007), 155–180.

[65] A. CONSTANTIN AND B. KOLEV, Geodesic flow on the diffeomorphism group of the circle, Comment. Math. Helv. 78 (2003), 787–804.

[66] A. CONSTANTIN AND D. LANNES, The hydrodynamical relevance of the Camassa-Holm and Degasperis-Procesi equations, Arch. Ration. Mech. Anal. 192 (2009), 165–186.

[67] A. CONSTANTIN AND H. P. MCKEAN, A shallow water equation on the circle, Comm. Pure Appl. Math. 52 (1999), 949–982.

[68] A. CONSTANTIN AND L. MOLINET, Global weak solutions for a shallow water equation, Comm. Math. Phys. 211 (2000), 45–61.

[69] A. CONSTANTIN, D. SATTINGER, AND W. STRAUSS, Variational formulations for steady water waves with vorticity, J. Fluid Mech. 548 (2006), 151–163.

[70] A. CONSTANTIN AND W. STRAUSS, Stability of the Camassa-Holm solitons, J. Nonlinear Sci. 12 (2002), 415–422.

[71] A. CONSTANTIN AND W. STRAUSS, Exact steady periodic water waves with vorticity, Comm. Pure Appl. Math. 57 (2004), 481–527.

[72] A. CONSTANTIN AND W. STRAUSS, Stability properties of steady water waves with vorticity, Comm. Pure Appl. Math. 60 (2007), 911–950.

[73] A. CONSTANTIN AND W. STRAUSS, Rotational steady water waves near stagnation, Philos. Trans. R. Soc. Lond. Ser. A Math. Phys. Eng. Sci. 365 (2007), 2227–2239.

[74] A. CONSTANTIN AND W. STRAUSS, Pressure beneath a Stokes wave, Comm. Pure Appl. Math. 63 (2010), 533–557.

[75] A. CONSTANTIN AND W. STRAUSS, Periodic traveling gravity water waves with discontinuous vorticity, Arch. Ration. Mech. Anal. 202 (2011), 133–175.

[76] A. CONSTANTIN AND E. VARVARUCA, Steady periodic water waves with constant vorticity: Regularity and local bifurcation, Arch. Ration. Mech. Anal. 199 (2011), 33–67.

[77] A. CONSTANTIN AND G. VILLARI, Particle trajectories in linear water waves, J. Math. Fluid Mech. 10 (2008), 1–18.

[78] J. B. CONWAY, *A Course in Functional Analysis*, Springer-Verlag, New York, 1990.

[79] W. A. COPPEL, *Stability and Asymptotic Behavior of Differential Equations*, D. C. Heath and Co., Boston, 1965.

[80] D. COUTAND AND S. SHKOLLER, Well-posedness of the free-surface incompressible Euler equations with or without surface tension, J. Amer. Math. Soc. 20 (2007), 829–930.

[81] W. Craig, Non-existence of solitary water waves in three dimensions, Philos. Trans. R. Soc. Lond. Ser. A Math. Phys. Eng. Sci. 360 (2002), 2127–2135.

[82] W. Craig, Surface water waves and tsunamis, J. Dynam. Differential Equations 18 (2006), 525–549.

[83] W. Craig, P. Guyenne, J. Hammack, D. Henderson, and C. Sulem, Solitary water wave interactions, Phys. Fluids 18 (2006), Art. 057106.

[84] W. Craig and P. Sternberg, Symmetry of solitary waves, Comm. Partial Differential Equations 13 (1988), 603–633.

[85] A. D. D. Craik, The origins of water wave theory, Annu. Rev. Fluid Mech. 36 (2004), 1–28.

[86] M. G. Crandall and P. H. Rabinowitz, Bifurcation from simple eigenvalues, J. Funct. Anal. 8 (1971), 321–340.

[87] G. D. Crapper, *Introduction to Water Waves*, Ellis Horwood Ltd., Chichester, UK, 1984.

[88] E. N. Dancer, Global structure of the solutions of non-linear real analytic eigenvalue problems, Proc. London Math. Soc. (3) 27 (1973), 747–765.

[89] A. F. T. da Silva and D. H. Peregrine, Steep, steady surface waves on water of finite depth with constant vorticity, J. Fluid Mech. 195 (1988), 281–302.

[90] E. B. Davies, *Linear Operators and Their Spectra*, Cambridge University Press, Cambridge, UK, 2007.

[91] L. Debnath, *Nonlinear Water Waves*, Academic Press, Boston, 1994.

[92] A. Degasperis and M. Procesi, Asymptotic integrability, in *Symmetry and Perturbation Theory*, A. Degasperis and G. Gaeta, eds., World Scientific, Singapore, 1999, pp. 23–37.

[93] P. Deift, *Orthogonal Polynomials and Random Matrices: A Riemann-Hilbert Approach*, Courant Lecture Notes in Mathematics, 3, CIMS, New York, 1999.

[94] P. Deift and E. Trubowitz, Inverse scattering on the line, Comm. Pure Appl. Math. 32 (1979), 121–251.

[95] F. Dias, Dynamics of tsunami waves, in *Extreme Man-Made and Natural Hazards in Dynamics of Structures*, Proc. of NATO Adv. Res. Workshop, Opaijta, Croatia, 2006.

[96] J. Dieudonné, *Foundations of Modern Analysis*, Academic Press, New York, 1969.

[97] P. G. Drazin, *Solitons*, London Mth. Soc. Lecture Notes, Cambridge University Press, Cambridge, UK, 1983.

[98] P. G. Drazin and R. S. Johnson, *Solitons: An Introduction*, Cambridge University Press, Cambridge, UK, 1989.

[99] M.-L. Dubreil-Jacotin, Sur la détermination rigoureuse des ondes permanentes périodiques d'ampleur finie, J. Math. Pures Appl. 13 (1934), 217–291.

[100] B. A. Dubrovin and S. P. Novikov, A periodic problem for the Korteweg-de Vries and Sturm-Liouville equations. Their connection with algebraic geometry (Russian), Dokl. Akad. Nauk SSSR 219 (1974), 531–534.

[101] J. Dugundji, *Topology*, Allyn and Bacon, Boston, 1966.

[102] P. L. Duren, *Theory of H^p Spaces*, Dover, New York, 2000.

[103] H. Dym and H. P. McKean, *Fourier Series and Integrals*, Academic Press, New York, London, 1972.

[104] D. G. Ebin, The equations of motion of a perfect fluid with free boundary are not well posed, Comm. Partial Differential Equations 12 (1987), 1175–1201.

[105] M. Ehrnström, A note on surface profiles for symmetric gravity waves with vorticity, J. Nonlinear Math. Phys. 13 (2006), 1–8.

[106] M. Ehrnström, Deep-water waves with vorticity: Symmetry and rotational behaviour, Discrete Contin. Dyn. Syst., 19 (2007), 483–491.

[107] M. Ehrnström, On the streamlines and particle paths of gravitational water waves, Nonlinearity 21 (2008), 1141–1154.

[108] M. Ehrnström and G. Villari, Linear water waves with vorticity: Rotational features and particle paths, J. Differential Equations 244 (2008), 1888–1909.

[109] J. Escher and B. Kolev, The Degasperis-Procesi equation as a non-metric Euler equation, Math. Z. (to appear).

[110] J. Escher, Y. Liu, and Z. Yin, Global weak solutions and blow-up structure for the Degasperis-Procesi equation, J. Funct. Anal. 241 (2006), 457–485.

[111] L. C. Evans, *Partial Differential Equations*, Amer. Math. Soc., Providence, RI, 1998.

[112] L. C. Evans and R. F. Gariepy, *Measure Theory and Fine Properties of Functions*, CRC Press, Boca Raton, FL, 1992.

[113] W. A. B. Evans and M. J. Ford, An exact integral equation for solitary waves (with new numerical results for some "internal" properties), Proc. R. Soc. Lond. Ser. A Math. Phys. Eng. Sci. 452 (1996), 373–390.

[114] H. M. Farkas and I. Kra, *Riemann Surfaces*, Springer-Verlag, New York, 1992.

[115] J. D. Fenton, Nonlinear wave theories, in *The Sea*, B. Le Méhauté and D. M. Hanes, eds., Wiley, New York, 1990.

[116] L. E. Fraenkel, *An Introduction to Maximum Principles and Symmetry in Elliptic Problems*, Cambridge University Press, Cambridge, UK, 2000.

[117] N. C. FREEMAN AND R. S. JOHNSON, Shallow water waves on shear flows, J. Fluid Mech. 42 (1970), 401–409.

[118] F. G. FRIEDLANDER, *Introduction to the Theory of Distributions*, Cambridge University Press, Cambridge, UK, 1998.

[119] K. O. FRIEDRICHS AND D. H. HYERS, The existence of solitary waves, Comm. Pure Appl. Math. 7 (1954), 517–550.

[120] B. FUCHSSTEINER AND A. S. FOKAS, Symplectic structures, their Bäcklund transformations and hereditary symmetries. Phys. D 4 (1981), 47–66.

[121] F. D. GAKHOV, *Boundary Value Problems*, Pergamon Press, Oxford, New York, Paris, 1966.

[122] P. R. GARABEDIAN, Surface waves of finite depth, J. Anal. Math. 14 (1965), 161–169.

[123] C. S. GARDNER, J. M. GREENE, M. D. KRUSKAL, AND R. MIURA, Method for solving the Korteweg-de Vries equation, Phys. Rev. Lett. 19 (1967), 1095–1097.

[124] P. GERMAIN, N. MASMOUDI, AND J. SHATAH, Global solutions for the gravity water waves equation in dimension 3, C. R. Math. Acad. Sci. Paris 347 (2009), 897–902.

[125] F. GERSTNER, Theorie der Wellen, Abh. Königl. Böhm. Ges. Wiss. (1802).

[126] F. GERSTNER, Theorie der Wellen samt einer daraus abgeleiteten Theorie der Deichprofile, Ann. Phys. 2 (1809), 412–445.

[127] A. GEYER, On some background flows for tsunami waves, J. Math. Fluid Mech. (to appear).

[128] B. GIDAS, W. M. NI, AND L. NIRENBERG, Symmetry and related properties via the maximum principle, Comm. Math. Phys. 68 (1979), 209–243.

[129] D. GILBARG AND N. S. TRUDINGER, *Elliptic Partial Differential Equations of Second Order*, Springer-Verlag, Berlin, 2001.

[130] R. GOYON, Contribution à la théorie des houles, Ann. Fac. Sci. Univ. Toulouse 22 (1958), 1–55.

[131] A. E. GREEN AND P. M. NAGHDI, A derivation of equations for wave propagation in water of variable depth, J. Fluid Mech. 78 (1976), 237–246.

[132] P. A. GRIFFITHS, *Introduction to Algebraic Curves*, Amer. Math. Soc., Providence, RI, 1989.

[133] M. D. GROVES AND E. WAHLÉN, Small-amplitude Stokes and solitary gravity water waves with an arbitrary distribution of vorticity, Phys. D 237 (2008), 1530–1538.

[134] K. GRUNERT AND G. TESCHL, Long-time asymptotics for the Korteweg-de Vries equation via nonlinear steepest descent, Math. Phys. Anal. Geom. 12 (2009), 287–324.

[135] V. GUILLEMIN AND A. POLLACK, *Differential Topology*, Prentice–Hall, Englewood Cliffs, NJ, 1974.

[136] J. K. HALE *Ordinary Differential Equations*, R. E. Krieger, Huntington, NY, 1980.

[137] R. S. HAMILTON, The inverse function theorem of Nash and Moser, Bull. Amer. Math. Soc. 7 (1982), 65–222.

[138] J. L. HAMMACK AND H. SEGUR, The Korteweg-de Vries equation and water waves, J. Fluid Mech. 84 (1978), 337–358.

[139] T. J. HEALEY, *Global Continuation Methods in Three Dimensional Elasticity*, Lecture Notes, University of Puerto Rico, 2003.

[140] T. J. HEALEY AND H. SIMPSON, Global continuation in nonlinear elasticity, Arch. Ration. Mech. Anal. 143 (1998), 1–28.

[141] D. HENRY, On Gerstner's water wave, J. Nonlinear Math. Phys. 15 (2008), 87–95.

[142] N. HERMAS AND S. DJEBALI, Existence de l'application exponentielle riemannienne d'un groupe de difféomorphismes muni d'une métrique de Sobolev, J. Math. Pures Appl. 94 (2010), 433–446.

[143] M. HERVÉ, *Several Complex Variables. Local Theory*, Oxford University Press, London, 1963.

[144] E. HILLE, *Lectures on Ordinary Differential Equations*, Addison–Wesley, Reading, MA, 1969.

[145] P. D. HISLOP AND I. M. SIGAL, *Introduction to Spectral Theory. With Applications to Schrödinger Operators*, Springer-Verlag, New York, 1996.

[146] H. HOCHSTADT, On the determination of a Hill's equation from its spectrum, Arch. Ration. Mech. Anal. 19 (1965), 353–362.

[147] B. H. HODGES, *Numerical Simulation of Nonlinear Free-Surface Waves on a Turbulent Open-Channel Flow*, Ph.D. thesis, Stanford University, 1997.

[148] H. HOFER AND E. ZEHNDER, *Symplectic Invariants and Hamiltonian Dynamics*, Birkhäuser-Verlag, Basel, 1994.

[149] H. HOLDEN AND X. RAYNAUD, Periodic conservative solutions of the Camassa-Holm equation, Ann. Inst. Fourier (Grenoble) 58 (2008), 945–988.

[150] L. HÖRMANDER, *The Analysis of Linear Partial Differential Operators. I. Distribution Theory and Fourier Analysis*, Springer-Verlag, Berlin, 1983.

[151] V. M. HUR, Global bifurcation theory of deep-water waves with vorticity, SIAM J. Math. Anal. 37 (2006), 1482–1521.

[152] V. M. HUR, Symmetry of steady periodic water waves with vorticity, Philos. Trans. R. Soc. Lond. Ser. A Math. Phys. Eng. Sci. 365 (2007), 2203–2214.

[153] V. M. Hur, Exact solitary water waves with vorticity, Arch. Ration. Mech. Anal. 188 (2008), 213–244.

[154] V. M. Hur, Symmetry of solitary water waves with vorticity, Math. Res. Lett. 15 (2008), 491–509.

[155] D. Ionescu-Kruse, Particle trajectories in linearized irrotational shallow water flows, J. Nonlinear Math. Phys. 15 (2008), 13–27.

[156] G. Iooss and P. I. Plotnikov, Small divisor problem in the theory of three-dimensional water gravity waves, Mem. Amer. Math. Soc. 200 (2009), no. 940.

[157] S. Israwi, Variable depth KdV equations and generalizations to more nonlinear regimes, M2AN Math. Model. Numer. Anal. 44 (2010), 347–370.

[158] A. R. Its, The Riemann-Hilbert problem and integrable systems, Notices Amer. Math. Soc. 50 (2003), 1389–1400.

[159] A. R. Its and V. B. Matveev, Hill operators with a finite number of lacunae (Russian), Funk. Anal. Pril. 9 (1975), 69–70.

[160] R. I. Ivanov, On the integrability of a class of nonlinear dispersive wave equations, J. Nonlinear Math. Phys. 12 (2005), 462–468.

[161] R. S. Johnson, Nonlinear gravity waves on the surface of an arbitrary shear flow with variable depth, in *Nonlinear Instability Analysis*, Adv. Fluid Mech. Vol. 12, Comput. Mech., Southampton, UK, 1997, pp. 221–243.

[162] R. S. Johnson, *A Modern Introduction to the Mathematical Theory of Water Waves*, Cambridge University Press, Cambridge, UK, 1997.

[163] R. S. Johnson, *Singular Perturbation Theory. Mathematical and Analytical Techniques with Applications to Engineering*, Springer-Verlag, New York, 2005.

[164] R. S. Johnson, Camassa-Holm, Korteweg-de Vries and related models for water waves, J. Fluid Mech. 457 (2002), 63–82.

[165] R. S. Johnson, The Camassa-Holm equation for water waves moving over a shear flow, Fluid Dynam. Res. 33 (2003), 97–111.

[166] R. S. Johnson, The classical problem of water waves: A reservoir of integrable and nearly-integrable equations, J. Nonlinear Math. Phys. 10 (2003), 72–92.

[167] I. G. Jonsson, Wave-current interactions, in *The Sea*, B. Le Méhauté and D. M. Hanes, eds., Wiley, New York, 1990, pp. 65–120.

[168] T. Kato, Quasi-linear equations of evolution, with applications to partial differential equations, in *Spectral Theory and Differential Equations*, Lecture Notes in Math. 448, Springer-Verlag, Berlin, 1975, pp. 25–70.

[169] T. Kato, On the Cauchy problem for the (generalized) Korteweg-de Vries equation, in *Studies in Applied Mathematics*, Adv. Math. Suppl. Stud. 8, Academic Press, New York, 1983, pp. 93–128.

[170] T. KATO, On the Korteweg–de-Vries equation, Manuscripta Math. 28 (1979), 89–99.

[171] T. KATO, *Perturbation Theory for Linear Operators*, Springer-Verlag, Berlin, 1995.

[172] G. KEADY AND J. NORBURY, On the existence theory for irrotational water waves, Math. Proc. Cambridge Philos. Soc. 83 (1978), 137–157.

[173] W. THOMSON (LORD KELVIN), On a disturbing infinity in Lord Rayleigh's solution for waves in a plane vortex stratum, Nature 23 (1880), 45–46.

[174] H. KIELHÖFER, Multiple eigenvalue bifurcation for Fredholm operators, J. Reine Angew. Math. 358 (1985), 104–124.

[175] D. KINDERLEHRER AND L. NIRENBERG, Regularity in free boundary problems, Ann. Sc. Norm. Super. Pisa Cl. Sci. (4) 4 (1977), 373–391.

[176] D. KINDERLEHRER, L. NIRENBERG, AND J. SPRUCK, Regularity in elliptic free boundary problems, J. Anal. Math. 34 (1978), 86–119.

[177] B. KINSMAN, *Wind Waves: Their Generation and Propagation on the Ocean Surface*, Dover, New York, 2002.

[178] A. A. KIRILLOV AND A. D. GVISHIANI, *Theorems and Problems in Functional Analysis*, Springer-Verlag, New York, Berlin, 1982.

[179] F. KLEIN, *Development of Mathematics in the* 19*th Century*, Math. Sci. Press, Brookline, MA, 1979.

[180] J. KO AND W. STRAUSS, Large-amplitude steady rotational water waves, Eur. J. Mech. B Fluids 27 (2007), 96–109.

[181] J. KO AND W. STRAUSS, Effect of vorticity on steady water waves, J. Fluid Mech. 608 (2008), 197–215.

[182] B. KOLEV, Lie groups and mechanics: An introduction, J. Nonlinear Math. Phys. 11 (2004), 480–498.

[183] P. KOOSIS, *Introduction to H_p-spaces*, Cambridge University Press, Cambridge, UK, New York, 1998.

[184] D. J. KORTEWEG AND G. DE VRIES, On the change of form of long waves advancing in a rectangular canal and on a new type of long stationary waves, Phil. Mag. 39 (1895), 422–443.

[185] R. A. KRAENKEL, M. A. MANNA, AND J. P. PEREIRA, The Korteweg-de Vries hierarchy and long water-waves, J. Math. Phys. 36 (1995), 307–320.

[186] S. G. KRANTZ AND H. R. PARKS, *A Primer of Real Analytic Functions,* Birkhäuser-Verlag, Basel, 1992.

[187] J. P. KRASOVSKII, On the theory of steady-state waves of finite amplitude, USSR Comput. Math. Math. Phys. 1 (1961), 996–1018.

[188] W. KRAWCEWICZ AND J. WU, *Theory of Degrees with Applications to Bifurcations and Differential Equations*, Wiley, New York, 1997.

[189] M. D. KRUSKAL AND N. J. ZABUSKY, Interaction of solitons in a collisionless plasma and the recurrence of initial states, Phys. Rev. Lett. 15 (1965), 240–243.

[190] M. LAKSHMANAN, Integrable nonlinear wave equations and possible connections to tsunami dynamics, in *Tsunami and Nonlinear Waves*, A. Kundu, ed., Springer-Verlag, Berlin, 2007, pp. 31–49.

[191] M. LAKSHMANAN, Solitons, tsunamis and oceanographical applications of, in *Extreme Environmental Events*, R. A. Meyers, ed., Springer-Verlag, Berlin, 2011, pp. 873–888.

[192] H. LAMB, *Hydrodynamics*, Cambridge University Press, Cambridge, UK, 1895.

[193] G. L. LAMB, *Elements of Soliton Theory*, Wiley, New York, 1980.

[194] D. LANNES, Well-posedness of the water-waves equations, J. Amer. Math. Soc. 18 (2005), 605–654.

[195] P. D. LAX, Integrals of nonlinear equations of evolution and solitary waves, Comm. Pure Appl. Math. 21 (1968), 467–490.

[196] P. D. LAX, Outline of a theory of the KdV equation, in *Recent Mathematical Methods in Nonlinear Wave Propagation* (Montecatini Terme, 1994), Lecture Notes in Math. 1640, Springer-Verlag, Berlin, 1996, pp. 70–102.

[197] P. D. LAX, *Linear Algebra and Its Applications*, Wiley-Interscience, Hoboken, NJ, 2007.

[198] J. LEECH, *Classical Mechanics*, Wiley, New York, 1958.

[199] S. LEIBOVICH, The form and dynamics of Langmuir circulations, Annu. Rev. Fluid. Mech. 15 (1983), 391–427.

[200] G. LEONI, *A First Course in Sobolev Spaces*, Amer. Math. Soc., Providence, RI, 2009.

[201] J. LERAY AND J. SCHAUDER, Topologie et équations fonctionelles, Ann. Sci. École Norm. Sup. (3) 51 (1934), 45–78.

[202] T. LEVI-CIVITA, Determinazione rigorosa delle onde irrotazionali periodiche in acqua profonda, Rend. Accad. Lincei 33 (1924), 141–150.

[203] B. M. LEVITAN, *Inverse Sturm-Liouville Problems*, VNU Science Press, Utrecht, 1987.

[204] H. LEWY, A note on harmonic functions and a hydrodynamical application, Proc. Amer. Math. Soc. 3 (1952), 111–113.

[205] G. M. LIEBERMAN AND N. S. TRUDINGER, Nonlinear oblique boundary value problems for nonlinear elliptic equations, Trans. Amer. Math. Soc. 295 (1986), 509–546.

[206] J. LIGHTHILL, *Waves in Fluids*, Cambridge University Press, Cambridge, UK, New York, 1978.

[207] H. LINDBLAD, Well-posedness for the motion of an incompressible liquid with free surface boundary, Ann. of Math. (2) 162 (2005), 109–194.

[208] J. LINDENSTRAUSS AND L. TZAFRIRI, *Classical Banach Spaces*, Springer-Verlag, Berlin, New York, 1977.

[209] S. LOJASIEWICZ, Sur le problème de la division, Studia Math. 18 (1959), 87–136.

[210] M. S. LONGUET-HIGGINS, Mass transport in water waves, Philos. Trans. R. Soc. Lond. Ser. A Math. Phys. Eng. Sci. 245 (1953), 535–581.

[211] M. S. LONGUET-HIGGINS, Mass transport in the boundary layer at a free oscillating surface, J. Fluid Mech. 8 (1960), 293–306.

[212] M. S. LONGUET-HIGGINS, Eulerian and Lagrangian aspects of surface waves, J. Fluid Mech. 173 (1986), 683–707.

[213] YA. B. LOPATINSKI, On a method of reducing boundary problems for a system of differential equations of elliptic type to regular integral equations, Ukrain. Math. J. 5 (1953), 123–151 (Russian); Engl. transl., Amer. Math. Soc. Transl. Ser. 2 89 (1970), 149–183.

[214] T.-W. MA, *Classical Analysis on Normed Spaces*, World Scientific, River Edge, NJ, 1995.

[215] W. MAGNUS AND S. WINKLER, *Hill's Equation*, Dover, New York, 1979.

[216] A. J. MAJDA AND A. L. BERTOZZI, *Vorticity and Incompressible Flow*, Cambridge University Press, Cambridge, UK, 2002.

[217] V. A. MARCHENKO AND I. V. OSTROVSKIJ, Characterization of spectrum of Hill's operator (Russian), Matem. Sbornik 97 (1975), 540–586.

[218] Y. MARTEL, F. MERLE, AND T. MIZUMACHI, Description of the inelastic collision of two solitary waves for the BBM equation, Arch. Ration. Mech. Anal. 196 (2010), 517–574.

[219] W. S. MASSEY, Cross products of vectors in higher dimensional Euclidean spaces, Amer. Math. Monthly 90 (1983), 697–701.

[220] Y. MATSUNO, The N-soliton solution of the Degasperis-Procesi equation, Inverse Problems 21 (2005), 2085–2101.

[221] V. B. MATVEEV, 30 years of finite-gap integration theory, Philos. Trans. R. Soc. Lond. Ser. A Math. Phys. Eng. Sci. 366 (2008), 837–875.

[222] R. K. MAZOVA AND J. F. RAMIREZ, Tsunami waves with an initial negative wave on the Chilean coast, Natural Hazards 20 (1999), 83–92.

[223] H. P. McKean, Integrable systems and algebraic curves, in *Global analysis*, Lecture Notes in Math. 755, Springer-Verlag, Berlin, 1979, pp. 83–200.

[224] H. P. McKean and P. van Moerbeke, The spectrum of Hill's equation, Invent. Math. 30 (1975), 217–274.

[225] H. P. McKean and V. Moll, *Elliptic Curves: Function Theory, Geometry, Arithmetic*, Cambridge University Press, Cambridge, UK, 1997.

[226] H. P. McKean and E. Trubowitz, Hill's operator and hyperelliptic function theory in the presence of infinitely many branch points, Comm. Pure Appl. Math. 29 (1976), 143–226.

[227] J. B. McLeod, The Stokes and Krasovskii conjectures for the wave of greatest height, University of Wisconsin Mathematics Research Center, Report 2041, 1979.

[228] K. R. Meyer and G. R. Hall, *Introduction to Hamiltonian Dynamical Systems and the N-Body Problem*, Springer-Verlag, New York, 1992.

[229] P. Michor and D. Mumford, Vanishing geodesic distance on spaces of submanifolds and diffeomorphisms, Doc. Math. 10 (2005), 217–245.

[230] J. W. Milnor, *Singular Points of Complex Hypersurfaces*, Ann. of Math. Stud. 61, Princeton University Press, Princeton, NJ, 1968.

[231] J. W. Milnor, Remarks on infinite-dimensional Lie groups, in *Relativity, Groups and Topology* (Les Houches, 1983), North–Holland, Amsterdam, 1984, pp. 1007–1057.

[232] J. W. Milnor, *Topology from the differentiable viewpoint*, Princeton University Press, Princeton, NJ, 1997.

[233] L. M. Milne-Thomson, *Theoretical Hydrodynamics*, Macmillan, New York, 1960.

[234] G. Misiolek, A shallow water equation as a geodesic flow on the Bott-Virasoro group, J. Geom. Phys. 24 (1998), 203–208.

[235] S. Monismith, H. Nepf, E. A. Cowen, L. Thais, and J. Magnaudet, Laboratory observations of mean flows under surface gravity waves, J. Fluid Mech. 573 (2007), 131–147.

[236] C. B. Morrey, *Multiple Integrals in the Calculus of Variations*, Springer-Verlag, Berlin, 1966.

[237] D. Mumford, *Jacobian Theta Functions and Differential Equations*, Birkhäuser Boston, Boston, 1984.

[238] J. R. Munkres, *Topology: A first course*, Prentice–Hall, Englewood Cliffs, NJ, 1975.

[239] O. G. Mustafa and D. O'Regan, On an inverse scattering algorithm for the Camassa-Holm equation, J. Nonlinear Math. Phys. 15 (2008), 283–290.

[240] F. Müller, On the continuation of solutions for elliptic equations in two variables, Ann. Inst. H. Poincaré Anal. Non Linéaire 19 (2002), 745–776.

[241] R. NARASIMHAN, *Analysis on Real and Complex Manifolds*, North–Holland, Amsterdam, 1968.

[242] A. H. NAYFEH, *Introduction to Perturbation Techniques*, Wiley, New York, 1981.

[243] A. I. NEKRASOV, On steady waves, Izv. Ivanovo-Voznesenk. Politekhn. 3 (1921).

[244] S. P. NOVIKOV, A periodic problem for the Korteweg-de Vries equation (Russian), Funk. Anal. Pril. 8 (1974), 54–66.

[245] H. OKAMOTO AND M. SHOJI, *The Mathematical Theory of Permanent Progressive Water Waves*, World Scientific, London, Singapore, Hong Kong, 2001.

[246] V. OVSIENKO AND B. KHESIN, Korteweg–de Vries superequations as an Euler equation, Funct. Anal. Appl. 21 (1987), 81–82.

[247] A. PARKER, On the Camassa-Holm equation and a direct method of solution: N-soliton solutions, Proc. R. Soc. Lond. Ser. A Math. Phys. Eng. Sci. 461 (2005), 3893–3911.

[248] A. PAZY, *Semigroups of Linear Operators and Applications to Partial Differential Equations*, Springer-Verlag, New York, 1983.

[249] D. H. PEREGRINE, Calculations of the development of an undular bore, J. Fluid Mech. 25 (1966), 321–330.

[250] D. H. PEREGRINE, Book reviews, J. Fluid Mech. 367 (1998), 378–381.

[251] M. PERLIN AND W. W. SCHULTZ, Capillary effects on surface waves, Annu. Rev. Fluid Mech. 32 (2000), 241–274.

[252] Q. PLAYFAIR, Cremona's forgotten curve, The Strad 110 (1999), 1194–1197.

[253] P. I. PLOTNIKOV, A proof of the Stokes conjecture in the theory of surface waves, Dinamika Splosh. Sredy 57 (1982), 41–76 (Russian); Engl. transl., Stud. Appl. Math. 3 (2002), 217–244.

[254] P. I. PLOTNIKOV AND J. F. TOLAND, Convexity of Stokes waves of extreme form, Arch. Ration. Mech. Anal. 171 (2004), 349–416.

[255] CH. POMMERENKE, Conformal maps at the boundary, in *Handbook of Complex Analysis: Geometric Function Theory, Vol.* 1, R. Kühnau, ed., North–Holland, Amsterdam, 2002, pp. 37–74.

[256] F. QUINN AND A. SARD, Hausdorff conullity of critical images of Fredholm maps, Amer. J. Math. 94 (1972), 1101–1110.

[257] P. RABINOWITZ, Some global results for nonlinear eigenvalue problems, J. Funct. Anal. 7 (1971), 487–513.

[258] W. J. M. RANKINE, On the exact form of waves near the surface of deep water, Philos. Trans. R. Soc. Lond. Ser. A Math. Phys. Eng. Sci. 153 (1863), 127–138.

[259] R. J. RAPP AND W. K. MELVILLE, Laboratory measurements of deep-water breaking waves, Philos. Trans. R. Soc. Lond. Ser. A Math. Phys. Eng. Sci. 331 (1990), 735–800.

[260] B. RAYLEIGH, On waves, Philos. Mag. 5 (1876), 257–279.

[261] R. M. REDHEFFER, The theorems of Bony and Brézis on flow-invariant sets, Amer. Math. Monthly 79 (1972), 740–747.

[262] M. REED AND B. SIMON, *Methods of Modern Mathematical Physics. I. Functional Analysis*, Academic Press, New York, 1980.

[263] E. H. ROTHE, *Introduction to Various Aspects of Degree Theory in Banach Spaces*, Amer. Math. Soc., Providence, RI, 1986.

[264] W. RUDIN, *Real and Complex Analysis*, McGraw–Hill, New York, 1987.

[265] G. SANSONE AND R. CONTI, *Non-linear Differential Equations*, Pergamon Press, New York, 1964.

[266] M. SCHLICHENMAIER, *An Introduction to Riemann Surfaces, Algebraic Curves and Moduli Spaces*, Lecture Notes in Physics, 322, Springer-Verlag, Berlin, 1989.

[267] P. SCHUUR, *Asymptotic Analysis of Soliton Problems: An Inverse Scattering Approach*, Springer Lecture Notes in Mathematics 1232 Springer-Verlag, New York, 1986.

[268] B. F. SCHUTZ, *Geometrical Methods of Mathematical Physics*, Cambridge University Press, Cambridge, UK, 1980.

[269] L. W. SCHWARTZ AND J. D. FENTON, Strongly nonlinear waves, Annu. Rev. Fluid Mech. 14 (1982), 39–60.

[270] J. SCOTT RUSSELL, Report on waves, Fourteenth meeting of the British Association for the Advancement of Science, 1844.

[271] H. SEGUR, Waves in shallow water with emphasis on the tsunami of 2004, in *Tsunami and Nonlinear Waves*, A. Kundu, ed., Springer-Verlag, Berlin, 2007, pp. 3–29.

[272] H. SEGUR, Integrable models of waves in shallow water, in *Probability, Geometry and Integrable Systems*, M. Pinsky et al., eds., Math. Sci. Res. Inst. Publ., Cambridge University Press, Cambridge, UK, 2008, pp. 345–371.

[273] Z. SHAPIRO, On general boundary problems for equations of elliptic type (Russian), Izv. Akad. Nauk SSSR 17 (1953), 539–562.

[274] S. S. SHEN, *A Course on Nonlinear Waves*, Kluwer Academic Publishers, Dordrecht, The Netherlands, 1993.

[275] V. V. SHOKUROV, Riemann surfaces and algebraic curves, in *Algebraic Geometry*, I. R. Shafarevich, ed., Encyclopaedia Math. Sci. 23, Springer-Verlag, Berlin, 1994, pp. 1–166.

[276] C. L. SIEGEL, *Topics in Complex Function Theory. Vol. II. Automorphic Functions and Abelian Integrals*, John Wiley & Sons, New York, 1988.

[277] J. A. SIMMEN AND P. G. SAFFMAN, Steady deep-water waves on a linear shear current, Stud. Appl. Math. 73 (1985), 35–57.

[278] S. SMALE, An infinite dimensional version of Sard's theorem, Amer. J. Math. 87 (1965), 861–866.

[279] A. SOMMERFELD, *Mechanics of Deformable Bodies*, Academic Press, New York, 1950.

[280] A. SPICAK AND J. VANEK, The M_W 9.0 Tohoku earthquake (Japan, March 11, 2011), Stud. Geophys. Geod. 55 (2011), 389–395.

[281] E. R. SPIELVOGEL, A variational principle for waves of infinite depth, Arch. Ration. Mech. Anal. 39 (1970), 189–205.

[282] G. SPRINGER, *Introduction to Riemann Surfaces*, Addison–Wesley, Reading, MA, 1957.

[283] E. M. STEIN, *Harmonic Analysis: Real-Variable Methods, Orthogonality, and Oscillatory Integrals*, Princeton University Press, Princeton, NJ, 1993.

[284] E. M. STEIN AND G. WEISS, *Introduction to Fourier Analysis on Euclidean Spaces*, Princeton University Press, Princeton, NJ, 1971.

[285] I. STIMPSON, Japan's Tohoku earthquake and tsunami, Geology Today 27 (2011), 96–98.

[286] J. J. STOKER, *Water Waves. The Mathematical Theory with Applications*, Interscience, New York, 1957.

[287] G. STOKES, On the theory of oscillatory waves, Trans. Cambridge Phil. Soc. 8 (1847), 441–455.

[288] G. STOKES, Considerations relative to the greatest height of oscillatory irrotational waves which can be propagated without change of form, in *Mathematical and Physical Papers, Vol.* I, Cambridge University Press, Cambridge, UK, 1880, pp. 225–228.

[289] W. A. STRAUSS, Steady water waves, Bull. Amer. Math. Soc. 47 (2010), 671–694.

[290] R. S. STRICHARTZ, *A Guide to Distribution Theory and Fourier Transforms*, CRC Press, Boca Raton, FL, 1994.

[291] D. STRUIK, Détermination rigoureuse des ondes irrotationelles périodiques dans un canal à profondeur finie, Math. Ann. 95 (1926), 595–634.

[292] M. STRUWE, *Variational Methods. Applications to Nonlinear Partial Differential Equations and Hamiltonian Systems*, Springer-Verlag, Berlin, Heidelberg, 1990.

[293] R. STUHLMEIER, KdV theory and the Chilean tsunami of 1960, Discrete Contin. Dyn. Syst. Ser. B 12 (2009), 623–632.

[294] C. SWAN, I. CUMMINS, AND R. JAMES, An experimental study of two-dimensional surface water waves propagating on depth-varying currents, J. Fluid Mech. 428 (2001), 273–304.

[295] T. TAO, Low-regularity global solutions to nonlinear dispersive equations, in *Surveys in Analysis and Operator Theory*, Proc. Centre Math. Appl. Austral. Nat. Univ., 2002, pp. 19–48.

[296] G. P. THOMAS, Wave-current interactions: An experimental and numerical study—nonlinear waves, J. Fluid Mech. 216 (1990), 505–536.

[297] J. F. TOLAND, On the existence of a wave of greatest height and Stokes's conjecture, Proc. R. Soc. Lond. Ser. A Math. Phys. Eng. Sci. 363 (1978), 469–485.

[298] J. F. TOLAND, Stokes waves, Topol. Methods Nonlinear Anal. 7 (1996), 1–48.

[299] J. F. TOLAND, On the symmetry theory for Stokes waves of finite and infinite depth, in *Trends in Applications of Mathematics to Mechanics (Nice, 1998)*, Chapman & Hall/CRC, Boca Raton, FL, pp. 207–217.

[300] A. TORCHINSKY, *Real-Variable Methods in Harmonic Analysis*, Dover, Mineola, NY, 2004.

[301] E. TRUBOWITZ, The inverse problem for periodic potentials, Comm. Pure Appl. Math. 30 (1977), 321–337.

[302] M. UMEYAMA, T. SHINTANI, AND S. WATANABE, Measurement of particle velocities and trajectories in a wave-current motion using particle image velocimetry, Proc. 32nd Internat. Conf. Coastal Engineering (2010).

[303] F. URSELL, Mass transport in gravity waves, Proc. Cambridge Phil. Soc. 40 (1953), 145–150.

[304] J.-M. VANDEN-BROECK, Periodic waves with constant vorticity in water of infinite depth, IMA J. Appl. Math. 56 (1996), 207–217.

[305] E. VARVARUCA, Singularities of Bernoulli free boundaries, Comm. Partial Differential Equations 31 (2006), 1451–1477.

[306] E. VARVARUCA, On some properties of traveling water waves with vorticity, SIAM J. Math. Anal. 39 (2008), 1686–1692.

[307] E. VARVARUCA, Some geometric and analytic properties of solutions of Bernoulli free-boundary problems, Interfaces Free Bound. 9 (2007), 367–381.

[308] E. WAHLÉN, On rotational water waves with surface tension, Philos. Trans. R. Soc. Lond. Ser. A Math. Phys. Eng. Sci. 365 (2007), 2215–2225.

[309] E. WAHLÉN, Steady water waves with a critical layer, J. Differential Equations 246 (2009), 2468–2483.

[310] S. WALSH, Stratified steady periodic water waves, SIAM J. Math. Anal. 41 (2009), 1054–1105.

[311] S. WALSH, Some criteria for the symmetry of stratified water waves, Wave Motion 46 (2009), 350–362.

[312] S. WALSH, Global bifurcation for steady water waves with surface tension, submitted.

[313] G. B. WHITHAM, *Linear and Nonlinear Waves*, Wiley, New York, 1974.

[314] G. B. WHITHAM, *Lectures on Wave Propagation*, Tata Institute of Fundamental Research Lectures on Mathematics and Physics, Springer-Verlag, Berlin, New York, 1979.

[315] D. V. WIDDER, The Airy transform, Amer. Math. Monthly 86 (1979), 271–277.

[316] E. P. WIGNER, The unreasonable effectiveness of mathematics in the natural sciences, Comm. Pure Appl. Math. 13 (1960), 1–14.

[317] J. T. WLOKA, *Partial Differential Equations*, Cambridge University Press, Cambridge, UK, 1987.

[318] S. WU, Well-posedness in Sobolev spaces of the full water wave problem in 2-D, Invent. Math. 130 (1997), 39–72.

[319] T. YANAGI, *Coastal Oceanography*, Kluwer Academic Publishers, Dordrecht, The Netherlands, 2003.

[320] D. YANAGIMOTO, M. KAWABE, AND S. FUJIO, Direct velocity measurements of deep circulation southwest of the Shatsky Rise in the western North Pacific, Deep Sea Research 57 (2010), 328–337.

[321] K. YOSIDA, *Functional Analysis*, Springer-Verlag, Berlin, 1995.

[322] E. ZEIDLER, Existenzbeweis für permanente Kapillar-Schwerewellen mit allgemeinen Wirbelverteilungen, Arch. Ration. Mech. Anal. 50 (1973), 34–72.

[323] G. M. ZIEGLER, Zur Mathematik von Tsunamis, DMV Mitt. 13 (2005), 51–53.

Index

amplitude parameter, 136

Bernoulli's law, 43, 87, 133, 168
Bott–Virasoro group, 268
  Bott cocycle, 268
  Euler equation, 269
  Gelfand–Fuchs 2-cocycle, 268
  Riemannian exponential map, 269
  right-invariant metric, 268
bound states, 195
boundary conditions
  dynamic, 14
  kinematic, 14
boundary layer, 32, 33
breaking wave, 241, 255, 256
  plunging breaker, 241, 243
  spilling breaker, 241, 243

cnoidal waves, 214
compact operator, 97, 104
complementing condition, 77, 120, 123, 124
continuum hypothesis, 5
current, 27
  adverse, 69
  favorable, 69
  flood and ebb, 29
  mean, 134

diffeomorphism group, 269
  Lie group exponential map, 269
  Riemannian exponential map, 269
  right-invariant metric, 269
discriminant, 214
dispersion relation
  Burns condition, 71
  capillary-gravity, 31
  constant vorticity, 68
  Gerstner's wave, 160
  irrotational, 64, 65, 138
dispersive effect, 65
double spectrum, 217
drawdown, 277

elliptic regularity, 60, 61, 116
energy dissipation, 32
equation
  Euler, 13, 86
  Gelfand–Levitan–Marchenko, 206
Eulerian viewpoint, 131

Faddeev function, 190
finite-gap potential, 215, 216
Floquet theory, 214
flow-invariant set, 14, 34
Fredholm operator, 76, 77, 97, 99, 108, 110, 121

Gerstner's wave, 135, 156
global bifurcation, 105
  analytic, 74, 78, 110
  via generalized Leray–Schauder degree, 74, 78, 109
  via Leray–Schauder degree, 105
gravity water waves, 6, 12
group speed, 282

Hölder spaces, 115
Hamiltonian system, 180
  conjugate variables, 180
  Hamiltonian flow, 186
  Hamiltonian function, 184
  Hamiltonian functional, 188
  integrability, 186
  integral of motion, 180, 189
  Jacobi's identity, 181

Noether's theorem, 186
Poisson bracket, 181, 184, 188
symplectic map, 184
Hamiltonian vector field, 184
Hardy function class, 191, 202, 204, 206
outer function, 202
Hilbert–Schmidt operator, 207
homogeneity, 5, 11
homotopy, 102, 104
hydraulic head, 43, 134, 168
hyperelliptic curve, 222
homogeneous coordinates, 222
Weierstrass function, 214

irrotational flow, 27
isometry, 259
isospectral problem, 189
Lax pair, 199

Jost function, 189

Lagrangian viewpoint, 25, 37, 131, 156
laminar flow, 47, 69
Lax–Milgram, 60, 118
local bifurcation, 55, 72, 73, 79, 99, 100, 110, 114
Lyapunov–Schmidt reduction, 111, 114

manifold, 182
cotangent bundle, 226
Lie derivative, 183
tangent bundle, 224
vector field, 182
maximum principle, 125
Hopf, 75, 81, 86, 94, 97, 126, 134, 143, 168
Serrin, 81, 126
strong, 81–83, 91, 92, 125, 146
weak, 75, 81, 86, 126
method of stationary phase, 213
minimization problem, 49, 80
model, 237
BBM equation, 237, 247, 263
Camassa–Holm equation, 267
Degasperis–Procesi equation, 267
Green–Naghdi equation, 246
inviscid Burgers equation, 270
Johnson equation, 248, 249
KdV equation, 165, 247, 262
linearized KdV equation, 235
moderate-amplitude long waves, 250, 253, 256, 263
underlying current, 273
variable depth, 272, 273

nondimensional variables, 136
norming constants, 199, 206
numerical simulation
finite-difference method, 88, 211
spectral method, 211

Paley–Wiener theorem, 206
particle drift, 148
particle path
Gerstner's wave, 157
irrotational linear theory, 139
solitary wave, 169
Stokes wave, 149
plate tectonics, 277
power series, 142, 160
pressure, 12
Gerstner's wave, 158
hydrostatic, 152
monotonicity, 153, 173
nondimensional, 136
proper map, 74, 108

quasi-periodic function, 238

real-analytic, 52, 94, 97, 123, 124
real-analytic variety, 111
curve selection, 112
local uniformizing parameter, 112
structure theorem, 112
reflection coefficient, 196
relative mass flux, 42, 84, 86
Riemann surface, 223
differential of the first kind, 226
differentials of the first kind, 234
genus, 223
homology group, 227
Jacobi variety, 227, 232, 234
meromorphic differential, 226
Riemann–Hilbert problem, 213
Riesz projections, 197

scaling, 137
  moderate-amplitude long waves, 248
scattering
  data, 199, 200
  matrix, 201
scattering theory
  forward, 201
  inverse, 204
Schauder estimates, 27, 60, 74, 76, 78, 81, 83, 84, 116–119, 121, 122, 125
semianalytic set, 111
semigroup theory, 258
  evolution system, 261
  generators of hyperbolic type, 259
  Kato's approach, 261
  Lumer–Phillips theorem, 259
  quasi-m-accretive generators, 259
  Stone's theorem, 260
shallowness parameter, 136
simple spectrum, 217
Sobolev spaces, 16, 38, 115
solitary wave, 9, 10, 163, 164, 166
  greatest height, 170
soliton, 174
  elastic collision, 178
  two-soliton interaction for KdV, 178
  two-soliton solution for KdV, 175
spectral forbidden bands, 215
spectral lacunae, 217, 232
stagnation point, 43, 152, 170
  weak, 79, 84, 89
Stokes wave, 131, 132
  greatest height, 143, 151
stream function, 26, 27, 41, 86, 133, 143, 167
Sturm–Liouville problem, 51
  ground state, 79
surface tension, 6, 29, 31, 33
swell, 7–9, 29, 33, 135

thrust fault, 277, 296
topological degree, 100
  Brouwer degree, 101
  Leray–Schauder degree, 103
trace formula, 216
transmission coefficient, 196
tsunami, 66, 275
  1958 Alaska, 295
  1960 Chile, 287
  2004 Thailand, 285
  2011 Japan, 296

unbound states, 195

velocity potential, 26, 144, 170, 246
viscosity, 6, 32
  dissipation, 33
vorticity, 18
  deformation effect, 19
  equation of evolution, 24
  local/global effect, 21, 23
  negative, 51
  nonnegative, 55
  rotational effect, 19
  translational effect, 18

wave train, 6
  amplitude, 8
  crest, 6
  deep water, 66
  period, 8, 66
  shallow water, 65
  speed, 8, 33, 67, 86, 135
  trough, 8
  wave height, 8, 29, 140
  wavelength, 8, 66, 86
winding number, 101